The Bellman Function Technique in Harmonic Analysis

The Bellman function, a powerful tool originating in control theory, can be used successfully in a large class of difficult harmonic analysis problems and has produced some notable results over the last 30 years. This book by two leading experts is the first devoted to the Bellman function method and its applications to various topics in probability and harmonic analysis. Beginning with basic concepts, the theory is introduced step-by-step starting with many examples of gradually increasing sophistication, culminating with Calderón–Zygmund operators and endpoint estimates. All necessary techniques are explained in generality, making this book accessible to readers without specialized training in nonlinear PDEs or stochastic optimal control. Graduate students and researchers in harmonic analysis, PDEs, functional analysis, and probability will find this to be an incisive reference, and can use it as the basis of a graduate course.

Vasily Vasyunin is Leading Researcher at the St. Petersburg Department of Steklov Mathematical Institute of Russian Academy of Sciences and Professor at the Saint Petersburg State University. His research interests include linear and complex analysis, operator models, and harmonic analysis. Vasyunin has taught at universities in Europe and the United States. He has authored or coauthored over 60 articles.

Alexander Volberg is Distinguished Professor of Mathematics at Michigan State University. He was the recipient of the Onsager Medal as well as the Salem Prize, awarded to a young researcher in the field of analysis. Along with teaching at institutions in Paris and Edinburgh, Volberg also served as a Humboldt senior researcher, Clay senior researcher, and a Simons fellow. He has coauthored 179 papers, and is the author of *Calderón–Zygmund Capacities and Operators on Non-Homogenous Spaces.*

The Bellman Function Technique in Harmonic Analysis

VASILY VASYUNIN
Russian Academy of Sciences

ALEXANDER VOLBERG
Michigan State University

CAMBRIDGE
UNIVERSITY PRESS

University Printing House, Cambridge CB2 8BS, United Kingdom

One Liberty Plaza, 20th Floor, New York, NY 10006, USA

477 Williamstown Road, Port Melbourne, VIC 3207, Australia

314–321, 3rd Floor, Plot 3, Splendor Forum, Jasola District Centre, New Delhi – 110025, India

79 Anson Road, #06–04/06, Singapore 079906

Cambridge University Press is part of the University of Cambridge.

It furthers the University's mission by disseminating knowledge in the pursuit of education, learning, and research at the highest international levels of excellence.

www.cambridge.org
Information on this title: www.cambridge.org/9781108486897
DOI: 10.1017/9781108764469

First published 2020

Printed in the United Kingdom by TJ International, Padstow Cornwall

A catalogue record for this publication is available from the British Library.

Library of Congress Cataloging-in-Publication Data
Names: Vasyunin, Vasily I., 1948– author. | Volberg, Alexander, 1956– author.
Title: The Bellman function technique in harmonic analysis /
Vasily Vasyunin, Alexander Volberg.
Description: Cambridge ; New York, NY : Cambridge University Press, 2020. |
Includes bibliographical references and index.
Identifiers: LCCN 2019042603 (print) | LCCN 2019042604 (ebook) |
ISBN 9781108486897 (hardback) | ISBN 9781108764469 (epub)
Subjects: LCSH: Harmonic analysis. | Functional analysis. | Control theory.
Classification: LCC QA403 .V37 2020 (print) | LCC QA403 (ebook) |
DDC 515/.2433–dc23
LC record available at https://lccn.loc.gov/2019042603
LC ebook record available at https://lccn.loc.gov/2019042604

ISBN 978-1-108-48689-7 Hardback

To our parents

Contents

Introduction

I.1 Preface

The subject of this book is the use of the Bellman function technique in harmonic analysis. The Bellman function, in principle, is the creature of another area of mathematics: control theory. We wish to show that it can be used very successfully in a big class of harmonic analysis problems. In the last 25–30 years some outstanding problems in harmonic analysis were solved by this approach. Later, 10–15 years after, another solution has been found by more classical methods involving some highly nontrivial stopping time argument.

This is what happened with the A_2 conjecture and then very recently with the A_1 conjecture concerning weighted estimates of singular integrals. Some other problems solved by the Bellman function method still await their "de-Bellmanisation." Among such problems, we can list the celebrated solution by Burkholder of Pełczyński's problem about Haar basis, the best L^p estimates of the Ahlfors–Beurling operator, and many matrix weight estimates.

One of the main technical advantages of the Bellman function technique is that it does not require the invention of any sophisticated stopping time argument of the kind that is so pervasive in modern harmonic analysis. We can express this feature by saying that the Bellman function knows how to stop the time correctly, but it does not show us its secret.

The purpose of this book is to present a wide range of problems in harmonic analysis having the same underlying structure that allows us to look at them as problems of stochastic optimal control and, consequently, to treat them by the methods originated from this part of control theory.

We intend to show that a certain class of harmonic analysis problems can be reduced (often without any loss of information) to solving a special partial differential equation called the Bellman equation of a problem. For that purpose, we first cast a corresponding harmonic analysis problem as a stochastic optimization problem.

A quintessentially typical problem of harmonic analysis is to find (or estimate) the norm of this or that (singular) operator in function space L^p. If we think about the operator as a black box, we should think about the unit ball of L^p as its input. The unit ball of L^p is not compact in norm topology, so there is a priori no extremizer, and moreover, it seems to be very difficult to "list" all functions in the unit ball and "try" them as black box inputs one by one.

The stochastic point of view helps here, because we can think about input as a stochastic process stopped at a certain time. This point of view gives a very nice and powerful way to list all inputs as solutions of simple stochastic differential equations with some unknown stochastic control. Then the norm of the operator becomes a functional on solutions of stochastic differential equations that we need to optimize by choosing optimal control.

The technique of doing that is to consider the Bellman function of this control problem and to write the Hamilton–Jacobi–Bellman equation whose solution the Bellman function is supposed to be.

This book can be used as the basis of a graduate course, and it can also serve as a reference on many (but not all) applications of the Bellman function technique in harmonic analysis.

A certain number of very important results obtained with the use of the Bellman function technique stayed outside of the scope of this book. For example, these are the twisted paraproducts results of V. Kovac [**91, 92**], and, in general, the applications of the Bellman function to multilinear and nonlinear harmonic analysis. In the last category one finds the works of C. Muscalu, T. Tao, and C. Thiele [**119**] concerning nonlinear analogs of the Hausdorff–Young inequality that relates the norm of a function and the norm of its nonlinear Fourier transform. This book does not present the recent results of O. Dragicevic and A. Carbonaro [**33, 34**], where the authors study universal multiplier theorems in the setting of symmetric contraction semigroups. In particular, the authors solved a long-standing problem of finding the optimal sector, where generators of symmetric contraction semigroups always admit a H^∞-type holomorphic functional calculus on L^p. This is done by a subtle application of the Bellman function technique on a flow that is given by the semigroup. Numerous multiplier theorems are improved due to this result, and new results on pointwise convergence related to a symmetric contraction semigroup on a closed sector are obtained.

The corresponding papers can be found in the References, and the reader is encouraged to study these beautiful applications of the Bellman function ideology.

I.2 Acknowledgments

Our foremost acknowledgments are to our teachers M. S. Birman and N. K. Nikolski. We also learned most of what we know from our collaborators, colleagues, and our students: Fedor Nazarov, Sergei Treil, Igor Verbitskiy, Dmitriy Stolyarov, Pavel Zatitskiy, Stefanie Petermichl, Paata Ivanisvili, Alexander Reznikov, Oliver Dragicevic, Irina Holmes, Leonid Slavin, and many others. The ideas we caught while working and/or talking with them are incorporated in this book. Special thanks are due to Marina Kuznetsov who corrected our most horrible blunders in English.

We are very grateful to Oberwolfach Mathematics Research Institute, to Centre International de Rencontres Mathématiques in Luminy, and to Department of Mathematics of Michigan State University for excellent working conditions and invitations for the two of us together to stay during 2014–2017. We are also grateful to NSF support that allowed us to work together and meet regularly.

Finally, we would like to thank our friends and relatives who encouraged us during writing of this book, especially our wives, Nina Vasyunina and Olga Volberg.

I.3 The Short History of the Bellman Function

The Bellman equation and the Euler–Lagrange equation both deal with extremal values of functionals. Given a functional, the Euler–Lagrange approach gives us a differential equation that rules the behavior of extremizers of a functional in question. The Bellman equation has quite a different nature. At first glance, it does not give any information on the extremizers. Moreover, typically, there will be no extremal function. Only a sequence of almost extremizers typically exists, which is yet another difference with applications of the Euler–Lagrange equation.

In the Bellman paradigm, the (system of) differential equation(s) is given to us, and we do not need to find it as in the Euler–Lagrange approach. However, the given differential equation (or a system of equations) also has an unknown functional parameter called control. We have to find the best control in the sense that this control will optimize a functional applied to the solution of (an) already given (system of) differential equation(s).

So the idea of Bellman is amazingly striking and incredibly simple simultaneously. Let us denote by $\mathbf{B}(x)$ the extremal value of the functional that we want to optimize on solutions of a given system of differential equations with initial value at x at initial time 0. Using the fact that at time Δt, we "know" where the solution is, and that having started at $x(\Delta t)$ this solution is also extremal (if the control is chosen correctly), one can deduce a partial differential equation on $\mathbf{B}$.

This equation is called the Hamilton–Jacobi equation. Due to the functional in need of optimization, and depending on the system of differential equations that is given to us, the Hamilton–Jacobi equation varies, but it stays in a certain class of first-order nonlinear (usually) PDEs.

However, if the system of differential equations mentioned previously is not the usual system, but is a system of stochastic differential equations, and we are asked to optimize not just a functional of solutions of this system, but the expectation (the average) of this functional, then the scheme mentioned previously can be applied as well. The resulting PDE is often called the Hamilton–Jacobi–Bellman equation, and the presence of stochasticity makes it a second-order nonlinear (usually) PDE. It belongs to the class of equations called degenerate elliptic equations, see, e. g., [**120**].

Previously we presented a very short exposition of the Bellman function of stochastic optimal control. In this branch of mathematics, it is also often called the value function. The reader who wants to acquire good knowledge in this area is advised to read [**95**].

But the goal of the book is to show the deep (and almost perfect) analogy between the Bellman function technique in stochastic optimal control and the Bellman function technique in harmonic analysis, which is the branch of analysis dealing with the estimates of singular integrals.

It was arguably in [**130**], and especially in [**131, 193**], where this parallelism between a wide class of harmonic analysis problems and the stochastic optimal control got recognized at face value.

The observation of this parallelism between two different branches of mathematics is sort of important for this book, but the ideas that now are generally recognized as the "Bellman function technique" in harmonic analysis have been around long before those papers.

Without any claim of completeness, we can list several articles and their ideas that now can be recognized as instances of application of the Bellman function technique (without ever mentioning stochastic control, the value function, or anything like that).

In the area of probability theory that deals with optimal problems for Brownian motion or for martingales, D. Burkholder [**22–31**], B. Davis [**50**], and Burkholder–Gundy [**32**] used what we call now the Bellman function as their main tool of finding the constants of best behavior of stopping times and of martingales with various restrictions.

But to the best of our knowledge, the first use of the idea underlying the Bellman function technique is due to A. Beurling, who found the exact function of uniform convexity for the space $L^p(0,1)$. Strangely enough, his work was not published: Beurling just made an oral report in Uppsala in 1945, and the exposition of his idea can be found in the paper of O. Hanner [**66**] of 1956. However, Beurling used certain magic guesses. These guesses were explained in the paper of Ivanisvili–Stolyarov–Zatitskiy [**82**], who showed that Beurling's function method is nothing other than perhaps the first occasion of the application of the Bellman function technique in harmonic analysis.

We think that chronologically the next case of using the Bellman function technique in harmonic analysis was again related to uniform convexity. But this case deals with the general theory of Banach spaces. In 1972, P. Enflo in [**59**] proved that the Banach space X is super-reflexive if and only if it can be given an equivalent norm that is uniformly convex. In fact, the "if" part was proved by R. C. James in [**86**]. Enflo proved the "only if" part, and the proof of Lemma 2 of [**59**] now reads as a typical Bellman function technique proof.

In 1975, G. Pisier [**154**] gave another proof of the James–Enflo result that the Banach space X is super-reflexive if and only if it has an equivalent uniformly convex norm. His proof used X-valued martingale interpretation of super-reflexivity. The uniformly convex norm on X was constructed in the second line of the proof of Theorem 3.1 of [**154**]. In fact, for a vector $x \in X$ this equivalent norm $|x|$ is defined as the infimum of a certain functional on X-valued martingales starting at x, and this is a quintessential Bellman function definition.

Let us briefly explain why we associate such an approach (also used in all the papers of Burkholder, Gundy, Davis mentioned previously) with the Bellman function technique described previously.

Roughly speaking, any martingale is a solution of a controlled stochastic differential equation (with continuous or discrete time), where martingale differences play the role of control that should be optimized to give a prescribed functional on martingale the "best" value. In the case explained in [**154**], the functional is given at the beginning of the proof of Theorem 3.1, and its optimal value is precisely $|x|$ – the equivalent norm of the initial vector x, where a martingale (the solution of a stochastic differential equation in our interpretation) has started.

The fact that $x \to |x|$ is uniformly convex is exactly the Hamilton–Jacobi–Bellman PDE, as the reader will conclude after reading this book. It sounds strange: Why should a certain inequality be called a partial differential equation?

It will be explained repeatedly that the Bellman PDE pertinent to a harmonic analysis problem is quite often, in fact, a certain second-order finite difference inequality.

It is difficult to find the optimal solutions of inequalities, so the reader will see in a case-by-case study how we account for this difficulty and how we remedy it.

In probability theory, there was an interest in understanding the relationship between the various norms of the stopping time T and the corresponding norms of $W(T)$, where W is the Brownian motion. The exposition of these results (and related martingale results) of B. Davis [**50**], G. Wang [**196**], and [**197**] can be found in Chapter 5.

A huge amount of work has been done in the papers by D. Burkholder [**21, 22**] and by Burkholder and Gundy [**32**]. These are all Bellman function technique papers. In particular, this method (without mentioning any stochastic optimal control or Hamilton–Jacobi equation) was used by Burkholder (in his seminal articles cited previously) to solve problems of A. Pełczyński, concerning sharp constants for unconditional Haar basis in L^p. We adapt Burkholder's solution to our language of the Bellman function technique, which is done in Section 1.8. This is one of those cases when it is easy to write the Bellman equation but difficult to solve it.

I.4 The Plan of the Book

In Chapter 1, we give nine precise Bellman functions corresponding to several typical harmonic analysis problems. As we already mentioned, Section 1.8 of this chapter is devoted to Burkholder's Bellman function. The John–Nirenberg inequality presents a very nice model for the application of the Bellman function technique, which the reader will find in Section 1.3. Then, in Section 1.5 we extend the method of the Bellman function to rather general functionals on the space BMO.

In Chapter 2, we first list elements of stochastic calculus and introduce the Bellman function of stochastic optimal control. Then in Section 2.6, we collect examples that show the perfect analogy between stochastic optimal control and a wide class of harmonic analysis problems. After that, we turn our attention to a class of problems from complex analysis that also can be adapted to the Bellman method. One of these problems is finding Pichorides constants, yet another question is concerned with the solution of Gohberg–Krupnik problem by B. Hollenbeck and I. Verbitsky [**69**]. Our main goal is to show that all harmonic analysis problems in this chapter can be interpreted as problems of stochastic optimal control. An important disclaimer should be made: the stochastic optimal control point of view helps us to write down a correct Bellman partial differential equation, but it does not, in any sense, help to solve it. And solving it can be a major difficulty.

Chapter 3 is devoted to sharp estimates of conformal martingales. We then use these results to consider one particular singular integral, the Ahlfors–Beurling transform. We give the best up-to-date estimates of this transform.

Chapter 4 demonstrates an interesting and unexpected feature of the Bellman function technique. Namely, it has been noticed that the Bellman function built for one problem can be used in another problem, sometimes not too close to the original problem. This allows us to use the Bellman functions for the weighted martingale transform to have the right estimates for much more complicated dyadic singular operators, the so-called dyadic shifts. Moreover, one need not know the precise form of the Bellman function, one should just know of its existence. This idea for the Ahlfors–Beurling transform was used by S. Petermichl and A. Volberg in [**152**]. In that chapter, we follow the ideas of S. Treil [**181**] with a slight modification.

It has been noticed repeatedly that the Bellman function technique can be used not only to prove the conjectural estimates of singular integrals, but also to disprove the estimates. This is, roughly speaking, the consequence of the fact that the language of Bellman functions is often exactly adequate and equivalent to harmonic analysis problems for which these functions are built. This observation helps to find sharp constants in several endpoint estimates for singular integrals. That point of view also brings counterexamples to several well-known conjectures. We devote Section 5.2 to such counterexamples. The rest of this chapter is devoted to using the Bellman function technique to find sharp estimates in several classical problems concerning the square function operator. Even though the sharp constants for this operator have been studied since 1975, there are still open questions and we discuss them in Chapter 5.

I.5 Notation

We conclude this Introduction by a short list of notation that will be used throughout the whole book.

The average of a summable function w over an interval I will be denoted by the symbol $\langle w \rangle_{I}$:

$$\langle w \rangle_{I} \stackrel{\text{def}}{=} \frac{1}{|I|} \int_{I} w(t)\, dt\,,$$

where $|I|$ stands for the Lebesgue measure of I.

We introduce the Haar system, normalized in L^∞:

$$H_I(t) = \begin{cases} -1 & \text{if } t \in I_-, \\ 1 & \text{if } t \in I_+, \end{cases} \tag{0.1}$$

and another one, normalized in L^2:

$$h_I(t) = \frac{1}{\sqrt{|I|}} H_I(t).$$

Then

$$|I|(\langle w\rangle_{I_+} - \langle w\rangle_{I_-}) = 2(w, H_I)$$

and

$$\sqrt{|I|}(\langle w\rangle_{I_+} - \langle w\rangle_{I_-}) = 2(w, h_I).$$

The characteristic function of a measurable set E is denoted by $\mathbf{1}_E$.

The symbol $\mathcal{D}$ stands for a dyadic lattice, and $\mathcal{D}_n$ stands for the grid of intervals (or cubes) of length (or side-length) 2^{-n}, $n \in \mathbb{Z}$. The σ-algebra generated by $\mathcal{D}_n$ is denoted by $\mathcal{F}_n$.

The symbol $\mathbb{E}_n$ stands for the expectation with respect to the σ-algebra $\mathcal{F}_n$. Then Δ_n stands for $\mathbb{E}_{n+1} - \mathbb{E}_n$, $\Delta_I \stackrel{\text{def}}{=} \mathbf{1}_I \Delta_n$ for $I \in \mathcal{D}_n$, and thus,

$$\Delta_n = \sum_{I \in \mathcal{D}_n} \Delta_I .$$

Bellman functions are usually denoted by $\mathbf{B}$, but in Sections 5.4–5.7 they are denoted by $\mathbf{U}$ to follow the established tradition coming from probability.

The matrix of second derivatives of a function B on $\mathbb{R}^d$ (the Hessian matrix) is denoted by $\frac{d^2 B}{dx^2}$ or H_B. The symbol $d^2 B$ stands for the second differential form of B, namely, for the quadratic form $(H_B\, dx, dx)$.

1

Examples of Exact Bellman Functions

1.1 A Toy Problem

Let us start by considering the following simple problem. Suppose we have two positive functions f_1 and f_2 on an interval I, $I \subset \mathbb{R}$, bounded, say, by 1 and having prescribed averages: $\langle f_i \rangle_{_I} = x_i$. We are interested in their scalar product: how large or how small it can be. That is, we would like to find the following two functions:

$$\mathbf{B}^{\max}(x_1, x_2) \stackrel{\text{def}}{=} \sup\left\{\langle f_1 f_2 \rangle_{_I} : 0 \le f_i \le 1,\ \langle f_i \rangle_{_I} = x_i\right\} \tag{1.1.1}$$

$$\mathbf{B}^{\min}(x_1, x_2) \stackrel{\text{def}}{=} \inf\left\{\langle f_1 f_2 \rangle_{_I} : 0 \le f_i \le 1,\ \langle f_i \rangle_{_I} = x_i\right\} \tag{1.1.2}$$

These functions will be called the Bellman functions of the corresponding extremal problem. In this simple case, the functions can be found by elementary consideration without using any special techniques. Nevertheless, we approach this problem as "a serious one" and provide all the steps in its derivation that we will need in the future consideration of more serious problems.

In what follows, we will consider only the first of these functions, and it will be denoted simply by $\mathbf{B}$ rather than $\mathbf{B}^{\max}$. The first question is about the domain of definition of our function. It is natural to define it on the set of all $x = (x_1, x_2) \in \mathbb{R}^2$ for which there exists at least one pair of test functions f_1 and f_2 such that $\langle f_i \rangle_{_I} = x_i$.

Definition 1.1.1 For a pair of functions $\{f_1, f_2\}$ from $L^1(I)$, we call the point $\mathfrak{b}_{f_1,f_2} \in \mathbb{R}^2$,

$$\mathfrak{b} = \mathfrak{b}_I(f_1, f_2) \stackrel{\text{def}}{=} (\langle f_1 \rangle_{_I}, \langle f_2 \rangle_{_I}),$$

the *Bellman point* of this pair. Very often, the pair of functions is fixed and we are interested in the dependence of the Bellman point on the interval. Then we omit arguments and use only the interval as the index:

$$\mathfrak{b}_J = (\langle f_1\rangle_{_J}, \langle f_2\rangle_{_J}) \quad \text{for any interval } J,\, J \subset I\,.$$

Clearly, the Bellman points of all admissible pairs fill the square

$$\Omega = \{x = (x_1, x_2)\colon 0 \le x_i \le 1\}\,.$$

Of course, function $\mathbf{B}$ is formally defined outside the square Ω as well, but it is not interesting to consider this function there because the supremum of the empty set is $-\infty$. Let us state this assertion as a formal proposition. It is trivial in this case, but it might not be so trivial for a more serious problem.

Proposition 1.1.2 (Domain of Definition) *The function* $\mathbf{B}$ *is defined on the domain* Ω.

Proof On the one hand, for any pair of test functions f_1, f_2, we have $0 \le \langle f_i\rangle_{_I} \le 1$, i.e., $\mathfrak{b}_I(f_1, f_2) \in \Omega$. On the other hand, for any $x \in \Omega$, the pair of constant functions $f_i \equiv x_i$ is an admissible pair and $\mathfrak{b}_I(x_1, x_2) = x$. □

Proposition 1.1.3 (Independence on the Interval) *The function* $\mathbf{B}$ *does not depend on the interval* I*, where the test functions are defined.*

Proof Indeed, if we have two intervals I_1 and I_2, then the linear change of variables maps the set of test functions from one interval to another preserving all averages. Therefore, for both intervals, the supremum in the definition of the Bellman function is taken over by the same set. □

We know the values of our function on the boundary $\partial\Omega$.

Proposition 1.1.4 (Boundary Conditions)

$$\begin{aligned} \mathbf{B}(0, x_2) &= 0, \qquad & \mathbf{B}(1, x_2) &= x_2, \\ \mathbf{B}(x_1, 0) &= 0, \qquad & \mathbf{B}(x_1, 1) &= x_1. \end{aligned} \tag{1.1.3}$$

Proof We easily know the boundary values because for these points, the set, over which supremum in the definition of the Bellman function is taken, consists of only one element. Indeed, if $\langle f_i\rangle_{_I} = 0$, then $f_i = 0$ almost everywhere (because $f_i \ge 0$), and therefore, $\langle f_1 f_2\rangle_{_I} = 0$. If $\langle f_i\rangle_{_I} = 1$, then $f_i = 1$ almost everywhere (because $f_i \le 1$), and hence, $\langle f_i f_j\rangle_{_I} = x_j$. □

Our function possesses an additional symmetry property:

PROPOSITION 1.1.5 (Symmetry)

$$\mathbf{B}(x_1, x_2) = \mathbf{B}(x_2, x_1)\,. \tag{1.1.4}$$

PROOF We can interchange the roles of f_1 and f_2 without changing the value of $\langle f_1 f_2 \rangle_I$. Then we interchange x_1 and x_2 keeping the value of the Bellman function stable. □

PROPOSITION 1.1.6 (Main Inequality) *For every pair of points $x^\pm$ from Ω and every pair of positive numbers $\alpha^\pm$ such that $\alpha^- + \alpha^+ = 1$, the following inequality holds:*

$$\mathbf{B}(\alpha^- x^- + \alpha^+ x^+) \geq \alpha^- \mathbf{B}(x^-) + \alpha^+ \mathbf{B}(x^+). \tag{1.1.5}$$

PROOF Let us split the interval I into two parts: $I = I^- \cup I^+$ such that $|I^\pm| = \alpha^\pm |I|$. The integral in the definition of $\mathbf{B}$ can be presented as a sum of two integrals, the first over I^- and the second over I^+:

$$\int_I f_1(s) f_2(s)\, ds = \int_{I^-} f_1(s) f_2(s)\, ds + \int_{I^+} f_1(s) f_2(s)\, ds.$$

After dividing over $|I|$ we get

$$\langle f_1 f_2 \rangle_I = \alpha^- \langle f_1 f_2 \rangle_{I^-} + \alpha^+ \langle f_1 f_2 \rangle_{I^+}.$$

Now, using the independence of the Bellman function on the interval (Proposition 1.1.3), we choose functions $f_i^\pm$ on the intervals $I^\pm$ such that they almost give us the supremum in the definition of $\mathbf{B}(x^\pm)$, i.e.,

$$\langle f_1^\pm f_2^\pm \rangle_{I^\pm} \geq \mathbf{B}(x^\pm) - \eta,$$

for a fixed small $\eta > 0$. Then for the functions $f_i(s),\ \ i = 1, 2$, on I, defined as f_i^+ on I^+ and f_i^- on I^-, we obtain the inequality

$$\langle f_1 f_2 \rangle_I \geq \alpha^- \mathbf{B}(x^-) + \alpha^+ \mathbf{B}(x^+) - \eta. \tag{1.1.6}$$

Observe that the pair of the compounded functions f_i is an admissible pair of test function corresponding to the point $x = \alpha^- x^- + \alpha^+ x^+$. Indeed, $x^\pm = \mathfrak{b}_{I^\pm}(f_1^\pm, f_2^\pm) = \mathfrak{b}_{I^\pm}(f_1, f_2)$, and therefore,

$$\mathfrak{b}_I(f_1, f_2) = \alpha^- \mathfrak{b}_{I^-}(f_1, f_2) + \alpha^+ \mathfrak{b}_{I^+}(f_1, f_2) = \alpha^- x^- + \alpha^+ x^+ = x.$$

The inequality $0 \leq f_i \leq 1$ is clearly fulfilled as well. So, we can take supremum in (1.1.6) over all admissible pairs of functions. This yields

$$\mathbf{B}(x) \geq \alpha^- \mathbf{B}(x^-) + \alpha^+ \mathbf{B}(x^+) - \eta,$$

which proves the main inequality because η is arbitrarily small. □

PROPOSITION 1.1.7 (Obstacle Condition)

$$\mathbf{B}(x) \geq x_1 \cdot x_2. \tag{1.1.7}$$

PROOF Since the constant functions $f_i = x_i$ belong to the set of admissible test functions corresponding to the point x, we come to the desired inequality $\sup\{\langle f_1 f_2 \rangle : \langle f_i \rangle_I = x_i\} \geq \langle x_1 x_2 \rangle_I = x_1 x_2$. □

Before stating the next proposition, we introduce some notation. Let $\mathcal{I}$ be a family of subintervals of an interval I with the following properties:

- $I \in \mathcal{I}$;
- if $J \in \mathcal{I}$, then there is a couple of almost disjoint intervals $J^\pm$ (i.e., with the disjoint interiors), such that $J = J^- \cup J^+$;
- $\mathcal{I} = \cup_{n \geq 0} \mathcal{I}_n$, where $\mathcal{I}_0 = \{I\}$, $\mathcal{I}_{n+1} = \{J^-, J^+ : J \in \mathcal{I}_n\}$;
- $\lim_{n \to \infty} \max\{|J| : J \in \mathcal{I}_n\} = 0$.

If the family $\mathcal{I}$ satisfies the following additional condition

- $|J^-| = |J^+|$,

it is called *dyadic*. For the dyadic family of subintervals, we use notation $\mathcal{D}(I)$ instead of $\mathcal{I}$.

PROPOSITION 1.1.8 (Bellman Induction) *If B is a continuous function on the domain Ω satisfying the main inequality (that is just concavity condition) and obstacle condition* (1.1.7), *then* $\mathbf{B}(x) \leq B(x)$.

PROOF Fix an interval I and its splitting $\mathcal{I}$. Take an arbitrary point $x \in \Omega$ and two test function f_1 and f_2 on I, $0 \leq f_i \leq 1$, such that $x = \mathfrak{b}_I(f_1, f_2)$. We can rewrite the main inequality in the form

$$|J|B(\mathfrak{b}_J)| \geq |J^+|B(\mathfrak{b}_{J^+}) + |J^-|B(\mathfrak{b}_{J^-}).$$

Let us take the sum of the earlier inequalities when J runs over $\mathcal{I}_k$, the set of subintervals of kth generation. Then $J^\pm$ are all intervals of the set $\mathcal{I}_{k+1}$, and we get

$$\sum_{J \in \mathcal{I}_k} |J|B(\mathfrak{b}_J) \geq \sum_{J \in \mathcal{I}_{k+1}} |J|B(\mathfrak{b}_J).$$

Therefore,

$$|I|B(x)| = |I|B(\mathfrak{b}_I) = \sum_{J \in \mathcal{I}_0} |J|B(\mathfrak{b}_J) \geq \sum_{J \in \mathcal{I}_n} |J|B(\mathfrak{b}_J) = \int_I B(x^{(n)}(s))\, ds,$$

where $x^{(n)}$ is a step function defined in the following way: $x^{(n)}(s) = \mathfrak{b}_J$, when $s \in J$, $J \in \mathcal{I}_n$.

We know that $x^{(n)}(s) \to (f_1(s), f_2(s))$ almost everywhere by the Lebesgue differentiation theorem. Since B is continuous, we have $B(x^{(n)}(s)) \to B(f_1(s), f_2(s))$. Now, using the obstacle condition (1.1.7) and the Lebesgue dominated convergence theorem, we can pass to the limit in the obtained inequality as $n \to \infty$.

$$|I|B(x) \geq \int_I B(f_1(s), f_2(s))\, ds \geq \int_I f_1(s) f_2(s)\, ds = |I|\langle f_1 f_2 \rangle_I. \quad (1.1.8)$$

Taking supremum in this inequality over all admissible pairs f_1, f_2 with $\mathfrak{b}_I(f_1, f_2) = x$, we come to the desired estimate. □

According to this proposition, every concave function satisfying the obstacle condition gives us an upper estimate of the functional under consideration. If we are interested in a sharp estimate, we need to look for minimal possible such functions. Due to the symmetry (see Proposition 1.1.5), it is sufficient to consider $x_1 \leq x_2$.

On a triangle, we know our function at the vertices: $\mathbf{B}(0,0) = 0$, $\mathbf{B}(0,1) = 0$, and $\mathbf{B}(1,1) = 1$. The minimal possible concave function passing through the given three points is a linear function. In our case, it is the function $B(x) = x_1$. By the symmetry on the whole square Ω, we get the following *Bellman candidate*[1] $B(x) = \min\{x_1, x_2\}$.

In fact, we have already found the Bellman function.

THEOREM 1.1.9

$$\mathbf{B}(x) = \min\{x_1, x_2\}.$$

PROOF First of all, by Proposition 1.1.8, the upper estimate $\mathbf{B}(x) \leq B(x)$ is true because B is concave and $\min\{x_1, x_2\} \geq x_1 x_2$. Since there is no concave function satisfying the required boundary condition and that is less than B, we get $\mathbf{B} = B$.

However, in a more difficult problem, it is not so clear that the Bellman candidate cannot be diminished. By this reason, we demonstrate on this example how we will typically prove the lower estimate $\mathbf{B}(x) \geq B(x)$. To this end for every point $x \in \Omega$, we present an admissible test function, realizing the supremum in the definition of the Bellman function. In some papers, such a function (in our case, it is a pair of functions) is called an *extremizer*, but in other papers it is called an *optimizer*. We shall use both these words as synonyms. In our case, the possible pair of extremizers is very

[1] Such a term is used for a function possessing the necessary properties of the Bellman function, e.g., concavity, symmetry, boundary values, etc. After a Bellman candidate is presented, we need to check that it indeed is the desired Bellman function.

simple: $f_i = \mathbf{1}_{[0,x_i]}$. We evidently have $\langle f_i \rangle_{[0,1]} = x_i$ and $\langle f_1 f_2 \rangle_{[0,1]} = \min\{x_i\}$. Since by definition $\mathbf{B}(x)$ is the supremum of $\langle f_1 f_2 \rangle_{[0,1]}$, when f_i runs over all admissible pairs corresponding to the point x, $\mathbf{B}(x)$ is not less that this particular value, which is equal to $\min\{x_i\} = B(x)$. □

At the end of this section, we would like to explain how to find the extremizers mentioned earlier. Look at the proof of Proposition 1.1.8. Let us take $B = \mathbf{B}$ in this chain of inequalities choosing at the beginning f_1, f_2 to be a pair of extremizers. Since the first and the last terms in the chain of inequalities (1.1.8) are equal, namely, they are $|I|\mathbf{B}(x)$, we must have equalities in each step. In other words, we need to choose such a splitting $x = \alpha^- x^- + \alpha^+ x^+$ to have equality rather than inequality in (1.1.5). In our case, it is easy to do because our Bellman candidate is a concatenation of two linear functions, and if we deal only with one of these linear functions, we always have equality in (1.1.5). Based on this reason, in this simple situation, we can choose extremizers in an almost arbitrary way; the only condition is that all three points x and $x^\pm$ must be in the same triangle: either in $\{x\colon x_1 \le x_2\}$ or in $\{x\colon x_1 \ge x_2\}$.

Let us construct a pair of optimizers for some point x with $x_1 \le x_2$. First we draw the straight line passing through the points x and $x^- \stackrel{\text{def}}{=} (1,1)$. It intersects the boundary of Ω at the point $(0, \frac{x_2 - x_1}{1 - x_1}) \stackrel{\text{def}}{=} x^+$. So, we have $x = x_1 \cdot x^- + (1 - x_1) \cdot x^+$, i.e., $\alpha^- = x_1$, $\alpha^+ = 1 - x_1$, and we need to split our initial interval I (take $I = [0,1]$) in the union $I^- = [0, x_1]$ and $I^+ = [x_1, 1]$. The point $x^- = (1,1)$ is the Bellman point of the only pair $f_1 = f_2 = 1$, hence on $[0, x_1]$ we take both extremal functions equal identically to 1. The point $x^+ = (0, \frac{x_2 - x_1}{1 - x_1})$ is the Bellman point, for example, the pair of constant functions, and we can put $f_1 = 0$ and $f_2 = \frac{x_2 - x_1}{1 - x_1}$ on $[x_1, 1]$. It is easy to check whether this pair of functions gives us an extremizer. However, the second function of this extremizer differs from that presented earlier. What to do to get that extremizer? We only have to split I^+ once more, presenting x^+ as the convex combination of $(0,1)$ and $(0,0)$:

$$x^+ = \frac{x_2 - x_1}{1 - x_1}(0,1) + \frac{1 - x_2}{1 - x_1}(0,0), \qquad I^+ = [x_1, 1] = [x_1, x_2] \cup [x_2, 1].$$

The function f_1 is, as before, the zero function on both subintervals, but we have to take f_2 equal to 1 on $[x_1, x_2]$ and equal to 0 on $[x_2, 1]$. In this way, we come to the pair of functions presented earlier.

We would like to provide support now to the readers for whom the latter paragraph remains unclear: you meet such kind of construction (splitting the interval and representing a Bellman point as a convex combination of two (or more) other Bellman points) many times on the pages of this book. We hope that after several repetitions, the construction becomes absolutely clear.

Exercises

PROBLEM 1.1.1 Find the function $\mathbf{B}$ defined for a similar problem, where the restriction $0 \le f_i \le 1$ is replaced by $|f_i| \le 1$

PROBLEM 1.1.2 Find the function $\mathbf{B}^{\min}$ defined in (1.1.2).

PROBLEM 1.1.3 Find the function $\mathbf{B}^{\min}$ for the set of test functions described in Problem 1.1.1.

1.2 Buckley Inequality

For an interval I and a number $r > 1$, the symbol $A_\infty(I,r)$ denotes the r-"ball" in the Muckenhoupt class A_∞:

$$A_\infty(I,r) \stackrel{\text{def}}{=} \Big\{ w\colon w \in L^1(I),\ w \ge 0,\ \langle w \rangle_{\scriptscriptstyle J} \le r e^{\langle \log w \rangle_{\scriptscriptstyle J}}\ \forall J \subset I \Big\}. \tag{1.2.1}$$

We denote by $\mathcal{D}(I)$ the set of all dyadic subintervals of I and by $A^d_\infty(I,r)$ the dyadic analog of (1.2.1), i.e., in the definition of $A^d_\infty(I,r)$, we consider only $J \in \mathcal{D}(I)$.

THEOREM (Buckley [**19**]) *There exists a constant $c = c(r)$ such that*

$$\sum_{J \in \mathcal{D}(I)} |J| \Big(\frac{\langle w \rangle_{\scriptscriptstyle J+} - \langle w \rangle_{\scriptscriptstyle J-}}{\langle w \rangle_{\scriptscriptstyle J}} \Big)^2 \le c(r)|I|$$

for any weight w from $A^d_\infty(I,r)$.

Now, we are ready to introduce the main object of our consideration, the so-called Bellman function of the problem.

$$\begin{aligned} \mathbf{B}(x) &= \mathbf{B}(x_1,x_2;r) \\ &\stackrel{\text{def}}{=} \sup_{w \in A^d_\infty(I,r)} \Bigg\{ \frac{1}{|I|} \sum_{J \in \mathcal{D}(I)} |J| \left(\frac{\langle w \rangle_{\scriptscriptstyle J+} - \langle w \rangle_{\scriptscriptstyle J-}}{\langle w \rangle_{\scriptscriptstyle J}} \right)^2 : \\ &\qquad\qquad \langle w \rangle_{\scriptscriptstyle I} = x_1,\ \langle \log w \rangle_{\scriptscriptstyle I} = x_2 \Bigg\}. \end{aligned} \tag{1.2.2}$$

Let us note that we did not assign the index I to $\mathbf{B}$ despite the fact that all test functions w in its definition are considered on I. This omission is not due to our desire to simplify notation, but rather an indication of the very important fact that the function $\mathbf{B}$ does not depend on I; Proposition 1.1.3 holds in this situation by the same reason.

For a given weight $w \in A_\infty^d(I,r)$, we introduce a Bellman point $\mathfrak{b}_I(w)$ in the following way: $\mathfrak{b}_I(w) = (\langle w \rangle_I, \langle \log w \rangle_I)$. Note that for all admissible weights and for any dyadic subinterval $J \subset I$, the corresponding Bellman point $\mathfrak{b}_J(w)$ is in the following domain Ω_r:

$$\Omega_r \stackrel{\text{def}}{=} \left\{ x = (x_1, x_2) \colon \log \frac{x_1}{r} \le x_2 \le \log x_1 \right\}.$$

Indeed, the right bound is simply Jensen's inequality and the left one is fulfilled because our weight w is from $A_\infty^d(I,r)$.

To show that Ω_r is the domain of the function $\mathbf{B}$, we need to check that for any point $x \in \Omega_r$ there exists an admissible weight with $\mathfrak{b}_I(w) = x$. However, we leave this for the reader as an exercise (see Problem 1.2.1).

Now we prove the crucial property of the function $\mathbf{B}$ that follows directly from its definition.

Lemma 1.2.1 (Main Inequality) *For every pair of points $x^\pm$ from Ω_r such that their mean $x = (x^+ + x^-)/2$ is also in Ω_r, the following inequality holds:*

$$\mathbf{B}(x) \ge \frac{\mathbf{B}(x^+) + \mathbf{B}(x^-)}{2} + \left(\frac{x_1^+ - x_1^-}{x_1} \right)^2. \tag{1.2.3}$$

Proof Let us split the sum in the definition of $\mathbf{B}$ into three parts: the sum over $\mathcal{D}(I^+)$, the sum over $\mathcal{D}(I^-)$, and an additional term corresponding to I itself:

$$\begin{aligned}
&\frac{1}{|I|} \sum_{J \in \mathcal{D}(I)} |J| \left(\frac{\langle w \rangle_{J+} - \langle w \rangle_{J-}}{\langle w \rangle_J} \right)^2 \\
&\quad = \frac{1}{2|I^+|} \sum_{J \in \mathcal{D}(I^+)} |J| \left(\frac{\langle w \rangle_{J+} - \langle w \rangle_{J-}}{\langle w \rangle_J} \right)^2 \\
&\quad + \frac{1}{2|I^-|} \sum_{J \in \mathcal{D}(I^-)} |J| \left(\frac{\langle w \rangle_{J+} - \langle w \rangle_{J-}}{\langle w \rangle_I} \right)^2 \\
&\quad + \left(\frac{\langle w \rangle_{I+} - \langle w \rangle_{I-}}{\langle w \rangle_I} \right)^2.
\end{aligned}$$

Using the fact that $\mathbf{B}$ does not depend on the interval where the test functions are defined, we can choose two weights $w^\pm$ on the intervals $I^\pm$ that almost give us the supremum in the definition of $\mathbf{B}(x^\pm)$, i.e.,

$$\frac{1}{|I^\pm|}\sum_{J\in\mathcal{D}(I^\pm)}|J|\left(\frac{\langle w^\pm\rangle_{J+}-\langle w^\pm\rangle_{J-}}{\langle w^\pm\rangle_J}\right)^2\geq \mathbf{B}(x^\pm)-\eta,$$

for an arbitrary fixed small $\eta>0$. Then for the weight w on I, defined as w^+ on I^+ and w^- on I^-, we obtain the inequality

$$\frac{1}{|I|}\sum_{J\in\mathcal{D}(I)}|J|\left(\frac{\langle w\rangle_{J+}-\langle w\rangle_{J-}}{\langle w\rangle_J}\right)^2\geq\frac{\mathbf{B}(x^+)+\mathbf{B}(x^-)}{2}-\eta+\left(\frac{x_1^+-x_1^-}{x_1}\right)^2. \tag{1.2.4}$$

Observe that the compound weight w is an admissible weight, corresponding to the point x. Indeed, $x^\pm=\mathfrak{b}_{I^\pm}(w)$ and by the construction of $w^\pm$ we have $w^\pm\in A_\infty^d(I^\pm,r)$. Therefore, the weight w satisfies the inequality $\langle w\rangle_J\leq re^{\langle\log w\rangle_J}$ for all $J\in\mathcal{D}(I^+)$, since w^+ does, and for all $J\in\mathcal{D}(I^-)$, since w^- does. Lastly, $\langle w\rangle_I\leq re^{\langle\log w\rangle_I}$, because, by assumption, $x\in\Omega_r$.

We can now take supremum in (1.2.4) over all admissible weights w, which yields

$$\mathbf{B}(x)\geq\frac{\mathbf{B}(x^+)+\mathbf{B}(x^-)}{2}-\eta+\left(\frac{x_1^+-x_1^-}{x_1}\right)^2.$$

This proves the main inequality because η is arbitrarily small. □

Lemma 1.2.2 (Boundary Condition)

$$\mathbf{B}(x_1,\log x_1)=0.$$

Proof Let us take a boundary point x of our domain Ω_r, that is a point with $x_2=\log x_1$. Since the equality in Jensen's inequality $e^{\langle w\rangle}\leq\langle e^w\rangle$ occurs only for constant functions w, the only test function corresponding to x is the constant (up to a set of measure zero) weight $w=x_1$. So, on this boundary, we have $\mathbf{B}(x)=0$. □

Lemma 1.2.3 (Homogeneity) *There is a function g on $[1,r]$ satisfying $g(1)=0$ and such that*

$$\mathbf{B}(x)=\mathbf{B}(x_1e^{-x_2},0)=g(x_1e^{-x_2}).$$

Proof For a weight w on an interval I and a positive number τ, consider a new weight $\tilde{w}=\tau w$. If $x=\mathfrak{b}_I(w)$, i.e., $x_1=\langle w\rangle_I,\ \ x_2=\langle\log w\rangle_I$, then for the point $\mathfrak{b}_I(\tilde{w})=\tilde{x}$ we have $\tilde{x}_1=\tau x_1$, $\tilde{x}_2=x_2+\log\tau$. Note that the expression in the definition of $\mathbf{B}$ is homogeneous of order 0 with respect to w, i.e., it does not depend on τ. Since the weights w and $\tilde{w}$ run over the whole set $A_\infty^d(I,r)$ simultaneously, we get $\mathbf{B}(x) = \mathbf{B}(\tilde{x})$. Choosing $\tau = e^{-x_2}$, we obtain

$$\mathbf{B}(x) = \mathbf{B}(x_1 e^{-x_2}, 0).$$

To complete the proof, it suffices to take $g(s) = \mathbf{B}(s,0)$. The boundary condition $g(1) = 0$ holds due to Lemma 1.2.2. □

We are now ready to demonstrate how the Bellman induction works in this case.

LEMMA 1.2.4 (Bellman Induction) *Let B be a nonnegative function on Ω_r satisfying the main inequality in Ω_r (Lemma 1.2.1). Then*

$$\mathbf{B}(x) \le B(x).$$

PROOF Fix an interval I and a point $x \in \Omega_r$. Take an arbitrary weight $w \in A_\infty^d(I,r)$ such that $\mathfrak{b}_I(w) = x$. Let us repeatedly use the main inequality in the form

$$|J|\,B(\mathfrak{b}_J) \ge |J^+|\,B(\mathfrak{b}_{J^+}) + |J^-|\,B(\mathfrak{b}_{J^-}) + |J|\left(\frac{\langle w\rangle_{J^+} - \langle w\rangle_{J^-}}{\langle w\rangle_J}\right)^2,$$

applying it first to I, then to the intervals of the first generation (that is $I^\pm$), and so on until $\mathcal{D}_n(I)$:

$$\begin{aligned}|I|\,B(\mathfrak{b}_I) &\ge |I^+|\,B(\mathfrak{b}_{I^+}) + |I^-|\,B(\mathfrak{b}_{I^-}) + |I|\left(\frac{\langle w\rangle_{I^+} - \langle w\rangle_{I^-}}{\langle w\rangle_I}\right)^2 \\ &\ge \sum_{J\in\mathcal{D}_n(I)} |J|\,B(\mathfrak{b}_J) + \sum_{k=0}^{n-1}\sum_{J\in\mathcal{D}_k(I)} |J|\left(\frac{\langle w\rangle_{J^+} - \langle w\rangle_{J^-}}{\langle w\rangle_J}\right)^2.\end{aligned}$$

Therefore,

$$\sum_{k=0}^{n-1}\sum_{J\in\mathcal{D}_k(I)} |J|\left(\frac{\langle w\rangle_{J^+} - \langle w\rangle_{J^-}}{\langle w\rangle_J}\right)^2 \le |I|\,B(\mathfrak{b}_I),$$

and passing to the limit as $n \to \infty$, we get

$$\sum_{J\in\mathcal{D}(I)} |J|\left(\frac{\langle w\rangle_{J^+} - \langle w\rangle_{J^-}}{\langle w\rangle_J}\right)^2 \le |I|\,B(x).$$

Taking supremum over all admissible weight w corresponding to the point x, we come to the desired estimate. □

Corollary 1.2.5 *Let g be a nonnegative function on $[1,r]$ such that the function $B(x) \stackrel{\text{def}}{=} g(x_1 e^{-x_2})$ satisfies inequality (1.2.3) in Ω_r. Then Buckley's inequality holds with the constant $c(r) = \|g\|_{L^\infty([1,r])}$.*

A natural question arises: How to find such a function g? To answer it, we first replace our main inequality, which is an inequality in finite differences, by a differential inequality. Let us denote the difference between x^+ and x^- by 2Δ, then $x^\pm = x \pm \Delta$ and the Taylor expansion around the point x gives us

$$\begin{aligned} B(x^\pm) &= B(x) \pm \frac{\partial B}{\partial x_1}\Delta_1 \pm \frac{\partial B}{\partial x_2}\Delta_2 \\ &\quad + \frac{1}{2}\frac{\partial^2 B}{\partial x_1^2}\Delta_1^2 + \frac{\partial^2 B}{\partial x_1 \partial x_2}\Delta_1\Delta_2 + \frac{1}{2}\frac{\partial^2 B}{\partial x_2^2}\Delta_2^2 + o(|\Delta|^2), \end{aligned}$$

and, therefore,

$$\begin{aligned} &\frac{B(x^+) + B(x^-)}{2} + \left(\frac{x_1^+ - x_1^-}{x_1}\right)^2 - B(x) \\ &= \frac{1}{2}\frac{\partial^2 B}{\partial x_1^2}\Delta_1^2 + \frac{\partial^2 B}{\partial x_1 \partial x_2}\Delta_1\Delta_2 + \frac{1}{2}\frac{\partial^2 B}{\partial x_2^2}\Delta_2^2 + 4\left(\frac{\Delta_1}{x_1}\right)^2 + o(|\Delta|^2). \end{aligned}$$

Thus, under the assumption that our candidate B is sufficiently smooth, the main inequality (1.2.3) implies the following matrix differential inequality:

$$\begin{pmatrix} \dfrac{\partial^2 B}{\partial x_1^2} + \dfrac{8}{x_1^2} & \dfrac{\partial^2 B}{\partial x_1 \partial x_2} \\[2ex] \dfrac{\partial^2 B}{\partial x_1 \partial x_2} & \dfrac{\partial^2 B}{\partial x_2^2} \end{pmatrix} \leq 0. \tag{1.2.5}$$

That is, this matrix has to be nonpositively defined.

By the preceding two lemmata, we can restrict our search to functions B of the form $B(x_1, x_2) = g(x_1 e^{-x_2})$, where g is a function on the interval $[1,r]$. In terms of g, our condition (1.2.5) can be rewritten as follows:

$$\begin{pmatrix} e^{-2x_2}\left(g'' + \dfrac{8}{s^2}\right) & -e^{-x_2}(sg')' \\[2ex] -e^{-x_2}(sg')' & s(sg')' \end{pmatrix} \leq 0,$$

where $g = g(s)$ and $s = x_1 e^{-x_2}$. From this matrix inequality, we conclude that

$$g'' + \frac{8}{s^2} \leq 0, \tag{1.2.6}$$

$$(sg')' \leq 0, \tag{1.2.7}$$

and that the determinant of the matrix must be nonnegative. However, we replace the last requirement by a stronger one – we require the determinant to be identically zero. This requirement comes from our desire to find the best possible estimate: If we take an extremal weight w, i.e., a weight on which the supremum in the definition of the Bellman function is attained, then we must have equalities on each step of the Bellman induction; therefore, on each step the main inequality (1.2.3) becomes an equality. Thus, for each dyadic subinterval J of I, there exists a direction through the point $\mathfrak{b}_J$ in Ω_r along which the quadratic form given by (1.2.5) is identically zero. Hence, the matrix (1.2.5) has a nontrivial kernel and so must have a zero determinant.[2]

Calculating the determinant, we get the equation

$$\Big(g' - \frac{8}{s}\Big)(sg')' = 0.$$

The general solution of this equation is $g(s) = c\log s + c_1$. Due to the boundary condition $g(1) = 0$, we have to take $c_1 = 0$.

Now we need to choose another constant, c. To this end, we return to the necessary conditions (1.2.6–1.2.7). The second inequality is fulfilled for all c because the expression is identically zero, while the first one gives $c \geq 8$. Since we would like to have g as small as possible (as it gives the upper bound in Buckley's inequality), it is natural to take $c = 8$. Finally, we get

$$g(s) = 8\log s \qquad \text{and} \qquad B(x_1, x_2) = 8(\log x_1 - x_2).$$

Lemma 1.2.6 *The function*

$$B(x_1, x_2) = 8(\log x_1 - x_2)$$

satisfies the main inequality (1.2.3).

Proof Put, as before, $\Delta = \frac{1}{2}(x^+ - x^-)$, so $x^\pm = x \pm \Delta$. Then

$$B(x) - \frac{B(x^+) + B(x^-)}{2} - \left(\frac{x_1^+ - x_1^-}{x_1}\right)^2$$
$$= 8\log x_1 - 8x_2 - 4\log(x_1^+ x_1^-) + 4(x_2^+ + x_2^-) - \left(\frac{x_1^+ - x_1^-}{x_1}\right)^2$$

[2] This is not a proof, the arguments are not absolutely correct; for example, the existence of an extremal weight w is not guaranteed. The supremum in the definition of the Bellman function can be not attainable and only an extremal sequence of weights can realize it. (By the way, for the Buckley inequality it is just the case.) Nevertheless, in the process of searching for a Bellman candidate, we may assume whatever we want (e.g., its smoothness to replace a finite difference condition by a differential one), but a rigorous proof starts after a candidate is found and we check that it is the true Bellman function.

$$= 4\log\frac{x_1^2}{(x_1+\Delta_1)(x_1-\Delta_1)} - 4\left(\frac{\Delta_1}{x_1}\right)^2$$

$$= -4\left[\log\left(1-\left(\frac{\Delta_1}{x_1}\right)^2\right) + \left(\frac{\Delta_1}{x_1}\right)^2\right] \geq 0.$$

□

Now we can apply Lemma 1.2.4 to $g(s) = 8\log s$, which yields the following:

THEOREM *The estimate*

$$\sum_{J\in\mathcal{D}(I)} |J|\Big(\frac{\langle w\rangle_{J+} - \langle w\rangle_{J-}}{\langle w\rangle_J}\Big)^2 \leq 8\log r\,|I|$$

holds for any weight $w \in A_\infty^d(I,r)$.

We would like to emphasize that we still have not found the Bellman function $\mathbf{B}$. The theorem just proved guarantees only the estimate

$$\mathbf{B}(x) \leq 8(\log x_1 - x_2).$$

To prove that this Bellman candidate is the true Bellman function (what proves sharpness of the earlier estimate), we need to find extremizers for every point of the domain Ω_r. However, this is a much more difficult task than it was in our previous example. For this reason, we now stop our investigation of the Bellman function for the Buckley inequality. The more experienced reader interested in completing investigation of this Bellman function can refer to Section 1.10, where we not only present the extremizers for the discussed Bellman function but also find the minimal Bellman function with completely different extremizers.

Exercises

PROBLEM 1.2.1 Check that the function $\mathbf{B}$ defined on the whole domain Ω_r, i.e., for every point x, $x \in \Omega_r$, there exists a function $w \in A_\infty^d(I,r)$ such that $x = \mathfrak{b}_I(w)$.

PROBLEM 1.2.2 Try to repeat the earlier procedure for finding the function $\mathbf{B}^{\min}$ defined by (1.2.2), where sup is replaced by inf.

PROBLEM 1.2.3 Try to find the following Bellman function

$$\mathbf{B}(x; m, M) \stackrel{\text{def}}{=} \sup_{u,v}\Big\{\frac{1}{|I|}\sum_{J\in\mathcal{D}(I)} |(u,h_J)|\,|(v,h_J)|\Big\},$$

where the supremum is taken over the set of all pairs of weights $u,\ v$ such that $\langle u\rangle_I = x_1,\ \langle v\rangle_I = x_2$, and $m^2 \le \langle u\rangle_J\langle v\rangle_J \le M^2,\ \forall J\in\mathcal{D}(I)$.

As a result you have to prove the following theorem:

THEOREM *If two weights $u, v \in L^1(I)$ satisfy the condition*

$$\sup_{J\in\mathcal{D}(I)} \langle u\rangle_J\langle v\rangle_J \le M^2,$$

then

$$\frac{1}{|I|}\sum_{J\in\mathcal{D}(I)} |J|\,|\langle u\rangle_{J+} - \langle u\rangle_{J-}|\,|\langle v\rangle_{J+} - \langle v\rangle_{J-}| \le 16M\sqrt{\langle u\rangle_I\langle v\rangle_I}.$$

1.3 John–Nirenberg Inequality

A function $\varphi\in L^1(I)$ is said to belong to the space $\mathrm{BMO}(I)$ if

$$\sup_J \langle|\varphi(s) - \langle\varphi\rangle_J|\rangle_J < \infty$$

for all subintervals $J\subset I$. If this condition holds only for the dyadic subintervals $J\in\mathcal{D}(I)$, we will write $\varphi\in\mathrm{BMO}^d(I)$. In fact, the following is true for any $p,\ p\in(0,\infty)$:

$$\varphi\in\mathrm{BMO}(I) \iff \Big(\sup_{J\subset I}\frac{1}{|J|}\int_J|\varphi(s)-\langle\varphi\rangle_J|^p\,ds\Big)^{\frac{1}{p}} < \infty.$$

If we factor over the constants, we get a normed space, where the expression on the right-hand side can be taken as one of the equivalent norm for any $p\in[1,\infty)$. In what follows, we will use the L^2-based norm:

$$\|\varphi\|^2_{\mathrm{BMO}(I)} = \sup_{J\subset I}\frac{1}{|J|}\int_J|\varphi(s)-\langle\varphi\rangle_J|^2\,ds = \sup_{J\subset I}\left(\langle\varphi^2\rangle_J - \langle\varphi\rangle_J^2\right).$$

The BMO ball of radius ε centered at 0 will be denoted by BMO_ε. Using the Haar decomposition

$$\varphi(s) = \langle\varphi\rangle_I + \sum_{J\in\mathcal{D}(I)}(\varphi,h_J)h_J(s),$$

we can write down the expression for the norm in the following way:

$$\|\varphi\|^2_{\mathrm{BMO}(I)} = \sup_{J\subset I}\frac{1}{|J|}\sum_{L\in\mathcal{D}(J)}|(\varphi,h_L)|^2$$
$$= \frac{1}{4}\sup_{J\subset I}\frac{1}{|J|}\sum_{L\in\mathcal{D}(J)}|L|\left(\langle\varphi\rangle_{L+} - \langle\varphi\rangle_{L-}\right)^2.$$

THEOREM (John–Nirenberg **[88]**) *There exist absolute constants c_1 and c_2 such that*

$$\left|\{s \in I\colon |\varphi(s) - \langle\varphi\rangle_{_I}| \geq \lambda\}\right| \leq c_1 e^{-c_2 \frac{\lambda}{\|\varphi\|}} |I|$$

for all $\varphi \in \mathrm{BMO}_\varepsilon(I)$.

An equivalent, integral form of the same assertion is the following:

THEOREM *There exists an absolute constant ε_0 such that for any $\varphi \in \mathrm{BMO}_\varepsilon(I)$ with $\varepsilon < \varepsilon_0$, the inequality*

$$\langle e^{\varphi}\rangle_{_I} \leq c\, e^{\langle\varphi\rangle_I}$$

holds with a constant $c = c(\varepsilon)$ not depending on φ.

We shall prove the theorem in this integral form and find the sharp constant $c(\varepsilon)$. Our Bellman function

$$\mathbf{B}(x;\varepsilon) \stackrel{\text{def}}{=} \sup_{\varphi\in\mathrm{BMO}_\varepsilon(I)} \left\{\langle e^{\varphi}\rangle_{_I} : \mathfrak{b}_I(\varphi) = x\right\},$$

where $\mathfrak{b}_I(\varphi) \stackrel{\text{def}}{=} (\langle\varphi\rangle_{_I}, \langle\varphi^2\rangle_{_I})$ is the Bellman point corresponding to the test function φ and the interval I. It is clear that the set of all Bellman points is the domain

$$\Omega_\varepsilon \stackrel{\text{def}}{=} \left\{x = (x_1, x_2)\colon x_1^2 \leq x_2 \leq x_1^2 + \varepsilon^2\right\},$$

i.e., Ω_ε is the domain where $\mathbf{B}$ is defined. Let us note from the beginning that we will consider $\varepsilon < 1$ only because $\varphi(s) = -\log s \in \mathrm{BMO}_1([0,1])$ and $\langle e^{\varphi}\rangle_{[0,1]} = \infty$.

First, we will consider the dyadic problem and deduce the main inequality for the dyadic Bellman function.

LEMMA 1.3.1 (Main Inequality) *For every pair of points $x^\pm$ from Ω_ε such that their mean $x = (x^+ + x^-)/2$ is also in Ω_ε, the following inequality holds*

$$\mathbf{B}(x) \geq \frac{\mathbf{B}(x^+) + \mathbf{B}(x^-)}{2}. \tag{1.3.1}$$

PROOF The proof repeats almost verbatim the proof of the main inequality for the Buckley's Bellman function. We split the integral in the definition of $\mathbf{B}$ into two parts, the integral over I^+ and the one over I^- :

$$\int_I e^{\varphi(s)}\,ds = \int_{I^+} e^{\varphi(s)}\,ds + \int_{I^-} e^{\varphi(s)}\,ds.$$

Now we choose such functions $\varphi^\pm$ on the intervals $I^\pm$ that they almost give us the supremum in the definition of $\mathbf{B}(x^\pm)$, i.e,

$$\frac{1}{|I^\pm|}\int_{I^\pm} e^{\varphi^\pm(s)}\,ds \geq \mathbf{B}(x^\pm) - \eta,$$

for a fixed small $\eta > 0$. Then for the function φ on I, defined as φ^+ on I^+ and φ^- on I^-, we obtain the inequality

$$\frac{1}{|I|}\int_I e^{\varphi(s)}\,ds \geq \frac{\mathbf{B}(x^+) + \mathbf{B}(x^-)}{2} - \eta. \tag{1.3.2}$$

Observe that the compound function φ is an admissible test function corresponding to the point x. Indeed, $x^\pm = \mathfrak{b}_{I^\pm}(\varphi)$ and by construction $\varphi^\pm \in \mathrm{BMO}_\varepsilon^d(I^\pm)$; therefore, the function φ satisfies the inequality $\langle\varphi^2\rangle_{_J} - \langle\varphi\rangle_{_J}^2 \leq \varepsilon^2$ for all $J \in \mathcal{D}(I^+)$, since φ^+ does, and for all $J \in \mathcal{D}(I^-)$, since φ^- does. Lastly, $\langle\varphi^2\rangle_{_I} - \langle\varphi\rangle_{_I}^2 \leq \varepsilon^2$, because, by assumption, $x \in \Omega_\varepsilon$.

We can now take supremum in (1.3.2) over all admissible functions φ, which yields

$$\mathbf{B}(x) \geq \frac{\mathbf{B}(x^+) + \mathbf{B}(x^-)}{2} - \eta.$$

This proves the main inequality because η is arbitrarily small. □

As in the case of the Buckley inequality, the next step is to derive a boundary condition for $\mathbf{B}$.

LEMMA 1.3.2 (Boundary Condition)

$$\mathbf{B}(x_1, x_1^2) = e^{x_1}. \tag{1.3.3}$$

PROOF The function $\varphi(s) = x_1$ is the only test function corresponding to the point $x = (x_1, x_1^2)$, because the equality in the Hölder inequality $x_2 \geq x_1^2$ occurs only for constant functions. Hence, $e^\varphi = e^{x_1}$. □

Now we are ready to describe super-solutions as functions verifying the main inequality and the boundary conditions.

LEMMA 1.3.3 (Bellman Induction) *If B is a continuous function on the domain Ω_ε, satisfying the main inequality* (1.3.1) *for any pair $x^\pm$ of points from Ω_ε such that $x \stackrel{\text{def}}{=} \frac{x^+ + x^-}{2} \in \Omega_\varepsilon$, as well as the boundary condition* (1.3.3), *then* $\mathbf{B}(x) \leq B(x)$.

PROOF Fix a bounded function $\varphi \in \mathrm{BMO}_\varepsilon(I)$ and put $x = \mathfrak{b}_I(\varphi)$. As in the case of Buckley inequality, we rewrite the main inequality in the form

$$|J|\,B(\mathfrak{b}_J) \geq |J^+|\,B(\mathfrak{b}_{J^+}) + |J^-|\,B(\mathfrak{b}_{J^-}),$$

applying it first to I, then to the intervals of the first generation (that is $I^{\pm}$), and so on until $\mathcal{D}_n(I)$:

$$\begin{aligned}|I|B(\mathfrak{b}_I)| &\geq |I^+|B(\mathfrak{b}_{I^+}) + |I^-|B(\mathfrak{b}_{I^-})\\ &\geq \sum_{J\in\mathcal{D}_n(I)} |J|B(\mathfrak{b}_J) = \int_I B(x^{(n)}(s))\,ds,\end{aligned}$$

where $x^{(n)}(s) = \mathfrak{b}_J$, when $s \in J,\ \ J \in \mathcal{D}_n(I)$. (Recall that $\mathcal{D}_n(I)$ stands for the set of subintervals of n-th generation.) By the Lebesgue differentiation theorem, we have $x^{(n)}(s) \to (\varphi(s), \varphi^2(s))$ almost everywhere. Now, we can pass to the limit in this inequality as $n \to \infty$. Since φ is assumed to be bounded, $x^{(n)}(s)$ runs in a bounded (and, therefore, compact) subdomain of Ω_ε. Since B is continuous, it is bounded on any compact set and so, by the Lebesgue dominated convergence theorem, we can pass to the limit in the integral using the boundary condition (1.3.3):

$$|I|B(\mathfrak{b}_I(\varphi)) \geq \int_I B(\varphi(s), \varphi^2(s))\,ds = \int_I e^{\varphi(s)} ds = |I|\langle e^{\varphi}\rangle_{I}. \tag{1.3.4}$$

To complete the proof of the lemma, we need to pass from bounded to arbitrary BMO test functions. To this end, we will use the following result:

LEMMA 1.3.4 (Cut-Off Lemma) *Fix $\varphi \in \mathrm{BMO}(I)$ and two real numbers c, d such that $c < d$. Let $\varphi_{c,d}$ be the cut-off of φ at heights c and d :*

$$\varphi_{c,d}(s) = \begin{cases} c, & if\ \varphi(s) \leq c;\\ \varphi(s), & if\ c < \varphi(s) < d;\\ d, & if\ \varphi(s) \geq d.\end{cases} \tag{1.3.5}$$

Then

$$\langle\varphi_{c,d}^2\rangle_J - \langle\varphi_{c,d}\rangle_J^2 \leq \langle\varphi^2\rangle_J - \langle\varphi\rangle_J^2, \quad \forall J,\ J \subset I,$$

and, consequently,

$$\|\varphi_{c,d}\|_{\mathrm{BMO}} \leq \|\varphi\|_{\mathrm{BMO}}.$$

PROOF If we integrate the evident inequality

$$|\varphi_{c,d}(s) - \varphi_{c,d}(t)|^2 \leq |\varphi(s) - \varphi(t)|^2$$

over J with respect to s and once more with respect to t, then dividing the result over $2|J|^2$, we get the desired estimate. □

Now, let $\varphi \in \mathrm{BMO}_\varepsilon(I)$ be a function bounded from above. Then, by the earlier lemma, $\varphi_n \stackrel{\text{def}}{=} \varphi_{-n,\infty} \in \mathrm{BMO}_\varepsilon(I)$. For the bounded function φ_n, inequality (1.3.4) is true, i.e.,

$$B(\langle\varphi_n\rangle_{I}, \langle\varphi_n^2\rangle_{I}) \geq \langle e^{\varphi_n}\rangle_{I}.$$

Since e^{φ_0} is a summable majorant for e^{φ_n} and B is continuous, we can pass to the limit and obtain the estimate (1.3.4) for any function φ bounded from above. Finally, we repeat this approximation procedure for an arbitrary φ. Now, we take $\varphi_n = \varphi_{-\infty,n}$ and we can pass to the limit in the right-hand side of the inequality by the monotone convergence theorem.

So, we have proved the inequality

$$B(\mathfrak{b}_I(\varphi)) \geq \langle e^{\varphi}\rangle_{I}$$

for arbitrary $\varphi \in \mathrm{BMO}_\varepsilon(I)$. Taking supremum over all admissible test functions corresponding to the point x, i.e., over all φ such that $\mathfrak{b}_I(\varphi) = x$, we get $B(x) \geq \mathbf{B}(x)$. □

As before, to come up with a candidate for the Bellman function, we pass from the finite difference inequality (1.3.1) to the infinitesimal one:

$$\frac{d^2B}{dx^2} \stackrel{\text{def}}{=} \begin{pmatrix} \dfrac{\partial^2 B}{\partial x_1^2} & \dfrac{\partial^2 B}{\partial x_1 \partial x_2} \\[2ex] \dfrac{\partial^2 B}{\partial x_1 \partial x_2} & \dfrac{\partial^2 B}{\partial x_2^2} \end{pmatrix} \leq 0, \tag{1.3.6}$$

and we will require this Hessian matrix to be degenerate, i.e., $\det(\frac{d^2B}{dx^2}) = 0$. Again, to solve this PDE, we use a homogeneity property to reduce the problem to an ODE.

Lemma 1.3.5 (Homogeneity) *There exists a function G on the interval $[0, \varepsilon^2]$ such that*

$$\mathbf{B}(x;\varepsilon) = e^{x_1} G(x_2 - x_1^2), \qquad G(0) = 1.$$

Proof Let φ be an arbitrary test function defined on an interval I and $x = (\langle\varphi\rangle_{I}, \langle\varphi^2\rangle_{I})$ its Bellman point. Then the function $\tilde{\varphi} \stackrel{\text{def}}{=} \varphi + \tau$ is also a test function with the same norm, and its Bellman point is

$$\tilde{x} = (x_1 + \tau, x_2 + 2\tau x_1 + \tau^2).$$

If φ runs over the set of all test functions corresponding to x, then $\tilde{\varphi}$ runs over the set of all test functions corresponding to $\tilde{x}$ and vice versa. Therefore,

$$\mathbf{B}(\tilde{x}) = \sup_{\tilde{\varphi}} \langle e^{\tilde{\varphi}} \rangle_{_I} = e^{\tau} \sup_{\varphi} \langle e^{\varphi} \rangle_{_I} = e^{\tau} \mathbf{B}(x).$$

Choosing $\tau = -x_1$, we get

$$\mathbf{B}(x) = e^{-\tau} \mathbf{B}(x_1 + \tau, x_2 + 2\tau x_1 + \tau^2) = e^{x_1} \mathbf{B}(0, x_2 - x_1^2).$$

Setting $G(s) = \mathbf{B}(0, s)$ completes the proof. □

Since $G > 0$, we can introduce $g(s) = \log G(s)$ and look for a function B of the form

$$B(x_1, x_2) = e^{x_1 + g(x_2 - x_1^2)}.$$

By direct calculation, we get

$$\begin{aligned}
\frac{\partial^2 B}{\partial x_1^2} &= \left(1 - 4x_1 g' + 4x_1^2 (g')^2 - 2g' + 4x_1^2 g''\right) B, \\
\frac{\partial^2 B}{\partial x_1 \partial x_2} &= \left(g' - 2x_1 (g')^2 - 2x_1 g''\right) B, \\
\frac{\partial^2 B}{\partial x_2^2} &= \left((g')^2 + g''\right) B.
\end{aligned} \tag{1.3.7}$$

The partial differential equation $\det(\frac{d^2 B}{dx^2}) = 0$ then turns into the following ordinary differential equation:

$$\begin{aligned}
\left(1 - 4x_1 g' + 4x_1^2 (g')^2 - 2g' + 4x_1^2 g''\right)&\left((g')^2 + g''\right) \\
&= \left(g' - 2x_1 (g')^2 - 2x_1 g''\right)^2,
\end{aligned}$$

which reduces to

$$g'' - 2g'g'' - 2(g')^3 = 0. \tag{1.3.8}$$

Dividing by $2(g')^3$ (since we are not interested in constant solutions), we get

$$\left(\frac{1}{g'} - \frac{1}{4(g')^2}\right)' = 1,$$

which yields

$$\frac{1}{g'} - \frac{1}{4(g')^2} = s + \text{const}$$

or, equivalently,

$$-\left(1 - \frac{1}{2g'}\right)^2 = s + \text{const}, \qquad \forall s \in [0, \varepsilon^2].$$

Since the left-hand side is nonpositive, the constant cannot be greater than $-\varepsilon^2$. Let us denote it by $-\delta^2$, where $\delta \geq \varepsilon$.

Thus, we have two possible solutions:

$$1 - \frac{1}{2g'_\pm} = \pm\sqrt{\delta^2 - s}. \tag{1.3.9}$$

Using the boundary condition $g(0) = 0$, we obtain

$$g_\pm(s) = \frac{1}{2}\int_0^s \frac{dt}{1 \mp \sqrt{\delta^2 - t}} = \log\frac{1 \mp \sqrt{\delta^2 - s}}{1 \mp \delta} \pm \sqrt{\delta^2 - s} \mp \delta.$$

These functions are well defined for $\delta \in [\varepsilon, 1)$ and give us two solutions for B:

$$B_\pm(x;\delta) = \frac{1 \mp \sqrt{\delta^2 - x_2 + x_1^2}}{1 \mp \delta}\exp\left\{x_1 \pm \sqrt{\delta^2 - x_2 + x_1^2} \mp \delta\right\}. \tag{1.3.10}$$

For the function $B = B_+(x;\delta)$, the Hessian $\frac{d^2B}{dx^2}$ is nonpositive and it is nonnegative for $B = B_-(x;\delta)$. This is possible to check either by the direct calculation (see Problem 1.3.1) or using formula (1.3.9) for g' and expression (1.3.7) for $B_{x_2x_2}$ together with relation (1.3.8) between g'' and g'. Therefore, we have to choose our Bellman candidate among the family of functions $B = B_+(x;\delta),\ \ \delta \in [\varepsilon, 1)$. Since

$$\frac{\partial B}{\partial \delta} = \frac{\delta^2}{(1-\delta)^2}\exp\left\{x_1 \pm \sqrt{\delta^2 - x_2 + x_1^2} \mp \delta\right\},$$

i.e., B increases in δ, and we are interested in the minimal possible majorant, it is natural to choose $\delta = \varepsilon$. However, this choice does not give us the dyadic Bellman function because for this function, the main inequality (1.3.1) is not fulfilled (see Problem 1.3.3). This is in contrast to the Buckley's Bellman function from Section 1.2.

To choose the proper δ for a given ε is not a simple task and for this reason, we refer the reader to [**171**]. Now we concentrate our attention on the function $B(x;\varepsilon)$. As already mentioned, this function is not concave in the strip Ω_ε (i.e., inequality (1.3.1) is not fulfilled everywhere in Ω_ε), but it is *locally concave*. This means that it is concave in every convex subset of Ω_ε because its Hessian is negative.

The latter statement is simple, but for the reader who is not familiar with the fact that a smooth function is locally concave in a domain if and only if its Hessian is nonpositive in this domain, we present a proof. Let us parametrize the interval $[x^-, x^+]$ as follows: $x(s) = (1-s)x^- + sx^+$,

$0 \le s \le 1$, and put $b(s) \stackrel{\text{def}}{=} B(x(s))$. Then for arbitrary $\alpha^\pm$, $\alpha^+ + \alpha^- = 1$, we have

$$B(\alpha^- x^- + \alpha^+ x^+) - \alpha^- B(x^-) - \alpha^+ B(x^+)$$
$$= b(\alpha^+) - \alpha^- b(0) - \alpha^+ b(1) = -\int_0^1 k(\alpha, s) b''(s)\, ds,$$

where

$$k(\alpha, s) = \begin{cases} \alpha^- s & \text{for } \ 0 \le s \le \alpha^+, \\ \alpha^+(1 - s) & \text{for } \ \alpha^+ \le s \le 1. \end{cases}$$

And since

$$b''(s) = \sum_{i,j=1}^{2} B_{x_i x_j}(x(s))(x_i^+ - x_i^-)(x_j^+ - x_j^-) \le 0,$$

we have the required concavity condition

$$B(\alpha^- x^- + \alpha^+ x^+) \ge \alpha^- B(x^-) + \alpha^+ B(x^+).$$

Our next goal is to show that $B_+(x, \varepsilon)$ is the Bellman function for classical (non-dyadic) BMO. We need to modify the Bellman induction for this situation because now we have significant freedom: We need to choose how to split the interval on each step of induction. For a given test function φ, we will split the interval I not in two equal halves as for dyadic case, but try to split it in two parts $I = I^+ \cup I^-$ in such a way that the segment $[\mathfrak{b}_{I^-}, \mathfrak{b}_{I^+}]$ is entirely if not in Ω_ε then in a slightly larger domain Ω_δ. If this were possible for some δ, we could run the Bellman induction for $B(x; \delta)$ and get $\mathbf{B}(x; \varepsilon) \le B(x; \delta)$. If such a procedure were possible for every $\delta > \varepsilon$, we could pass to the limit $\delta \to \varepsilon$ to get the final upper estimate $\mathbf{B}(x; \varepsilon) \le B(x; \varepsilon)$.

To realize this plan, we prove the following purely geometric result that is crucial to applying the Bellman function method to the usual, non-dyadic BMO.

Lemma 1.3.6 (Splitting Lemma) *Fix two positive numbers ε and δ with $\varepsilon < \delta$. For an arbitrary interval I and any function $\varphi \in \mathrm{BMO}_\varepsilon(I)$, there exists a splitting $I = I^+ \cup I^-$ such that the whole straight-line segment $[\mathfrak{b}_{I^-}(\varphi), \mathfrak{b}_{I^+}(\varphi)]$ is inside Ω_δ. Moreover, the parameters of splitting $\alpha^\pm \stackrel{\text{def}}{=} |I^\pm|/|I|$ are separated from 0 and 1 by constants depending on ε and δ only, i.e., uniformly with respect to the choice of I and φ.*

Proof Fix an interval I and a function $\varphi \in \mathrm{BMO}_\varepsilon(I)$. We now demonstrate an algorithm to find a splitting $I = I^- \cup I^+$ (i.e., choose the splitting parameters $\alpha^\pm = |I^\pm|/|I|$) so that the statement of the lemma holds. For simplicity, put $x^0 = \mathfrak{b}_I$ and $x^\pm = \mathfrak{b}_{I^\pm}$.

First, we take $\alpha^- = \alpha^+ = \frac{1}{2}$ (see Figure 1.1). If the whole segment $[x^-, x^+]$ is in Ω_δ, we fix this splitting. Assuming it is not the case, i.e., there exists a point x on this segment with $x_2 - x_1^2 > \delta^2$. Observe that only one of the segments, either $[x^-, x^0]$ or $[x^+, x^0]$, contains such points. Denote the corresponding endpoint (x^- or x^+) by ξ and define a function ρ by

$$\rho(\alpha^+) = \max_{x \in [x^-, x^+]} \{x_2 - x_1^2\} = \max_{x \in [\xi, x^0]} \{x_2 - x_1^2\}.$$

By assumption, $\rho\left(\frac{1}{2}\right) > \delta^2$.

Recall that our test function φ is fixed and the position of the Bellman points $x^\pm$ depends on the splitting parameter α^+ only. We will now change α^+ so that ξ approaches x^0, i.e., we will increase α^+ if $\xi = x^+$ and decrease it if $\xi = x^-$. We stop when $\rho(\alpha^+) = \delta^2$ and fix that splitting. It remains to check that such a moment occurs and that the corresponding α^+ is separated from 0 and 1.

Without loss of generality, assume that $\xi = x^+$. Since the function $x^+(\alpha^+)$ is continuous on the interval $(0, 1]$ and $x^+(1) = x^0$, ρ is continuous on $[\frac{1}{2}, 1]$. We have $\rho\left(\frac{1}{2}\right) > \delta^2$ and we also know that $\rho(1) \le \varepsilon^2 < \delta^2$ (because $x^0 \in \Omega_\varepsilon$). Therefore, there is a point $\alpha^+ \in \left[\frac{1}{2}, 1\right]$ with $\rho(\alpha^+) = \delta^2$ (Figure 1.2).

Having just proved that the desired point exists, we need to check that the corresponding α^+ is not too close to 0 or 1. If $\xi = x^+$, we have $\alpha^+ > \frac{1}{2}$ and $\xi_1 - x_1^0 = x_1^+ - x_1^0 = \alpha^-(x_1^+ - x_1^-)$. Similarly, if $\xi = x^-$, we have $\alpha^- > \frac{1}{2}$ and $\xi_1 - x_1^0 = x_1^- - x_1^0 = \alpha^+(x_1^- - x_1^+)$. Thus, $|\xi_1 - x_1^0| = \min\{\alpha^\pm\}|x_1^- - x_1^+|$. For the stopping value of α^+, the straight line through the points x^-, x^+, and

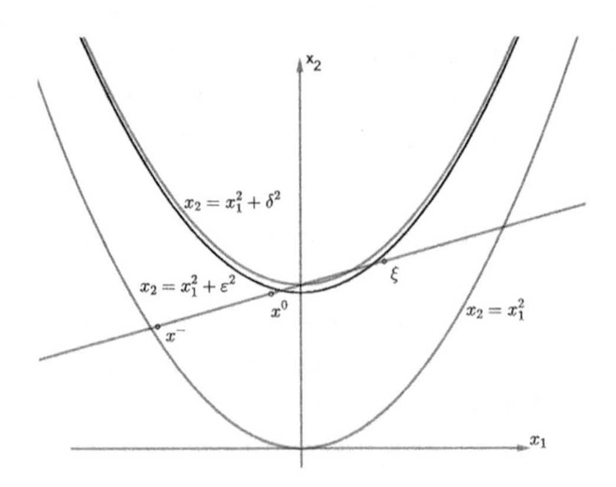

Figure 1.1 The initial splitting: $\alpha^- = \alpha^+ = \frac{1}{2}$, $\xi = x^+$.

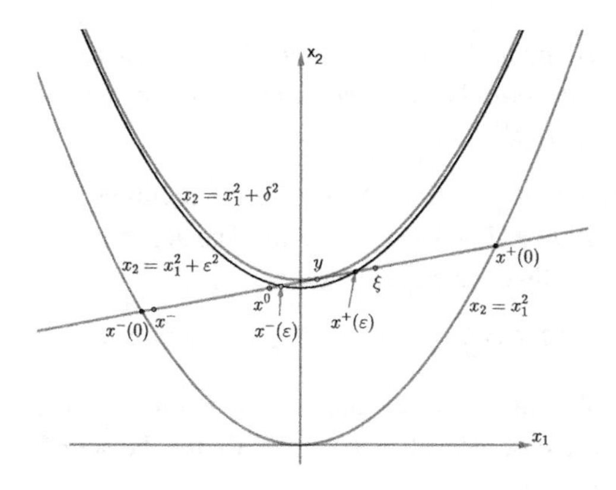

Figure 1.2 The stopping time: $[x^-, \xi]$ is tangent to the parabola $x_2 = x_1^2 + \varepsilon^2$.

x^0 is tangent to the parabola $x_2 = x_1^2 + \delta^2$ at some point y. The equation of this line is, therefore, $x_2 = 2x_1y_1 - y_1^2 + \delta^2$. The line intersects the graph of $x_2 = x_1^2 + s^2$ at the points

$$x^{\pm}(s) = \left(y_1 \pm \sqrt{\delta^2 - s^2},\, y_2 \pm 2y_1\sqrt{\delta^2 - s^2}\right).$$

Let us focus on the points $x^{\pm}(0)$ and $x^{\pm}(\varepsilon)$. We have

$$[x^-(\varepsilon), x^+(\varepsilon)] \subset [x^0, \xi] \subset [x^-, x^+] \subset [x^-(0), x^+(0)]$$

and, therefore,

$$\begin{aligned} 2\sqrt{\delta^2 - \varepsilon^2} &= |x_1^+(\varepsilon) - x_1^-(\varepsilon)| \le |x_1^0 - \xi_1| = \min\{\alpha^{\pm}\}|x_1^+ - x_1^-| \\ &\le \min\{\alpha^{\pm}\}|x_1^+(0) - x_1^-(0)| = \min\{\alpha^{\pm}\}2\delta, \end{aligned}$$

which implies

$$\sqrt{1 - \left(\frac{\varepsilon}{\delta}\right)^2} \le \alpha^+ \le 1 - \sqrt{1 - \left(\frac{\varepsilon}{\delta}\right)^2}.$$

As promised, this estimate does not depend on φ or I. □

From now on, we shall consider not the dyadic Bellman function $\mathbf{B}$, but the "true" one:

$$\mathbf{B}(x; \varepsilon) \stackrel{\text{def}}{=} \sup_{\varphi \in \mathrm{BMO}_\varepsilon(J)} \left\{ \langle e^{\varphi} \rangle_J : \langle \varphi \rangle_J = x_1,\ \langle \varphi^2 \rangle_J = x_2 \right\}.$$

The test functions now run over the ε-ball of the non-dyadic BMO.

Using the splitting lemma, we are able to make the Bellman induction work in the non-dyadic case.

LEMMA 1.3.7 (Bellman Induction) *If B is a continuous, locally concave function on the domain Ω_δ, satisfying the boundary condition* (1.3.3), *then* $\mathbf{B}(x;\varepsilon) \le B(x)$ *for all* $\varepsilon < \delta$.

PROOF Fix a function $\varphi \in \mathrm{BMO}_\varepsilon(I)$. By the splitting lemma, we can split every subinterval $J \subset I$ in such a way that the segment $[\mathfrak{b}_{J^-}, \mathfrak{b}_{J^+}]$ is inside Ω_δ. Since B is locally concave, we have

$$|J|B(\mathfrak{b}_J) \ge |J^+|B(\mathfrak{b}_{J^+}) + |J^-|B(\mathfrak{b}_{J^-})$$

for any such splitting. Now we can repeat, word for word, the arguments used in the dyadic case. Recall that $\mathcal{I}_n$ stands for the set of intervals of n-th generation, then

$$\begin{aligned}|I|B(\mathfrak{b}_I) &\ge |I^+|B(\mathfrak{b}_{I^+}) + |I^-|B(\mathfrak{b}_{I^-})\\ &\ge \sum_{J\in\mathcal{I}_n} |J|B(\mathfrak{b}_J) = \int_I B(x^{(n)}(s))\,ds,\end{aligned}$$

where $x^{(n)}(s) = \mathfrak{b}_J$, when $s \in J$, $J \in \mathcal{I}_n$. By the Lebesgue differentiation theorem, we have $x^{(n)}(s) \to (\varphi(s), \varphi^2(s))$ almost everywhere. (We have used here the fact that we split the intervals so that all coefficients $\alpha^\pm$ are uniformly separated from 0 and 1, and, therefore, $\max\{|J|\colon J \in \mathcal{I}_n\} \to 0$ as $n \to \infty$.) Now, we can pass to the limit in this inequality as $n \to \infty$. Again, first we assume φ to be bounded and, by the Lebesgue dominated convergence theorem, pass to the limit in the integral using the boundary condition (1.3.3)

$$|I|B(\mathfrak{b}_I(\varphi)) \ge \int_I B(\varphi(s), \varphi^2(s))\,ds = \int_I e^{\varphi(s)}ds = |I|\langle e^\varphi\rangle_I.$$

Then using the cut-off approximation, we get the same inequality for an arbitrary $\varphi \in \mathrm{BMO}_\varepsilon(I)$ such that $\mathfrak{b}_I(\varphi) = x$ for any given $x \in \Omega_\varepsilon$. □

COROLLARY 1.3.8

$$\mathbf{B}(x;\varepsilon) \le B(x;\delta), \qquad \varepsilon < \delta < 1.$$

PROOF The function $B(x;\delta)$ was constructed as a locally concave function satisfying boundary condition (1.3.3). □

COROLLARY 1.3.9

$$\mathbf{B}(x;\varepsilon) \le B(x;\varepsilon). \tag{1.3.11}$$

PROOF Since the function $B(x;\delta)$ is continuous with respect to the parameter $\delta \in (0,1)$, we can pass to the limit $\delta \to \varepsilon$ in the preceding corollary. □

Now, we would like to prove the inequality converse to (1.3.11). To this end, for every point x of Ω_ε, we construct a test function φ with BMO-norm ε, satisfying $\langle e^\varphi\rangle = B(x;\varepsilon)$, and such that its Bellman point is x. This would imply the inequality $\mathbf{B}(x;\varepsilon) \geq B(x;\varepsilon)$. Recall that such a test function that realizes the extremal value for the functional under investigation is called an *extremizer*.

First, we construct an extremizer φ_0 for the point $(0,\varepsilon^2)$. Without loss of generality, we can work on $I = [0,1]$. Note that the function $\varphi_a \stackrel{\text{def}}{=} \varphi_0 + a$ will then be an extremizer for the point $(a, a^2+\varepsilon^2)$. Indeed, φ_a has the same norm as φ_0, and if

$$\langle e^{\varphi_0}\rangle = B(0,\varepsilon^2;\varepsilon) = \frac{e^{-\varepsilon}}{1-\varepsilon},$$

then

$$\langle e^{\varphi_a}\rangle = \frac{e^{a-\varepsilon}}{1-\varepsilon} = B(a,a^2+\varepsilon^2;\varepsilon).$$

The point $(0,\varepsilon^2)$ is on the extremal line starting at $(-\varepsilon,\varepsilon^2)$. To keep equality on each step of the Bellman induction, when we split I into two subintervals I^- and I^+, the segment $[x^-, x^+]$ has to be contained in the extremal line along which our function B is linear. Since x is a convex combination of x^- and x^+, one of these points, say x^+, has to be to the right of x. However, the extremal line ends at $x = (0,\varepsilon^2)$, and so there seems to be nowhere to place that point. We circumvent this difficulty by placing x^+ infinitesimally close to x and using an approximation procedure. Where should x^- be placed? We already know extremizers for points on the lower boundary $x_2 = x_1^2$, since the only test function there are constants. Thus, it is convenient to put x^- there. Therefore, we set

$$x^- = (-\varepsilon,\varepsilon^2) \qquad \text{and} \qquad x^+ = (\Delta\varepsilon,\varepsilon^2),$$

for small Δ. To get these two points, we have to split I in proportion $1:\Delta$, that is, we take $I^+ = [0,\frac{1}{1+\Delta}]$ and $I^- = [\frac{1}{1+\Delta},1]$. To get the point x^-, we have

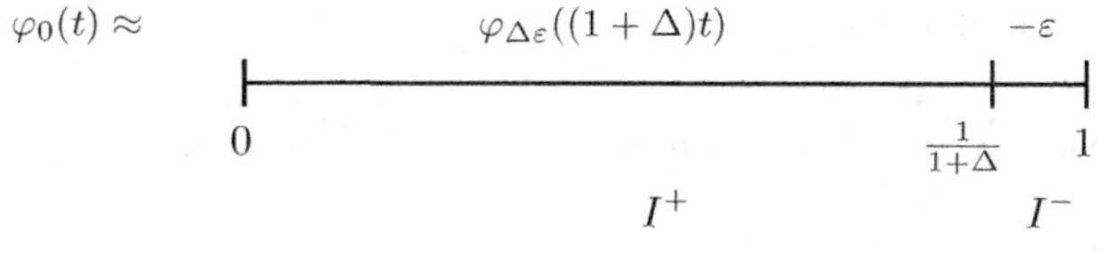

$\varphi_0(t) \approx$ $\varphi_{\Delta\varepsilon}((1+\Delta)t)$ $-\varepsilon$

0 $\frac{1}{1+\Delta}$ 1

I^+ I^-

to put $\varphi_0(t) = -\varepsilon$ on I^-. On I^+, we put a function corresponding not to the point x^+, but to the point $(\Delta\varepsilon,(1+\Delta^2)\varepsilon^2)$ on the upper boundary, which is close to x^+ (the distance between these two points is of order Δ^2). For such a point, the extremal function is $\varphi_{\Delta\varepsilon}(t) = \varphi_0(t) + \Delta\varepsilon$. Therefore, this function, when properly rescaled, can be placed on I^+. As a result, we obtain

$$\varphi_0(t) \approx \varphi_0\big((1+\Delta)t\big) + \Delta\varepsilon \approx \varphi_0(t) + \varphi_0'(t)\Delta t + \Delta\varepsilon,$$

which yields

$$\varphi_0'(t) = -\frac{\varepsilon}{t}.$$

Taking into account the boundary condition $\varphi_0(1) = -\varepsilon$, we get

$$\varphi_0(t) = \varepsilon \log\frac{1}{t} - \varepsilon.$$

Let us check whether we have found what we need. By the direct calculation, we get: $\langle\varphi_0\rangle_{[0,1]} = 0$, $\langle\varphi_0^2\rangle_{[0,1]} = \varepsilon^2$, and

$$\langle e^{\varphi_0}\rangle_{[0,1]} = \int_0^1 e^{-\varepsilon}\frac{dt}{t^\varepsilon} = \frac{e^{-\varepsilon}}{1-\varepsilon} = B(0,\varepsilon^2;\varepsilon).$$

It is easy now to get an extremal function for an arbitrary point x in Ω_ε. First of all, we draw the extremal line through x. It touches the upper boundary at the point $(a, a^2+\varepsilon^2)$ with $a = x_1 + \sqrt{\varepsilon^2 - x_2 + x_1^2}$ and intersects the lower boundary at the point (u, u^2) with $u = a - \varepsilon$. Now, we split the interval $[0,1]$ in the ratio $(x_1 - u) : (a - x_1)$ and concatenate the two known extremizers, $\varphi = u$ for the $x^- = (u, u^2)$ and $\varphi = \varphi_a$ for $x^+ = (a, a^2+\varepsilon^2)$. This gives the following function:

$$\varphi(t) = \begin{cases} \varepsilon\log\frac{x_1-u}{\varepsilon t} + u & \text{for } 0 \le t \le \frac{x_1-u}{\varepsilon} \\ u & \text{for } \frac{x_1-u}{\varepsilon} \le t \le 1 \end{cases}, \tag{1.3.12}$$

where

$$u = x_1 + \sqrt{\varepsilon^2 - x_2 + x_1^2} - \varepsilon.$$

This is a function from BMO_ε satisfying the required property $\langle e^\varphi\rangle_{[0,1]} = B(x;\varepsilon)$ (see Problem 1.3.4 later on).

This completes the proof of the following theorem.

THEOREM 1.3.10 *If $\varepsilon < 1$, then*

$$\mathbf{B}(x;\varepsilon) = \frac{1 - \sqrt{\varepsilon^2 - x_2 + x_1^2}}{1-\varepsilon}\exp\Big\{x_1 + \sqrt{\varepsilon^2 - x_2 + x_1^2} - \varepsilon\Big\};$$

if $\varepsilon \ge 1$, then $\mathbf{B}(x;\varepsilon) = \infty$.

Indeed, the second statement can be verified by the same extremal function φ because e^φ is not summable on $[0,1]$ for $\varepsilon \ge 1$.

Historical Remarks

The first proof of the theorem mentioned earlier was independently presented in [**168**] and [**183**]; a complete proof of this result together with the estimate from below (i.e., the lower Bellman function) and consideration of the dyadic version of the problem can be found in [**171**]. The Bellman function for the classical weak form of the John–Nirenberg inequality can be found in [**185**] or in [**191**].

Exercises

PROBLEM 1.3.1 Calculate the quadratic form of the Hessian $\sum_{i,j} B_{x_i x_j}\Delta_i\Delta_j$ for the function $B = B_+(x,\delta)$.

PROBLEM 1.3.2 Find the extremal trajectories along which the Hessian degenerates. Check that these are the tangent line to the parabola $y_2 = y_1^2 + \delta^2$.

PROBLEM 1.3.3 Check that for the function $B = B_+(x;\delta)$ from (1.3.10), the main inequality (1.3.1) is not true for some points. In particular,

$$B(u, u^2+\delta^2;\delta) \le \frac{B(u-\frac{\delta}{\sqrt{2}}, (u-\frac{\delta}{\sqrt{2}})^2;\delta) + B(u+\frac{\delta}{\sqrt{2}}, (u+\frac{\delta}{\sqrt{2}})^2+\delta^2;\delta)}{2}.$$

PROBLEM 1.3.4 Verify the following properties of the extremal function φ :

- $\langle\varphi\rangle_{[0,1]} = x_1$;
- $\langle\varphi^2\rangle_{[0,1]} = x_2$;
- $\langle e^{\varphi}\rangle_{[0,1]} = B(x_1,x_2;\varepsilon)$;
- $\varphi \in \mathrm{BMO}_\varepsilon$.

PROBLEM 1.3.5 Recall that we also obtained a second solution in (1.3.10):

$$B_-(x;\varepsilon) = \frac{1+\sqrt{\varepsilon^2 - x_2 + x_1^2}}{1+\varepsilon}\,\exp\Big\{x_1 - \sqrt{\varepsilon^2 - x_2 + x_1^2} + \varepsilon\Big\}.$$

Check that this is the solution of the following extremal problem:

$$\mathbf{B}_{\min}(x;\varepsilon) \stackrel{\text{def}}{=} \inf_{\varphi\in\mathrm{BMO}_\varepsilon(I)} \left\{\langle e^{\varphi}\rangle_I : \langle\varphi\rangle_I = x_1,\ \langle\varphi^2\rangle_I = x_2\right\},$$

that is, check that the Bellman induction works and construct an extremal function for every $x \in \Omega_\varepsilon$.

1.4 Homogeneous Monge–Ampère Equation

Now, we change the subject of our consideration for a while and look for the solutions of the equation

$$B_{x_1x_1}B_{x_2x_2} = (B_{x_1x_2})^2 \tag{1.4.1}$$

in a general setting. Section 1.3 motivates our interest in these solutions.

Linear functions always satisfy (1.4.1). Since we are looking for the smallest possible concave function B, it will always be linear, if there exists a linear function satisfying the required boundary conditions. It is a simple case, and in what follows we assume that B is not linear. This means that in each point x of the domain, there exists a unique (up to a scalar coefficient) vector, say, $\Theta(x)$, from the kernel of the Hessian matrix $\frac{d^2B}{dx^2}$.

Let us check that functions B_{x_i} are constant along the vector field Θ. The tangent vector field to the level set $f(x_1, x_2) = \text{const}$ has the form $\begin{pmatrix} -f_{x_2} \\ f_{x_1} \end{pmatrix}$ (it is orthogonal to $\operatorname{grad} f = \begin{pmatrix} f_{x_1} \\ f_{x_2} \end{pmatrix}$). Thus, we need to check that both vectors $\begin{pmatrix} -(B_{x_i})_{x_2} \\ (B_{x_i})_{x_1} \end{pmatrix}$ are in the kernel of the Hessian (i.e., proportional to the kernel vector Θ). This is a direct consequence of (1.4.1). For example, for $i = 1$ we have

$$\begin{pmatrix} B_{x_1x_1} & B_{x_1x_2} \\ B_{x_2x_1} & B_{x_2x_2} \end{pmatrix} \begin{pmatrix} -(B_{x_1})_{x_2} \\ (B_{x_1})_{x_1} \end{pmatrix} = \begin{pmatrix} -B_{x_1x_1}B_{x_1x_2} + B_{x_1x_2}B_{x_1x_1} \\ -B_{x_2x_1}B_{x_1x_2} + B_{x_2x_2}B_{x_1x_1} \end{pmatrix} = 0.$$

If we parameterize the integral curves of the field Θ by some parameter s, we can write $B_{x_i} \stackrel{\text{def}}{=} t_i(s),\quad s = s(x_1, x_2)$. Any B_{x_i} that is not identically constant can itself be taken as s. However, usually it is more convenient to parameterize the integral curves by some other parameter with a clear geometrical meaning.

Now, we check that function $t_0 \stackrel{\text{def}}{=} B - x_1t_1 - x_2t_2$ is also constant along the integral curves. Since

$$-\frac{\partial t_0}{\partial x_2} = -B_{x_2} + x_1\frac{\partial t_1}{\partial x_2} + x_2\frac{\partial t_2}{\partial x_2} + t_2 = x_1B_{x_1x_2} + x_2B_{x_2x_2}$$

and

$$\frac{\partial t_0}{\partial x_1} = B_{x_1} - t_1 - x_1\frac{\partial t_1}{\partial x_1} - x_2\frac{\partial t_2}{\partial x_1} = -x_1B_{x_1x_1} - x_2B_{x_1x_2},$$

we have

$$\begin{pmatrix} -(t_0)_{x_2} \\ (t_0)_{x_1} \end{pmatrix} = -x_1 \begin{pmatrix} -(t_1)_{x_2} \\ (t_1)_{x_1} \end{pmatrix} - x_2 \begin{pmatrix} -(t_2)_{x_2} \\ (t_2)_{x_1} \end{pmatrix} \in \operatorname{Ker} \frac{d^2B}{dx^2}.$$

So, we have proved that in the representation

$$B = t_0 + x_1 t_1 + x_2 t_2 \tag{1.4.2}$$

of a solution to the homogeneous Monge–Ampère equation, the coefficients t_i are constant along the vector field generated by the kernel of the Hessian. Now we prove that the integral curves of this vector field are, in fact, straight lines given by the equation

$$dt_0 + x_1 dt_1 + x_2 dt_2 = 0. \tag{1.4.3}$$

This is, indeed, the equation of a straight line because all the differentials are constant along the trajectory. If some parametrization of the trajectories is chosen, this equation can be rewritten as a usual linear equation with constant coefficients. For example, let us take $s = t_0$; then (1.4.3) turns into

$$1 + x_1 \frac{dt_1}{dt_0} + x_2 \frac{dt_2}{dt_0} = 0,$$

where the coefficients $\frac{dt_i}{dt_0}$, being functions of t_0, are constant on each trajectory.

Now, let us deduce equation (1.4.3). On the one hand, we have

$$dB = B_{x_1} dx_1 + B_{x_2} dx_2 = t_1 dx_1 + t_2 dx_2. \tag{1.4.4}$$

On the other hand, representation (1.4.2) yields

$$dB = dt_0 + t_1 dx_1 + x_1 dt_1 + t_2 dx_2 + x_2 dt_2. \tag{1.4.5}$$

Comparing (1.4.4) and (1.4.5) we get (1.4.3).

The homogeneous Monge–Ampère equation on an m-dimensional domain

$$\det\left(\{B_{x_i x_j}\}_{i,j=1}^m\right) = 0$$

has a similar solution of the form

$$B(x) = t_0 + \sum_{i=1}^m x_i t_i,$$

if the Hessian has a one-dimensional kernel. The extremal lines are the straight lines as well, determined by the equation

$$dt_0 + \sum_{i=1}^m x_i dt_i = 0. \tag{1.4.6}$$

This is now a system of $m - 1$ equations because each integral line is determined by $m - 1$ parameters $s_1, \ldots, s_{m-1}$. If we took $s_i = t_i$, $1 \le i \le m-1$, we would get the following system of linear equations:

$$\frac{\partial t_0}{\partial t_i} + x_i + x_m \frac{\partial t_m}{\partial t_i} = 0, \qquad 1 \le i \le m-1.$$

Each of these equations determines an affine hyperplane and the intersection of these hyperplanes gives us our extremal line.

Historical Remarks

Probably for the first time, it was noted in [**169**] that if the main inequality gives us purely concavity (or convexity) property, then the Bellman equation is, in fact, the homogeneous Monge–Ampère equation. After preprint [**169**] appeared, many papers solving some specific Bellman function problem by using the Monge–Ampère equation followed, for example, [**80, 81, 160, 170, 172, 189–191**].

1.5 Bellman Function for General Integral Functionals on BMO

Now, when having some information about solutions of the Monge–Ampère equation, we can repeat solving the Bellman problem for the John–Nirenberg inequality once more from a new point of view. However, we try to solve a more general problem on BMO. Namely, if to prove the John–Nirenberg inequality , we estimate the average of the function $e^{\varphi(s)}$, then we take more or less arbitrary function f instead of e^t and look for maximal value of the average of $f(\varphi(s))$, i.e., we try to calculate the Bellman function

$$\mathbf{B}(x;\varepsilon) \stackrel{\text{def}}{=} \sup_{\varphi \in \mathrm{BMO}_\varepsilon(I)} \left\{ \langle f(\varphi) \rangle_{I} : \mathfrak{b}_I(\varphi) = x \right\}, \tag{1.5.1}$$

where, as before $\mathfrak{b}_I(\varphi) = (\langle I \rangle_\varphi, \langle I \rangle_\varphi^2)$. We will not specify the condition on the function f now. We only need that all the expressions be correctly defined.

As in the case of John–Nirenberg inequality, we will look for our Bellman candidate among locally concave functions, i.e., we assume the Hessian to be nonpositive.[3] Due to the extremality of our function, it is natural to assume that

[3] When trying to find the Bellman function for the John–Nirenberg inequality we were able to prove the main inequality (i.e., concavity) in the dyadic case only. Nevertheless, we have looked for a Bellman candidate among locally concave functions. We justified our assumptions after proving that the found candidate is indeed the required Bellman function. We made also other assumptions that are a priori not evident. For example, a priori it is not clear why the Bellman function must be smooth enough to be a solution of the Monge–Ampère equation.

the Hessian is degenerated – thus we come to the Monge–Ampère equation. The same arguments that proved Lemma 1.3.3 yield the boundary condition in general situation

$$\mathbf{B}(x_1, x_1^2; \varepsilon) = \mathbf{B}(x) = f(x_1). \tag{1.5.2}$$

(We will omit index ε when it is assumed to be fixed.)

1.5.1 Tangent Domains

For the John–Nirenberg Bellman function, we had a family of segments of straight lines that fill in the domain of definition and do not intersect each other. Bellman function was linear on each of these segments. This was an example of *foliation* and these segments were called *extremal lines*. Let us try to construct the same foliation by extremal lines as we have for the John–Nirenberg Bellman function: these are the segments of the lines

$$x_2 = 2(u + \varepsilon)(x_1 - u) + u^2, \qquad u \le x_1 \le u + \varepsilon, \tag{1.5.3}$$

which are tangent to the upper parabola at the point $\big(u + \varepsilon, (u + \varepsilon)^2 + \varepsilon^2\big)$ and go from this point to the left till the lower boundary, intersecting it at the point (u, u^2). We call such extremal lines the left tangent lines. We have parametrized all of them by the first coordinate u of the common point with the lower parabola, where the boundary value is given, and for our Bellman candidate B, we have to take $B(u, u^2) = f(u)$. We can solve equation (1.5.3) with respect to u and get an explicit formula for the function $x \mapsto u(x)$:

$$u(x) = x_1 - \varepsilon + \sqrt{x_1^2 + \varepsilon^2 - x_2}. \tag{1.5.4}$$

Since we would like to have our tangent lines as the extremal lines, B has to be linear along each such line, i.e.,

$$B(x) = k(u)(x_1 - u) + f(u), \qquad u \le x_1 \le u + \varepsilon, \tag{1.5.5}$$

However, when searching for a possible Bellman candidate, we can make any assumption simplifying our job if, as a result, we can prove that the found function is just what we need. In Section 1.9, we show an example where the Bellman function is smooth and satisfies the differential Bellman equation only on a part of its domain, and on another part of its domain is not differentiable and cannot be a solution of the Bellman equation. Nevertheless, when we speak about concavity, i.e., when the Bellman function must be concave for the dyadic problem, the solution of the corresponding extremal problem in the "usual" non-dyadic setting is necessarily locally concave. We shall not prove this rather technical assertion here, but the interested reader can find the proof in [**174**].

at any point $x = (x_1, x_2)$ on the line (1.5.3). We remember that the derivatives of the solutions of Monge–Ampère equations are constant on the extremal lines. Let us check when the function given by (1.5.5) possesses this property.

Let us calculate the derivative with respect to x_2:

$$B_{x_2}(x) = \big(k'(u)(x_1 - u) - k(u) + f'(u)\big)u_{x_2}. \qquad (1.5.6)$$

We differentiate the extremal line equation (1.5.3) to get the expression for u_{x_2}:

$$1 = \big(2(x_1 - u) - 2(u + \varepsilon) + 2u\big)u_{x_2},$$

i.e.,

$$u_{x_2} = \frac{1}{2(x_1 - u - \varepsilon)}.$$

Note that this derivative is always negative, since $x_1 < u + \varepsilon$ at any point of our line except the tangency point. This fact is very clear from the geometrical point of view: if we move the point x straight up, then the point of the intersection of the left tangent line passing through x goes to the left.

Now, plugging this expression for u_{x_2} into (1.5.6), we get

$$B_{x_2} = \frac{k'(u)(x_1 - u) - k(u) + f'(u)}{2(x_1 - u - \varepsilon)} = \frac{1}{2}k'(u) + \frac{k'(u)\varepsilon - k(u) + f'(u)}{2(x_1 - u - \varepsilon)}.$$

Since the derivative of the solution of the Monge–Ampère equation could depend only on the extremal line (i.e., on u) but not on the place of the point on this line (i.e., there is no dependence on x_1), we have

$$t_2 = B_{x_2} = \frac{1}{2}k'(u),$$

and we need to satisfy the following equation:

$$\varepsilon k'(u) - k(u) + f'(u) = 0.$$

This equation determines the coefficient $k(u)$ up to a constant

$$k(u) = e^{u/\varepsilon}\left[c - \frac{1}{\varepsilon}\int e^{-u/\varepsilon} f'(u)\,du\right]. \qquad (1.5.7)$$

If we assume that some part of our domain ($a \le u \le b$) is foliated by the left tangent lines (1.5.3), then we can determine the constant c, if we know the slope at the point b. We rewrite (1.5.7) with a definite integral

$$k(u) = e^{u/\varepsilon}\left[c + \frac{1}{\varepsilon}\int_u^b e^{-t/\varepsilon} f'(t)\,dt\right], \qquad (1.5.8)$$

where $c = e^{-b/\varepsilon}k(b)$.

Now let us find when the Hessian of our solution is nonpositive. For this, it is necessary and sufficient to have

$$B_{x_2x_2} = t_2'(u)u_{x_2} = \frac{1}{2}k''(u)u_{x_2} \leq 0.$$

Since we already know that $u_{x_2} \leq 0$, we come to the condition

$$k''(u) \geq 0.$$

Consider now the case when the whole domain can be foliated by the left tangent lines. (Recall that this was the case in Section 1.3; see Exercises 1.2.2, 1.11.5). Then we can determine the integration constant from the concavity and minimality only.

Twice differentiating formula (1.5.8) with $b = \infty$, we get

$$k''(u) = e^{u/\varepsilon}\left[\frac{c}{\varepsilon^2} + \frac{1}{\varepsilon}\int_u^{\infty} e^{-t/\varepsilon} f'''(t)\,dt\right]. \tag{1.5.9}$$

This formula can be deduced if one first differentiates (1.5.8) twice and then integrates twice by parts assuming that all integrals converge and therefore all terms vanish at infinity. Alternatively one can replace the variable t in (1.5.8) by $t + u$ and then differentiate under the integral sign.

We assume, of course, that the integral converges at infinity, and hence the condition $k''(u) \geq 0$ holds only if $c \geq 0$. Moreover, we see that if $f''' \geq 0$ then the necessary condition $k'' \geq 0$ is fulfilled. Now we recall that we are interested in the minimal possible B of the form (1.5.5). Since the term $x_1 - u$ is positive here, from all possible k we have to take the minimal one, i.e., we have to put $c = 0$. Finally, we have in this case

$$k(u) = \frac{1}{\varepsilon}\int_u^{\infty} e^{(u-t)/\varepsilon} f'(t)\,dt.$$

Applying this to the case of the John–Nirenberg Bellman function, when $f(t) = e^t$, we obtain that $f''' > 0$ and

$$k(u) = \frac{1}{\varepsilon}e^{u/\varepsilon}\int_u^{\infty} e^{\frac{\varepsilon-1}{\varepsilon}t}\,dt = \frac{e^u}{1-\varepsilon}, \qquad 0 < \varepsilon < 1.$$

and

$$B(x) = \Big(\frac{x_1 - u}{1-\varepsilon} + 1\Big)e^u.$$

If we plug expression (1.5.4) here for u, we finally get the formula that was deduced in Section 1.3:

$$B(x;\varepsilon) = \frac{1 - \sqrt{\varepsilon^2 - x_2 + x_1^2}}{1-\varepsilon}\exp\Big\{x_1 - \varepsilon + \sqrt{\varepsilon^2 - x_2 + x_1^2}\Big\}.$$

Any boundary function f with $f''' \leq 0$ supply us with an example of the foliation by the tangent lines going into opposite directions. The necessary and sufficient condition for such foliation is $k'' \leq 0$. For the slope k, we can obtain a similar equation

$$\varepsilon k'(u) + k(u) + f'(u) = 0$$

and its general solution

$$k(u) = e^{-u/\varepsilon}\left[c + \frac{1}{\varepsilon}\int e^{u/\varepsilon} f'(u)\,du\right]. \tag{1.5.10}$$

If the whole parabolic strip is foliated by these right tangents, then we can specify the constant: $c = 0$, and then

$$k(u) = \frac{1}{\varepsilon}\int_{-\infty}^{u} e^{(t-u)/\varepsilon} f'(t)\,dt.$$

Such a situation occurs for the Bellman function $\mathbf{B}_{\min}$ from Exercise 1.3.5. Let us make a general remark: if you wish to obtain an estimate from below for a functional f, it is the same as taking with the opposite sign an estimate from above for a functional $-f$, i.e., $\mathbf{B}_{\min}(x; f) = -\mathbf{B}_{\max}(x; -f)$. So, if we want to estimate the quantity $\langle e^{\varphi}\rangle$ from below as in the John–Nirenberg case, we need to consider $f(t) = -e^t$, in which case we have $f''' < 0$ and, therefore, the foliation by the right tangent lines. So, we have

$$k(u) = -\frac{1}{\varepsilon}e^{-u/\varepsilon}\int_{-\infty}^{u} e^{\frac{1+\varepsilon}{\varepsilon}t} f'(t)\,dt = -\frac{e^u}{1+\varepsilon}.$$

Finally, we get

$$B(x) = -\Big(\frac{x_1 - u}{1+\varepsilon} + 1\Big)e^u,$$

and since now

$$u(x) = x_1 + \varepsilon - \sqrt{x_1^2 + \varepsilon^2 - x_2},$$

we come to the solution from Exercise 1.3.5:

$$-B(x;\varepsilon) = \frac{1 + \sqrt{\varepsilon^2 - x_2 + x_1^2}}{1+\varepsilon}\exp\left\{x_1 + \varepsilon - \sqrt{\varepsilon^2 - x_2 + x_1^2}\right\}.$$

1.5.2 Chordal Domains

We have described the case when one end of an extremal line is on the lower boundary and the second one belongs to the upper boundary. Now we consider the second possibility: when an extremal line connects two points on the lower boundary. Let $A = (a, a^2)$ and $B = (b, b^2)$[4] be the endpoints of such an extremal line. If x is a point on the chord $[A, B]$, then we immediately can write down

$$B(x) = \frac{f(b) - f(a)}{b - a} x_1 + \frac{bf(a) - af(b)}{b - a}. \tag{1.5.11}$$

However, the function built in this way is a Bellman candidate only if its derivatives B_{x_1} and B_{x_2} are constant along this chord. To calculate these derivatives, we assume that we have a family of such chords $[A, B]$ parametrized by their horizontal length $\ell = b - a$. It is clear that the right endpoint of the chord moves to the right and the left endpoint moves to the left as ℓ increases. So, $b(\ell)$ is an increasing function and $a(\ell)$ is a decreasing one. If boundary function f is not sufficiently smooth, it can happen that one endpoint is stable and only the second one is moving. But for simplicity, we shall not consider such a situation and assume that the functions a and b are differentiable and then $a' < 0$ and $b' > 0$. [5]

If x is a point of the domain foliated by the chords, we can consider ℓ as a function of x. The equation of the line passing through the points A and B is

$$x_2 = (a + b)x_1 - ab.$$

For a fixed x_1, we can consider x_2 as a function of ℓ. Then

$$x_2' = x_1(a' + b') - (ab' + ba'),$$

or

$$\ell_{x_2} = \frac{1}{x_1(a' + b') - (ab' + ba')}. \tag{1.5.12}$$

Let us show that this expression is always positive. In fact, it is an obvious from the picture: a bigger chord is above a smaller one. But it is easy to see that formally:

$$x_2' = b'(x_1 - a) + a'(x_1 - b),$$

where the both summands are positive.

Now we calculate the partial derivative B_{x_2}. Differentiating the identity (1.5.11) in x_2, we get

$$B_{x_2}(x_1, x_2) = \frac{\alpha x_1 + \beta}{(b - a)^2} \ell_{x_2}, \tag{1.5.13}$$

[4] We denote a point on the boundary and a Bellman candidate by the same letter B. We hope this will not lead to misunderstanding

[5] Example of domains foliated by extremal chords that share a common point on the boundary can be found, for example, in **[172]** or **[191]**).

where

$$\alpha = \big(f'(b)b' - f'(a)a'\big)(b-a) - \big(f(b) - f(a)\big)(b' - a');$$
$$\beta = \big(b'f(a) + bf'(a)a' - a'f(b) - af'(b)b'\big)(b-a)$$
$$\qquad - \big(bf(a) - af(b)\big)(b' - a'),$$

and ℓ_{x_2} is given by (1.5.12).

Since B_{x_2} has to be constant along the chords, it does not depend on x_1 if ℓ is fixed. In our case, the quotient of two linear functions does not depend on the variable, therefore their coefficients are proportional, i.e.,

$$\alpha(ab' + ba') = -\beta(a' + b').$$

After elementary calculations, we come to the following identity:

$$a'b'\left(\frac{f'(a) + f'(b)}{2} - \frac{f(b) - f(a)}{b - a}\right) = 0.$$

Dividing by $a'b'$, we get

$$\langle f' \rangle_{[a,b]} = \frac{f'(a) + f'(b)}{2}. \tag{1.5.14}$$

Thus, under the assumption $a'b' \neq 0$, the derivatives of B are constant on the chords if and only if the values of the boundary function at their ends satisfy equation (1.5.14). This relation is usually called the *cup equation*, because the subdomain situated under some chord is called a *cup*. A cup is called *full* if the horizontal length of the upper chord (i.e., $\ell = b - a$) is 2ε. This means that the full cup is maximal possible, because the upper chord of such cup is at tangent to the upper parabola. A cup is called *simple* if it is foliated by the chords only, i.e., there are no linearity domains inside the cup (i.e., no domains where the Hessian is zero identically). If there are such domains of linearity, the cup is called *composite*. For a simple cup, the functions $a(\ell)$ and $b(\ell)$ are defined on some interval $[0, l]$ and the point $c = a(0) = b(0)$ is called the *origin* of the cup. (It would be more correct to speak about the point $C = (c, c^2)$ on the lower boundary, as about the origin.) Now we would like two questions answered: the first, if a subdomain is foliated by chords, when does this foliation produce a concave function, and the second, what points can be the origins of cups.

However, before we turn to these questions, we deduce the cup equation using a much simpler (and more general!) geometrical reasoning. If we have a concave function being linear along some chord, then we can imagine the tangent plane to the graph of our function at any point of this chord. Then the whole chord (more precisely, the graph of the function above this chord) lies in this tangent plane. Moreover, the vectors tangent to the boundary curve at the ends of the chord belong to the same tangent plane. If three vectors are in one plane, then the matrix whose columns are made up of these vectors is degenerate, i.e., its determinant is zero. But this is just our condition (1.5.14):

$$\begin{vmatrix} b-a & 1 & 1 \\ b^2-a^2 & 2a & 2b \\ f(b)-f(a) & f'(a) & f'(b) \end{vmatrix} = 0. \tag{1.5.15}$$

Now we turn to the concavity of the constructed function. Since our function is linear in one direction, it is sufficient to verify its concavity along any other direction. Chords never go vertically and therefore it is enough to check the sign of $B_{x_2x_2}$. First, using (1.5.14), we simplify formula (1.5.13); Since the expression for B_{x_2} does not depend on x_1, we have

$$\begin{aligned} B_{x_2}(x_1,x_2) &= \frac{\alpha}{(a'+b')(b-a)^2} \\ &= \frac{\big(f'(b)b'-f'(a)a'\big)(b-a)-\big(f(b)-f(a)\big)(b'-a')}{(a'+b')(b-a)^2} \\ &= \frac{2f'(b)b'-2f'(a)a'-\big(f'(b)+f'(a)\big)(b'-a')}{2(a'+b')(b-a)} \\ &= \frac{f'(b)-f'(a)}{2(b-a)}. \end{aligned}$$

As we already know, ℓ strictly increases as x_2 grows, therefore it is sufficient to study the sign of $B_{x_2\ell}$. By direct calculations, we have

$$\begin{aligned} 2B_{x_2\ell} &= \frac{f''(b)b'-f''(a)a'}{b-a} - \frac{f'(b)-f'(a)}{(b-a)^2}(b'-a') \\ &= \frac{b'D_{\mathrm{R}}(a,b)-a'D_{\mathrm{L}}(a,b))}{b-a}, \end{aligned} \tag{1.5.16}$$

where

$$D_{\mathrm{L}}(a,b)=f''(a)-\langle f''\rangle_{[a,b]} \quad \text{and} \quad D_{\mathrm{R}}(a,b)=f''(b)-\langle f''\rangle_{[a,b]}.$$

These two functions the left one and the right one, are called *differentials*, by the following reason. Let us write the cup equation as $\Phi(a,b)=0$, where

$$\Phi(a,b)=f'(b)+f'(a)-2\langle f'\rangle_{[a,b]}.$$

Then on the manifold $\Phi(a,b)=0$ we have

$$d\Phi(a,b)=D_{\mathrm{R}}(a,b)db+D_{\mathrm{L}}(a,b)da$$

and in particular for the pair (a, b) connected by the cup equation, we have

$$D_{\mathrm{R}}(a,b)b' + D_{\mathrm{L}}(a,b)a' = 0.$$

Since $b' > 0$ and $a' < 0$, differentials have the same sign. And looking on the expression (1.5.16), we see that our function is convex if the differentials are positive and concave if they are negative.

Finally, we would like to consider a question about the possible place of the origins of cups. We shall prove that any origin c is a root of the third derivative of the boundary function f. Moreover, we show that if f''' is positive in some left neighborhood of this c and negative in some right neighborhood, then we can solve the cup equation in a small neighborhood of c and obtain a concave function B.

Let us take any point $c \in [a, b]$. We can write down the Taylor decomposition in a neighborhood of the point (c, c):

$$\Phi(a,b) = \frac{1}{6} f'''(c)(b-a)^2 + O\big((b-a)^3\big).$$

Therefore, if $a \to c$ and $b \to c$, then

$$f'''(c) = 6 \lim \frac{\Phi(a,b)}{(b-a)^2} = 0.$$

Now we show that for any $\ell = b - a$ sufficiently small, the cup equation has a solution $\big(a(\ell), b(\ell)\big)$ such that $a < c < b$. We check that $\Phi(c-\ell, c) > 0$ and $\Phi(c, c+\ell) < 0$. Whence the function $t \mapsto \Phi(t-\ell, t)$ has a root on $(0, \ell)$. This root will be $b(\ell)$ and $a(\ell) = b(\ell) - \ell$.

Let us check the first inequality. If $f''' > 0$ on $(c-\ell, c)$ for some ℓ, then the function f' is convex on this interval, i.e., its graph is below the straight line segment with the same endpoints:

$$f'(t) < f'(c) - \frac{f'(c) - f'(c-\ell)}{\ell}(c-t).$$

Averaging this inequality over $(c-\ell, c)$, we come to $\Phi(c-\ell, c) > 0$. The inequality $\Phi(c, c+\ell) < 0$ follows from concavity of f' in a right neighborhood of c in the same manner.

To be sure that we have constructed a concave function B we need to check that differentials $D_{\mathrm{L}}(a,b)$ and $D_{\mathrm{R}}(a,b)$ are negative. Let us consider the following auxiliary function:

$$q(t) = f'(t) - f'(a) - \frac{f'(b) - f'(a)}{b-a}(t-a).$$

Note that $q(a) = q(b) = 0$ and $\langle q \rangle_{[a,b]} = 0$ due to the cup equation. Since $q'' = f'''$, the function q is strictly convex on (a, c) and strictly concave on (c, b). We shall see that this implies that $q'(a) < 0$ and $q'(b) < 0$.

Indeed, assume that $q'(a) \geq 0$. Since $q'' > 0$ on (a, c), the function q' is increasing and therefore positive on (a, c). So, q in its turn increases from 0 to some positive value $q(c)$. Since q is concave on (c, b) it is strictly positive there as well because it vanishes at the end of the interval (recall that $q(b) = 0$). But this is impossible because $\langle q \rangle_{[a,b]} = 0$. Therefore, $q'(a) < 0$. In a similar way, we can check that $q'(b) < 0$. It remains to note that $q'(a) = D_{\mathrm{L}}(a, b)$ and $q'(b) = D_{\mathrm{R}}(a, b)$.

We conclude this section by the following remark. If a full cup is constructed, then it can be glued C^1-smoothly to a left tangent domain on the left and to a right tangent domain on the right (see Figure 1.3). The constants of integration for the tangent domains are determined by the continuity condition on the common tangents, which are two halves of the upper chord of the cup that touch the upper parabola. If the first coordinates of the ends of this chord are a_0 and b_0, then $b_0 - a_0 = 2\varepsilon$. The slope on this chord of our Bellman function is

$$\frac{f(b_0) - f(a_0)}{2\varepsilon} = \frac{f'(a_0) + f'(b_0)}{2}.$$

Therefore, choosing for the left chordal domain this initial condition for the slope, we get from (1.5.7)

$$k(u) = \frac{f'(a_0) + f'(b_0)}{2} e^{\frac{u-a_0}{\varepsilon}} + \frac{1}{\varepsilon} \int_u^{a_0} e^{\frac{u-t}{\varepsilon}} f'(t)\, dt, \qquad u \leq a_0.$$

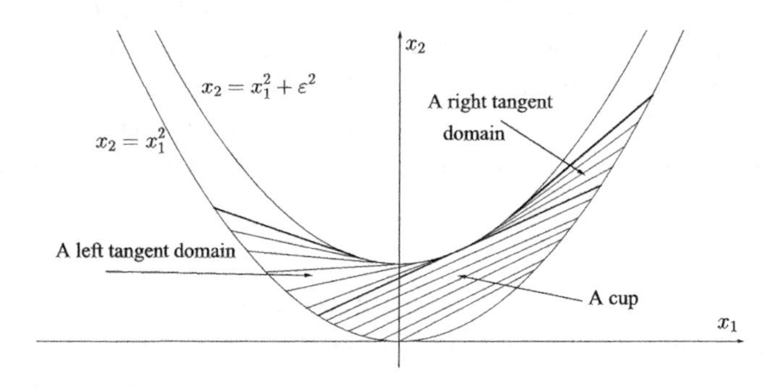

Figure 1.3 A cup surrounded by tangent domains.

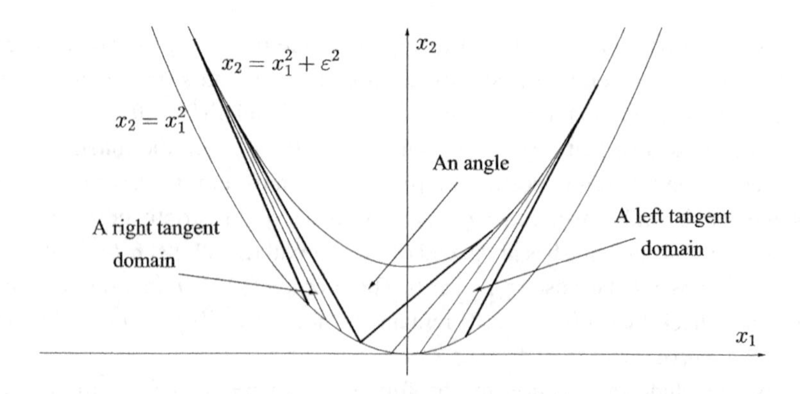

Figure 1.4 An angle surrounded by tangent domains.

For the right chordal domain we get

$$k(u) = \frac{f'(a_0) + f'(b_0)}{2} e^{\frac{b_0 - u}{\varepsilon}} + \frac{1}{\varepsilon} \int_{b_0}^{u} e^{\frac{t-u}{\varepsilon}} f'(t)\,dt, \qquad u \geq b_0.$$

Calculating the second derivatives of the slopes we get

$$k''(u) = \frac{1}{\varepsilon} D_{\mathrm{L}}(a_0, b_0) e^{\frac{u-a_0}{\varepsilon}} + \frac{1}{\varepsilon} \int_{u}^{a_0} e^{\frac{u-t}{\varepsilon}} f'''(t)\,dt, \qquad u \leq a_0\,;$$

$$k''(u) = -\frac{1}{\varepsilon} D_{\mathrm{R}}(a_0, b_0) e^{\frac{b_0-u}{\varepsilon}} + \frac{1}{\varepsilon} \int_{b_0}^{u} e^{\frac{t-u}{\varepsilon}} f'''(t)\,dt, \qquad u \geq b_0.$$

Since both differentials are negative, we see that the requirement $k'' \leq 0$ is fulfilled in some left neighborhood of a_0 and the requirement $k'' \geq 0$ is fulfilled in some right neighborhood of b_0. As a result, we have obtained a solution of the Monge–Ampère equation with given boundary values in some neighborhood of the cup (see Figure 1.3). Using a cup, we glue two tangent foliations together. To glue these foliation in the opposite order we need to apply another construction, which is called an *angle*. This is a domain between two tangent lines of different directions with a common point on the lower boundary (see Figure 1.4). The Bellman function on this domain is linear. In the next section, we consider briefly all possible domains of linearity.

1.5.3 Linearity Domains

We classify all possible domains of linearity by the number of their points common with the lower boundary. It is clear that the unique possibility of such a domain to have only one point on the lower boundary is an *angle* described at the end of the of the preceding section. If we have two points of a linearity domain on the lower boundary, then the chord connecting these points is, of course, the part of the boundary of the linearity domain. Except for the arc of the upper parabola, such a domain has to be bounded by two tangent lines. If these tangent lines go in one direction, then the corresponding is called a *trolleybus* (a *right trolleybus* or a *left trolleybus* depending on the direction of the tangent lines); see Figure 1.5 for a left trolleybus and Figure 1.6 for a right one.

Otherwise, it is called a *birdie* (see Figure 1.7).

All domains with more than two points on the lower parabola are called *multifigures*. The lower boundary of a multifigure consists of a union of chords and, maybe, some arcs of the lower parabola, whereas the upper

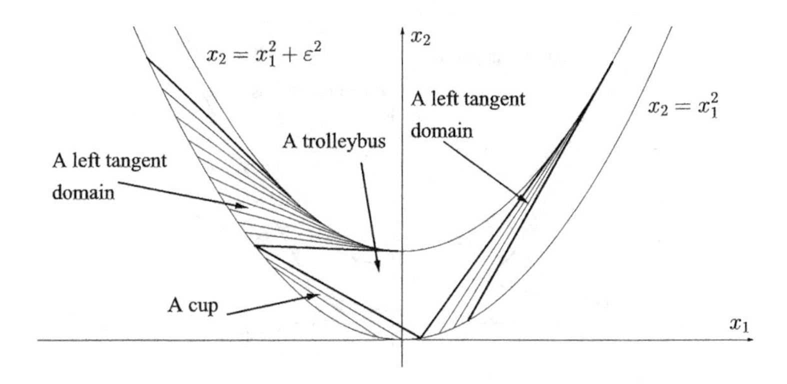

Figure 1.5 A left trolleybus together with surrounding tangent domains and an underlying cup.

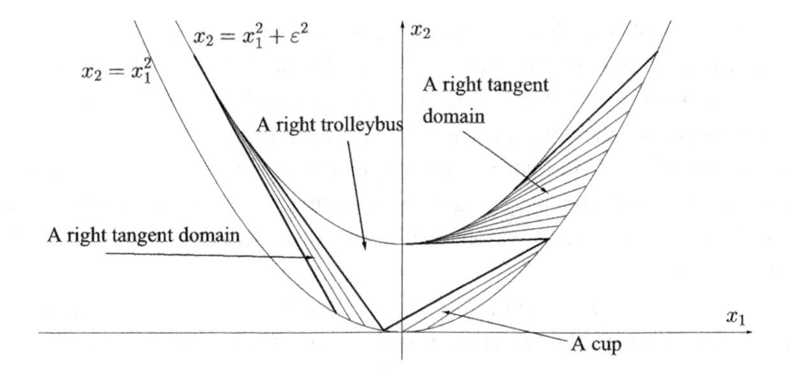

Figure 1.6 A right trolleybus together with surrounding tangent domains and an underlying cup.

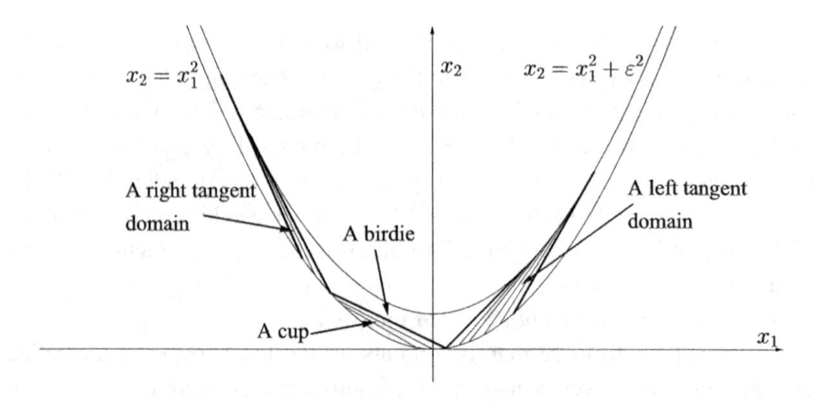

Figure 1.7 A birdie together with surrounding tangent domains and an underlying cup.

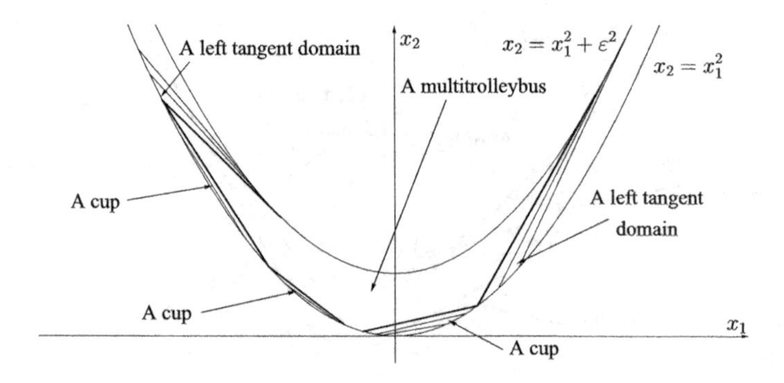

Figure 1.8 A multitrolleybus together with surrounding tangent domains and an underlying cups.

boundary is either a chord (then such a domain is called a *closed multicup domain*) or an arc of the upper parabola. In the last case, the multifigure is called a *multitrolleybus domain* if the left and the right boundary of this domain are the tangent lines of the same direction (see Figure 1.8). Otherwise it is called either a *multicup* (if the left boundary is a left tangent line and the right boundary is a right tangent line), or a *multibirdie* (if the boundary tangent lines have opposite directions), see Figures 1.9 and 1.10, respectively.

Using the described figures, it is possible to construct a foliation for the Bellman function that corresponds to an arbitrary integral functional on BMO.

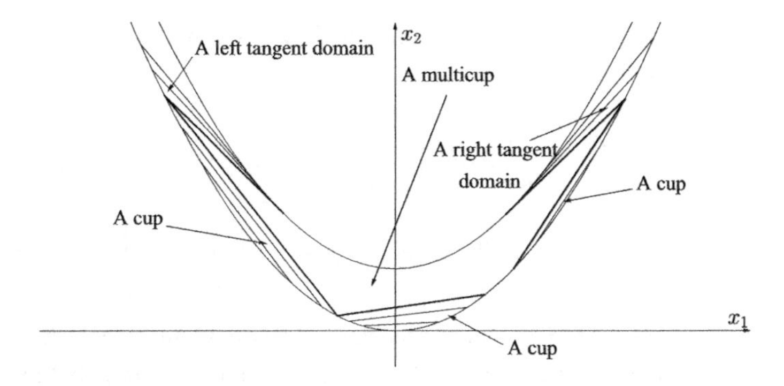

Figure 1.9 A multicup together with surrounding tangent domains and an underlying cup.

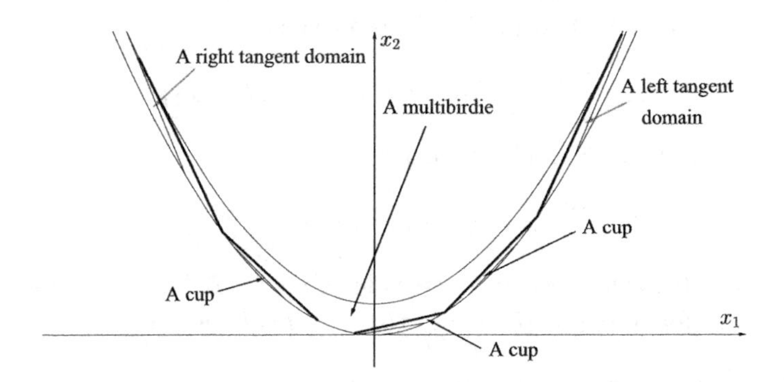

Figure 1.10 A multibirdie together with surrounding tangent domains and an underlying cup.

1.6 Dyadic Maximal Operator

Let us define the dyadic maximal operator on the set of positive, locally summable functions w, as follows:

$$(Mw)(t) \stackrel{\text{def}}{=} \sup_{I \in \mathcal{D}(\mathbb{R}),\, t \in I} \langle w \rangle_I .$$

We would like to estimate the norm of M as an operator acting from $L^p(\mathbb{R})$ to $L^p(\mathbb{R})$. Even though the operator is defined on the whole line, we first localize its action to a fixed dyadic interval I; we will get to all of $\mathbb{R}$ at the end. Thus, we are looking for function

$$\mathbf{B}(x_1, x_2; L) \stackrel{\text{def}}{=} \sup_{w \geq 0} \Big\{ \langle (Mw)^p \rangle_I : \ \langle w \rangle_I = x_1, \\ \langle w^p \rangle_I = x_2, \quad \sup_{J \supset I,\, J \in \mathcal{D}(\mathbb{R})} \langle w \rangle_J = L \Big\}. \tag{1.6.1}$$

We need the "external" parameter L because M is not truly local: the value of Mw on interval I depends not only on the behavior of w on I, but also on that on the whole line $\mathbb{R}$. Function $\mathbf{B}$ depends on three variables, and each of them can change when we split the interval of definition. Nevertheless, we will consider L as a parameter. The reason will become clear a bit later.

As before, $\mathbf{B}$ does not depend on I. Its domain is

$$\Omega \stackrel{\text{def}}{=} \{(x_1, x_2; L) \colon 0 < x_1 \leq L,\ x_1^p \leq x_2\},$$

or, if we consider L to be a fixed parameter,

$$\Omega_L \stackrel{\text{def}}{=} \{(x_1, x_2) \colon 0 < x_1 \leq L,\ x_1^p \leq x_2\}.$$

It is not difficult to check that every point of Ω_L is a Bellman point for some weight, i.e., for every $x \in \Omega_L$, there exists $w \in L^p(I)$ such that

$$\langle w \rangle_I = x_1, \quad \langle w^p \rangle_I = x_2, \quad \text{and} \quad \sup_{J \supset I,\, J \in \mathcal{D}(\mathbb{R})} \langle w \rangle_J = L.$$

Note that the boundary $x_1 = 0$ does not belong to the domain: if $x_1 = 0$, then w must be identically zero on I, which means that $x_2 = 0$. So, formally speaking, the points $(0, 0; L)$ should be included into our domain.

LEMMA 1.6.1 (Main Inequality) *Take $(x; L) \in \Omega$, and let the points $(x^\pm; L^\pm) \in \Omega$ be such that $x = \frac{1}{2}(x^+ + x^-)$ and $L^\pm = \max\{x_1^\pm, L\}$. Then the following inequality holds:*

$$\mathbf{B}(x; L) \geq \frac{\mathbf{B}(x^+; L^+) + \mathbf{B}(x^-; L^-)}{2}. \tag{1.6.2}$$

PROOF The proof should feel standard by now. Fixing an interval I and a small number $\eta > 0$, we take a pair of test functions $w^\pm$ on intervals $I^\pm$ such that their Bellman points are $\mathfrak{b}_{I^\pm}(w^\pm) = (x^\pm, L^\pm)$ and they almost realize the supremum in the definition of the Bellman function:

$$\langle (Mw^\pm)^p \rangle_{I^\pm} \geq \mathbf{B}(x^\pm; L^\pm) - \eta.$$

We set

$$w(t) = \begin{cases} w^{\pm}(t), & \text{if } t \in I^{\pm}, \\ L, & \text{if } t \notin I. \end{cases}$$

Then $\mathfrak{b}_I(w) = (x; L)$ and $(Mw)(t) = (Mw^{\pm})(t)$ for $t \in I_{\pm}$. Therefore,

$$\begin{aligned} \mathbf{B}(x;L) \geq \langle (Mw)^p \rangle_{_I} &= \tfrac{1}{2}\big(\langle (Mw^+)^p \rangle_{_{I_+}} + \langle (Mw^-)^p \rangle_{_{I_-}}\big) \\ &\geq \tfrac{1}{2}\big(\mathbf{B}(x^+;L^+) + \mathbf{B}(x^-;L^-)\big) - \eta, \end{aligned}$$

which proves the lemma. □

Corollary 1.6.2 (Concavity) *For a fixed L, the function $\mathbf{B}$ is concave on Ω_L.*

Proof Domain Ω_L is obviously convex. For any pair $x^{\pm} \in \Omega_L$, we have $L^{\pm} = L$, and (1.6.2) becomes the usual concavity condition. □

Corollary 1.6.3 (Boundary Condition) *If the function $\mathbf{B}$ is sufficiently smooth, then*

$$\frac{\partial \mathbf{B}}{\partial L}(x; x_1) = 0. \tag{1.6.3}$$

Proof First of all, we note that the definition of $\mathbf{B}$ immediately yields the inequality $\frac{\partial \mathbf{B}}{\partial L} \geq 0$. Now, take an arbitrary point x on the boundary $x_1 = L$ and a pair $x^{\pm}$ such that $x = \frac{1}{2}(x^+ + x^-)$. Let $\Delta = \frac{1}{2}(x^+ - x^-)$ and assume, without loss of generality that $x_1^+ > x_1^-$, i.e., $\Delta_1 > 0$. Then $x^{\pm} = x \pm \Delta$ and $x_1^- < x_1 = L < x_1^+$; therefore, $L^+ = x_1^+$ and $L^- = L$. Writing the main inequality up to the terms of first order in Δ, we get

$$\begin{aligned} 0 &\leq \mathbf{B}(x;L) - \tfrac{1}{2}\big(\mathbf{B}(x^+;L^+) + \mathbf{B}(x^-;L^-)\big) \\ &= \mathbf{B}(x_1,x_2;x_1) - \tfrac{1}{2}\big(\mathbf{B}(x_1+\Delta_1, x_2+\Delta_2; x_1+\Delta_1) \\ &\quad + \mathbf{B}(x_1-\Delta_1, x_2-\Delta_2; x_1)\big) \\ &\approx \mathbf{B} - \tfrac{1}{2}\big(\mathbf{B} + \mathbf{B}_{x_1}\Delta_1 + \mathbf{B}_{x_2}\Delta_2 + \mathbf{B}_L\Delta_1 + \mathbf{B} - \mathbf{B}_{x_1}\Delta_1 - \mathbf{B}_{x_2}\Delta_2\big) \\ &= -\tfrac{1}{2}\mathbf{B}_L(x;x_1)\Delta_1. \end{aligned}$$

Since $\mathbf{B}_L(x;x_1) \geq 0$, the last inequality is possible only if $\mathbf{B}_L(x;x_1) = 0$. □

Lemma 1.6.4 (Homogeneity) *If the function $\mathbf{B}$ is sufficiently smooth, then*

$$\mathbf{B}(x;L) = \frac{1}{p}x_1\mathbf{B}_{x_1} + x_2\mathbf{B}_{x_2} + \frac{1}{p}L\,\mathbf{B}_L. \tag{1.6.4}$$

Proof As before, along with a test function w, we consider the function $\tilde{w} = \tau w$ for $\tau > 0$. Comparing the Bellman functions at the corresponding Bellman points gives us the equality

$$\mathbf{B}(\tau x_1, \tau^p x_2; \tau L) = \tau^p \mathbf{B}(x_1, x_2; L).$$

Differentiating this identity with respect to τ at point $\tau = 1$ proves the lemma. □

Before we start looking for a Bellman candidate, let us state one more boundary condition – in fact, the principal one.

LEMMA 1.6.5 (Boundary Condition)

$$\mathbf{B}(u, u^p; L) = L^p. \tag{1.6.5}$$

PROOF The only test function corresponding to point $x = (u, u^p)$ is the function identically equal to u on interval J. Hence, Mw is identical to L on that interval. □

REMARK 1.6.6 Note that the boundary $x_1 = 0$ is not accessible, it does not belong to the domain. Therefore, we have no boundary condition on the boundary $x_1 = 0$.

We are now ready to search for a Bellman candidate. To this end, we will, as before, solve a Monge–Ampère boundary value problem. Our arguments for looking for a solution of the Monge–Ampère equation are the same as before: the concavity condition forces us to look for a function whose Hessian is negative, and the optimality condition requires the Hessian to be degenerate. And, of course, all these "natural" assumptions on the Bellman function get justified at the very end when we prove that a Bellman candidate obtained with the help of these assumptions is truly the Bellman function defined at the beginning of this section.

Again, we are looking for a solution in the form

$$B(x) = t_0 + x_1 t_1 + x_2 t_2, \tag{1.6.6}$$

that is linear along extremal trajectories given by

$$dt_0 + x_1 dt_1 + x_2 dt_2 = 0.$$

Let us parameterize the extremal lines by the first coordinate of their points of intersection with boundary $x_2 = x_1^p$. Since the boundary $x_1 = 0$ is not accessible, such an extremal line can either be vertical (i.e., parallel to the x_2-axis) or slanted to the right in which case, it intersects the boundary $x_1 = L$ at a point, say, (L, v). We leave it to the reader to answer the question: Why can an extremal line not have both ends on the boundary $x_2 = x_1^p$? (See Problem 1.6.1.)

The former case is very simple. Since B is linear on each vertical line and satisfies the boundary condition (1.6.5), it has the form

$$B(x;L) = k(x_1, L)(x_2 - x_1^p) + L^p.$$

Since $B_{x_2x_2} = 0$ and the matrix $B_{x_ix_j}$ must be nonpositive, we have $B_{x_1x_2} = k_{x_1} = 0$, i.e., k does not depend on x_1, $k = k(L)$. Now we use the second boundary condition $B_L(x_1, x_2; x_1) = 0$, which turns into $k'(x_1)(x_2 - x_1^p) + px_1^{p-1} = 0$. The last equation has no solution; therefore, this case (i) impossible, at least in the whole domain Ω_L.

Consider the latter case, where the extremal line goes from the bottom boundary to the right boundary, as shown in the picture (see Figure 1.11). The boundary condition (1.6.5) on the bottom boundary gives us

$$t_0 + ut_1 + u^p t_2 = L^p, \tag{1.6.7}$$

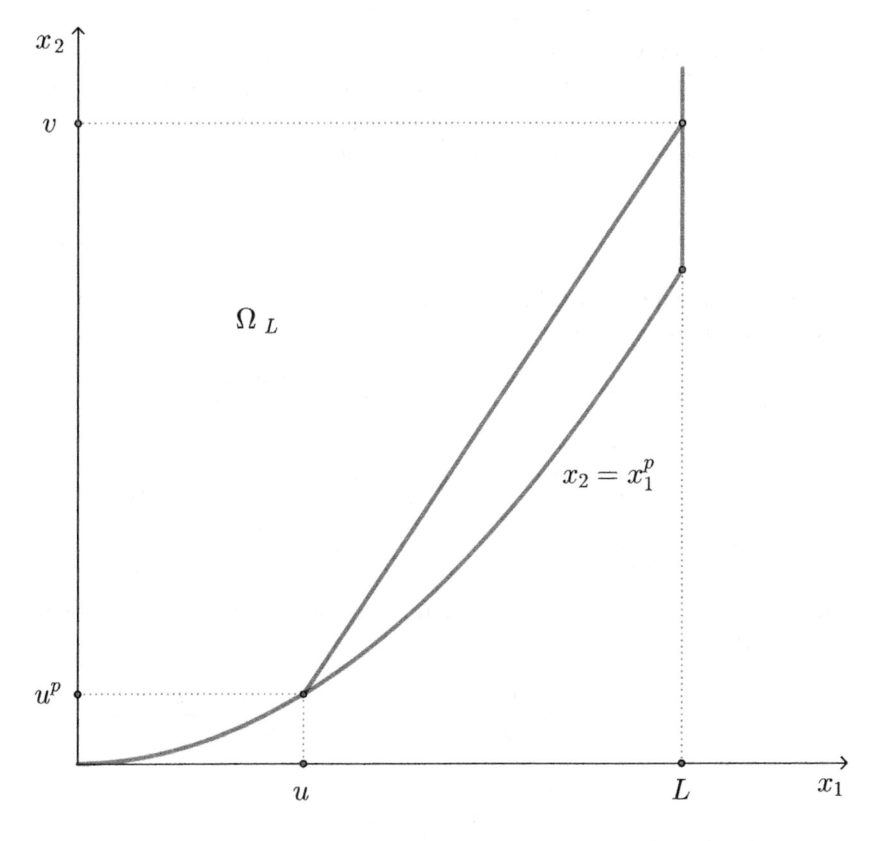

Figure 1.11 The extremal trajectory passing through (u, u^p) and (L, v).

and the condition (1.6.3) on the right boundary, along with (1.6.4), yields

$$t_0 + Lt_1 + vt_2 = \frac{1}{p}Lt_1 + vt_2. \tag{1.6.8}$$

From the last equation, we get

$$t_0 + \frac{p-1}{p}Lt_1 = 0. \tag{1.6.9}$$

Now we differentiate (1.6.7),

$$(dt_0 + udt_1 + u^p dt_2) + (t_1 + pu^{p-1}t_2)du = 0,$$

and use the fact that the point (u, u^p) is on the trajectory, i.e., $dt_0 + udt_1 + u^p dt_2 = 0$. Thus,

$$(t_1 + pu^{p-1}t_2)du = 0.$$

That equation gives us two possibilities. The first option is $u = \text{const}$, producing a family of trajectories, all passing through the point (u, u^p), or $t_1 + pu^{p-1}t_2 = 0$. The first option cannot give us a foliation of the whole Ω_L, since it would result in trajectories connecting two points of the bottom boundary (see Problem 1.6.1). Therefore, let us consider the second possibility, i.e.,

$$t_1 + pu^{p-1}t_2 = 0. \tag{1.6.10}$$

Solving the system of three linear equations, (1.6.7), (1.6.9), and (1.6.10), of three variables t_0, t_1, and t_2, we obtain:

$$t_0 = \frac{L^{p+1}}{L-u}, \qquad t_0' = \frac{L^{p+1}}{(L-u)^2},$$

$$t_1 = -\frac{pL^p}{(p-1)(L-u)}, \qquad t_1' = -\frac{pL^p}{(p-1)(L-u)^2},$$

$$t_2 = \frac{L^p}{(p-1)u^{p-1}(L-u)}, \qquad t_2' = \frac{L^p[pu-(p-1)L]}{(p-1)u^p(L-u)^2}.$$

Now we can plug the derivatives of t_i into the equation of extremal trajectories $dt_0 + x_1 dt_1 + x_2 dt_2 = 0:$

$$\frac{L^{p+1}}{(L-u)^2} - x_1\frac{pL^p}{(p-1)(L-u)^2} + x_2\frac{L^p[pu-(p-1)L]}{(p-1)u^p(L-u)^2} = 0$$

or

$$x_2 = \frac{px_1 - (p-1)L}{pu-(p-1)L}u^p. \tag{1.6.11}$$

If we plug the found expressions for t_i and this expression for x_2 into the initial formula (1.6.6) for B, we get

$$B(x;L) = \frac{px_1 - (p-1)L}{pu - (p-1)L} L^p = \frac{L^p}{u^p} x_2, \tag{1.6.12}$$

where the function $u = u(x_1, x_2)$ is implicitly given by (1.6.11).

We see that the extremals form a "fan" of lines passing through point $(\frac{p-1}{p}L, 0)$. However, those elements of the fan that intersect the "forbidden" boundary $x_1 = 0$ cannot be extremal trajectories. Therefore, the acceptable lines foliate not the whole domain Ω_L, but only the subdomain $x_1 \ge \frac{p-1}{p}L$. To foliate the rest, we return to considering vertical lines. Earlier, we had refused this type of trajectories for the whole domain Ω_L since the foliation produced in such a way would not give us a function satisfying the boundary condition on line $x_1 = L$. However, such trajectories are perfectly suited for foliating the subdomain $x_1 \le \frac{p-1}{p}L$, especially because the boundary of the two subdomains, the vertical line $x_1 = \frac{p-1}{p}L$, fits as an element of both foliations. On that line we have $x_1 = u = \frac{p-1}{p}L$, and so

$$B(\tfrac{p-1}{p}L, x_2; L) = \frac{p^p}{(p-1)^p} x_2.$$

As we have seen, the Bellman candidate on the vertical trajectories must be of the form

$$B(x;L) = k(L)(x_2 - x_1^p) + L^p.$$

To get $B = \frac{p^p}{(p-1)^p} x_2$ on line $x_1 = \frac{p-1}{p}L$, we have to take $k(L) = \frac{p^p}{(p-1)^p}$, which gives the following Bellman candidate in the left part of Ω_L :

$$B(x;L) = \tfrac{p^p}{(p-1)^p}(x_2 - x_1^p) + L^p. \tag{1.6.13}$$

So, we have got a Bellman candidate on the whole domain Ω_L:

$$B(x;L) = \begin{cases} \frac{p^p}{(p-1)^p}(x_2 - x_1^p) + L^p, & 0 < x_1 \le \frac{p-1}{p}L,\ x_2 \ge x_1^p, \\ \\ \frac{L^p}{u^p} x_2, & \frac{p-1}{p}L \le x_1 \le L,\ x_2 \ge x_1^p, \end{cases} \tag{1.6.14}$$

where $u = u(x)$ is defined by (1.6.11).

We can solve equation (1.6.11) explicitly only for $p = 2$. In that case, we have

$$u = \frac{\sqrt{x_2}L}{\sqrt{x_2} + \sqrt{x_2 - L(2x_1 - L)}}.$$

That yields

$$B(x;L) = \begin{cases} 4(x_2 - x_1^2) + L^2, & 0 < x_1 \le \frac{L}{2},\ x_2 \ge x_1^2, \\ \left(\sqrt{x_2} + \sqrt{x_2 - L(2x_1 - L)}\right)^2, & \frac{L}{2} \le x_1 \le L,\ x_2 \ge x_1^2. \end{cases} \tag{1.6.15}$$

Now, we start proving that the Bellman candidate just found is indeed the Bellman function of our problem.

LEMMA 1.6.7 *The function defined by* (1.6.14) *satisfies the main inequality* (1.6.2).

PROOF Let us define a family of functions $\tilde{B}$ in the domain

$$\tilde{\Omega} \stackrel{\text{def}}{=} \{x = (x_1, x_2)\colon\ x_1 > 0,\ x_2 \ge x_1^p\}$$

by the following formula:

$$\tilde{B}(x;L) \stackrel{\text{def}}{=} \begin{cases} B(x;L), & 0 < x_1 \le L,\ x_2 \ge x_1^p, \\ B(x;x_1), & x_1 \ge L,\ x_2 \ge x_1^p. \end{cases}$$

In other words, for $x_1 \ge L$ the function $\tilde{B}$ is defined by

$$\tilde{B}(x;L) = \frac{x_1^p}{u^p} x_2,$$

where $u = u(x)$ is implicitly determined by the relation

$$x_2 = \frac{x_1 u^p}{pu - (p-1)x_1}.$$

Let us calculate the first partial derivatives:

$$\tilde{B}_{x_1}(x;L) = \begin{cases} -\dfrac{p^{p+1}}{(p-1)^p} x_1^{p-1}, & 0 < x_1 \le \frac{p-1}{p} L, \\[2ex] -\dfrac{pL^p}{(p-1)(L-u)}, & \frac{p-1}{p} L \le x_1 \le L, \\[2ex] -\dfrac{px_1^p}{(p-1)(x_1-u)}, & x_1 \ge L. \end{cases}$$

$$\tilde{B}_{x_2}(x;L) = \begin{cases} \dfrac{p^p}{(p-1)^p}, & 0 < x_1 \le \frac{p-1}{p} L, \\[2ex] \dfrac{L^p}{(p-1)(L-u)u^{p-1}}, & \frac{p-1}{p} L \le x_1 \le L, \\[2ex] \dfrac{x_1^p}{(p-1)(x_1-u)u^{p-1}}, & x_1 \ge L. \end{cases}$$

From these expressions, we see that our function $\tilde{B}$ is C^1-smooth. Further, we calculate the second derivative

$$\tilde{B}_{x_1x_1}(x;L) = \begin{cases} -\dfrac{p^{p+1}}{(p-1)^{p-1}}x_1^{p-2}, & 0 < x_1 \le \frac{p-1}{p}L, \\ -\dfrac{pL^p u^{p+1}}{(p-1)^2(L-u)^3x_2}, & \frac{p-1}{p}L \le x_1 \le L, \\ -\dfrac{px_1^{p-1}}{(p-1)^2(x_1-u)^3}Q, & x_1 \ge L, \end{cases}$$

where $Q = (p-1)^2x_1^2 - (p-1)(2p-1)x_1u + (p^2-p+1)u^2$, and check that it is strictly negative. In the first two subdomains it is evident, for the third one we calculate the discriminant of the quadratic form Q, which is $-3(p-1)^2$, and conclude that $Q > 0$. Therefore, to prove concavity of $\tilde{B}$ in the domain $\tilde{\Omega}$ it is sufficient to show that the determinant of the Hessian matrix is nonnegative. We know that this determinant is zero in Ω_L and, therefore, we need to calculate the second derivatives of $\tilde{B}$ only in the domain $x_1 > L$, where we have

$$\tilde{B}_{x_1x_2} = \frac{x_1^{p+1}u^{p+1}}{(p-1)^2(x_1-u)^3x_2^2},$$
$$\tilde{B}_{x_2x_2} = -\frac{x_1^{p+2}u^{p+1}}{p(p-1)^2(x_1-u)^3x_2^3},$$

which yields

$$\tilde{B}_{x_1x_1}\tilde{B}_{x_2x_2} - \tilde{B}_{x_1x_2}^2 = \frac{x_1^{2p+1}u^{p+1}}{(p-1)^2(x_1-u)^4x_2^3}.$$

The concavity just proved immediately implies (1.6.2). Indeed, we have proved that the function $\tilde{B}$ is locally concave in each subdomain of $\tilde{\Omega}$, and it is C^1-smooth on the whole domain; therefore, it is concave everywhere on $\tilde{\Omega}$. Furthermore, relation (1.6.2) is a special case of the concavity condition on the function $\tilde{B}$, when x^- and x are in the subdomain Ω_L. □

LEMMA 1.6.8 (Bellman Induction) *For any continuous function B satisfying the main inequality* (1.6.2) *and the boundary condition* (1.6.5), *we have*

$$\mathbf{B}(x;L) \le B(x;L).$$

PROOF The proof is standard. First, we fix a test function w on $\mathbb{R}$ and a dyadic interval I. This gives us a Bellman point $\mathfrak{b}_I = (x;L)$. Then we start splitting the interval I, while repeatedly applying the main inequality:

$$|I|B(\mathfrak{b}_I) \geq |I^+|B(\mathfrak{b}_{I^+}) + |I^-|B(\mathfrak{b}_{I^-})$$
$$\geq \sum_{J\in\mathcal{D}_n} |J|B(\mathfrak{b}_J) = \int_I B\big(x^{(n)}(s); L^{(n)}(s)\big)\, ds,$$

where $(x^{(n)}(s); L^{(n)}(s)) = \mathfrak{b}_J$, when $s \in J$, $J \in \mathcal{D}_n$. By the Lebesgue differentiation theorem, we have $x^{(n)}(s) \to \big(w(s), w^p(s)\big)$, and by the definition of the maximal function, $L^{(n)}(s) \to (Mw)(s)$ almost everywhere. For bounded w, we can pass to the limit under the integral

$$B\big(x^{(n)}(s); L^{(n)}(s)\big) \to B\big(w(s), w^p(s); (Mw)(s)\big) = (Mw)^p(s)$$

and obtain

$$B(x; L) \geq \langle (Mw)^p(s)\rangle_I.$$

Then, approximating, as before, an arbitrary test function w by its bounded cut-offs, we get the same inequality for all w, which immediately gives us the required property: $B(x; L) \geq \mathbf{B}(x; L)$. □

COROLLARY 1.6.9 *For the function B, given by* (1.6.14), *the inequality*

$$B(x; L) \geq \mathbf{B}(x; L)$$

holds.

To prove the converse inequality, we need to construct an optimizer. However, in the present setting, we have no test function realizing the supremum in the definition of the Bellman function. Thus, an optimizer will be given by a sequence of test functions.

First, we choose an arbitrary $u \in (\frac{p-1}{p}L, L)$, and consider the extremal line passing through point (u, u^p) given by equation (1.6.11). This extremal line intersects the boundary $x_1 = L$ at the point (L, v), with

$$v = \frac{Lu^p}{pu - (p-1)L}. \tag{1.6.16}$$

Note that when u increases from $\frac{p-1}{p}L$ to L, this point of intersection runs over the whole right boundary, i.e., it runs from $+\infty$ to L^p. Now we start constructing the extremal sequence of test functions on the interval $(0, 1)$ for point $x = (L, v)$. We need to split the interval $(0, 1)$ into two halves, which splits the Bellman point $x = (L, v)$ into a pair of points $x^{\pm}$, $x = (x^- + x^+)/2$. We use the homogeneity of the problem in our construction. We know that the set of test functions for the point $\tilde{x} = (\tau x_1, \tau^p x_2)$ is the same as the set of test functions for the point x, each multiplied by τ. Therefore, if w is an optimizer for

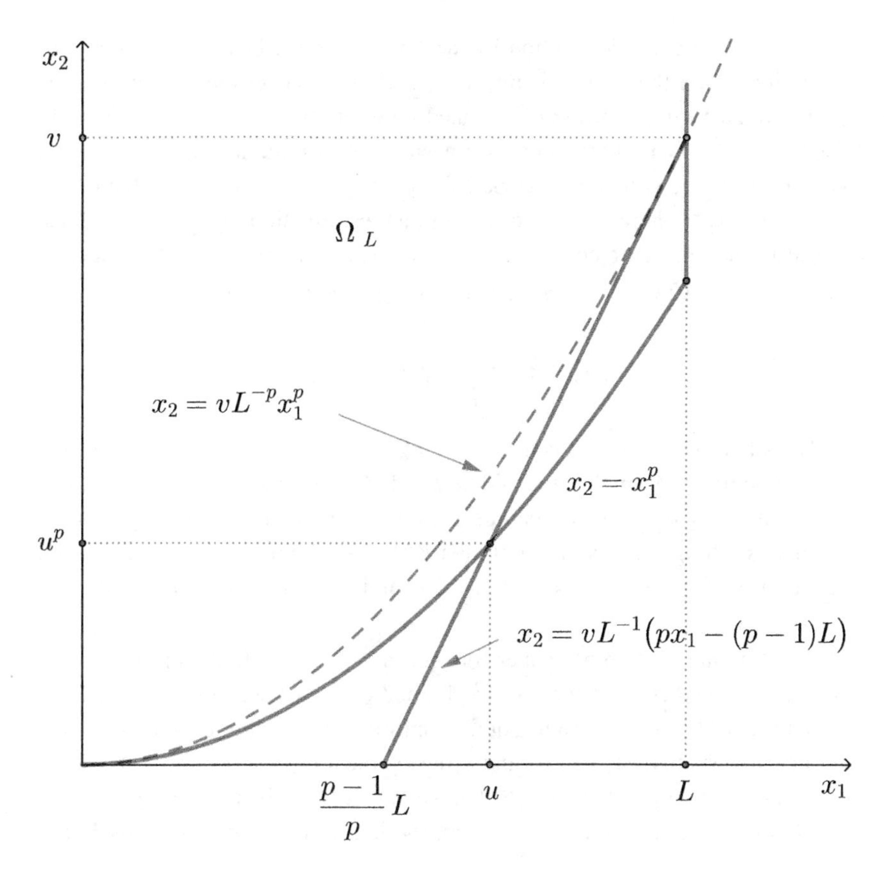

Figure 1.12 The extremal trajectory passing through (u, u^p) and (L, v) is tangent to the curve $x_2 = vL^{-p}x_1^p$.

the point x, then τw is an optimizer for $\tilde{x}$. For that reason, for the first splitting of $x = (L, v)$, it is more convenient to take the right point x^+ not on the continuation of the extremal line, but on the curve $x_2 = vL^{-p}x_1^p$, which is tangent to our extremal line at the point x. So, we shall split our Bellman points not along the extremal line being tangent to the mentioned curve at that point, but along the secant line, which tends to the tangent when we choose x^+ closer and closer to x. Moving the point x^+ along the curve $x_2 = vL^{-p}x_1^p$, we can make any ratio of the length of the segment $[x, x^+]$ to the length of the part of our secant line lying inside the domain (i.e., $[x^-, x]$). For our sequence of test functions, we choose that ratio to run the values 2^{-n}. Let u_n be the first coordinate of the point where our secant line intersects the lower boundary. We shall not specify the value u_n. The only fact that is important to us is that $u_n \to u$.

Now we start the description of the test functions. Due to our choice of the point x^+ on the right half-interval $(\frac{1}{2}, 1)$, we can set the optimizer to be proportional to the appropriately scaled copy of itself: $w(t) = \tau w(2t-1)$ for $t \in (\frac{1}{2}, 1)$. What function do we need to take on the left half-interval? In case $n = 1$, the situation is trivial: the point x^- is on the lower boundary, $x^- = (u_1, u_1^p)$, where we have only constant test function $w|_{(0,\frac{1}{2})} = u_1$. Since our test function has to correspond to point (v, L), we deduce that we need to take $\tau = \tau_1 = (2 - \frac{u_1^p}{v})^{1/p}$. As a result, we get the following test function:

$$w_1(t) = u_1 \sum_{k=0}^{\infty} \tau_1^k \chi_{[1-2^{-k}, 1-2^{-k-1}]}$$

The situation for $n = 2$ is not much more difficult: After the first splitting we get a point x^-, which is the middle point of the segment with the endpoints (L, v) and (u_2, u_2^p), and we can take these points as the Bellman points for the next splitting, i.e., on the right interval $(\frac{1}{4}, \frac{1}{2})$, we put a copy of the scaled function w_2 $\big(w_2|_{(\frac{1}{4},\frac{1}{2})}(t) = w_2(4t-1)\big)$ and we put constant u_2 on the left interval $(0, \frac{1}{4})$.

We continue in such a manner for arbitrary n: In the first splitting we put $w_n(t) = \tau_n w_n(2t-1)$ for $t \in (\frac{1}{2}, 1)$, and after that we always choose the initial point (L, v) as the right point of splitting. The length of the secant line is chosen in such a way that for the n-th step the left point occurs on the lower boundary, so on the interval $(0, 2^{-n})$, we put $w_n(t)$ to be the constant u_n and on all preceding intervals we put the appropriate scaled copies of w_n itself. So, we get

$$w_n(t) = \begin{cases} u_n & 0 < t < 2^{-n}, \\ w_n(2^k t - 1) & 2^{-k} < t < 2^{-k+1},\ 1 < k < n, \\ \tau_n w_n(2t-1) & \frac{1}{2} < t < 1. \end{cases} \tag{1.6.17}$$

Let us verify that this recurrent relation defines the sequence $\{w_n\}$ correctly. To that end, let us introduce a sequence $\{w_{n,m}\}$ by induction:

$$w_{n,0}(t) = \begin{cases} u_n & 0 < t < 2^{-n}, \\ 0 & 2^{-n} < t < 1; \end{cases}$$

$$w_{n,m}(t) = \begin{cases} u_n & 0 < t < 2^{-n}, \\ w_{n,m-1}(2^k t - 1) & 2^{-k} < t < 2^{-k+1},\ 1 < k < n, \\ \tau_n w_{n,m-1}(2t-1) & \frac{1}{2} < t < 1. \end{cases}$$

We see that $w_{n,m}(t) = w_{n,m-1}(t)$ for all t, such that $w_{n,m-1}(t) \neq 0$, and the measure of the set, where $w_{n,m-1}(t) = 0$, is $(1 - 2^{-n})^m$, i.e., it tends to zero as $m \to \infty$. Therefore, $w_{n,m}$ stabilizes almost everywhere as a sequence in m, and its limit w_n satisfies the recurrent relation (1.6.17).

Now we need to calculate the maximal function of w_n, which is a simple matter:

$$Mw_n = \frac{L}{u_n} w_n.$$

This follows from recursive formula (1.6.17) that defines a unique function w_n, if the constants u_n and τ_n are given. It is clear that the function Mw_n satisfies the same recursive relation, where u_n is replaced by L. Since the formula is linear with respect to that parameter, we come to the earlier formula for Mw_n.

Finally, we have

$$\langle (Mw_n)^p \rangle_{[0,1]} = \frac{L^p}{u_n^p} \langle w_n^p \rangle_{[0,1]} \longrightarrow \frac{L^p}{u^p} v = B(L, v; L),$$

and, therefore,

$$\mathbf{B}(L, v; L) \geq B(L, v; L).$$

We have described the earlier construction of extremizers for pedagogical reasons: It is the usual method of proving the inverse inequality. Moreover, sometimes the knowledge of extremizers helps us in the investigations of the corresponding operator. Here, for example, we see that the extremal test functions are, in a sense, the eigenfunctions of the maximal operator. However, in this case, we can prove the inverse inequality, using only homogeneity and the main inequality. In fact, we shall use the same idea: We have "to go" along the extremals and use the homogeneity of our Bellman function.

We write down the main inequality

$$(x_1^+ - x_1^-)\mathbf{B}(x; L) \geq (x_1 - x_1^-)\mathbf{B}(x^+; L^+) + (x_1^+ - x_1)\mathbf{B}(x^-; L^-)$$

for the same point $x = (L, v)$. As before, the point x^+ will be placed on the curve $x_2^+ = \frac{v}{L^p}(x_1^+)^p$, and the point x^- will be placed on the lower boundary $x_2^- = (x_1^-)^p$. It is clear that $L^+ = x_1^+$ and $L^- = L$. For points chosen in such a way, we have

$$\mathbf{B}(x^+; L^+) = \left(\frac{x_1^+}{L}\right)^p \mathbf{B}(x; L) = \frac{x_2^+}{v} \mathbf{B}(x; L) \qquad \text{and} \qquad \mathbf{B}(x^-; L^-) = L^p.$$

Then the earlier inequality turns into

$$\mathbf{B}(x;L) \geq \frac{(x_1^+ - L)L^p}{(x_1^+ - x_1^-) - \frac{x_2^+}{v}(L - x_1^-)} = \frac{L^p v}{x_2^-}.$$

In the latter equality, we have used the fact that the points x and $x^\pm$ are placed on a straight line:

$$\frac{L - x_1^-}{x_1^+ - L} = \frac{v - x_2^-}{x_2^+ - v}.$$

Since $x_2^- = (x_1^-)^p \to u^p$ as $x_1^+ \to L$ in the limit, we have the desired inequality

$$\mathbf{B}(L, v; L) \geq \frac{L^p}{u^p} v = B(L, v; L).$$

Thus, we have proved the inequality $\mathbf{B}(x; L) \geq B(x; L)$ for x on line $x_1 = L$. Now let us take an arbitrary $x \in \Omega_L$, with $x_1 > \frac{p-1}{p} L$. Let the extremal line passing through this point intersect the two boundaries of Ω_L at points (u, u^p) and (L, v). Also let us assume that the point x splits the segment between those two points in the ratio $\alpha : (1 - \alpha)$. Using the main inequality for $\mathbf{B}$, linearity of B on the extremal line, and the just-proved inequality $\mathbf{B}(L, v; L) \geq B(L, v; L)$, we can write down the following chain of estimates:

$$\begin{aligned}
\mathbf{B}(x; L) &\geq \alpha \mathbf{B}(L, v; L) + (1 - \alpha)\mathbf{B}(u, u^p; L) \\
&= \alpha \mathbf{B}(L, v; L) + (1 - \alpha)L^p \\
&\geq \alpha B(L, v; L) + (1 - \alpha)L^p \\
&= \alpha B(L, v; L) + (1 - \alpha)B(u, u^p; L) = B(x; L).
\end{aligned}$$

We use the same trick to prove inequality $\mathbf{B}(x; L) \geq B(x; L)$ for $x_1 \leq \frac{p-1}{p} L$, except now, instead of the vertical extremal line, we use a nearby line with a large slope. Take a number ξ close to x_1, $\xi < x_1$, and take the line passing through x and $(\xi, 0)$. Let (a, a^p) and (L, v) be the points where that line intersects two boundaries of Ω_L. Let (u, u^p) be the point of intersection of the extremal line, passing through (L, v), with the boundary. Then from similar triangles, we have the following proportions:

$$\frac{v}{L - \xi} = \frac{x_2}{x_1 - \xi} = \frac{a^p}{a - \xi}.$$

If $\xi \to x_1$, we have

$$a \to x_1, \qquad u \to \frac{p-1}{p} L,$$

and

$$\frac{x_1 - a}{x_1 - \xi} = 1 - \frac{a - \xi}{x_1 - \xi} = 1 - \frac{a^p}{x_2} \xrightarrow[\xi \to x_1]{} 1 - \frac{x_1^p}{x_2}.$$

Therefore,

$$\begin{aligned}
\mathbf{B}(x; L) &\geq \frac{x_1 - a}{L - a}\mathbf{B}(L, v) + \frac{L - x_1}{L - a}\mathbf{B}(a, a^p) \\
&= \frac{x_1 - a}{L - a} \cdot \frac{L^p}{u^p} v + \frac{L - x_1}{L - a} L^p \\
&= \frac{L^p}{u^p} \cdot \frac{L - \xi}{L - a} \cdot \frac{x_1 - a}{x_1 - \xi} x_2 + \frac{L - x_1}{L - a} L^p \\
&\xrightarrow[\xi \to x_1]{} \frac{p^p}{(p-1)^p}(x_2 - x_1^p) + L^p = B(x; L).
\end{aligned}$$

Thus, we have proved the following lemma.

Lemma 1.6.10

$$\mathbf{B}(x; L) \geq B(x; L).$$

Taken together, this lemma and Corollary 1.6.9 prove the following theorem:

Theorem 1.6.11

$$\mathbf{B}(x; L) = \begin{cases} \frac{p^p}{(p-1)^p}(x_2 - x_1^p) + L^p, & 0 < x_1 \leq \frac{p-1}{p} L,\ x_2 \geq x_1^p, \\ \\ \frac{L^p}{u^p} x_2, & \frac{p-1}{p} L \leq x_1 \leq L,\ x_2 \geq x_1^p, \end{cases}$$

where $u = u(x)$ is defined by (1.6.11).

Reducing the Number of Variables

In this section, we explain how to find the same Bellman function in two steps: first, we find a family of auxiliary Bellman functions, which are much simpler and depend only on variable x_1 and parameter L; second, we reconstruct the desired function, roughly speaking, as the pointwise minimum of the functions from the found family.

We shall look for the following auxiliary Bellman function:

$$\mathbb{B}_c(x_1; L) \stackrel{\text{def}}{=} \sup_{w \geq 0} \Big\{ \langle (Mw)^p \rangle_{_J} - c^p \langle w^p \rangle_{_I} :$$

$$\langle w \rangle_{_I} = x_1, \quad \sup_{J \supset I,\ J \in \mathcal{D}(\mathbb{R})} \langle w \rangle_{_J} = L \Big\}.$$

Its domain of definition is the set $\{(x_1; L) \colon 0 < x_1 \leq L\}$. The same main inequality (1.6.2) holds for the same reason, but now we have a one-dimensional variable $x = x_1$, instead of the two-dimensional ones earlier. So, $\mathbb{B}_c$ is a concave function in x_1 on the interval $(0, L]$. The same boundary condition at the end $x_1 = L$

$$\frac{\partial \mathbb{B}_c}{\partial L}(x_1; x_1) = 0 \tag{1.6.18}$$

holds for the same reason, as before. The homogeneous condition will be used now in another way: We will remove parameter L and reduce the problem of finding a function on the interval $(0, 1]$. Since

$$\mathbb{B}_c(\tau x_1; \tau L) = \tau^p \mathbb{B}_c(x_1; L),$$

we can introduce a concave function

$$\beta_c(y) \stackrel{\text{def}}{=} \mathbb{B}_c(y; 1), \qquad y \in (0, 1],$$

and then

$$\mathbb{B}_c(x_1; L) = L^p \beta_c\Big(\frac{x_1}{L}\Big).$$

For β_c the boundary condition (1.6.18) turns into

$$\beta_c'(1) = p\beta_c(1). \tag{1.6.19}$$

But now, we no longer have the other boundary condition. Instead of (1.6.5), we have an obstacle condition

$$\beta_c(y) \geq 1 - c^p y^p. \tag{1.6.20}$$

Indeed, if we take a constant test function $w = x_1$, then the maximal function is $Mw = L$, and we get the necessary condition

$$\mathbb{B}_c(x_1; L) \geq L^p - c^p x_1^p,$$

which coincides with (1.6.20).

As before, at this point, we start to search for a Bellman candidate, but now it will be much easier, because the one-dimensional “Monge–Ampère equation" means simply that the second derivative is zero. Therefore, our function has to be linear, where it is bigger than the obstacle function. The boundary condition (1.6.19) is not fulfilled for the obstacle function itself for any c, therefore in all cases $\mathbb{B}_c$ should be linear on some interval $[u, L]$, and now our task is to find the value $u = u(c)$.

A linear function satisfies (1.6.19) if and only if it is zero at point $y = \frac{p-1}{p}$. Note that the obstacle condition vanishes at point $y = \frac{1}{c}$. So, if $c < \frac{p}{p-1}$ (see Figure 1.13), then any function we can get by concatenating the part of the obstacle function with any straight line passing through point $(\frac{p-1}{p}, 0)$ will not be concave. Therefore, for such c we have $\beta_c = +\infty$ (we leave the rigorous proof of this fact as an exercise, see Problem 1.6.3). If c is bigger than $\frac{p}{p-1}$, then our only choice is to draw the tangent line to the obstacle curve, starting at point $(\frac{p-1}{p}, 0)$, and if $y = \frac{u}{L}$ is the tangency point, then take this linear function on the interval $[\frac{u}{L}, 1]$ (see Figure 1.14), i.e., for $c \geq \frac{p}{p-1}$ our Bellman candidate is

$$\beta_c = \begin{cases} 1 - c^p y^p & \text{for } 0 < yL < u(c), \\ \left(1 - \frac{c^p u^p}{L^p}\right) \frac{p-1-py}{(p-1)L-pu} L & \text{for } u(c) \leq yL \leq L. \end{cases}$$

A few more words about the function $u(c)$. Since we have chosen our straight line to be tangent to the obstacle curve, function β_c is C^1-smooth. Comparing the left and the right derivatives at point u, we come to the equation

$$(p-1)(L-u)c^p u^{p-1} - L^p = 0, \tag{1.6.21}$$

whose unique solution on interval $(\frac{p-1}{p}L, L]$ supplies us with the desired function $u(c)$.

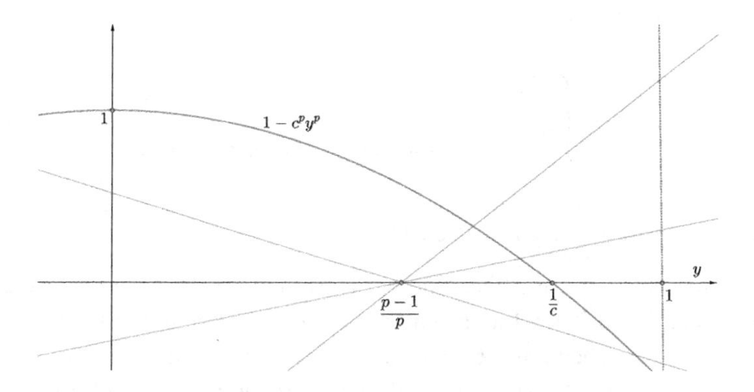

Figure 1.13 No concave solution.

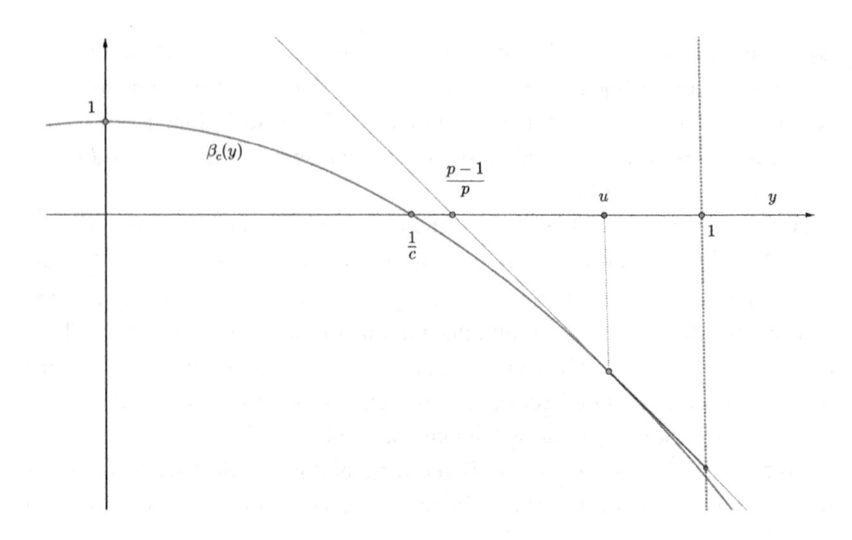

Figure 1.14 An appropriate tangent line to the obstacle curve.

Using relation (1.6.21), we can rewrite the formula for β_c in a simpler form as follows:

$$\beta_c = \begin{cases} 1 - c^p y^p & \text{for } 0 < yL < u, \\ \frac{p-1-py}{(p-1)(L-u)} L & \text{for } u \le yL \le L. \end{cases}$$

After all these preparations, we can state the assertion similar to Theorem 1.6.11:

Theorem 1.6.12 *If $c < \frac{p}{p-1}$, then $\mathbb{B}_c = +\infty$. If $c \le \frac{p}{p-1}$, then $\mathbb{B}_c$ is given by the following formula*:

$$\mathbb{B}_c(x_1; L) = \begin{cases} L^p - c^p x_1^p & \text{for } 0 < x_1 < u, \\ \frac{L^p}{L-u}\left(L - \frac{p}{p-1} x_1\right) & \text{for } u \le x_1 \le L, \end{cases} \tag{1.6.22}$$

where $u = u(c)$ is defined by (1.6.21).

We shall not prove this theorem, but leave that for the reader as exercises (see Problems 1.6.3–1.6.4 at the end of this section) that repeat the reasoning in the proof of Theorem 1.6.11 in a simpler setting. And now we discuss how to reconstruct the function $\mathbf{B}$ of three variable by means of a family of functions $\mathbb{B}_c$.

Reconstruction of the Function $\mathbf{B}$ from the Family of $\mathbb{B}_c$

Take any test function w with the Bellman point $(x; L)$. Since

$$\langle (Mw)^p \rangle = \big[\langle (Mw)^p \rangle - c^p \langle w^p \rangle\big] + c^p \langle w^p \rangle \le \mathbb{B}_c(x_1; L) + c^p x_2,$$

we have

$$\mathbf{B}(x; L) \le \inf_{c \in \mathbb{R}} \big\{\mathbb{B}_c(x_1; L) + c^p x_2\big\}.$$

Let us calculate the function in the right-hand side of this inequality to see that, in fact, it coincides with $\mathbf{B}$.

First, consider the case $x_1 \le \frac{p-1}{p} L$. Then $x_1 \le u$ (recall that $u = u(c)$ is defined by (1.6.21)) and

$$\begin{aligned}\inf_{c \in \mathbb{R}} \big\{\mathbb{B}_c(x_1; L) + c^p x_2\big\} &= \inf_{c \ge \frac{p}{p-1}} \big\{L^p - c^p x_1^p + c^p x_2\big\} \\ &= L^p + \tfrac{p^p}{(p-1)^p}\big(x_2 - x_1^p\big).\end{aligned}$$

Now let $\frac{p-1}{p} L < x_1 \le L$. Here it is more convenient to consider u as a parameter, rather than c. The inverse function $c = c(u)$ is monotonously increasing, and c runs over the whole ray $\big[\frac{p}{p-1}, \infty\big)$, when u runs over $\big[\frac{p-1}{p} L, L\big)$. Since in the domain $0 < x_1 \le u$ function $\mathbb{B}_c(x_1; L) + c^p x_2 = L^p + c^p(x_2 - x_1^p)$ is increasing in c (and therefore in u), it takes its minimum at the minimal value of u, for $u = x_1$. Therefore, it is sufficient to consider the infimum of the expression, when $x_1 \ge u$:

$$\begin{aligned}\mathbb{B}_c(x_1; L) + c^p x_2 &= \frac{L^p}{L-u}\Big(L - \frac{p}{p-1} x_1\Big) + \frac{L^p}{(p-1)(L-u)u^{p-1}} x_2 \\ &\overset{\text{def}}{=} \psi(u).\end{aligned}$$

over $u \in \big[\frac{p-1}{p} L, x_1\big]$. Direct calculation gives us

$$\psi'(u) = \frac{L^p}{(p-1)u^p(L-u)^2}\big([(p-1)L - p x_1]u^p - [(p-1)L - pu]x_2\big).$$

Since $\psi'\big(\frac{p-1}{p} L\big) < 0$ and $\psi'(x_1) > 0$, the equation $\psi'(u) = 0$ has a solution on the interval $\big(\frac{p-1}{p} L, x_1\big)$ (it is not difficult to see that the solution is unique), where the function ψ attains its local minimum, which is equal to $\frac{L^p}{u^p} x_2$. Here u is the solution of the equation $\psi'(u) = 0$, or

$$[(p-1)L - p x_1]u^p - [(p-1)L - pu]x_2 = 0,$$

which is just equation (1.6.11). So, we have reconstructed the function $\mathbf{B}$, starting from the family $\mathbb{B}_c$.

Historical Remarks

The Bellman setup of the problem discussed in this section was first stated in [**127**], where a supersolution was found, which gives the correct norm of the maximal operator. Theorem 1.6.11 supplying us with the analytic expression for the Bellman function was found by A. Melas in [**116**]. The Monge–Ampère approach to calculation of the Bellman function was presented in [**170**] (an initial version in [**169**]). The method of reducing it to a family of functions of two variables and Theorem 1.6.12, together with recovering from that family the initial Bellman function (1.6.14) can be found in [**143**].

Exercises

PROBLEM 1.6.1 Show that an extremal line of the function $\mathbf{B}$ cannot connect two points of the boundary $x_2 = x_1^p$.

PROBLEM 1.6.2 Show that Theorem 1.6.11 implies the following estimate:

$$\|Mw\|_{L^p(\mathbb{R})} \le \frac{p}{p-1}\|w\|_{L^p(\mathbb{R})}.$$

The estimate is sharp, i.e., $\|M\|_{L^p\to L^p} = \frac{p}{p-1}$.

PROBLEM 1.6.3 Prove that $\mathbb{B}_c = +\infty$ for $c < \frac{p}{p-1}$.

PROBLEM 1.6.4 Prove that $\mathbb{B}_c$ is given by (1.6.22) if $c \ge \frac{p}{p-1}$.

1.7 Weak Estimate of the Martingale Transform

We will work not on the whole $\mathbb{R}$, but on a finite interval. The result for the whole axis can be obtained by enlarging the underlying interval and taking into consideration that the estimates will not depend on the interval.

Let φ be a dyadic martingale starting at x_1 and ψ its martingale transform, starting at x_2, i.e.,

$$\varphi = x_1 + \sum_{J\in\mathcal{D}(I)} (\varphi, h_J)h_J, \qquad \psi = x_2 + \sum_{J\in\mathcal{D}(I)} \varepsilon_J(\varphi, h_J)h_J. \quad (1.7.1)$$

Until the very end of this section, we consider the class of $\pm$transforms, i.e., we assume that $\varepsilon_J = \pm 1$. Recall that h_J are normalized in $L^2(\mathbb{R})$ Haar function of the interval J:

$$h_J(t) = \begin{cases} \frac{1}{\sqrt{|J|}}, & t \in J^+ \\ -\frac{1}{\sqrt{|J|}}, & t \in J^- \end{cases},$$

where $J^{\pm}$ are two halves of the interval J. With every pair of functions φ, ψ on I, we associate the Bellman point $\mathfrak{b}_I = \mathfrak{b}_I(\varphi, \psi) = x = (x_1, x_2, x_3)$ with coordinates

$$x_1 = \langle \varphi \rangle_I, \qquad x_2 = \langle \psi \rangle_I, \qquad x_3 = \langle |\varphi| \rangle_I,$$

and introduce the following function:

$$\mathbf{B}(x) = \mathbf{B}(x_1, x_2, x_3) \stackrel{\text{def}}{=} \sup \frac{1}{|I|} \big|\{t \in I : \psi(t) \geq 0\}\big|, \tag{1.7.2}$$

where the sup is taken over all admissible pairs of the test functions φ and ψ, corresponding to the point x. We will consider a pair of functions $\varphi, \psi \in L^1(I)$ to be admissible for point x, if they are as described earlier, i.e., if φ is a dyadic martingale starting at x_1; ψ is its martingale transform, starting at x_2; and x_3 is the average of $|\varphi|$. The function $\mathbf{B}$ is defined in convex domain $\Omega \subset \mathbb{R}^3$:

$$\Omega \stackrel{\text{def}}{=} \{x = (x_1, x_2, x_3) \in \mathbb{R}^3 : |x_1| \leq x_3\}.$$

Note that the function $\mathbf{B}$ should not be indexed by I because, as before, it does not depend on I.

1.7.1 Properties of B

- **Range.** By definition, it is immediately clear that

$$0 \leq \mathbf{B}(x) \leq 1. \tag{1.7.3}$$

- **Symmetry.** Function $\mathbf{B}$ is invariant under reflection with respect to x_1:

$$\mathbf{B}(-x_1, x_2, x_3) = \mathbf{B}(x_1, x_2, x_3), \tag{1.7.4}$$

  because if $x = (x_1, x_2, x_3) = \mathfrak{b}_I(\varphi, \psi)$, then $\widetilde{\varphi} = -\varphi$, $\widetilde{\psi} = \psi$ is an admissible pair, corresponding to $\widetilde{x} = (-x_1, x_2, x_3)$.
- **Homogeneity.**

$$\mathbf{B}(\tau x_1, \tau x_2, \tau x_3) = \mathbf{B}(x_1, x_2, x_3), \qquad \tau > 0, \tag{1.7.5}$$

  because if $x = (x_1, x_2, x_3) = \mathfrak{b}_I(\varphi, \psi)$, then $\widetilde{\varphi} = \tau\varphi$, $\widetilde{\psi} = \tau\psi$ is an admissible pair, corresponding to $\widetilde{x} = (\tau x_1, \tau x_2, \tau x_3)$ and functions ψ and $\widetilde{\psi}$ are positive simultaneously.
- **Boundary Condition.**

$$\mathbf{B}(0, x_2, 0) = \begin{cases} 1, & \text{if } x_2 \geq 0, \\ 0, & \text{if } x_2 < 0, \end{cases} \tag{1.7.6}$$

  because the only admissible pair for point $x = (0, x_2, 0)$ is $\varphi = 0$, $\psi = x_2$.

- **Obstacle Condition.**

$$\mathbf{B}(x_1, x_2, |x_1|) \geq \begin{cases} 1, & \text{if } x_2 \geq 0, \\ 0, & \text{if } x_2 < 0, \end{cases} \tag{1.7.7}$$

because the pair of constant functions $\varphi = x_1, \ \psi = x_2$ is an admissible pair for point $x = (x_1, x_2, |x_1|)$.

By the way, since $\mathbf{B} \leq 1$ by definition, the obstacle condition supplies us with function $\mathbf{B}$ on half of the boundary, namely, $\mathbf{B}(x) = 1$ for $x_2 \geq 0$. We shall see soon that this is not the whole part of the boundary, where $\mathbf{B}(x) = 1$. However, first we shall derive the main inequality.

1.7.2 Main Inequality

LEMMA 1.7.1 *Let $x^{\pm}$ be two points in Ω, such that $|x_1^+ - x_1^-| = |x_2^+ - x_2^-|$ and $x = \frac{1}{2}(x^+ + x^-)$. Then*

$$\mathbf{B}(x) - \frac{\mathbf{B}(x^+) + \mathbf{B}(x^-)}{2} \geq 0. \tag{1.7.8}$$

In particular, this means that if we change variables $x_1 = y_1 - y_2, \ x_2 = y_1 + y_2, \ x_3 = y_3$ and introduce function $\mathbf{M}(y) \stackrel{\text{def}}{=} \mathbf{B}(x)$, defined in the domain $G \stackrel{\text{def}}{=} \{y \in \mathbb{R}^3 \colon |y_1 - y_2| \leq y_3\}$, then we get that for each fixed y_2, $\mathbf{M}(y_1, \cdot, y_3)$ is concave and for each fixed y_1, $\mathbf{M}(\cdot, y_2, y_3)$ is concave.

PROOF Fix $x^{\pm} \in \Omega$ and let $\varphi^{\pm}, \ \psi^{\pm}$ be two pairs of test functions, giving the supremums in $\mathbf{B}(x^+)$ and $\mathbf{B}(x^-)$, respectively, up to a small number $\eta > 0$. Using the fact that the function $\mathbf{B}$ does not depend on the interval, where test functions are defined, we assume that $\varphi^+, \ \psi^+$ live on I^+ and $\varphi^-, \ \psi^-$ live on I^-, i.e.,

$$\varphi^{\pm} = x_1^{\pm} + \sum_{J \in \mathcal{D}(I^{\pm})} a_J h_J, \qquad \psi^{\pm} = x_2^{\pm} + \sum_{J \in \mathcal{D}(I^{\pm})} \varepsilon_J a_J h_J.$$

Consider

$$\varphi(t) := \begin{cases} \varphi^+(t), & \text{if } t \in I^+ \\ \varphi^-(t), & \text{if } t \in I^- \end{cases} = \frac{x_1^+ + x_1^-}{2} + \frac{x_1^+ - x_1^-}{2} H_I + \sum_{J \in \mathcal{D}(I) \setminus \{I\}} a_J h_J$$

and

$$\psi(t) := \begin{cases} \psi^+(t), \text{ if } t \in I^+ \\ \psi^-(t), \text{ if } t \in I^- \end{cases} = \frac{x_2^+ + x_2^-}{2} + \frac{x_2^+ - x_2^-}{2} H_I + \sum_{J \in \mathcal{D}(I) \setminus \{I\}} \varepsilon_I a_I h_I.$$

Since $|x_1^+ - x_1^-| = |x_2^+ - x_2^-|$, function ψ is a martingale transform of φ, and the pair $\varphi,\ \psi$ is an admissible pair of test functions, corresponding to point x. Therefore,

$$\begin{aligned}\mathbf{B}(x) &\geq \frac{1}{|I|}\big|\{t \in I\colon \psi(t) \geq 0\}\big| \\ &= \frac{1}{2|I^+|}\big|\{t \in I^+\colon \psi(t) \geq 0\}\big| + \frac{1}{2|I^-|}\big|\{t \in I^-\colon \psi(t) \geq 0\}\big| \\ &\geq \tfrac{1}{2}\mathbf{B}(x^+) + \tfrac{1}{2}\mathbf{B}(x^-) - \eta.\end{aligned}$$

Since this inequality holds for an arbitrarily small η, we can pass to the limit $\eta \to 0$, which confirms the required assertion. □

1.7.3 Supersolution

LEMMA 1.7.2 *Let B be a continuous function on Ω, satisfying the main inequality* (1.7.8) *and the obstacle condition* (1.7.7). *Then $\mathbf{B}(x) \leq B(x)$.*

PROOF Let us fix a point $x \in \Omega$ and a pair of admissible functions $\varphi,\ \psi$ on I, corresponding to x, i.e., $\mathfrak{b}_I(\varphi,\psi) = x$. Since the pair $\varphi,\ \psi$ is fixed, we will write $\mathfrak{b}_J$ instead of $\mathfrak{b}_J(\varphi,\psi)$. Using repeatedly the main inequality for function B, we can write down the following chain of inequalities:

$$\begin{aligned}|I|B(x) &\geq |I^+|B(\mathfrak{b}_{I^+}) + |I^-|B(\mathfrak{b}_{I^-}) \\ &\geq \sum_{J \in \mathcal{D}_n(I)} |J|B(\mathfrak{b}_J) = \int_I B(x^{(n)}(t))dt,\end{aligned}$$

where $x^{(n)}(t) = \mathfrak{b}_J$, if $t \in J,\ \ J \in \mathcal{D}_n(I)$.

Note that $x^{(n)}(t) \to (\varphi(t), \psi(t), |\varphi(t)|)$ almost everywhere (at any Lebesgue point t). Therefore, since B is continuous and bounded, we can pass to the limit in the integral. So, we come to the inequality

$$|I|B(x) \geq \int_I B(\varphi(t), \psi(t), |\varphi(t)|)dt \geq \int_{\{t\in I\colon \psi(t)\geq 0\}} dt = \big|\{t \in I\colon \psi(t) \geq 0\}\big|, \tag{1.7.9}$$

where we have used the property $B(x_1, x_2, |x_1|) = 1$ for $x_2 \geq 0$. Now, after dividing by $|I|$, we take supremum in (1.7.9) over all admissible pairs $\varphi,\ \psi$, and get the required estimate $B(x) \geq \mathbf{B}(x)$. □

Now we explain how we will apply this lemma. Let us denote

$$T\varphi \stackrel{\text{def}}{=} \sum_{J \in \mathcal{D}} \varepsilon_J(\varphi, h_J) h_J(x).$$

It is a dyadic singular operator (actually, it is a family of operators enumerated by sequences ε of ± 1). To prove that it is of weak type $(1,1)$ is the same as to prove that

$$\mathbf{B}(x) \le \frac{C\,x_3}{|x_2|}. \tag{1.7.10}$$

Indeed, if $\varphi,\ \psi$ is an admissible pair, corresponding to point x, then $T\varphi = \psi - x_2$. Therefore, for a given φ with $\langle\varphi\rangle = x_1$ and $\langle|\varphi|\rangle = x_3$, the best estimate of the value of $|\{t\colon T\varphi \ge \lambda\}|$ gives us function $\mathbf{B}(x)$ with $x_2 = -\lambda$. Thus, were we to find any function B with the required estimate and satisfying conditions of Lemma 1.7.2, we would immediately get the needed weak type $(1,1)$, and in fact, more precise information on the level set of $T\varphi$.

1.7.4 Bellman Function on the Boundary

First of all, we note that the boundary $\partial\Omega$ consists of two independent parts:

$$\partial\Omega_+ \stackrel{\text{def}}{=} \{x = (x_1, x_2, x_1)\colon x_1 \ge 0, -\infty < x_2 < +\infty\} \qquad \text{and}$$
$$\partial\Omega_- \stackrel{\text{def}}{=} \{x = (x_1, x_2, -x_1)\colon x_1 \le 0, -\infty < x_2 < +\infty\}.$$

They are independent in the following sense. If we have a pair of test functions $\varphi,\ \psi$, whose Bellman point $x = \mathfrak{b}(\varphi, \psi)$ is on the boundary (whence the sign of $\varphi(t)$ is constant on the whole interval), then after splitting the interval we get a pair of Bellman points $x^{\pm}$ from the same part of the boundary. So, the main inequality (1.7.8) has to be fulfilled separately on $\partial\Omega_+$ and $\partial\Omega_-$. Due to the symmetry condition, it is sufficient to find the function, say, on $\partial\Omega_+$. And further, we assume that $x_1 \ge 0$.

So we look for a minimal function on the half-plane $\{x_1 \ge 0\}$, satisfying the main inequality and the boundary condition (1.7.6). We pass to the variable y ($x_1 = y_1 - y_2$ and $x_2 = y_1 + y_2$) and look for a function M in the half-plane $y_2 < y_1$, which satisfies the main inequality (i.e., it is concave in each variable: in y_1, when y_2 is fixed, and in y_2, when y_1 is fixed) and with the given values on the boundary $y_2 = y_1$: $\mathbf{M} = 1$ if $y_1 = y_2 \ge 0$ and $\mathbf{M} = 0$ if $y_1 = y_2 < 0$.

First, we use concavity of $\mathbf{M}$ with respect to y_2 for some fixed $y_1 \ge 0$. A concave function bounded from below cannot decrease, therefore it has to be identically 1 on any such ray due to fixed boundary condition. It remains for us to find $\mathbf{M}$ in domain $y_2 < y_1 < 0$. Here we use concavity along y_1. We know that our function is 0 at $y_1 = y_2$ and it is 1 at $y_1 = 0$. Therefore, between these two points, it is at least the linear function $M = 1 - \frac{y_1}{y_2}$, i.e., $\mathbf{M} \geq M$, where

$$M = \begin{cases} 1, & \text{if } y_1 \geq 0, \\ 1 - \frac{y_1}{y_2}, & \text{if } y_1 < 0. \end{cases}$$

To prove the opposite inequality, we note that M is concave in each variable and it satisfies the obstacle condition. Therefore, Lemma 1.7.2 guarantees us the required inequality.

Returning to variables x, we can write $\mathbf{B} = \frac{2x_1}{x_1 - x_2}$ in the half-plane $x_1 \geq 0$. As a result, we have proved the following:

PROPOSITION 1.7.3

$$\mathbf{B}(x_1, x_2, |x_1|) = B(x_1, x_2, |x_1|) \stackrel{\text{def}}{=} \begin{cases} 1, & \text{if } x_2 \geq -|x_1|, \\ \frac{2|x_1|}{|x_1| - x_2}, & \text{if } x_2 \leq -|x_1|. \end{cases} \tag{1.7.11}$$

1.7.5 Full Bellman Function for the Weak Type Estimate

Now we present the full Bellman function:

THEOREM 1.7.4 *For function* $\mathbf{B}$, *defined by* (1.7.2), *we have the following analytic expression*:

$$\mathbf{B}(x) = B(x) \stackrel{\text{def}}{=} \begin{cases} 1, & \text{if } x_3 + x_2 \geq 0, \\ 1 - \dfrac{(x_3 + x_2)^2}{x_2^2 - x_1^2}, & \text{if } x_3 + x_2 < 0. \end{cases} \tag{1.7.12}$$

PROOF As earlier, we change variables

$$x_1 = y_1 - y_2, \quad x_2 = y_1 + y_2, \quad x_3 = y_3,$$
$$\text{i.e.,} \quad y_1 = \frac{x_1 + x_2}{2}, \quad y_2 = \frac{x_2 - x_1}{2}, \tag{1.7.13}$$

and will be looking for a function M

$$M(y) \stackrel{\text{def}}{=} B(x),$$

which is defined on $\Omega \stackrel{\text{def}}{=} \{y = (y_1, y_2, y_3) : y_3 \geq |y_1 - y_2|\}$, is biconcave in variables (y_1, y_3) and (y_2, y_3), and satisfies boundary condition (1.7.11), or, in terms of M,

$$M(y_1, y_2, |y_1 - y_2|) = \begin{cases} 1, & \text{if either } y_1 \geq 0 \text{ or } y_2 \geq 0, \\ 1 - \dfrac{\max\{y_1, y_2\}}{\min\{y_1, y_2\}}, & \text{if } y_1 < 0 \text{ and } y_2 < 0. \end{cases} \tag{1.7.14}$$

Since function $\mathbf{B}$ is even with respect to x_1, as before, it is sufficient to consider the half-space $\{x_1 > 0\}$, or the half-space $\{y_2 < y_1\}$ in y-variable. But, in fact, we can restrict ourselves to the cone $\{x_2 < -x_1 < 0, x_3+x_1 > 0\}$ (or $\{y_2 < y_1 < 0, y_3 > y_1 - y_2\}$) because in all other points, our function is identically 1 for the same reason as before: it is concave, nonnegative, and bounded by 1 on every ray $\{y_1 = \text{const}, y_2 = \text{const}, y_3 > y_1 - y_2\}$.

The boundary function is not smooth because the boundary itself is not smooth at the line $\{x_1 = x_3 = 0\}$ and, moreover, the boundary condition on this line has a jump. But inside the domain, we can look for a smooth candidate B. Then it has to satisfy the boundary condition $\frac{\partial B}{\partial x_1}|_{x_1=0} = 0$, or, in terms of M,

$$\frac{\partial M}{\partial y_1}\bigg|_{y_1=y_2} = \frac{\partial M}{\partial y_2}\bigg|_{y_1=y_2}. \tag{1.7.15}$$

Our function has to be concave in each plane $\{y_1 = \text{const}\}$ and in each plane $\{y_2 = \text{const}\}$, and we have to decide in which plane concavity is degenerate, i.e., in which plane M satisfies the Monge–Ampère equation. Looking on the boundary, we see that the extremals are segments of lines $\{y_2 = \text{const}\}$, and therefore it is natural to look for a solution of the Monge–Ampère equation

$$M_{y_1y_1}M_{y_3y_3} - M^2_{y_1y_3} = 0$$

in this plane. (The intersection of our domain Ω with this plane is shown on Figure 1.15.)

Note that the half-lines $\{y_3+y_1 = \text{const}\colon y_3 > y_1-y_2\}$ are in that domain if const $\geq y_2$. Moreover, if const $\geq -y_2$ (recall that $y_2 < 0$), then the boundary value on this ray is 1, and hence, it is identically 1 for $y_3 + y_2 + y_1 \geq 0$. Therefore, we need to solve the Monge–Ampère equation only in the triangle with the vertices $(0, y_2, -y_2)$, $(y_2, y_2, 0)$, and $(y_2, y_2, -2y_2)$:

$$\{y = (y_1, y_2, y_3)\colon y_2 = \text{const}, y_1 > y_2, y_1 - y_2 < y_3 < -y_1 - y_2\}$$

with the boundary conditions

$$M(y_1, y_2, y_1 - y_2) = 1 - \frac{y_1}{y_2}, \qquad M(y_1, y_2, -y_1 - y_2) = 1,$$
$$M_{y_1}(y_2, y_2, y_3) = M_{y_2}(y_2, y_2, y_3).$$

Our function is linear on two sides of the triangle, so the minimal concave function linear on two sides is the linear function in the whole triangle. However, this function does not satisfy the boundary condition on the side $y_1 = y_2$. Therefore, the extremal lines cannot intersect inside the triangle, and

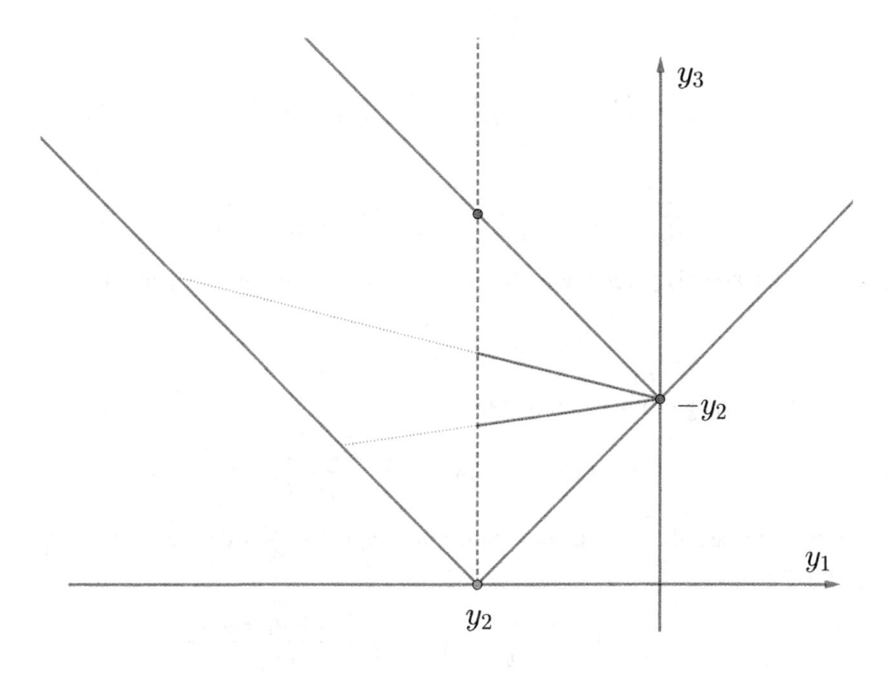

Figure 1.15 Section of the domain by a plane $y_2 = \text{const}$.

the only way to foliate this triangle without singularities inside the domain is a fan of straight-line segments, starting at point $(0, y_2, -y_2)$, which we parametrize by the slope k of each extremal line:

$$y_3 = ky_1 - y_2. \tag{1.7.16}$$

The slope runs over the interval $[-1, 1]$. For $k = -1$, we get the upper side of the triangle $y_1 + y_2 + y_3 = 0$, where $M = 1$; for $k = 1$, we get the lower side $y_3 = y_1 - y_2$, where $M = 1 - \frac{y_1}{y_2}$. On all other extremal lines M is linear in y_1 as well:

$$M = 1 + m(k, y_2)y_1.$$

Now our task is to find its slope $m = m(k, y_2)$ with the prescribed values at the points $k = \pm 1$: $m(-1, y_2) = 0$ and $m(1, y_2) = -\frac{1}{y_2}$. We find this function from the boundary condition (1.7.15) on the third side of the triangle.

First, we deduce from (1.7.16) that $k = k(y_1, y_2, y_3) = \frac{y_3 + y_2}{y_1}$, and hence,

$$\frac{\partial k}{\partial y_1} = -\frac{k}{y_1} \quad \text{and} \quad \frac{\partial k}{\partial y_2} = \frac{1}{y_1}.$$

Therefore,

$$\frac{\partial M}{\partial y_1} = m + y_1 \frac{\partial m}{\partial k}\frac{\partial k}{\partial y_1} = m - k\frac{\partial m}{\partial k},$$

$$\frac{\partial M}{\partial y_2} = y_1\Big(\frac{\partial m}{\partial y_2} + \frac{\partial m}{\partial k}\frac{\partial k}{\partial y_2}\Big) = y_1\frac{\partial m}{\partial y_2} + \frac{\partial m}{\partial k}.$$

Thus, the boundary condition (1.7.15) turns into the following equation:

$$m - (k+1)\frac{\partial m}{\partial k} = y_2\frac{\partial m}{\partial y_2},$$

which has the general solution of the form

$$m(k, y_2) = (k+1)\Phi\Big(\frac{k+1}{y_2}\Big),$$

where Φ is an arbitrary function. Since $m(1, y_2) = -\frac{1}{y_2}$, we have $\Phi(t) = -\frac{t}{4}$. And finally,

$$M(y) = 1 - \frac{(k+1)^2}{4y_2}y_1 = 1 - \frac{(y_1+y_2+y_3)^2}{4y_1y_2},$$

or

$$B(x) = 1 - \frac{(x_2+x_3)^2}{x_2^2 - x_1^2}.$$

Now it is an easy task to check that the found function M is concave in each of the planes $\{y_1 = \text{const}\}$ and $\{y_2 = \text{const}\}$. Since M is symmetric with respect to interchanging variables y_1 and y_2, it is sufficient to check concavity in variables, say, y_1, y_3. Hessian

$$\begin{pmatrix} M_{y_1y_1} & M_{y_1y_3} \\ M_{y_3y_1} & M_{y_3y_3} \end{pmatrix} = \begin{pmatrix} -\dfrac{(y_2+y_3)^2}{2y_1^3y_2} & \dfrac{y_2+y_3}{2y_1^2y_2} \\[2ex] \dfrac{y_2+y_3}{2y_1^2y_2} & -\dfrac{1}{2y_1y_2} \end{pmatrix}$$

is a negative degenerate matrix, and function M is C^1-smooth near the plane $y_1 + y_2 + y_3 = 0$, and, therefore, we have the desired concavity in the whole domain.

Since B is not a continuous function, we cannot apply Lemma 1.7.2 directly to get the desired inequality. Nevertheless, the conclusion of Lemma 1.7.2 is true in our case, i.e., $\mathbf{B}(x) \le B(x)$. This can be justified in many different ways. For example, we can use a convolution of our solution with a carefully chosen smooth approximate identity to get a family of continuous supersolutions, which converges to B. However, we leave the details for the reader (see Exercise 1.7.1).

Now we have to prove the opposite inequality. For this Bellman function, it is very easy due to its following a special property. Note that it is linear on the extremal lines not only in the domain, where we were looking for them, but also on the continuation of each extremal line (see Figure 1.15). Indeed, all extremal lines in the triangle under investigation are parametrized by their slope k, $-1 < k < 1$, and have the form

$$y_3 = ky_1 - y_2, \qquad y_2 \le y_1 \le 0,$$

and the found function on this line is

$$M(y_1, y_2, ky_1 - y_2) = 1 - \frac{(k+1)^2}{4y_2} y_1.$$

Thus, we see that this function is linear not only on the interval $(y_2, 0)$ but also for $y_1 < y_2$. So, we can continue this extremal line up to its second point of intersection with the boundary $y_3 = |y_2 - y_1|$ where it coincides with $\mathbf{M}$. As a result, we have two points where the concave function $\mathbf{M}$ coincides with the linear function M; therefore, between these two points we have $\mathbf{M}(y) \ge M(y)$. Since the described continued extremal lines foliate the whole domain $y_1 + y_2 + y_3 < 0$, we have the desired inequality for arbitrary point y from Ω. □

1.7.6 Two-Sided Estimate

As already mentioned at the end of section 1.7.3, the found Bellman function supplies us with the estimate $|\{t \colon T\varphi \ge \lambda\}| \le \mathbf{B}(x_1, -\lambda, x_3)$. If we are interested in estimating the martingale transform by modulus, or in a more general estimate of the measure of the set, where $\lambda_- \le T\varphi \le \lambda_+$, we need to look for a bit more difficult Bellman function

$$\mathbf{B}(x) = \mathbf{B}(x_1, x_2, x_3; \lambda_-, \lambda_+) \stackrel{\text{def}}{=} \sup \frac{1}{|I|} \big|\{t \in I \colon \lambda_- \le \psi(t) \le \lambda_+\}\big|, \tag{1.7.17}$$

where, as before, the sup is taken over all admissible pairs of test functions φ, ψ, corresponding to point x. For the new Bellman function, we have the same domain and the same range. Symmetry condition (1.7.4) holds true, but in the homogeneity condition (1.7.5), we have to change parameters $\lambda_\pm$:

$$\mathbf{B}(\tau x_1, \tau x_2, \tau x_3; \tau\lambda_-, \tau\lambda_+) = \mathbf{B}(x; \lambda_-, \lambda_+), \qquad \tau > 0. \tag{1.7.18}$$

In addition, we now have the following translation invariance:

$$\mathbf{B}(x_1, x_2+\tau, x_3; \lambda_- +\tau, \lambda_+ +\tau) = \mathbf{B}(x; \lambda_-, \lambda_+), \qquad -\infty<\tau<\infty, \tag{1.7.19}$$

which immediately follows from replacing function ψ by $\psi+\tau$. Shifting it by $\tau = -\frac{1}{2}(\lambda_- + \lambda_+)$, we reduce the problem to a symmetrical one. Therefore, without loss of generality, we can consider the following function:

$$\mathbf{B}(x) = \mathbf{B}(x_1, x_2, x_3; \lambda) \stackrel{\text{def}}{=} \sup \frac{1}{|I|}\big|\{t\in I\colon |\psi(t)|\le\lambda\}\big|. \tag{1.7.20}$$

This function evidently possesses the symmetry condition with respect to both x_1 and x_2:

$$\mathbf{B}(-x_1, x_2, x_3; \lambda) = \mathbf{B}(x_1, -x_2, x_3; \lambda) = \mathbf{B}(x; \lambda). \tag{1.7.21}$$

Therefore, it is sufficient to find this function for $x_1 \ge 0$ and $x_2 \ge 0$. Using the homogeneity property (1.7.18), it would be possible to reduce the situation to the case $\lambda = 1$, but this does not simplify the formulas too much, and for that reason we leave an arbitrary λ.

Boundary condition for this function takes the following form:

$$\mathbf{B}(0, x_2, 0) = \begin{cases} 1, & \text{if } |x_2| \le \lambda, \\ 0, & \text{if } |x_2| < \lambda, \end{cases} \tag{1.7.22}$$

and the obstacle condition is

$$\mathbf{B}(x_1, x_2, |x_1|) \ge \begin{cases} 1, & \text{if } |x_2| \le \lambda, \\ 0, & \text{if } |x_2| < \lambda. \end{cases} \tag{1.7.23}$$

The main inequality (Lemma 1.7.1) holds without any changes. The same can be said about Lemma 1.7.2 (where, of course, we have to consider the proper obstacle condition, namely, (1.7.23) rather than (1.7.7)).

As before, first we consider the Bellman function on the boundary, and as it was already noted, due to symmetry, it is sufficient to find function $\mathbf{B}$ in quadrant $x_1 \ge 0,\ \ x_2 \ge 0$, or function $\mathbf{M}$ in $y_1 \ge |y_2|$.

On the boundary, we have

$$\mathbf{M}(y_1, y_1, 0) = \begin{cases} 1, & \text{if } |y_1| \le \frac{1}{2}\lambda, \\ 0, & \text{if } |y_1| > \frac{1}{2}\lambda. \end{cases} \tag{1.7.24}$$

First, we construct the domain, where $\mathbf{M} = 1$. As before, if we can draw a ray parallel to an axis and starting at a point, where $\mathbf{M} = 1$, then $\mathbf{M} = 1$ on the whole ray. So, in the quadrant $y_1 \ge |y_2|$ we get a subdomain

$$\{y = (y_1, y_2)\colon y_1 \ge |y_2|,\ y_2 \le \tfrac{1}{2}\lambda\},$$

where $\mathbf{M} = 1$. The remaining sector

$$\{y = (y_1, y_2) \colon \tfrac{1}{2}\lambda \le y_2 \le y_1\}$$

we fill, as before, with straight-line segments of the linear function. As a result, we come to the following function on the boundary (compare with (1.7.14)):

$$M(y_1, y_2, y_1 - y_2) = \begin{cases} 1 - \dfrac{y_2 - \frac{1}{2}\lambda}{y_1 - \frac{1}{2}\lambda}, & \text{if } \frac{1}{2}\lambda \le y_2 \le y_1, \\ 1, & \text{if } y_1 \ge |y_2|,\ y_2 \le \frac{1}{2}\lambda. \end{cases} \tag{1.7.25}$$

The three-dimensional function is recovered from this boundary condition in the same way. Thus, in the subdomain $\{y_1 \ge |y_2|\}$, we obtain

$$M(y_1, y_2, y_3) = \begin{cases} 1 - \dfrac{(y_1 + y_2 - y_3 - \lambda)^2}{(2y_1 - \lambda)(2y_2 - \lambda)}, & \text{if } y_3 < y_1 + y_2 - \lambda, \\ 1, & \text{otherwise}, \end{cases} \tag{1.7.26}$$

and in terms of variables x in the whole domain Ω, we have

$$B(x_1, x_2, x_3; \lambda) = \begin{cases} 1 - \dfrac{(|x_2| - x_3 - \lambda)^2}{(|x_2| - \lambda)^2 - x_1^2}, & \text{if } x_3 < |x_2| - \lambda, \\ 1, & \text{otherwise}. \end{cases} \tag{1.7.27}$$

The estimate of tails is more interesting. If we measure the set, where $|\psi(t)| \ge \lambda$ (the more general case when either $\psi(t) \le \lambda_-$ or $\psi(t) \ge \lambda_+$ is reduced to this one by a shift with respect to variable x_2), then we need to look for the following Bellman function:

$$\mathbf{B}(x; \lambda) \stackrel{\text{def}}{=} \sup \frac{1}{|I|} \big|\{t \in I \colon |\psi(t)| \ge \lambda\}\big|. \tag{1.7.28}$$

The boundary condition for this function is

$$\mathbf{B}(0, x_2, 0; \lambda) = \begin{cases} 1, & \text{if } |x_2| \ge \lambda, \\ 0, & \text{if } |x_2| < \lambda, \end{cases} \tag{1.7.29}$$

or, in terms of $\mathbf{M}$

$$\mathbf{M}(y_1, y_1, 0; \lambda) = \begin{cases} 1, & \text{if } |y_1| \ge \frac{1}{2}\lambda, \\ 0, & \text{if } |y_1| < \frac{1}{2}\lambda. \end{cases} \tag{1.7.30}$$

Consider Bellman function on the boundary, and due to symmetry, it is sufficient to find function $\mathbf{B}$ in quadrant $x_1 \ge 0,\ \ x_2 \ge 0$, or function $\mathbf{M}$ in $y_1 \ge |y_2|$.

Now we have $\mathbf{M}(y_1, y_2, |y_1 - y_2|) = 1$ outside the square $|y_i| < \frac{1}{2}\lambda$. Therefore, we have to consider the triangle $|y_2| \le y_1 < \frac{1}{2}\lambda$. We split it into two smaller triangles:

$$\{0 \le y_2 \le y_1 < \tfrac{1}{2}\lambda\} \qquad \text{and} \qquad \{0 \le -y_2 \le y_1 < \tfrac{1}{2}\lambda\}.$$

The minimal possible candidate in the first triangles is linear with respect to y_1:

$$M = \frac{y_1 - y_2}{\frac{1}{2}\lambda - y_2}.$$

Note that this function is concave with respect to y_2, as we need.

In the second triangle, we can try to check a function either linear with respect to y_1, or linear with respect to y_2. But in any case, we come to the same result:

$$M = \frac{4y_1y_2 + 2\lambda(y_1 - y_2)}{\lambda^2}.$$

This function is linear in both directions.

So, for the Bellman candidate, we have the following boundary value:

$$M(y_1, y_2, y_1 - y_2; \lambda) = \begin{cases} 1, & \text{if } y_1 \ge \frac{1}{2}\lambda, \\[2ex] \dfrac{4y_1y_2 + 2\lambda(y_1 - y_2)}{\lambda^2}, & \text{if } 0 \le -y_2 \le y_1 < \frac{1}{2}\lambda, \\[2ex] \dfrac{y_1 - y_2}{\frac{1}{2}\lambda - y_2}, & \text{if } 0 \le y_2 \le y_1 < \frac{1}{2}\lambda. \end{cases} \tag{1.7.31}$$

In terms of x, we rewrite this formula, as follows:

$$B(x_1, x_2, x_1; \lambda) = \begin{cases} 1, & \text{if } x_1 + x_2 \ge \lambda, \\[2ex] 1 - \dfrac{(x_1 - \lambda)^2 - x_2^2}{\lambda^2}, & \text{if } x_1 + x_2 < \lambda,\ x_2 \le x_1, \\[2ex] \dfrac{2x_1}{\lambda + x_1 - x_2}, & \text{if } x_1 + x_2 < \lambda,\ x_2 \ge x_1. \end{cases}$$

Using these boundary values, we begin to build a Bellman candidate in the interior of the domain. As before, we start by determining the subdomain, where our function is equal to 1. In the preceding discussion, the lower bound of the domain, where $M = 1$, was determined by the lowest infinite rays, starting from the points of the boundary, where $M = 1$. Now this surface will be generated by the straight-line segments connecting two such points on the boundary. Indeed, consider a section of our domain by plane $y_2 = \text{const}$. We know, that $M(y_1, y_2, |y_1 - y_2|) = 1$ if $|y_1| \geq \frac{1}{2}\lambda$. The two lowest such points are $(\frac{1}{2}\lambda, y_2, \frac{1}{2}\lambda - y_2)$ and $(-\frac{1}{2}\lambda, y_2, \frac{1}{2}\lambda + y_2)$. The straight line passing through these points is given by the equation $y_3 = \frac{\lambda^2 - 4y_1y_2}{2\lambda}$, and therefore,

$$M(y;\lambda) = 1 \qquad \text{if } y_3 \geq \frac{\lambda^2 - 4y_1y_2}{2\lambda}.$$

In terms of x, we get

$$B(1;\lambda) = 1 \qquad \text{if } x_3 \geq \frac{\lambda^2 + x_1^2 - x_2^2}{2\lambda}.$$

In the triangle $\{(y_1, y_2)\colon |y_2| \leq y_1 \leq \frac{1}{2}\lambda\}$, the boundary function is linear along the segments $y_2 = \text{const}$. Therefore, it is natural to assume that in the three-dimensional domain

$$\{(y_1, y_2, y_3)\colon 0 \geq y_1 \geq \tfrac{1}{2}\lambda,\ -y_1 \leq y_2 \leq y_1,\ y_3 \geq y_1 - y_2\},$$

the extremal lines of our Bellman function are in the plane $y_2 = \text{const}$. In this plane, we have to foliate only the triangle

$$\{(y_1, y_3)\colon |y_2| \leq y_1 \leq \tfrac{1}{2}\lambda,\ y_1 - y_2 \leq y_3 \leq \frac{\lambda^2 - 4y_1y_2}{2\lambda}\}$$

because we know that for $y_3 \geq \frac{\lambda^2 - 4y_1y_2}{2\lambda}$ we have $M = 1$.

As in the case of the one-sided problem, the only way to foliate this triangle is by a fan of extremal lines, starting from the vertex $(y_1, y_3) = (\frac{1}{2}\lambda, \frac{1}{2}\lambda - y_2)$. We parametrize the segments of this fan by their slopes k:

$$y_3 = (\tfrac{1}{2}\lambda - y_2) + k(y_1 - \tfrac{1}{2}\lambda).$$

We look for a function, linear along each such line, i.e.,

$$M(y_1, y_2, y_3) = 1 + m(k, y_2)(y_1 - \tfrac{1}{2}\lambda), \qquad \text{where} \qquad k = \frac{y_3 + y_2 - \frac{1}{2}\lambda}{y_1 - \frac{1}{2}\lambda}.$$

We know the boundary values of our function on two sides of the triangle. On the side with $k = -\frac{2y_2}{\lambda}$, we have $m(k, y_2) = 0$, since

$$M = 1 \quad \text{for} \quad y_3 = \frac{\lambda^2 - 4y_1y_2}{2\lambda}.$$

On the side $k = 1$, we have different boundary conditions for $y_2 > 0$ and $y_2 < 0$. From the expression (1.7.31), we see that

$$M = \frac{4y_1y_2 + 2\lambda(y_1 - y_2)}{\lambda^2}, \quad \text{i.e.,} \quad m(1, y_2) = \frac{4y_2 + 2\lambda}{\lambda^2} \quad \text{if} \quad y_2 \leq 0,$$

and

$$M = \frac{y_1 - y_2}{\frac{1}{2}\lambda - y_2}, \qquad \text{i.e.,} \qquad m(1, y_2) = \frac{1}{\frac{1}{2}\lambda - y_2} \quad \text{if} \quad y_2 \geq 0.$$

On the third side $y_1 = |y_2|$, we have the Neumann boundary condition due to symmetry of our function with respect to planes $x_1 = 0$ and $x_2 = 0$. Namely,

$$\left.\frac{\partial M}{\partial y_1}\right|_{y_1=y_2} = \left.\frac{\partial M}{\partial y_2}\right|_{y_1=y_2} \quad \text{and} \quad \left.\frac{\partial M}{\partial y_1}\right|_{y_1=-y_2} = -\left.\frac{\partial M}{\partial y_2}\right|_{y_1=-y_2}.$$

This gives us a differential equation for the unknown function $m(k, y_2)$. Since $k = \frac{y_3+y_2-\frac{1}{2}\lambda}{y_1-\frac{1}{2}\lambda}$, we have

$$\frac{\partial k}{\partial y_1} = -\frac{k}{y_1 - \frac{1}{2}\lambda} \quad \text{and} \quad \frac{\partial k}{\partial y_2} = \frac{1}{y_1 - \frac{1}{2}\lambda},$$

whence

$$\frac{\partial M}{\partial y_1} = m - k\frac{\partial m}{\partial k} \quad \text{and} \quad \frac{\partial M}{\partial y_2} = (y_1 - \tfrac{1}{2}\lambda)\frac{\partial m}{\partial y_2} + \frac{\partial m}{\partial k}.$$

Let us first consider the case, where $y_2 < 0$. Then on the boundary $y_1 = -y_2$ we have the following equation:

$$m = (k-1)\frac{\partial m}{\partial k} + (y_2 + \tfrac{1}{2}\lambda)\frac{\partial m}{\partial y_2}.$$

Its general solution has the following form:

$$m(k, y_2) = (y_2 + \tfrac{1}{2}\lambda)\Phi\Big(\frac{k-1}{y_2 + \frac{1}{2}\lambda}\Big).$$

The boundary condition at $k = 1$ yields $\Phi(0) = \frac{4}{\lambda^2}$. To satisfy the boundary condition at $k = -\frac{2y_2}{\lambda}$, we need $\Phi(-\frac{2}{\lambda}) = 0$. We still have a lot of freedom in choosing function Φ. We take a linear function Φ because it gives us a function M, linear both in y_1 and y_3, that is the minimal possible convex function. A linear function is uniquely determined by our boundary conditions, that is,

$$\Phi(t) = \frac{4}{\lambda^2} + \frac{2}{\lambda}t,$$

whence

$$m(k, y_2) = \frac{4y_2}{\lambda^2} + \frac{2k}{\lambda}. \tag{1.7.32}$$

Plugging $k = \frac{y_3+y_2-\frac{1}{2}\lambda}{y_1-\frac{1}{2}\lambda}$, we get the final formula for Bellman candidate M:

$$M(y_1, y_2, y_3) = \frac{4y_1y_2 + 2\lambda y_3}{\lambda^2}. \tag{1.7.33}$$

Let us turn to the case $y_2 > 0$. On the boundary $y_1 = y_2$, we now have the following equation:

$$m = (k+1)\frac{\partial m}{\partial k} + (y_2 - \tfrac{1}{2}\lambda)\frac{\partial m}{\partial y_2}$$

Its general solution has the form

$$m(k, y_2) = (y_2 - \tfrac{1}{2}\lambda)\Phi\Big(\frac{k+1}{y_2 - \frac{1}{2}\lambda}\Big).$$

The boundary condition at $k = 1$ turns into

$$(y_2 - \tfrac{1}{2}\lambda)\Phi\Big(\frac{2}{y_2 - \frac{1}{2}\lambda}\Big) = -\frac{1}{y_2 - \frac{1}{2}\lambda},$$

whence

$$\Phi(t) = -\frac{t^2}{4}$$

and

$$m(k, y_2) = -\frac{(k+1)^2}{4(y_2 - \frac{1}{2}\lambda)}. \tag{1.7.34}$$

This gives us the following expression for the Bellman candidate:

$$M(y_1, y_2, y_3) = 1 - \frac{(y_1 + y_2 + y_3 - \lambda)^2}{(\lambda - 2y_1)(\lambda - 2y_2)}.$$

Let us note that this function does not satisfy the second boundary condition, therefore, it cannot be a Bellman candidate in the whole domain. We try to glue this solution with the already found linear one, which satisfied the required boundary condition not only for $y_2 < 0$, but for all y_2.

Two solutions are equal on the extremal line $y_3 = (\frac{1}{2}\lambda - y_2) + k(y_1 - \frac{1}{2}\lambda)$, if they have equal slopes:

$$-\frac{(k+1)^2}{4(y_2 - \frac{1}{2}\lambda)} = \frac{4y_2}{\lambda^2} + \frac{2k}{\lambda}.$$

This equation has only one solution $k = 1 - \frac{4y_2}{\lambda}$. Note that this is a root of multiplicity two and this guarantees C_1-smoothness of the solution we obtain after gluing. So, for $k < 1 - \frac{4y_2}{\lambda}$, we take solution (1.7.32) and for $k > 1 - \frac{4y_2}{\lambda}$, we take solution (1.7.34). Note that for $y_2 < 0$, we have no k with $k > 1 - \frac{4y_2}{\lambda}$ and we have only solution (1.7.32) there, so it is not necessary to separate the cases $y_2 < 0$ and $y_2 > 0$.

The condition $k = 1 - \frac{4y_2}{\lambda}$ is the equation of the surface

$$y_3 = y_1 + y_2 - \frac{4y_1y_2}{\lambda}$$

and $k = -\frac{2y_2}{\lambda}$ is the equation of the surface

$$y_3 = \frac{\lambda^2 - 4y_1y_2}{2\lambda}.$$

Thus, we get the following Bellman candidate:

$$M(y_1, y_2, y_3; \lambda)$$
$$= \begin{cases} 1, & \text{if } y_3 \geq \frac{\lambda^2 - 4y_1y_2}{2\lambda}, \\ \dfrac{4y_1y_2 + 2\lambda y_3}{\lambda^2}, & \text{if } y_1 + y_2 - \dfrac{4y_1y_2}{\lambda} \leq y_3 \leq \dfrac{\lambda^2 - 4y_1y_2}{2\lambda}, \\ 1 - \dfrac{(y_1 + y_2 + y_3 - \lambda)^2}{(\lambda - 2y_1)(\lambda - 2y_2)}, & \text{if } y_3 \leq y_1 + y_2 - \dfrac{4y_1y_2}{\lambda}. \end{cases}$$

In terms of x, we rewrite this formula as follows:

$$B(x; \lambda) = \begin{cases} 1, & \text{if } x_3 \geq \dfrac{\lambda^2 + x_1^2 - x_2^2}{2\lambda}, \\ \dfrac{2\lambda x_3 - x_1^2 + x_2^2}{\lambda^2}, & \text{if } x_2 + \dfrac{x_1^2 - x_2^2}{\lambda} \leq x_3 \leq \dfrac{\lambda^2 + x_1^2 - x_2^2}{2\lambda}, \\ 1 - \dfrac{(x_3 + x_2 - \lambda)^2}{(\lambda - x_2)^2 - x_1^2}, & \text{if } x_3 \leq x_2 + \dfrac{x_1^2 - x_2^2}{\lambda}. \end{cases}$$

This candidate was obtained in subdomain

$$\{x \in \Omega\colon |x_1| + |x_2| < \lambda,\ x_i \geq 0\}.$$

Using symmetry, we get a Bellman candidate in the whole domain. We split Ω in three subdomains:

$$\Omega_1 = \{x \in \Omega\colon |x_1| + |x_2| \geq \lambda\} \cup \{x \in \Omega\colon |x_1| + |x_2| \leq \lambda,\ x_3 \geq \frac{\lambda^2 + x_1^2 - x_2^2}{2\lambda}\},$$
$$\Omega_2 = \{x \in \Omega\colon |x_1| + |x_2| < \lambda,\ |x_2| + \frac{x_1^2 - x_2^2}{\lambda} \leq x_3 \leq \frac{\lambda^2 + x_1^2 - x_2^2}{2\lambda}\},$$
$$\Omega_3 = \{x \in \Omega\colon |x_1| + |x_2| < \lambda,\ x_3 \leq |x_2| + \frac{x_1^2 - x_2^2}{\lambda}\}.$$

The final Bellman candidate in the whole domain is then given by the following formula:

$$B(x;\lambda)=\begin{cases}1, & x\in\Omega_1,\\[2mm] \dfrac{2\lambda x_3-x_1^2+x_2^2}{\lambda^2}, & x\in\Omega_2,\\[2mm] 1-\dfrac{(x_3+|x_2|-\lambda)^2}{(\lambda-|x_2|)^2-x_1^2}, & x\in\Omega_3.\end{cases} \tag{1.7.35}$$

Now we state the final result of this section.

THEOREM 1.7.5 *The explicit expression for the function defined in* (1.7.28) *is given by formula* (1.7.35), *i.e.,*

$$\mathbf{B}(x;\lambda)=B(x;\lambda).$$

To prove this theorem, we have to check the required concavity in order to apply an "improved" version of Lemma 1.7.2 (our function B is discontinuous at two points $(0,\pm\lambda,0)$) and then to check the reverse inequality by constructing an extremal pair of test functions for every point of Ω. However, we leave this task to the reader as very useful exercises (see Exercises 1.7.2 and 1.7.3).

Historical Remarks

Probably the formula (1.7.12) for one-sided estimate of martingale transform appeared in [**162**] for the first time. Independently this formula was found in [**142**]. Theorem 1.7.5 together with formula (1.7.35) for two-sided estimate was proved by A. Osękowski also in [**142**].

Exercises

PROBLEM 1.7.1 Prove that the conclusion of Lemma 1.7.2 is true for function $\mathbf{B}$, defined in (1.7.2) and function B, defined in (1.7.12).

PROBLEM 1.7.2 Prove inequality $\mathbf{B}(x;\lambda)\le B(x;\lambda)$ for function $\mathbf{B}$, defined in (1.7.28) and function B, defined in (1.7.35).

PROBLEM 1.7.3 Prove inequality $\mathbf{B}(x;\lambda)\ge B(x;\lambda)$ for function $\mathbf{B}$, defined in (1.7.28) and function B, defined in (1.7.35).

1.8 Burkholder's Bellman Function

Here we consider L^p-estimate for the martingale transform discussed in the preceding section. The corresponding Bellman function is the famous auxiliary function, introduced by Burkholder in [**23**]. It is possible to say that this particular function opened a new estimation method in analysis and probability. In this section, we essentially repeat our paper [**190**], where we tried to explain how to find this Burkholder's auxiliary function.

1.8.1 Notation, Definitions, and Statement of Main Result

Fix a real p, $1 < p < \infty$, and let $p' = \frac{p}{p-1}$, $p^* = \max\{p, p'\}$. Introduce the following domain in $\mathbb{R}^3$:

$$\Omega = \Omega(p) = \{x = (x_1, x_2, x_3) : \ x_3 \geq 0, \ |x_1|^p \leq x_3\}.$$

For a fixed partition $\mathcal{I}$ of an interval I, we define two functions on this domain

$$\mathbf{B}_{\max}(x) = \mathbf{B}_{\max}(x; p) = \sup_{\varphi, \psi}\{\langle\, |\psi|^p \rangle_{_I}\},$$

$$\mathbf{B}_{\min}(x) = \mathbf{B}_{\min}(x; p) = \inf_{\varphi, \psi}\{\langle\, |\psi|^p \rangle_{_I}\},$$

where the supremum and infimum were taken over all functions φ, ψ from $L^p(J)$, such that $\langle \varphi \rangle_{_I} = x_1$, $\langle \psi \rangle_{_I} = x_2$, $\langle\, |\varphi|^p \rangle_{_I} = x_3$, and $|(\varphi, h_{_J})| = |(\psi, h_{_J})|$ for all J, $J \in \mathcal{I}$. We shall refer to any such pair of functions φ, ψ, as an admissible pair.

Remark 1 As usual, function $\mathbf{B}$ does not depend on the interval I, where admissible pairs are defined. Moreover, it does not depend on the fixed partition $\mathcal{I}$, so we can assume that it is the dyadic partition: $\mathcal{I} = \mathcal{D}(I)$. Next, we work only with dyadic partitions.

Remark 2 The Bellman function for $\pm$ transform coincides with the Bellman function defined for admissible pairs, when ψ is differentially subordinated to φ, i.e., $|\varepsilon_J| \leq 1$ in (1.7.1). That was not trivial for the weak estimate, but for the norm estimate that is a common fact.

Remark 3 In the case $p = 2$, Bellman functions are evident:

$$\mathbf{B}_{\max}(x) = \mathbf{B}_{\min}(x) = x_2^2 + x_3 - x_1^2.$$

Indeed, since

$$\|\varphi\|_2^2 = |I|x_3 = |I|x_1^2 + \sum_{J \in \mathcal{I}} |(\varphi, h_{_J})|^2,$$

we have

$$\langle |\psi|^2 \rangle_{I} = \frac{1}{|I|}\|g\|_2^2 = x_2^2 + \frac{1}{|I|}\sum_{J\in\mathcal{I}} |(\psi, h_{J})|^2$$
$$= x_2^2 + \frac{1}{|I|}\sum_{J\in\mathcal{I}} |(\varphi, h_{J})|^2 = x_2^2 + x_3 - x_1^2.$$

Define the following function on $\mathbb{R}_+^2 = \{z = (z_1, z_2)\colon z_i > 0\}$:

$$F_p(z_1, z_2) = \begin{cases} z_1^p - (p^* - 1)^p z_2^p, & \text{if } z_1 \le (p^*-1)z_2, \\ p(1 - \frac{1}{p^*})^{p-1}(z_1+z_2)^{p-1}\big[z_1-(p^*-1)z_2\big], & \text{if } z_1 \ge (p^*-1)z_2. \end{cases} \tag{1.8.1}$$

Note that for $p = 2$, these expressions are reduced to $F_2(z_1, z_2) = z_1^2 - z_2^2$.

1.8.2 Main Result

Now we are ready to state the main result:

THEOREM 1.8.1 *The equation* $F_p(|x_1|, |x_2|) = F_p(x_3^{\frac{1}{p}}, \mathbf{B}^{\frac{1}{p}})$ *implicitly determines function* $\mathbf{B} = \mathbf{B}_{\min}(x; p)$, *and the equation* $F_p(|x_2|, |x_1|) = F_p(\mathbf{B}^{\frac{1}{p}}, x_3^{\frac{1}{p}})$ *implicitly determines function* $\mathbf{B} = \mathbf{B}_{\max}(x; p)$.

1.8.3 How to Find Bellman Functions

We start from deducing the main inequality for Bellman functions. As in the preceding section, we introduce the following new variables: $y_1 = \frac{1}{2}(x_2 + x_1)$, $y_2 = \frac{1}{2}(x_2 - x_1)$, and $y_3 = x_3$. In terms of the new variables, we define function $\mathbf{M}$,

$$\mathbf{M}(y_1, y_2, y_3) = \mathbf{B}(x_1, x_2, x_3) = \mathbf{B}(y_1 - y_2, y_1 + y_2, y_3),$$

on the domain

$$\Xi = \{y = (y_1, y_2, y_3)\colon\ y_3 \ge 0,\ |y_1 - y_2|^p \le y_3\}.$$

Since the point of the boundary $x_3 = |x_1|^p$ ($y_3 = |y_1 - y_2|^p$) occurs only for the constant test function $\varphi = x_1$ (and therefore then $\psi = x_2$ is a constant function as well), we have

$$\mathbf{B}(x_1, x_2, |x_1|^p) = |x_2|^p,$$

or

$$\mathbf{M}(y_1, y_2, |y_1 - y_2|^p) = |y_1 + y_2|^p. \tag{1.8.2}$$

Note that function $\mathbf{B}$ is even with respect to x_1 and x_2, i.e.,

$$\mathbf{B}(x_1, x_2, x_3) = \mathbf{B}(-x_1, x_2, x_3) = \mathbf{B}(x_1, -x_2, x_3).$$

This follows from the definition of $\mathbf{B}$ if we consider the test functions $\tilde{\varphi} = -\varphi$ for the first equality and $\tilde{\psi} = -\psi$ for the second one. For function $\mathbf{M}$, this means that we have symmetry with respect to lines $y_1 = \pm y_2$:

$$\mathbf{M}(y_1, y_2, y_3) = \mathbf{M}(y_2, y_1, y_3) = \mathbf{M}(-y_1, -y_2, y_3). \tag{1.8.3}$$

Therefore, it is sufficient to find function $\mathbf{B}$ in the domain

$$\Omega_+ = \Omega_+(p) = \{x = (x_1, x_2, x_3) : x_i \geq 0, \ |x_1|^p \leq x_3\}, \tag{1.8.4}$$

or the function $\mathbf{M}$ in the domain

$$\Xi_+ = \{y = (y_1, y_2, y_3) \colon \ y_1 \geq 0, \ -y_1 \leq y_2 \leq y_1, \ (y_1 - y_2)^p \leq y_3\}. \tag{1.8.5}$$

Then we get the solution in the whole domain by putting

$$\mathbf{B}(x_1, x_2, x_3) = \mathbf{B}(|\,x_1|, |\,x_2|, x_3).$$

Due to symmetry (1.8.3), we have the following boundary conditions on the "new part" of the boundary $\partial\Xi_+$:

$$\begin{aligned} \frac{\partial \mathbf{M}}{\partial y_1} &= \frac{\partial \mathbf{M}}{\partial y_2} && \text{on the hyperplane } y_2 = y_1, \\ \frac{\partial \mathbf{M}}{\partial y_1} &= -\frac{\partial \mathbf{M}}{\partial y_2} && \text{on the hyperplane } y_2 = -y_1. \end{aligned} \tag{1.8.6}$$

If we consider the family of test functions $\tilde{\varphi} = \tau\varphi, \ \ \tilde{\psi} = \tau\psi$ along with φ and ψ, we come to the following homogeneity condition:

$$\mathbf{B}(\tau x_1, \tau x_2, \tau^p x_3) = \tau^p \mathbf{B}(x_1, x_2, x_3)$$

or, in our other notation,

$$\mathbf{M}(\tau y_1, \tau y_2, \tau^p y_3) = \tau^p \mathbf{M}(y_1, y_2, y_3).$$

As usual, we shall use this property in the following form: take a derivative with respect to τ and put $\tau = 1$. Then we get

$$y_1 \frac{\partial \mathbf{M}}{\partial y_1} + y_2 \frac{\partial \mathbf{M}}{\partial y_2} + p y_3 \frac{\partial \mathbf{M}}{\partial y_3} = p\mathbf{M}(y_1, y_2, y_3). \tag{1.8.7}$$

The main inequality for the Bellman function $\mathbf{B} = \mathbf{B}_{\max}$ is the same as was stated in Lemma 1.7.1 for the weak problem. For function $\mathbf{B} = \mathbf{B}_{\min}$, we have the opposite inequality, i.e., the corresponding convexity condition. The proof is absolutely the same. We take the same test functions, but in the last step we estimate the required Bellman function. For $\mathbf{B} = \mathbf{B}_{\max}$, we have

$$\mathbf{B}(x) \geq \langle |\psi|^p \rangle_I = \tfrac{1}{2}\langle |\psi^+|^p \rangle_{I^+} + \tfrac{1}{2}\langle |\psi^-|^p \rangle_{I^-} \geq \tfrac{1}{2}\mathbf{B}(x^+) + \tfrac{1}{2}\mathbf{B}(x^-) - \eta.$$

Since η is arbitrary, we conclude that

$$\mathbf{B}(x) \geq \tfrac{1}{2}\mathbf{B}(x^+) + \tfrac{1}{2}\mathbf{B}(x^-). \tag{1.8.8}$$

In a similar way, for function $\mathbf{B} = \mathbf{B}_{\min}$, we can get

$$\mathbf{B}(x) \leq \tfrac{1}{2}\mathbf{B}(x^+) + \tfrac{1}{2}\mathbf{B}(x^-). \tag{1.8.9}$$

Recall that this is not quite a concavity (convexity) condition because we have the restriction $|x_1^+ - x_1^-| = |x_2^+ - x_2^-|$. But in terms of function $\mathbf{M}$

$$\mathbf{M}_{\max}(y) \geq \tfrac{1}{2}\mathbf{M}_{\max}(y^+) + \tfrac{1}{2}\mathbf{M}_{\max}(y^-),$$
$$\mathbf{M}_{\min}(y) \leq \tfrac{1}{2}\mathbf{M}_{\min}(y^+) + \tfrac{1}{2}\mathbf{M}_{\min}(y^-),$$

when either $y_1 = y_1^+ = y_1^-$ or $y_2 = y_2^+ = y_2^-$, we have concavity (convexity) of function $\mathbf{M}$ with respect to y_2, y_3 under a fixed y_1, and with respect to y_1, y_3 under a fixed y_2. Such a property is called "separate concavity" (convexity).

Since the domain is convex, under the assumption that function $\mathbf{B}$ is sufficiently smooth, these conditions of concavity (convexity) are equivalent to the following differential inequalities:

$$\begin{pmatrix} \mathbf{M}_{y_1y_1} & \mathbf{M}_{y_1y_3} \\ \mathbf{M}_{y_3y_1} & \mathbf{M}_{y_3y_3} \end{pmatrix} \leq 0, \quad \begin{pmatrix} \mathbf{M}_{y_2y_2} & \mathbf{M}_{y_2y_3} \\ \mathbf{M}_{y_3y_2} & \mathbf{M}_{y_3y_3} \end{pmatrix} \leq 0, \quad \forall y \in \Xi, \tag{1.8.10}$$

for $\mathbf{M} = \mathbf{M}_{\max}$ and

$$\begin{pmatrix} \mathbf{M}_{y_1y_1} & \mathbf{M}_{y_1y_3} \\ \mathbf{M}_{y_3y_1} & \mathbf{M}_{y_3y_3} \end{pmatrix} \geq 0, \quad \begin{pmatrix} \mathbf{M}_{y_2y_2} & \mathbf{M}_{y_2y_3} \\ \mathbf{M}_{y_3y_2} & \mathbf{M}_{y_3y_3} \end{pmatrix} \geq 0, \quad \forall y \in \Xi, \tag{1.8.11}$$

for $\mathbf{M} = \mathbf{M}_{\min}$.

Extremal properties of the Bellman function require for one of the matrices in (1.8.10) and (1.8.11) to be degenerated. So, we arrive at the Monge–Ampère equation:

$$\mathbf{M}_{y_iy_i}\mathbf{M}_{y_3y_3} = (\mathbf{M}_{y_iy_3})^2 \tag{1.8.12}$$

either for $i = 1$ or for $i = 2$. As usual, to find a candidate M for the role of the true Bellman function $\mathbf{M}$, we shall solve that equation. After finding a solution, we shall prove that $M = \mathbf{M}$.

As usual, we look for solution of the Monge–Ampère equation in the form

$$M = t_iy_i + t_3y_3 + t_0, \qquad i = 1, 2, \tag{1.8.13}$$

where $t_k = M_{y_k}$, $k = 1, 2, 3$. Solution M is linear along the *extremal trajectories*

$$y_i dt_i + y_3 dt_3 + dt_0 = 0. \tag{1.8.14}$$

One of the ends of the extremal trajectory has to be a point on the boundary $y_3 = |y_1 - y_2|^p$, where constant functions are the only test functions corresponding to these points. Denote that point by $U = (y_1, u, (y_1 - u)^p)$. Note that we write $(y_1 - u)^p$, instead of $|y_1 - u|^p$ because the domain Ξ_+ is under consideration. For the second end of the extremal trajectory, we have four possibilities:

1) it belongs to the same boundary $y_3 = (y_1 - y_2)^p$;
2) it is at infinity $(y_1, y_2, +\infty)$, i.e., the extremal lines goes parallel to the y_3-axis;
3) it belongs to the boundary $y_2 = y_1$;
4) it belongs to the boundary $y_2 = -y_1$.

The first possibility gives us no solution. More exactly, we have the following.

PROPOSITION 1.8.2 *If $p \neq 2$, then function $\mathbf{B}$ cannot be equal to $B(x) = M(y)$, where M is the solution of the Monge–Ampère equation (1.8.12), such that all of its extremal trajectories are of type* 1).

PROOF Let us write down the relation between the ends of possible extremal chords. Assume that the boundary is parametrized as $g(t) = (g_1(t), g_2(t))$, and $f(t)$ is the boundary value of our Bellman function. It was already explained in Section 1.5 that if segment $[g(a), g(b)]$ is an extremal chord, there exists a tangent plane to the graph of our Bellman function such that the graph over that chord is a straight-line segment $[(g_1(a), g_2(a), f(a)); (g_1(b), g_2(b), f(b))]$ in three-dimensional space, belonging to that tangent plane. The two tangent vectors to the boundary curve: $(g_1'(a), g_2'(a), f'(a))$ and $(g_1'(b), g_2'(b), f'(b))$ also belong to that plane. Hence, we have

$$\det \begin{pmatrix} g_1'(a) & g_2'(a) & f'(a) \\ g_1'(b) & g_2'(b) & f'(b) \\ g_1(b) - g_1(a) & g_2(b) - g_2(a) & f(b) - f(a) \end{pmatrix} = 0. \tag{1.8.15}$$

If we go along a family of shrinking chords $[g(a), g(b)]$ to the origin of the cup (let us denote it by $g(c)$), then from (1.8.15) we come to the following equation for such a point:

$$\det\begin{pmatrix} g_1'(c) & g_2'(c) & f'(c) \\ g_1''(c) & g_2''(c) & f''(c) \\ g_1'''(c) & g_2'''(c) & f'''(c) \end{pmatrix} = 0. \tag{1.8.16}$$

That is the point where the torsion of the boundary curve vanishes.

In our case, we can parametrize the boundary curve by its first coordinate, i.e., put $g_1(x_1) = x_1$, $x_3 = g_2(x_1) = x_1^p$, and $f(x_1) = x_2^p$, where $x_2 = \sigma x_1 + \text{const}$, $\sigma = \pm 1$. So matrix (1.8.16) turns into

$$\det\begin{pmatrix} 1 & px_1^{p-1} & px_2^{p-1}\sigma \\ 0 & p(p-1)x_1^{p-2} & p(p-1)x_2^{p-2} \\ 0 & p(p-1)(p-2)x_1^{p-3} & p(p-1)(p-2)x_2^{p-3}\sigma \end{pmatrix} = 0.$$

And we come to the equation $p^2(p-1)^2(p-2)x_1^{p-3}x_2^{p-3}(\sigma x_1 - x_2) = 0$. It is always fulfilled, if $p = 2$. The solution $x_2 = \sigma x_1$ in Ω_+ means that $\sigma = 1$, and we get a foliation in the plane $x_1 = x_2$ only. The torsion of the boundary curve is identically zero and this gives us the linear function $B = x_3$, and we can consider that plane foliated by the extremal lines going parallel to the y_3-axis and include this special plane into case 2).

Now we show that a cup cannot be originated in the symmetry planes $x_1 = 0$ and $x_2 = 0$. Such points could be only the endpoints of extremals. Indeed, by symmetry, the extremals from different sides of the plane of symmetry are in orthogonal planes: if one goes in, say, plane $x_1 = x_2$, the symmetrical extremal line goes in plain $x_1 = -x_2$, and these segments cannot be parts of one straight line.

So, the only possibility we need to consider is a singular cup with the origin on one of two axes: either $x_1 = 0$ or $x_2 = 0$. Let us consider the case when the possible origin of a singular cup is on line $x_1 = 0$. Denote the second coordinate of the origin by u. Then the extremal line, started at point $(0, u, 0)$ ends at some point $(b, u + \sigma b, b^p)$. The cup equation (1.8.15) turns into

$$\det\begin{pmatrix} 1 & 0 & p\sigma u^{p-1} \\ 1 & pb^{p-1} & p\sigma(u+\sigma b)^{p-1} \\ b & b^p & (u+\sigma b)^p - u^p \end{pmatrix} = 0,$$

i.e.,

$$pub^{p-1}\big[(u+\sigma b)^{p-1} - u^{p-1} - (p-1)\sigma b u^{p-2}\big] = 0.$$

This equation has no solution, except for $b = 0$, if $p \neq 2$. Indeed, denoting the expression in brackets by $h(b)$, we see that $h(0) = 0$ and

$$h'(b) = (p-1)u\sigma\big((u+\sigma b)^{p-2} - u^{p-2}\big).$$

This expression does not change its sign, and therefore h is monotonic. We conclude that h has no more zeroes, except for $b = 0$. □

Now we check the second possibility among the possibilities 1)–4) listed earlier.

PROPOSITION 1.8.3 *If function* $\mathbf{B}$ *has foliation of type* 2) *in a subdomain, then it is of the form* $B(x) = x_2^p + C(x_3 - x_1^p)$ *with a constant* C.

PROOF Since the extremal line is parallel to the y_3-axis, the Bellman function has to be of the form

$$M(y) = A(y_1, y_2) + C(y_1, y_2)y_3.$$

Any pair of inequalities, both (1.8.10) and (1.8.11), implies $M_{y_i y_i} M_{y_3 y_3} - (M_{y_i y_3})^2 \geq 0$. Since $M_{y_3 y_3} = 0$, this yields $M_{y_i y_3} = \frac{\partial C}{\partial y_i} = 0, \ \ i = 1, 2$, i.e., C is a constant. From the boundary condition (1.8.2) we get

$$A(y_1, y_2) + C(y_1 - y_2)^p = (y_1 + y_2)^p,$$

whence

$$A(y_1, y_2) = (y_1 + y_2)^p - C(y_1 - y_2)^p,$$

and

$$M(y) = (y_1 + y_2)^p + C(y_3 - (y_1 - y_2)^p), \tag{1.8.17}$$

or

$$B(x) = x_2^p + C(x_3 - x_1^p). \tag{1.8.18}$$

□

Let us note that this solution cannot satisfy the necessary conditions in the whole domain Ξ_+, except for the case when $p = 2$. Constant C must be positive. Otherwise, the extremal lines cannot tend to infinity along y_3-axes because M must be a nonnegative function. Therefore, the straight line

$$y_1 + y_2 = C^{\frac{1}{p-2}}(y_1 - y_2) \text{ or } x_2 = C^{\frac{1}{p-2}} x_1$$

splits Ξ_+ into two subdomains in one of which the derivatives

$$\frac{\partial^2 M}{\partial y_1^2} = \frac{\partial^2 M}{\partial y_2^2} = p(p-1)\Big((y_1 + y_2)^{p-2} - C(y_1 - y_2)^{p-2})\Big)$$

are positive (i.e., it could be a candidate for $\mathbf{B}_{\min}$), and in another subdomain those derivatives are negative (i.e., it could be a candidate for $\mathbf{B}_{\max}$).

Thus, this simple solution cannot give us the whole Bellman function and we need to continue the consideration of the possibilities 3) and 4). Until now, we have not fixed which of the two matrices in (1.8.10) or in (1.8.11) is degenerated, i.e., what is i in the Monge–Ampère equation (1.8.12), because for the vertical extremal lines both these equations are fulfilled. Now, when considering possibility 3) or 4), we need to investigate separately both Monge–Ampère equations (1.8.12). We shall refer to these cases as 3_i) and 4_i).

Let us start with simultaneous consideration of the cases 3_1) and 4_1) (we recall that this means that y_2 is fixed). We look for a function

$$M = t_1 y_1 + t_3 y_3 + t_0$$

on the domain Ξ_+, which is linear along the extremal lines

$$y_1 dt_1 + y_3 dt_3 + dt_0 = 0.$$

Now, one endpoint of our extremal line $V = (v, y_2, (v - y_2)^p)$ belongs to the boundary $y_3 = |y_1 - y_2|^p$ and the second endpoint $W = (|y_2|, y_2, w)$ is on the boundary $y_1 = |y_2|$, where we have boundary condition (1.8.6). Due to symmetry (1.8.3), on the boundary $y_1 = y_2$ (which means that our fixed y_2 is positive) we have

$$\frac{\partial M}{\partial y_2} = \frac{\partial M}{\partial y_1} = t_1,$$

and on the boundary $y_1 = -y_2$ (which means that our fixed y_2 is negative) we have

$$\frac{\partial M}{\partial y_2} = -\frac{\partial M}{\partial y_1} = -t_1.$$

In both cases,

$$y_2 \frac{\partial M}{\partial y_2} = y_1 \frac{\partial M}{\partial y_1} = |y_2| t_1,$$

and therefore (1.8.7) and (1.8.13) imply

$$2t_1|y_2| + pwt_3 = pM(W) = pt_1|y_2| + pwt_3 + pt_0,$$

whence we conclude that

$$t_0 = (\tfrac{2}{p} - 1) t_1 |y_2|.$$

This gives us the formula for $t_0(t_1)$ (remember that y_2 is fixed, as we are considering the cases 3_1), 4_1) now). Thus, we get

$$M(y) = \Big[y_1 + (\tfrac{2}{p} - 1)|y_2| \Big] t_1 + y_3 t_3. \tag{1.8.19}$$

Since $dt_0 = (\frac{2}{p} - 1)|y_2|\,dt_1$, the equation of the extremal trajectories (1.8.14) takes the form:

$$\Big[y_1 + (\tfrac{2}{p} - 1)|y_2|\Big]dt_1 + y_3\,dt_3 = 0, \tag{1.8.20}$$

and we can rewrite (1.8.19), as follows:

$$M(y) = \Big(t_3 - t_1\tfrac{dt_3}{dt_1}\Big)y_3.$$

We see that the expression $M(y)/y_3$ is constant along the trajectory and we can find it, evaluating at the point V, where the boundary condition (1.8.2) is known:

$$M(y) = \left(\frac{v+y_2}{v-y_2}\right)^p y_3, \tag{1.8.21}$$

where $v = v(y_1, y_2, y_3)$ satisfies the following equation:

$$\frac{y_1 + (\frac{2}{p} - 1)|y_2|}{y_3} = \frac{v + (\frac{2}{p} - 1)|y_2|}{(v-y_2)^p} \tag{1.8.22}$$

because the point $V = (v, y_2, (v - y_2)^p)$ is on the extremal line (1.8.20). We shall not even check under what conditions equation (1.8.22) has a solution and when it is unique. Later, we shall show that regardless, this function M that we have just found cannot be the Bellman function we are interested in because neither condition (1.8.10) nor (1.8.11) can be fulfilled: the matrix $\{M_{y_iy_j}\}_{i,j=2,3}$ is neither negative definite nor positive definite. We postpone the verification because the calculation of the sign of the Hessian matrices is the same for this solution as for another solution of the Monge–Ampère equation that supplies us with the true Bellman function. These calculations will be made simultaneously a bit later. And now, we shall rewrite our solution in an implicit form, more convenient for calculation.

We introduce

$$\omega \stackrel{\text{def}}{=} \left(\frac{M(y)}{y_3}\right)^{\frac{1}{p}}, \tag{1.8.23}$$

then (1.8.21) yields

$$v = \frac{\omega+1}{\omega-1}\,y_2. \tag{1.8.24}$$

Recall that we are considering $V = \big(v, y_2, (v - y_2)^p\big) \in \Xi_+$, i.e., $v \ge |y_2| \ge 0$ and hence,

$$\operatorname{sgn} y_2 = \operatorname{sgn}(\omega - 1). \tag{1.8.25}$$

After substitution of (1.8.24) in (1.8.22), we get

$$\left(\frac{2y_2}{\omega-1}\right)^p\left[y_1+\left(\tfrac{2}{p}-1\right)|y_2|\right]=y_3\left[\frac{\omega+1}{\omega-1}y_2+\left(\tfrac{2}{p}-1\right)|y_2|\right]$$

or

$$2^p|y_2|^{p-1}\left[py_1+(2-p)|y_2|\right]=y_3|\omega-1|^{p-1}\left[(\omega+1)p+(2-p)|\omega-1|\right]$$

For the case 3_1), we have $y_2>0$ (i.e., $x_2>x_1$, as we are looking for $\omega>1$ or $B>y_3$), and the latter equation can be rewritten in the initial coordinates, as follows:

$$(x_2-x_1)^{p-1}\left[x_2+(p-1)x_1\right]=(B^{\frac{1}{p}}-x_3^{\frac{1}{p}})^{p-1}\left[B^{\frac{1}{p}}+(p-1)x_3^{\frac{1}{p}}\right].$$

For the case 4_1), we have $y_2<0$ (i.e., $x_2<x_1$, as we are looking for $\omega<1$ or $B<y_3$) and the equation takes the form

$$(x_1-x_2)^{p-1}\left[x_1+(p-1)x_2\right]=(x_3^{\frac{1}{p}}-B^{\frac{1}{p}})^{p-1}\left[x_3^{\frac{1}{p}}+(p-1)B^{\frac{1}{p}}\right].$$

Let us introduce the following function:

$$G(z_1,z_2)\stackrel{\text{def}}{=}(z_1+z_2)^{p-1}\left[z_1-(p-1)z_2\right], \tag{1.8.26}$$

defined on the half-plane $z_1+z_2\geq 0$. Then in the case 3_1), we have the relation

$$G(x_2,-x_1)=G(B^{\frac{1}{p}},-x_3^{\frac{1}{p}}), \tag{1.8.27}$$

or

$$G(y_2+y_1,y_2-y_1)=y_3G(\omega,-1).$$

In the case 4_1), we have

$$G(x_1,-x_2)=G(x_3^{\frac{1}{p}},-B^{\frac{1}{p}}), \tag{1.8.28}$$

or

$$G(y_1-y_2,-y_1-y_2)=y_3G(1,-\omega).$$

Now we have to consider the Monge–Ampère equation (1.8.12) for the cases 3_2) and 4_2). This means that we fix y_1 now. Let us begin with the case 3_2), where an extremal line starts from a point $U=(y_1,u,(y_1-u)^p)$ and ends at a point $W=(y_1,y_1,w)$. Again, the symmetry condition at the point W is

$$\frac{\partial M}{\partial y_1}=\frac{\partial M}{\partial y_2}=t_2,$$

and the homogeneity condition (1.8.7) plus condition (1.8.13) at W yield

$$2y_1t_2+pwt_3=pM(W)=py_1t_2+pwt_3+pt_0,$$

whence

$$t_0 = (\tfrac{2}{p} - 1)y_1 t_2,$$

and therefore

$$M(y) = \Big[y_2 + (\tfrac{2}{p} - 1)y_1\Big]t_2 + y_3 t_3. \tag{1.8.29}$$

Since $dt_0 = (\frac{2}{p} - 1)y_1\, dt_2$, the equation of the extremal trajectories takes the form

$$\Big[y_2 + (\tfrac{2}{p} - 1)y_1\Big]dt_2 + y_3\, dt_3 = 0, \tag{1.8.30}$$

and we can rewrite (1.8.29), as follows:

$$M(y) = \Big(t_3 - t_2\frac{dt_3}{dt_2}\Big)y_3.$$

Again, from this equality, we conclude that the expression $M(y)/y_3$ is constant along the trajectory and we can find it, evaluating at the point U, where the boundary condition (1.8.2) is known:

$$M(y) = \Big(\frac{y_1 + u}{y_1 - u}\Big)^p y_3, \tag{1.8.31}$$

where $u = u(y_1, y_2, y_3)$ can be found from (1.8.30):

$$\frac{y_2 + (\frac{2}{p} - 1)y_1}{y_3} = \frac{u + (\frac{2}{p} - 1)y_1}{(y_1 - u)^p}. \tag{1.8.32}$$

This extremal line intersects the plane $y_1 = y_2$ at the point with the third coordinate

$$y_3 = \frac{2y_1}{p} \cdot \frac{(y_1 - u)^p}{u + (\frac{2}{p} - 1)y_1}.$$

Therefore, this point is in the domain if $y_3 > 0$, i.e., $u > -(\frac{2}{p} - 1)y_1$, and the starting points U with $u \le -(\frac{2}{p} - 1)y_1$ cannot be acceptable for the case under consideration (see Figure 1.16).

Let us check that equation (1.8.32) has exactly one solution $u = u(y_1, y_2, y_3)$ in the sector $-(\frac{2}{p} - 1)y_1 < y_2 < y_1$. It is easy to see that the function

$$u \mapsto y_3\Big[u + (\tfrac{2}{p} - 1)y_1\Big] - (y_1 - u)^p\Big[y_2 + (\tfrac{2}{p} - 1)y_1\Big]$$

is monotonously increasing for $u < y_1$. Since it has negative value $-(\frac{2}{p}y_1)^p\Big[y_2 + (\frac{2}{p} - 1)y_1\Big]$ at point $u = -(\frac{2}{p} - 1)y_1$ and positive value

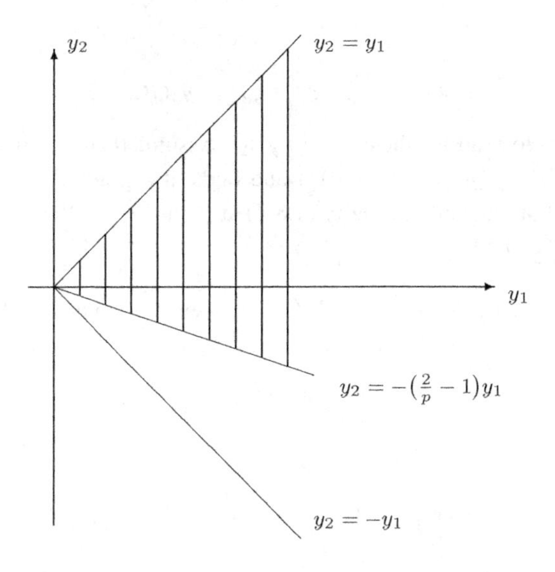

Figure 1.16 Acceptable sector for the case 3_2).

$\frac{2}{p}y_1y_3$ at point $u = y_1$, it has a root in between, and due to monotonicity, this root is unique.

Now we rewrite the solution (1.8.31) in an implicit form, as before, using notation (1.8.23): $\omega = \left(\frac{M(y)}{y_3}\right)^{\frac{1}{p}}$. From (1.8.31), we have

$$u = \frac{\omega - 1}{\omega + 1} y_1, \tag{1.8.33}$$

therefore, from (1.8.32), we obtain

$$2^{-p}y_3(\omega+1)^{p-1}[p(\omega-1)+(2-p)(\omega+1)] = y_1^{p-1}\big[py_2+(2-p)y_1\big]$$

or

$$2^{-p+1}y_3(\omega+1)^{p-1}(\omega-p+1) = y_1^{p-1}\big[py_2+(2-p)y_1\big],$$

which is (again using notation (1.8.23))

$$(B^{\frac{1}{p}}+x_3^{\frac{1}{p}})^{p-1}\big[B^{\frac{1}{p}}-(p-1)x_3^{\frac{1}{p}}\big] = (x_1+x_2)^{p-1}\big[x_2-(p-1)x_1\big].$$

In terms of function G (see (1.8.26)), this can be rewritten, as follows:

$$G(x_2,x_1) = G(B^{\frac{1}{p}}, x_3^{\frac{1}{p}}),$$

or

$$G(y_1 + y_2, y_1 - y_2) = y_3 G(\omega, 1).$$

It remains to examine the possibility 4_2). Assume that an extremal line starts at a point $U = (y_1, u, (y_1 - u)^p)$ and ends at a point $W = (y_1, -y_1, w)$. Again, the homogeneity property (1.8.7) at point W and the symmetry $\frac{\partial M}{\partial y_1} = -\frac{\partial M}{\partial y_2} = -t_2$ yield

$$-2y_1 t_2 + pwt_3 = pM(W) = -py_1 t_2 + pwt_3 + pt_0,$$

whence

$$t_0 = \left(1 - \tfrac{2}{p}\right) y_1 t_2,$$

and therefore

$$M(y) = \left[y_2 + \left(1 - \tfrac{2}{p}\right) y_1\right] t_2 + y_3 t_3. \tag{1.8.34}$$

Since $dt_0 = (1 - \frac{2}{p}) y_1 \, dt_2$, the equation of the extremal trajectories takes the form

$$\left[y_2 + \left(1 - \tfrac{2}{p}\right) y_1\right] dt_2 + y_3 \, dt_3 = 0, \tag{1.8.35}$$

and we can rewrite (1.8.34), as follows:

$$M(y) = \left(t_3 - t_2 \tfrac{dt_3}{dt_2}\right) y_3.$$

Again, the expression $M(y)/y_3$ is constant along the trajectory and from the boundary condition (1.8.2), we get the same expression:

$$M(y) = \left(\frac{y_1 + u}{y_1 - u}\right)^p y_3. \tag{1.8.36}$$

Now $u = u(y_1, y_2, y_3)$ is a solution of the equation

$$\frac{y_2 - (\frac{2}{p} - 1) y_1}{y_3} = \frac{u - (\frac{2}{p} - 1) y_1}{(y_1 - u)^p} \tag{1.8.37}$$

that we get from (1.8.35). As before, we get trajectories ending at the plane $y_2 = -y_1$ not in the whole domain Ξ_+ but only in the sector $-y_1 < y_2 < (\frac{2}{p} - 1) y_1$ (see Figure 1.17), and equation (1.8.37) has a unique solution for every point from this sector. As before, relation (1.8.33) allows us to rewrite the equation of extremal trajectories (1.8.37) as an implicit expression for ω (and hence for M):

$$G(x_1, x_2) = G(x_3^{\frac{1}{p}}, B^{\frac{1}{p}}),$$

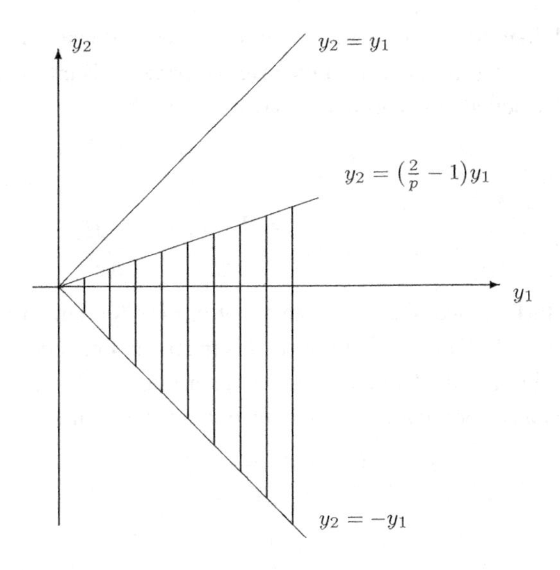

Figure 1.17 Acceptable sector for the case 4_2).

or

$$G(y_1 - y_2, y_1 + y_2) = y_3 G(1, \omega).$$

Now we start the verification of which of the obtained solutions satisfies conditions (1.8.10) or (1.8.11). We need to calculate $D_i \stackrel{\text{def}}{=} M_{y_i y_i} M_{y_3 y_3} - M^2_{y_i y_3}$, $i = 1, 2$, for the following four cases:

$$3_1)\ G(y_1 + y_2, -y_1 + y_2) = y_3 G(\omega, -1)\,; \tag{1.8.38}$$

$$4_1)\ G(y_1 - y_2, -y_1 - y_2) = y_3 G(1, -\omega)\,; \tag{1.8.39}$$

$$3_2)\ G(y_1 + y_2, y_1 - y_2) = y_3 G(\omega, 1)\,; \tag{1.8.40}$$

$$4_2)\ G(y_1 - y_2, y_1 + y_2) = y_3 G(1, \omega), \tag{1.8.41}$$

where $M = y_3 \omega^p$. In all situations, we have a relation of the form

$$\Phi(\omega) = \frac{H(y_1, y_2)}{y_3}.$$

At this time, we will not specify the expressions either for Φ and H, or for their derivatives. Instead, we will plug in the specific expressions only in the final result, after numerous cancellations. Moreover, let us introduce two more auxiliary functions:

$$R_1 = R_1(\omega) \stackrel{\text{def}}{=} \frac{1}{\Phi'} \quad \text{and} \quad R_2 = R_2(\omega) \stackrel{\text{def}}{=} R_1' = -\frac{\Phi''}{\Phi'^2}.$$

We would like to mention here that this idea, allowing us to make the calculation shorter, is taken from the original paper of Burkholder [**22**].

First, we calculate the partial derivatives of ω:

$$\Phi'\omega_{y_3} = -\frac{H}{y_3^2} \quad \Longrightarrow \quad \omega_{y_3} = -\frac{R_1 H}{y_3^2},$$

$$\Phi'\omega_{y_i} = \frac{H_{y_i}}{y_3} \quad \Longrightarrow \quad \omega_{y_i} = \frac{R_1 H_{y_i}}{y_3} = \frac{R_1 H'}{y_3}, \qquad i = 1, 2.$$

Here and further, we shall use the notation H' for any partial derivative H_{y_i}, $i = 1, 2$. This cannot cause misunderstanding because only one i participates in the calculation of Hessian determinants D_i. Moreover, we shall not mention anymore that index i can take two values: either $i = 1$, or $i = 2$.

$$\omega_{y_3 y_3} = -\frac{R_2 \omega_{y_3} H}{y_3^2} + 2\frac{R_1 H}{y_3^3} = \frac{R_1 H}{y_3^4}(R_2 H + 2y_3),$$

$$\omega_{y_3 y_i} = -\frac{R_2 \omega_{y_i} H}{y_3^2} - \frac{R_1 H'}{y_3^2} = -\frac{R_1 H'}{y_3^3}(R_2 H + y_3),$$

$$\omega_{y_i y_i} = \frac{R_2 \omega_{y_i} H'}{y_3} + \frac{R_1 H''}{y_3} = \frac{R_1}{y_3^2}(R_2 (H')^2 + y_3 H'').$$

Now we pass to the calculation of derivatives of $M = y_3 \omega^p$:

$$\begin{aligned} M_{y_3} &= p y_3 \omega^{p-1} \omega_{y_3} + \omega^p, \\ M_{y_i} &= p y_3 \omega^{p-1} \omega_{y_i}; \end{aligned}$$

$$\begin{aligned} M_{y_3 y_3} &= p y_3 \omega^{p-1} \omega_{y_3 y_3} + 2p\omega^{p-1}\omega_{y_3} + p(p-1) y_3 \omega^{p-2} \omega_{y_3}^2 \\ &= \frac{p\omega^{p-2} R_1 H^2}{y_3^3}[\omega R_2 + (p-1)R_1], \qquad (1.8.42) \\ M_{y_3 y_i} &= p y_3 \omega^{p-1} \omega_{y_3 y_i} + p\omega^{p-1}\omega_{y_i} + p(p-1) y_3 \omega^{p-2} \omega_{y_3}\omega_{y_i} \\ &= -\frac{p\omega^{p-2} R_1 H H'}{y_3^2}[\omega R_2 + (p-1)R_1], \\ M_{y_i y_i} &= p y_3 \omega^{p-1} \omega_{y_i y_i} + p(p-1) y_3 \omega^{p-2} \omega_{y_i}^2 \\ &= \frac{p\omega^{p-2} R_1}{y_3}\left([\omega R_2 + (p-1)R_1](H')^2 + \omega y_3 H''\right). \end{aligned}$$

This yields

$$D_i = M_{y_3 y_3} M_{y_i y_i} - M_{y_3 y_i}^2 = \frac{p^2 \omega^{2p-3} R_1^2 H^2 H''}{y_3^3}[\omega R_2 + (p-1)R_1]. \qquad (1.8.43)$$

Notice that H' disappeared completely.

Now we need to calculate second derivatives of

$$H(y_1, y_2) = G(\alpha_1 y_1 + \alpha_2 y_2, \beta_1 y_1 + \beta_2 y_2),$$

where $\alpha_i, \beta_i = \pm 1$:

$$\begin{aligned} H'' &= \frac{\partial^2}{\partial y_i^2} G(\alpha_1 y_1 + \alpha_2 y_2, \beta_1 y_1 + \beta_2 y_2) \\ &= \alpha_i^2 G_{z_1 z_1} + 2\alpha_i \beta_i G_{z_1 z_2} + \beta_i^2 G_{z_2 z_2} \\ &= G_{z_1 z_1} + G_{z_2 z_2} \pm 2G_{z_1 z_2}, \end{aligned}$$

where the "+" sign has to be chosen if the coefficients in front of y_i are equal, and the "−" sign in the opposite case.

The derivatives of G are simple:

$$\begin{aligned} G_{z_1} &= p(z_1 + z_2)^{p-2}\big[z_1 - (p-2)z_2\big], \\ G_{z_2} &= -p(p-1)z_2(z_1 + z_2)^{p-2}; \\ G_{z_1 z_2} &= p(p-1)(z_1 + z_2)^{p-3}\big[z_1 - (p-3)z_2\big], \\ G_{z_1 z_2} &= -p(p-1)(p-2)z_2(z_1 + z_2)^{p-3}, \\ G_{z_2 z_2} &= -p(p-1)(z_1 + z_2)^{p-3}\big[z_1 + (p-1)z_2\big]. \end{aligned}$$

Note that $G_{z_1 z_1} + G_{z_2 z_2} = 2G_{z_1 z_2}$, and therefore, $H'' = 4G_{z_1 z_2}$, if $\alpha_i = \beta_i$; and $H'' = 0$, if $\alpha_i = -\beta_i$. The first case occurs for $H_{y_2 y_2}$ in cases 3_1), 4_1) and for $H_{y_1 y_1}$ in cases 3_2), 4_2). The second case occurs for $H_{y_1 y_1}$ in cases 3_1), 4_1) and for $H_{y_2 y_2}$ in cases 3_2), 4_2). In fact, we know that the equality $D_i = 0$ has to be fulfilled in the cases 3_i) and 4_i) because it is just the Monge–Ampère equation we have been solving.

So, we have:

$$\begin{aligned} &3_1) \quad z_1 = y_1 + y_2, \\ &\qquad z_2 = -y_1 + y_2, \qquad G_{z_1 z_2} = p(p-1)(p-2)(y_1 - y_2)(2y_2)^{p-3}, \\ &4_1) \quad z_1 = y_1 - y_2, \\ &\qquad z_2 = -y_1 - y_2, \qquad G_{z_1 z_2} = p(p-1)(p-2)(y_1 + y_2)(-2y_2)^{p-3}, \\ &3_2) \quad z_1 = y_1 + y_2, \\ &\qquad z_2 = y_1 - y_2, \qquad G_{z_1 z_2} = -p(p-1)(p-2)(y_1 - y_2)(2y_1)^{p-3}, \end{aligned}$$

$4_2)$ $z_1 = y_1 - y_2,$
$$z_2 = y_1 + y_2, \qquad G_{z_1 z_2} = -p(p-1)(p-2)(y_1+y_2)(2y_1)^{p-3}.$$

By the way, we would like to call the attention of the reader to the fact that all expressions of the form $(\pm 2y_i)^{p-3}$ make sense because y_1 is always positive, $y_2 > 0$ in the case 3_1), and $y_2 < 0$ in the case 4_1).

To complete the investigation of sgn D_i, we need to calculate the sign of the expression in brackets in (1.8.43):

$$\omega R_2 + (p-1)R_1 = R_1^2[(p-1)\Phi' - \omega\Phi''] \tag{1.8.44}$$

$3_1)$
$$\begin{aligned}\Phi(\omega) &= G(\omega,-1),\\ \Phi'(\omega) &= G_{z_1}(\omega,-1) = p(\omega-1)^{p-2}(\omega+p-2),\\ \Phi''(\omega) &= G_{z_1 z_1}(\omega,-1) = p(p-1)(\omega-1)^{p-3}(\omega+p-3),\end{aligned}$$

$$(p-1)\Phi' - \omega\Phi'' = -p(p-1)(p-2)(\omega-1)^{p-3};$$

$4_1)$
$$\begin{aligned}\Phi(\omega) &= G(1,-\omega),\\ \Phi'(\omega) &= -G_{z_2}(1,-\omega) = -p(p-1)\omega(1-\omega)^{p-2},\\ \Phi''(\omega) &= G_{z_2 z_2}(1,-\omega) = p(p-1)(1-\omega)^{p-3}[1-(p-1)\omega],\end{aligned}$$

$$(p-1)\Phi' - \omega\Phi'' = -p(p-1)(p-2)\omega(1-\omega)^{p-3};$$

$3_2)$
$$\begin{aligned}\Phi(\omega) &= G(\omega,1),\\ \Phi'(\omega) &= G_{z_1}(\omega,1) = p(\omega+1)^{p-2}(\omega-p+2),\\ \Phi''(\omega) &= G_{z_1 z_1}(\omega,1) = p(p-1)(\omega+1)^{p-3}(\omega-p+3),\end{aligned}$$

$$(p-1)\Phi' - \omega\Phi'' = -p(p-1)(p-2)(\omega+1)^{p-3};$$

$4_2)$
$$\begin{aligned}\Phi(\omega) &= G(1,\omega),\\ \Phi'(\omega) &= G_{z_1}(1,\omega) = -p(p-1)\omega(\omega+1)^{p-2},\\ \Phi''(\omega) &= G_{z_1 z_1}(\omega,-1) = -p(p-1)(\omega-1)^{p-3}[1+(p-1)\omega],\end{aligned}$$

$$(p-1)\Phi' - \omega\Phi'' = -p(p-1)(p-2)\omega(\omega+1)^{p-3}.$$

Note once more that we always take a power of nonnegative expressions. For example, in 4_1) above we must have $y_2 < 0$, so by (1.8.25), $\omega < 1$, hence $(1-\omega)^{p-3}$ is fine there. The same type of observation works for the other cases above. We see that in all cases, $\operatorname{sgn}[(p-1)\Phi' - \omega\Phi''] = -\operatorname{sgn}(p-2)$.

Therefore, in the first two cases, we have $D_2 < 0$ and this solution satisfies neither requirement (1.8.10) nor requirement (1.8.11). In the second two cases, we have $D_1 > 0$, and function M can be a candidate either for $\mathbf{M}_{\max}$ or for $\mathbf{M}_{\min}$, depending on the sign of the second derivative $M_{y_3 y_3}$.

Recall that (see (1.8.42))

$$M_{y_3 y_3} = \frac{p\omega^{p-2} R_1 H^2}{y_3^3}[\omega R_2 + (p-1)R_1].$$

In case 3_2), we have $\operatorname{sgn}[\omega R_2 + (p-1)R_1] = -\operatorname{sgn}(p-2)$, and therefore, we only need to know $\operatorname{sgn} R_1 = \operatorname{sgn} \Phi' = \operatorname{sgn} \frac{d}{d\omega} G(\omega, 1)$. Since this solution is considered only in sector $\frac{p-2}{p} y_1 < y_2 < y_1$ (see Figure 1.16), we have

$$G(y_1 + y_2, y_1 - y_2) = (2y_1)^{p-1}[py_2 - (p-2)y_1] > 0, \tag{1.8.45}$$

and ω, being the unique positive solution of the equation

$$G(\omega, 1) = (\omega + 1)^{p-1}[\omega - p + 1] = \frac{1}{y_3} G(y_1 + y_2, y_1 - y_2), \tag{1.8.46}$$

satisfies the condition $\omega > p - 1$. Therefore, $\operatorname{sgn} R_1 = \operatorname{sgn} \frac{d}{d\omega} G(\omega, 1) = \operatorname{sgn} p(\omega + 1)^{p-2}(\omega - p + 2) > 0$, and so $\operatorname{sgn} M_{y_3 y_3} = -\operatorname{sgn}(p-2)$, i.e., for $p > 2$ this is a candidate for $\mathbf{M}_{\max}$, and for $p < 2$ this is a candidate for $\mathbf{M}_{\min}$.

We are still considering the case 3_2). Recall that this function is defined not in the whole domain Ξ_+, but only in the sector $\frac{p-2}{p} y_1 < y_2 < y_1$. To get a solution everywhere, we need to "glue" this solution with the one we obtained when considering case 2) (see (1.8.17)):

$$M(y) = (y_1 + y_2)^p + C(y_3 - (y_1 - y_2)^p). \tag{1.8.47}$$

To glue this solution along the plane $y_2 = \frac{p-2}{p} y_1$ with the one we just obtained earlier, let us require that the resulting function be continuous everywhere. From (1.8.46) and (1.8.45), we see that $G(\omega, 1) = 0$ on this plane. Therefore, $\omega = p - 1$ and $M = \omega^p y_3 = (p-1)^p y_3$. Solution (1.8.47) has the same value on this plane for $C = (p-1)^p$.

Now we need to check that we have the correct continuation, in the sense that if the solution satisfies (1.8.10), then its continuation satisfies the same condition as well and if the solution satisfies (1.8.11), then the same is true for its continuation. The Hessian determinants will have the right sign automatically (actually $D_2 = 0$ identically). We only need to check the sign of

$$M_{y_1 y_1} = M_{y_2 y_2} = p(p-1)\big((y_1 + y_2)^{p-2} - (p-1)^p (y_1 - y_2)^{p-2}\big)$$

in the domain $-y_1 < y_2 < \frac{p-2}{p} y_1$ or, in the initial coordinates, for $0 < x_2 < (p-1)x_1$.

For $p > 2$, we have

$$(y_1 + y_2)^{p-2} = x_2^{p-2} < (p-1)^{p-2} x_1^{p-2} \\ < (p-1)^p x_1^{p-2} = (p-1)^p (y_1 - y_2)^{p-2},$$

and for $p < 2$, we have

$$(y_1 + y_2)^{p-2} = x_2^{p-2} > (p-1)^{p-2} x_1^{p-2} \\ > (p-1)^p x_1^{p-2} = (p-1)^p (y_1 - y_2)^{p-2}.$$

This means that M is a candidate for $\mathbf{M}_{\max}$ if $p > 2$, and a candidate for $\mathbf{M}_{\min}$, if $p < 2$, as it should be.

Let us rewrite expression (1.8.47) in the same form as was done in (1.8.46).

$$M - Cy_3 = (y_1 + y_2)^p - C(y_1 - y_2)^p = x_2^p - Cx_1^p. \tag{1.8.48}$$

Whence, if we change the definition of G a bit, defining it on the quadrant $z_i \geq 0$, as follows:

$$G_p(z_1, z_2) = \begin{cases} z_1^p - (p-1)^p z_2^p, & \text{if } z_1 \leq (p-1)z_2, \\ (z_1 + z_2)^{p-1}\big[z_1 - (p-1)z_2\big], & \text{if } z_1 \geq (p-1)z_2, \end{cases} \tag{1.8.49}$$

then we can write our two solutions M on Ξ_+ in an implicit form, as before:

$$G(y_1 + y_2, y_1 - y_2) = y_3 G(\omega, 1).$$

or solutions B on Ω_+

$$G(x_2, x_1) = G(B^{\frac{1}{p}}, x_3^{\frac{1}{p}}), \tag{1.8.50}$$

In case 4_2), we again consider exactly the same $G = G_p$ from (1.8.49). In a similar way, we can continuously glue the solution for the case 4_2) found in sector $-y_1 < y_2 < \frac{2-p}{p} y_1$:

$$G(1, \omega) = (\omega + 1)^{p-1}[1 - (p-1)\omega] = \frac{1}{y_3} G(y_1 - y_2, y_1 + y_2), \tag{1.8.51}$$

which is the same as

$$G(x_1, x_2) = G(x_3^{\frac{1}{p}}, B^{\frac{1}{p}}), \tag{1.8.52}$$

with the solution (1.8.47) along the line $y_2 = \frac{2-p}{p} y_1$. Here we have to take $C = (p' - 1)^p$ because on the line $y_2 = \frac{2-p}{p} y_1$, we have $G(1, \omega) = 0$, i.e., $\omega = p' - 1$. Now, in the sector $-y_1 < y_2 < \frac{2-p}{p} y_1$, we have

$$M_{y_3y_3} = \frac{p\omega^{p-2}R_1^2H^2}{y_3^3} \cdot \frac{p-2}{\omega+1}.$$

Therefore, $\operatorname{sgn} M_{y_3y_3} = \operatorname{sgn}(p-2)$, i.e., for $p < 2$ this is a candidate for $\mathbf{M}_{\max}$, and for $p > 2$ this is a candidate for $\mathbf{M}_{\min}$.

In the "dual" sector $x_2 > (p'-1)x_1$ (or $y_2 > \frac{2-p}{p}y_1$) for $p > 2$, we have

$$\begin{aligned}(y_1+y_2)^{p-2} = x_2^{p-2} &> (p'-1)^{p-2}x_1^{p-2} \\ &> (p'-1)^p x_1^{p-2} = (p'-1)^p(y_1-y_2)^{p-2},\end{aligned}$$

and for $p < 2$, we have

$$\begin{aligned}(y_1+y_2)^{p-2} = x_2^{p-2} &< (p'-1)^{p-2}x_1^{p-2} \\ &< (p'-1)^p x_1^{p-2} = (p'-1)^p(y_1-y_2)^{p-2}.\end{aligned}$$

This means that M is a candidate for $\mathbf{M}_{\max}$, if $p < 2$, and a candidate for $\mathbf{M}_{\min}$, if $p > 2$.

Using the same "generalized" definition (1.8.49) of function G, we can write our solutions M on Ξ_+ in an implicit form, as before:

$$G(y_1-y_2, y_1+y_2) = y_3G(1,\omega)$$

or solutions B on Ω_+

$$G(x_1, x_2) = G(x_3^{\frac{1}{p}}, B^{\frac{1}{p}}), \tag{1.8.53}$$

which should give us, as we said earlier, the candidate for $\mathbf{B}_{\max}$ for $p < 2$ and $\mathbf{B}_{\min}$ for $p > 2$. Notice that for $p > 2$ the candidate for, say, $\mathbf{B}_{\max}$ is given by equation (1.8.50).

It is a bit inconvenient to use one equation for, say, $B_{\max}$, if $p > 2$ (this will be (1.8.50)), and another one for the same $B_{\max}$, if $p < 2$ (this will be (1.8.53)). We note that after interchanging the roles of z_i and replacing p by p', we get the scalar multiple of the original expression in both lines of (1.8.49). This allows us to give one expression for $B_{\max}$ for all p, using notation of $p^* = \max\{p, p'\}$. In such a way, we come to formula (1.8.1) for F_p, where we introduce additional scalar coefficients to make this function not only continuous, but C^1-smooth everywhere in Ω_+. This smoothness guarantees us that the solution B is C^1-smooth as well.

1.8.4 Proof of Theorem 1.8.1. Verification Theorem

From now on, we shall denote by $B_{\max}$ the unique positive solution of the equation $F(|x_2|, |x_1|) = F(B^{\frac{1}{p}}, x_3^{\frac{1}{p}})$, and by $B_{\min}$ we shall denote the unique positive solution of the equation $F(|x_1|,|x_2|) = F(x_3^{\frac{1}{p}}, B^{\frac{1}{p}})$, where function $F = F_p$ is defined in (1.8.1). Existence and uniqueness of these solutions follow from the fact that $F(z_1, z_2)$ is strictly increasing in z_1 from $-p^{*(p-1)}(p^* - 1)^p z_2^p$ to $+\infty$, as z_1 runs from 0 to $+\infty$, and it is strictly decreasing in z_2 from $p(p^* - 1)^{p-1} z_2^p$ to $-\infty$, as z_2 runs from 0 to $+\infty$. Indeed, the first partial derivatives of F are

$$F_{z_1} = \begin{cases} p z_1^{p-1}, & \text{if } z_1 \le (p^*-1)z_2, \\ p(1-\frac{1}{p^*})^{p-1}(z_1+z_2)^{p-2}\big[pz_1 - ((p-1)(p^*-1)-1)z_2\big], & \text{if } z_1 \ge (p^*-1)z_2; \end{cases} \tag{1.8.54}$$

$$F_{z_2} = \begin{cases} -(p^*-1)^p p z_2^{p-1}, & \text{if } z_1 \le (p^*-1)z_2, \\ -p(1-\frac{1}{p^*})^{p-1}(z_1+z_2)^{p-2}\big[(p^*-p)z_1 + p(p^*-1)z_2\big], & \text{if } z_1 \ge (p^*-1)z_2. \end{cases} \tag{1.8.55}$$

Note that both derivatives are continuous everywhere, even at the origin, where they vanish. Moreover, $F_{z_1} > 0$ if $z_1 > 0$, and $F_{z_2} < 0$ if $z_2 > 0$, i.e., F is strictly increasing in z_1 and strictly decreasing in z_2.

In the case of $B_{\max}$, we look for a solution of the equation

$$F(B^{\frac{1}{p}}, x_3^{\frac{1}{p}}) = F(|x_2|, |x_1|)$$

or

$$F(\omega, 1) = \frac{1}{x_3} F(|x_2|, |x_1|).$$

Thus, we get a continuous solution $\omega(x)$ everywhere, except in the plane $x_3 = 0$, where ω is not defined. But we can easily estimate the behavior of ω near the line $x_3 = x_1 = 0$. Note that the intersection of plane $x_3 = 0$ with the domain Ω consists of that line only.

Since F is decreasing in z_2 and $0 \le |x_1| \le x_3^{\frac{1}{p}}$, we have

$$F(\omega, 1) = F\Big(\frac{|x_2|}{x_3^{1/p}}, \frac{|x_1|}{x_3^{1/p}}\Big) \ge F\Big(\frac{|x_2|}{x_3^{1/p}}, 1\Big).$$

Using the fact that F is increasing in z_1, we get

$$\omega \ge \frac{|x_2|}{x_3^{1/p}}.$$

To estimate ω from above, we introduce ω_0, which is the unique solution of the equation

$$(\omega_0+1)^{p-1}(\omega_0-p^*+1)=\frac{|x_2|^p}{x_3}$$

on interval $(p^*-1,+\infty)$. In other words, ω_0 is a solution of the equation

$$F(\omega_0,1)=F\Big(\frac{|x_2|}{x_3^{1/p}},0\Big).$$

Once again, using the fact that F is decreasing in z_2, we get

$$F(\omega_0,1)=F\Big(\frac{|x_2|}{x_3^{1/p}},0\Big)\geq F\Big(\frac{|x_2|}{x_3^{1/p}},\frac{|x_1|}{x_3^{1/p}}\Big)=F(\omega,1)$$

whence $\omega<\omega_0$, because F is increasing in z_1. Furthermore,

$$(\omega_0-p^*+1)^p\leq(\omega_0+1)^{p-1}(\omega_0-p^*+1)=\frac{|x_2|^p}{x_3},$$

i.e.,

$$\omega_0\leq p^*-1+\frac{|x_2|}{x_3^{1/p}}.$$

Therefore, for $B=\omega^p x_3$, we have the following estimate:

$$|x_2|^p\leq B\leq\big(|x_2|+(p^*-1)x_3^{1/p}\big)^p,$$

which gives us the continuity near $x_3=0$. Thus, the solution $B_{\max}$ is continuous in the closure of Ω.

Similar considerations give us the continuity of $B_{\min}$. In that case, we have the equation

$$F(1,\omega)=\frac{1}{x_3}F(|x_1|,|x_2|),$$

and hence,

$$F\Big(0,\frac{|x_2|}{x_3^{1/p}}\Big)\leq F(1,\omega)=F\Big(\frac{|x_1|}{x_3^{1/p}},\frac{|x_2|}{x_3^{1/p}}\Big)\leq F\Big(1,\frac{|x_2|}{x_3^{1/p}}\Big).$$

Now $F(1,\omega)$ is decreasing in ω, therefore,

$$\frac{|x_2|}{x_3^{1/p}}\leq\omega\leq\omega_0,$$

where ω_0 is the solution of the equation

$$1-(p^*-1)^p\omega_0^p=-(p^*-1)^p\frac{|x_2|^p}{x_3},$$

i.e., $\omega_0^p = (p^* - 1)^{-p} + |x_2|^p/x_3$, and for $B = \omega^p x_3$, we have the following estimate:

$$|x_2|^p \le B \le |x_2|^p + (p^* - 1)^{-p} x_3,$$

which gives us continuity near $x_3 = 0$. Thus, the solution $B_{\min}$ is continuous in the closed domain Ω as well.

The first step of the proof is to check that the main inequality (i.e., concavity (1.8.8) for candidate $B_{\max}$, and convexity (1.8.9) for candidate $B_{\min}$) is fulfilled, if points x^+, x^- satisfy the extra condition on their coordinates:

$$|x_1^+ - x_1^-| = |x_2^+ - x_2^-|. \tag{1.8.56}$$

This was almost done in the preceding section, when we constructed these candidates. We know that the Hessians of our candidates have the required signs everywhere in the domain Ω, except, possibly, in the planes $x_1 = 0$, $x_2 = 0$, and, either $|x_2| = (p^* - 1)|x_1|$ for $B_{\max}$, or $|x_1| = (p^* - 1)|x_2|$ for $B_{\min}$. On these hyperplanes, our solutions are not C^2-smooth, but they are C^1-smooth, and that is quite sufficient for global convexity/concavity.

Let us consider three points x, x^+, x^-, satisfying the following relations:

$$\begin{gathered} |x_1^+ - x_1^-| = |x_2^+ - x_2^-|, \quad x_3^+ \ge |x_1^+|^p, \quad x_3^- \ge |x_1^-|^p, \\ x = \alpha^- x^- + \alpha^+ x^+, \quad \alpha^- + \alpha^+ = 1, \quad \alpha^\pm > 0. \end{gathered} \tag{1.8.57}$$

We know that for our solution $B_{\max}$, the following main inequality (zigzag concavity) holds:

$$B(x) - \alpha^- B(x^-) - \alpha^+ B(x^+) \ge 0, \tag{1.8.58}$$

and the opposite main inequality (zigzag convexity):

$$B(x) - \alpha^- B(x^-) - \alpha^+ B(x^+) \le 0 \tag{1.8.59}$$

is true for the solution $B_{\min}$.

LEMMA 1.8.4 *If function B, continuous in Ω, satisfies the main inequality* (1.8.58) *and the boundary restriction $B(x_1, x_2, |x_1|^p) \ge |x_2|^p$, then $B \ge \mathbf{B}_{\max}$. If it satisfies* (1.8.59) *and $B(x_1, x_2, |x_1|^p) \le |x_2|^p$, then $B \le \mathbf{B}_{\min}$.*

PROOF Let $I = [0, 1]$ and let $\mathcal{I}$ be a tree of subintervals that is obtained by splitting each interval into two of its children, not obligatorily of equal lengths. The only requirement for the splitting is that its ratio $|J_\pm|/|J|$ should be uniformly separated from 0 and from 1 over the whole tree. The dyadic tree, where this ratio is always $\frac{1}{2}$, is a partial case of the general tree. We denote the set of subintervals of the n-th generation by $\mathcal{I}_n$, i.e., $\mathcal{I}_0 = \{I\}$. Thus, $\mathcal{I}_n$ is the set of children of elements from $\mathcal{I}_{n-1}$. For a given tree $\mathcal{I}$, we fix two bounded measurable test functions φ, ψ on I such that $|(\psi, h_J)| = |(\varphi, h_J)|$ for any

$J \in \mathcal{I}$. Since the test functions are fixed, we omit them in notation of the Bellman point:

$$\mathfrak{b}_J = (\langle\varphi\rangle_J, \langle\psi\rangle_J, \langle|\varphi|^p\rangle_J).$$

The fact that $|(\psi, h_J)| = |(\varphi, h_J)|$ guarantees that $\mathfrak{b}_{J\pm}$ satisfies assumptions of (1.8.57), and we can rewrite inequalities (1.8.58) and (1.8.59) in the form

$$|J|\,B(\mathfrak{b}_J) - |J^-|\,B(\mathfrak{b}_{J^-}) - |J^+|\,B(\mathfrak{b}_{J^+}) \geq 0, \tag{1.8.60}$$

$$|J|\,B(\mathfrak{b}_J) - |J^-|\,B(\mathfrak{b}_{J^-}) - |J^+|\,B(\mathfrak{b}_{J^+}) \leq 0. \tag{1.8.61}$$

So, adding up all our inequalities (1.8.60) with $J \in \mathcal{I}_{k-1}$, we get

$$\sum_{J\in\mathcal{I}_{k-1}} |J|\,B(\mathfrak{b}_J) \geq \sum_{J\in\mathcal{I}_k} |J|\,B(\mathfrak{b}_J).$$

Adding up these inequalities over k from 1 to n, we get

$$B(\mathfrak{b}_I) \geq \sum_{J\in\mathcal{I}_n} |J|\,B(\mathfrak{b}_J) = \int_0^1 B(x^{(n)}(t))\,dt,$$

where $x^{(n)}(t)$ is the step function equal to $\mathfrak{b}_J$ for all $t \in J$, $J \in \mathcal{I}_n$.

Notice that $\langle f\rangle_J \to f(t)$ almost everywhere, when J runs over a family of nested intervals, shrinking to point t for an arbitrary summable function f. Therefore, almost everywhere

$$x^{(n)}(t) = (\langle\varphi\rangle_J, \langle\psi\rangle_J, \langle|\varphi|^p\rangle_J) \to (\varphi(t), \psi(t), |\varphi(t)|^p) \qquad \text{as } n \to \infty,$$

and since B is continuous, we have

$$B(x^{(n)}(t)) \to B(\varphi(t), \psi(t), |\varphi(t)|^p) \geq |\psi(t)|^p.$$

Now, using the Lebesgue dominant convergence theorem, we come to the estimate

$$\langle|\psi|^p\rangle_I \leq B(x)$$

for every pair of bounded measurable test functions φ, ψ. And finally, approximating arbitrary φ, $\psi \in L^p(I)$ by simple functions (i.e., constant on all $J \in \mathcal{I}_n$ for some n), we can extend this inequality to the set of arbitrary possible test functions φ and ψ, which means that $\mathbf{B}_{\max}(x) \leq B(x)$.

In all these considerations, for the case of $\mathbf{B}_{\min}$, we only need to change the sign of inequalities, in order to get $\mathbf{B}_{\min}(x) \geq B(x)$ for B, satisfying (1.8.59).

□

We are left to prove the opposite inequalities:

$$\mathbf{B}_{\max} \geq B_{\max}(x) \quad \text{and} \quad \mathbf{B}_{\min} \leq B_{\min}(x).$$

This can be done by reversing the reasoning in the lemma as we have just proved. Using the fact that domain $\Omega = \{x = (x_1, x_2, x_3) \colon x_3 \geq |x_1|^p\}$

is foliated by the straight-line segments (extremal trajectories), it is possible to construct the sequence of test functions φ_n, ψ_n, corresponding to any given point $x \in \Omega$ such that $\langle |\psi_n|^p \rangle_I \to B(x)$. This supplies us with the required inequality. The main idea is to travel along the extremal trajectories, starting from $x \in \Omega$, to build a net $\mathcal{N} \stackrel{\text{def}}{=} \{x^+, x^-, x^{++}, x^{+-}, x^{--}, x^{-+}, \ldots\}$. All points of the net should belong to Ω, and for a while we put them on the same extremal trajectory on which x lies. If one of them, say z, hits the boundary $\partial\Omega$, we stop building children z^+, z^-. If a point, say ζ, hits the special hyperplanes $x_1 = 0$ or $x_2 = 0$, we have to leave the extremal trajectory. In that case, we shall choose children ζ^+, ζ^- in such a way that they lie in different quadrants, very close to ζ. Then we start anew the building of the net for ζ^+ and ζ^- separately. The closer ζ^+, ζ^- are to ζ, the smaller will be the difference $\langle |\psi_n|^p \rangle_I - B(x)$. In this way, for arbitrary ε, we obtain the inequalities

$$\mathbf{B}_{\max}(x) \geq B_{\max}(x) - \varepsilon \quad \text{and} \quad \mathbf{B}_{\min}(x) \leq B_{\min}(x) + \varepsilon.$$

Following the described procedure, the reader can construct a net $\mathcal{N}$ that generates a required pair of functions φ, ψ, approximating the value of the Bellman function. In the proof of the following lemma, we only state the result of the described construction that supplies us with a recursive definition of φ and ψ.

LEMMA 1.8.5 *Functions $B_{\max}$ and $B_{\min}$ satisfy the inequalities*:

$$\mathbf{B}_{\max}(x) \geq B_{\max}(x) \quad \textit{and} \quad \mathbf{B}_{\min}(x) \leq B_{\min}(x).$$

PROOF We construct an extremal sequence of pairs φ, ψ for the case 3_2) on the plane $x_1 = 0$, i.e., this will be an extremizer for function $\mathbf{B}_{\max}(x)$, if $p > 2$, and for function $\mathbf{B}_{\min}(x)$, if $p < 2$. The similar construction for the case 4_2) works on the plane $x_2 = 0$.

We will not split the points strictly along the extremal trajectories, but close to them such that they are tending to the extremal trajectories in the limit. However, our splitting has to go strictly either in the plane (y_1, y_3) or in (y_2, y_3) to get an admissible pair. Take a small ε and make the first splitting $x_1^\pm = x_1 \pm \varepsilon = \pm\varepsilon$ (because $x_1 = 0$) and $x_2^\pm = x_2 \mp \varepsilon$. After that, we jump from x^- and from all its right successors to the initial point x, and make it so that the n-th right successor hits the boundary $x_3 = |x_1|^p$. Thus we come to the boundary point $(c, x_2 - c, c^p)$, where $c = 2^n\varepsilon$. Indeed, since $x_1^+ = \varepsilon$ for the first coordinate of the right successors are 2ε, $4\varepsilon, \ldots, 2^k\varepsilon, \ldots$, and finally, $2^n\varepsilon$ for the n-th successor. We have to go strictly in the plane $x_1 + x_2 = \text{const}$; therefore, the second coordinates of the right successors are $x_2 - 2\varepsilon$, $x_2 - 4\varepsilon$, $\ldots, x_2 - 2^n\varepsilon = x_2 - c \stackrel{\text{def}}{=} d_+$. As a result, we get the following picture on the right half of the interval:

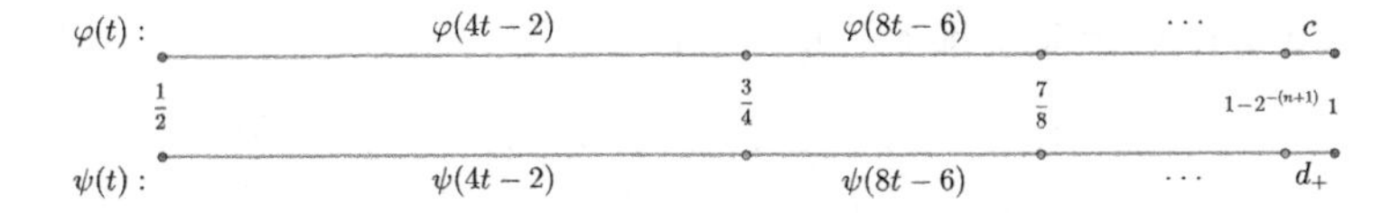

Figure 1.18 Construction of an extremal pair on I^+.

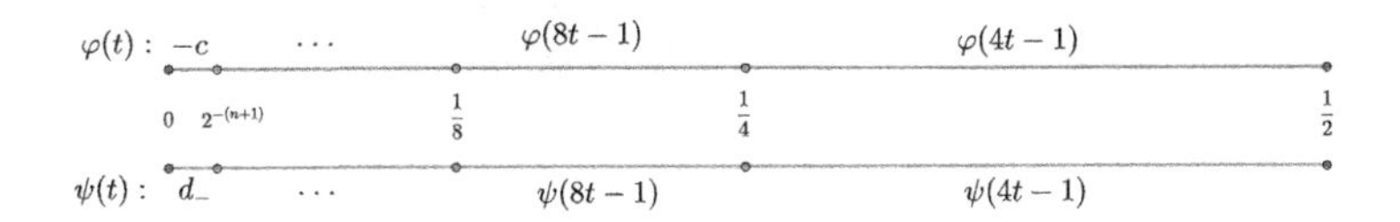

Figure 1.19 Construction of an extremal pair on I^-.

The left half-interval I^- we split symmetrically. The first coordinates of the left successors of the point x^- will be $-2\varepsilon,\ \ -4\varepsilon, \ldots, -2^k\varepsilon, \ldots$, and finally $-2^n\varepsilon = -c$. Their second coordinates will be 2ε times bigger than the second coordinates of their counterparts, i.e., $x_2,\ \ x_2 - 2\varepsilon,\ \ x_2 - 6\varepsilon, \ldots,$ $x_2-(2^n-2)\varepsilon = x_2-c+2\varepsilon \stackrel{\text{def}}{=} d_-$. As on the left half interval, the right jump for all successors will be the same: it will be the point $(0, x_2 + 2\varepsilon, *)$ on the plane $x_1 = 0$. The third coordinate is at our disposal. We choose it in such a way that a pair of test functions for it is proportional to the pair for the initial point x, i.e., $\big(0, x_2 + 2\varepsilon, \big(1 + \frac{2\varepsilon}{x_2}\big)^p x_3\big)$. The result of this splitting is shown in Figure 1.19.

Now only parameter c is left for us to specify. To choose it, we recall that the equality $\langle\, |\varphi|^p\rangle = x_3$ should be fulfilled (the conditions $\langle\varphi\rangle = x_1 = 0$ and $\langle\psi\rangle = x_2$ are fulfilled by construction):

$$x_3 = \langle\, |\varphi|^p\rangle = 2^{-n}c^p + \big(\tfrac{1}{2} - 2^{-(n+1)}\big)\Big[x_3 + \big(1 + \tfrac{2\varepsilon}{x_2}\big)^p x_3\Big].$$

Therefore, c has to depend on n and to satisfy the following equation:

$$c^p = \left[1 - (1 - 2^{-n})\frac{(1 + \frac{2^{1-n}c}{x_2})^p - 1}{2^{1-n}}\right]x_3. \tag{1.8.62}$$

In the limit as $n \to \infty$, this equation turns into

$$c^p = \left[1 - \frac{cp}{x_2}\right]x_3, \tag{1.8.63}$$

which clearly has a unique positive solution. Therefore, for n big enough, equation (1.8.62) also has a unique solution $c = c_n$, close to the solution $c = c_\infty$ of (1.8.63).

Let us now calculate the average $\langle\,|\psi|^p\rangle$:

$$\langle\,|\psi|^p\rangle = 2^{-(n+1)}(d_+^p + d_-^p) + \left(\tfrac{1}{2} - 2^{-(n+1)}\right)\Big[1 + \big(1 + \frac{2\varepsilon}{x_2}\big)^p\Big]\langle|\psi|^p\rangle,$$

whence

$$\langle\,|\psi|^p\rangle = \frac{2^{-(n+1)}(d_+^p + d_-^p)}{1 - \left(\tfrac{1}{2} - 2^{-(n+1)}\right)\Big[1 + \big(1 + \frac{2\varepsilon}{x_2}\big)^p\Big]}.$$

Let $n \to \infty$, then we have $d_\pm \to x_2 - c_\infty$, $\varepsilon \to 0$, and

$$\langle\,|\psi|^p\rangle \to \frac{(x_2 - c_\infty)^p}{1 - \frac{pc_\infty}{x_2}} = \frac{(x_2 - c_\infty)^p}{c_\infty^p} x_3.$$

If we introduce $\omega \stackrel{\text{def}}{=} \frac{x_2 - c_\infty}{c_\infty}$ and plug it into (1.8.63), then we see that ω verifies the equation

$$\left(\frac{x_2}{\omega+1}\right)^p = \left(1 - \frac{p}{\omega+1}\right) x_3,$$

or

$$x_2^p = (\omega+1)^{p-1}(\omega - p + 1)x_3.$$

Thus, we see that ω is just the solution of the equation

$$G(x_2, 0) = x_3 G(\omega, 1),$$

and

$$\lim_{n\to\infty} \langle\,|\psi|^p\rangle = \omega^p x_3 = B(0, x_2, x_3).$$

Therefore, we have constructed a sequence of pairs of test functions supporting the value $B(0, x_2, x_3)$, which proves that $\mathbf{B}_{\max}(0, x_2, x_3) \geq B_{\max}(0, x_2, x_3)$, if $p > 2$, and $\mathbf{B}_{\min}(0, x_2, x_3) \leq B_{\min}(0, x_2, x_3)$, if $p < 2$.

Now we consider the inner points, i.e., $x_1 > 0$. If $x_2 < (p-1)x_1$, then there exists an extremal line passing through x with the endpoints on the boundary $x_3 = x_1^p$ and on the plane $x_1 = 0$. We know that at both endpoints, the functions $\mathbf{B}$ and B coincide. Therefore, concavity/convexity of $\mathbf{B}$ and linearity of B on this line guarantees us the desired inequality along the whole line.

If $x_2 > (p-1)x_1$, then the extremal passing through x is infinite, it goes parallel to x_3 axis. And we apply here the same trick that had been used in Section 1.6 when working with vertical extremal lines of the Bellman function for the maximal operator. Namely, we take a number ξ, $\xi > x_1$, and draw a line in the plane $x_1 + x_2 = \text{const}$ through the points x and $(\xi, x_1 + x_2 - \xi, \xi^p)$. This line hits the plane $x_1 = 0$ at the point $\bar{x} = (0, x_1 + x_2, \bar{x}_3)$, where $\bar{x}_3 = \frac{\xi x_3 - \xi^p x_1}{\xi - x_1}$.

To be precise, let us assume that we deal with the function $\mathbf{B} = \mathbf{B}_{\max}$ (for $\mathbf{B}_{\min}$, we will simply have the opposite inequality). Then due to concavity of $\mathbf{B}_{\max}$ we have

$$\mathbf{B}(x) \geq (1 - \frac{x_1}{\xi})\mathbf{B}(\bar{x}) + \frac{x_1}{\xi}(x_1 + x_2 - \xi)^p.$$

For the point $\bar{x}$, the extremizer has been constructed, and we know that $\mathbf{B}(\bar{x}) = \omega^p \bar{x}_3$, where ω is the solution of the following equation:

$$G(\bar{x}_2, \bar{x}_1) = G(\omega, 1)\bar{x}_3.$$

Since $\bar{x}_1 = 0$ and $\bar{x}_2 = x_1 + x_2$, i.e., $G(\bar{x}_2, \bar{x}_1) = (x_1 + x_2)^p$ and $G(\omega, 1) = (\omega + 1)^{p-1}(\omega - p + 1)$, we conclude that ω satisfies the following equation:

$$(x_1 + x_2)^p = (\omega + 1)^{p-1}(\omega - p + 1)\frac{\xi x_3 - \xi^p x_1}{\xi - x_1}.$$

Therefore, $\omega \to p - 1$ as $\xi \to x_1$.

Let us return to our estimate:

$$\begin{aligned}\mathbf{B}(x) &\geq \frac{\xi - x_1}{\xi} \cdot \mathbf{B}(\bar{x}) + \frac{x_1}{\xi} \cdot (x_1 + x_2 - \xi)^p \\ &= \frac{\xi - x_1}{\xi} \cdot \omega^p \cdot \frac{\xi x_3 - \xi^p x_1}{\xi - x_1} + \frac{x_1}{\xi}(x_1 + x_2 - \xi)^p \\ &= \omega^p(x_3 - \xi^{p-1}x_1) + \frac{x_1}{\xi}(x_1 + x_2 - \xi)^p \\ &\xrightarrow[\xi \to x_1]{} (p-1)^p(x_3 - x_1^p) + x_2^p = B(x).\end{aligned}$$

Thus, we have proved that $\mathbf{B}_{\max} \geq B_{\max}$ for $p > 2$ in the domain $x_1 \geq 0$, $x_2 \geq 0$. By symmetry, the inequality is true everywhere. All other cases are absolutely similar. □

1.8.5 Function u_p from Function B

We found Burkholder's functions $\mathbf{B}_{\max}$ and $\mathbf{B}_{\min}$, as claimed in Theorem 1.8.1. As an immediate corollary, we get the sharp constant in Burkholder's inequality:

THEOREM 1.8.6 *Let $\langle \varphi \rangle_I = x_1$, $\langle \psi \rangle_I = x_2$, and let ψ be a martingale transform of φ, and $|x_2| \leq |x_1|$. Then*

$$\langle\, |\psi|^p \rangle_I \le (p^* - 1)^p \langle\, |\varphi|^p \rangle_I .$$

The constant $p^* - 1$ is sharp.

Proof If we analyze the form of function $\mathbf{B}_{\max}$ from Theorem 1.8.1, we immediately see that

$$\sup_{x \in \Omega,\, |x_2| \le |x_1|} \frac{\mathbf{B}_{\max}(x_1, x_2, x_3)}{x_3} = (p^* - 1)^p .$$

□

Theorem 1.8.7 *Let $\langle \varphi \rangle_I = x_1$, $\langle \psi \rangle_I = x_2$, and let ψ be a martingale transform of φ, and $|x_2| \le |x_1|$. Then*

$$\langle\, |\varphi|^p \rangle_I \le (p^* - 1)^p \langle\, |\psi|^p \rangle_I .$$

The constant $p^* - 1$ is sharp.

Proof If we analyze the form of function $\mathbf{B}_{\min}$ from Theorem 1.8.1, we immediately see that

$$\inf_{x \in \Omega,\, |x_2| \ge |x_1|} \frac{\mathbf{B}_{\min}(x_1, x_2, x_3)}{x_3} = (p^* - 1)^{-p} .$$

□

Remark The same analysis shows that $\langle\, |\psi|^p \rangle_I \le (p^* - 1)^p \langle\, |\varphi|^p \rangle_I$ if and only if $|x_2| \le (p^* - 1)|x_1|$ in Theorem 1.8.6, and $\langle\, |f|^p \rangle_I \le (p^* - 1)^p \langle\, |g|^p \rangle_I$ if and only if $|x_2| \ge (p^* - 1)^{-1}|x_1|$ in Theorem 1.8.7.

Notation Next, we use $\beta_p \stackrel{\text{def}}{=} (p^* - 1)^p$. Put

$$\phi_{\max}(x_1, x_2) \stackrel{\text{def}}{=} \sup_{x_3 :\, (x_1, x_2, x_3) \in \Omega} \big[\mathbf{B}_{\max}(x_1, x_2, x_3) - \beta_p x_3\big],$$

$$\phi_{\min}(x_1, x_2) \stackrel{\text{def}}{=} \inf_{x_3 :\, (x_1, x_2, x_3) \in \Omega} \big[\mathbf{B}_{\min}(x_1, x_2, x_3) - \beta_p^{-1} x_3\big].$$

These functions are defined on the whole $\mathbb{R}^2$.

Definition If for all pairs of points $x^\pm \in \mathbb{R}^2$, such that

$$|x_1^+ - x_1^-| = |x_2^+ - x_2^-| \quad \text{and} \quad x = \frac{x^+ + x^-}{2}, \tag{1.8.64}$$

function ϕ on $\mathbb{R}^2$ satisfies the condition

$$\phi(x) - \frac{\phi(x^-) + \phi(x^+)}{2} \ge 0, \tag{1.8.65}$$

then it is called zigzag concave. If the opposite inequality holds

$$\phi(x) - \frac{\phi(x^-) + \phi(x^+)}{2} \leq 0, \tag{1.8.66}$$

function ϕ is called zigzag convex. The next theorem gives us an independent description of $\phi_{\max}$ and $\phi_{\min}$.

THEOREM 1.8.8 *Function $\phi_{\max}$ is the least zigzag concave majorant of function $h_{\max}(x) \stackrel{\text{def}}{=} |x_2|^p - \beta_p|x_1|^p$. Function $\phi_{\min}$ is the greatest zigzag convex minorant of function $h_{\min}(x) \stackrel{\text{def}}{=} |x_2|^p - \beta_p^{-1}|x_1|^p$.*

REMARK Notice that this is slightly counterintuitive: $\mathbf{B}_{\max}(x) - \beta_p x_3$ is a zigzag concave for any fixed x_3, and the supremum of concave functions usually is *not* concave. The same is true about infimum of convex functions.

PROOF Let $x^\pm$ and x be as in (1.8.64). It is obvious that $\phi_{\max}$ is a zigzag concave. One verifies that just by definition. In fact, if for any $x^\pm \in \mathbb{R}^2$, we can choose $x_3^\pm$, such that the supremum in the definition of $\phi_{\max}$ is almost attained, i.e., $\mathbf{B}_{\max}(x^\pm) - \beta_p x_3^\pm > \phi_{\max}(x^\pm) - \varepsilon$ for a given ε, then we define $x_3 = \frac{x_3^- + x_3^+}{2}$ and $\tilde{x} = (x, x_3)$. Now, using (1.8.58) we can write

$$\begin{aligned}
\phi_{\max}(x) &\geq \mathbf{B}_{\max}(\tilde{x}) - \beta_p x_3 \\
&\geq \frac{\mathbf{B}_{\max}(\tilde{x}^-) + \mathbf{B}_{\max}(\tilde{x}^+)}{2} - \beta_p \frac{x_3^- + x_3^+}{2} \\
&\geq \frac{\phi_{\max}(x^-) + \phi_{\max}(x^+)}{2} - \varepsilon,
\end{aligned}$$

which yields (1.8.65). Inequality (1.8.66) is totally similar.

As sup is not smaller than lim, we conclude

$$\phi_{\max}(x) \geq \lim_{x_3 \to |x_1|^p} \left[\mathbf{B}_{\max}(\tilde{x}) - \beta_p x_3\right] = |x_2|^p - \beta_p|x_1|^p = h_{\max}(x).$$

Since inf is not bigger than lim, we can analogously get

$$\phi_{\min}(x) \leq \lim_{x_3 \to |x_1|^p} \left[\mathbf{B}_{\min}(\tilde{x}) - \beta_p^{-1} x_3\right] = |x_2|^p - \beta_p^{-1}|x_1|^p = h_{\min}(x).$$

That is because the boundary values of $\mathbf{B}_{\max}$ and $\mathbf{B}_{\min}$ are $|x_2|^p$.

We are left to see that $\phi_{\max}$ is the *least* of such a majorant (and a symmetric claim for $\phi_{\min}$). Let λ be a zigzag concave function such that

$$\phi_{\max} \geq \lambda \geq h_{\max}. \tag{1.8.67}$$

Consider function $\Lambda(\tilde{x}) \stackrel{\text{def}}{=} \lambda(x) + \beta_p x_3$. It immediately follows that Λ satisfies (1.8.58). On the boundary of Ω, we have $\Lambda(x) \geq |x_2|^p$.

That is just by the right-hand side of (1.8.67). Then Lemma 1.8.4 yields

$$\Lambda(\tilde{x}) \geq \mathbf{B}_{\max}(\tilde{x}).$$

Then, obviously,

$$\lambda(x) = \sup_{x_3:\, \tilde{x}\in\Omega} \big[\Lambda(\tilde{x}) - \beta_p x_3\big] \geq \sup_{x_3:\, \tilde{x}\in\Omega} \big[\mathbf{B}_{\max}(\tilde{x}) - \beta_p x_3\big] = \phi_{\max}(x).$$

So we proved that $\phi_{\max}$ is the least zigzag concave majorant of $h_{\max}$. Symmetric consideration will bring us to the conclusion that $\phi_{\min}$ is the largest zigzag convex minorant of $h_{\min}$. □

The reader should now look at function F_p from Theorem 1.8.1. It would be interesting to obtain the formulas for $\phi_{\max}$ and $\phi_{\min}$, especially using this F_p. It would also be interesting to understand the role of function

$$u_p(x_1, x_2) \stackrel{\text{def}}{=} p(1 - \tfrac{1}{p^*})^{p-1}(|x_1| + |x_2|)^{p-1}(|x_2| - (p^* - 1)|x_1|), \tag{1.8.68}$$

used repeatedly by Burkholder. Maybe it is equal to $\phi_{\max}$? The answer is "no," but it coincides with the nontrivial part of $\phi_{\max}$, where $\phi_{\max}$ does not coincide with the obstacle function. Namely, we prove the following:

THEOREM 1.8.9

$$\phi_{\max}(x_1, x_2) = F_p(|x_2|, |x_1|). \tag{1.8.69}$$

PROOF We shall consider only the case $p > 2$, the case $p < 2$; is similar. Due to symmetry between the change of x_1 to $-x_1$ and x_2 to $-x_2$, it is enough to check equality (1.8.69) in the quadrant $x_1 > 0,\ \ x_2 > 0$. If $x_2 \leq (p-1)x_1$, we get an explicit formula for $\mathbf{B}_{\max}$ from Theorem 1.8.1: $\mathbf{B}_{\max}(\tilde{x}) = x_2^p + (p-1)^p(x_3 - x_1^p)$, and therefore,

$$\phi_{\max}(x_1, x_2) = \sup_{x_3:\, \tilde{x}\in\Omega} \big[\mathbf{B}_{\max}(\tilde{x}) - (p-1)^p x_3\big] = x_2^p - (p-1)^p x_1^p = F_2(x_2, x_1).$$

So in the rest of the proof, we shall consider only the domain $\{x = (x_1, x_2)\colon 0 \leq (p-1)x_1 < x_2\}$. Moreover, since both functions $\phi_{\max}$ and F_p are p-homogeneous, it is sufficient to check (1.8.69) on the interval $S \stackrel{\text{def}}{=} \{x\colon 0 \leq px_1 < 1, x_1 + x_2 = 1\}$. (Indeed, condition $px_1 < 1$ on line $x_1 + x_2 = 1$ means $x_2 > (p-1)x_1$.)

Function F_p is linear on S: $F_p(1 - x_1, x_1) = p^{2-p}(p-1)^{p-1}(1 - px_1)$. Now we check that $\phi_{\max}$ is linear as well. To this end, we check the inequality

$$\phi_{\max}(x) \leq \frac{\phi_{\max}(x_1 - a, x_2 + a) + \phi_{\max}(x_1 + a, x_2 - a)}{2} \tag{1.8.70}$$

for all $x \in S$ and sufficiently small a, which just means linearity of $\phi_{\max}$ on S, because the opposite inequality follows from the zigzag concavity of $\phi_{\max}$ (1.8.65).

Fix $x \in S$ and $\varepsilon > 0$. Take x_3 such that $B(\tilde{x}) - \beta_p x_3 \geq \phi_{\max}(x) - \varepsilon$. Due to condition $x_2 > (p-1)x_1$, the extremal trajectory L_x of $\mathbf{B}_{\max}$, passing through point $\tilde{x} = (x, x_3)$ is not vertical: at some point, it hits the plane $x_1 = 0$. Therefore, we can take two different points $\tilde{x}^{\pm} = (x^{\pm}, x_3^{\pm})$ on L_x such that $\tilde{x} = \frac{1}{2}(\tilde{x}^+ + \tilde{x}^-)$. We know three things:

$$\begin{aligned}
\mathbf{B}_{\max}(\tilde{x}) - \beta_p x_3 &\geq \phi_{\max}(x) - \varepsilon, \\
\mathbf{B}_{\max}(\tilde{x}^+) - \beta_p x_3^+ &\leq \phi_{\max}(x^+), \\
\mathbf{B}_{\max}(\tilde{x}^-) - \beta_p x_3^- &\leq \phi_{\max}(x^-).
\end{aligned}$$

Since function $\mathbf{B}_{\max}$ is linear along L_x, we can write the following chain of inequalities:

$$\begin{aligned}
\phi_{\max}(x) - \varepsilon &\leq \mathbf{B}_{\max}(\tilde{x}) - \beta_p x_3 \\
&= \frac{\left[\mathbf{B}_{\max}(\tilde{x}^+) - \beta_p x_3^+\right] + \left[\mathbf{B}_{\max}(\tilde{x}^-) - \beta_p x_3^-\right]}{2} \\
&\leq \frac{\phi_{\max}(x^+) + \phi_{\max}(x^-)}{2}.
\end{aligned}$$

Since ε is arbitrary, we come to the desired convexity (1.8.70).

Function $F_p(x_2, x_1)$ is a concave C^1-smooth function majorizing $h_{\max}$ on S. This immediately follows from its formula. Functions $\phi_{\max}(x_1, x_2)$ and $F_p(x_2, x_1)$ are linear on S, and at point $x = w_p \stackrel{\text{def}}{=} (\frac{1}{p}, 1 - \frac{1}{p})$ both are equal to $h_{\max}(w_p) = 0$. Therefore, to prove that they are identical, it is sufficient to check that their derivatives at w_p along S are equal as well. Since $\phi_{\max}$ is a majorant of $h_{\max}$ and both functions are equal at w_p, the left derivative of $\phi_{\max}$ at w_p is not greater than the derivative of $h_{\max}$ at that point. On the other hand, since $\phi_{\max}$ is the least majorant, it is not greater than F_p, i.e., its left derivative at w_p is not less than the derivative of F_p there. The latter coincides with the derivative of $h_{\max}$. Hence, all three derivatives along S are equal at point w_p, and we proved $\phi_{\max}(x_1, x_2) = F_p(|x_2|, |x_1|)$. □

THEOREM 1.8.10

$$\phi_{\min}(x_1, x_2) = -(p^* - 1)^{-p} F_p(|x_1|, |x_2|). \tag{1.8.71}$$

The proof of this theorem is absolutely similar to the proof of Theorem 1.8.9.

Burkholder often used function u_p from (1.8.68). To demystify it, let us notice that it is also p-homogeneous and as such can be considered only on segment $x_1 + x_2 = 1,\ \ x_i > 0$. On that segment, function u_p becomes linear. It is a majorant of h_p, and its graph is tangential to the graph h_p exactly at point w_p on S, where h_p vanishes. It is not the least zigzag concave function greater than h_p (of course not, $\phi_{\max}$ is), but it is the least zigzag concave function larger than h_p and such that on all segments $\{x\colon x_i > 0,\, x_1 + x_2 = \text{const}\}$, it is not only concave, but also linear. (Keeping in mind the symmetries $x \to -x_1, x_2 \to -x_2$, we can consider first quadrant only.)

One more thing we want to mention is that we could have considered a slightly more general problem. Namely, instead of majorizing function $h_{\max}(x_1, x_2) = |x_2|^p - (p^* - 1)^p |x_1|^p$, we could have started with any function

$$h_c(x_1, x_2) \stackrel{\text{def}}{=} |x_2|^p - c|x_1|^p.$$

The reader can easily see that we have proved the following theorem (of course, Burkholder had already proved most of it).

COROLLARY 1.8.11 *The smallest c for which there exists a zigzag concave function ϕ_c, majorizing h_c, is equal to $(p^* - 1)^p$. For that c, the least zigzag concave majorant is $F_p(|x_1|, |x_2|)$. The smallest c for which there exists a zigzag concave function ϕ_c, majorizing h_c, such that it is linear on segment $\{x\colon x_i > 0,\, x_1 + x_2 = \text{const}\}$, symmetric and p-homogeneous, is equal to $(p^* - 1)^p$. For that c, the least zigzag concave majorant, linear on $\{x\colon x_i > 0,\, x_1 + x_2 = \text{const}\}$, symmetric and p-homogeneousis $u_p(x_1, x_2)$.*

REMARK Notice an interesting thing that we do not know how to explain. Given function $\mathbf{B}_{\max}$ from Theorem 1.8.1, we can easily diminish the number of variables and construct $\phi_{\max}$. But amazingly, we can also find $\mathbf{B}_{\max}$ if only $\phi_{\max}$ is given. In fact, Theorem 1.8.9 gives us the formula for $\phi_{\max}$ via F_p. Then F_p allows us to find $\mathbf{B}_{\max}$. If we now combine Theorem 1.8.1 and Theorems 1.8.9–1.8.10, we get the following:

COROLLARY 1.8.12 *Given a point $x \in \Omega$, if we know $\phi_{\max}$, we can find $\mathbf{B}_{\max}(x)$ by solving the equation:*

$$\phi_{\max}(x_1, x_2) = \phi_{\max}(x_3^{\frac{1}{p}}, \mathbf{B}_{\max}^{\frac{1}{p}}).$$

The symmetric formula allows us to find $\mathbf{B}_{\min}$, if $\phi_{\min}$ is known:

$$\phi_{\min}(x_2, x_1) = \phi_{\min}(\mathbf{B}_{\min}^{\frac{1}{p}}, x_3^{\frac{1}{p}}).$$

1.9 On the Bellman Function Generated by a Weak Type Estimate of a Square Function

The function we will discuss in this section differs very much from all Bellman functions described earlier. This is an example of the Bellman function that satisfies the corresponding differential equation only on some part of its domain, on another part of the domain it is not differentiable even at a dense subset. Here we try to show how it is possible to investigate the main inequality when it cannot be reduced to its infinitesimal form.

1.9.1 Level 0: What to Keep in Mind When Reading This Section

Organization of the Section. Since this section is quite technical in some places, we decided to write the text not in the usual "linear" manner where each statement is immediately followed by its proof and each proof contains all the needed auxiliary statements, but rather in a "tree-like" manner, where the top level is occupied by just statements of the main results, the second level is occupied by statements of the auxiliary results, and the proofs of the main results without some technical details, the third level is occupied by the technical details missing in the second level and so on. So, the reader who only wants to get a general impression of what has been done in this section can read just Level 1; the reader who wants, in addition, to get a general idea of how everything is proved can read through Level 2, and so on.

Such a structure means that at each level, we will freely use the results from the next levels and the notation from the previous ones. Within each level, we use just the usual linear structure.

1.9.1.1 Warning about Computer-Assisted Proofs. Many of our proofs of various "elementary inequalities" are computer assisted. On the other hand, our standards for using computers in the proofs are quite strict: We allow only algebraic symbolic manipulation of rational functions and basic integer arithmetic. All our computations were done using the Mathematica program by Wolfram Research run on the Windows XP platform. We believe that there were no bugs in the software that could affect our results but, of course, the reader is welcome to check the computations using different programs on different platforms.

1.9.1.2 Notation and Facts to Remember throughout the Entire Section. The following facts and notation are "global" and will be used freely throughout the text without any further references after their first occurrence.

Everything else is "local" to each particular section and can be safely forgotten when exiting that section.

- The definition of square function Sf (Section 1.9.2);
- The definition and the properties of nonlinear mean M (see Section 1.9.3.1);
- The definition and the properties of the dyadic suspension bridge $\mathbf{A}$ (see Section 1.9.3.1);
- The definition of function $\mathbf{B}$ (see Section 1.9.2);
- The notation $X(x,\tau) = \dfrac{x+\tau}{\sqrt{1-\tau^2}}$;
- The Bellman inequality in its standard form (1.9.1) (on page 120) and the inverse function form (1.9.2) (on page 124);
- The notation $\mathcal{B}(x) = \begin{cases} 1, & x \le 0 \\ \frac{1}{1+x^2}, & x \ge 0 \end{cases}$ and the fact that $\mathcal{B}$ satisfies the Bellman inequality;
- The definition of supersolution and the fact that $\mathbf{B}$ is the least supersolution (Section 1.9.3.3);
- The notation $\Phi(x) = \int_x^\infty e^{-y^2/2}\,dy$ and $\Psi = \Phi^{-1}$;
- The differential Bellman inequality $xB'(x)+B''(x) \le 0$ and its equivalence to the concavity of function $B \circ \Psi$ (Section 1.9.3.4);
- The increasing property of the ratio $\dfrac{\mathbf{B}(x)}{\Phi(x)}$ (Section 1.9.3.4).

This list is here to serve as a reminder to a reader who might otherwise occasionally get lost in this text or who might want to read its various parts in some nontrivial order. In addition to this list, it may be useful to keep in mind the statements in the titles of sections and the summary of results in Level 1, though it is not formally necessary.

1.9.2 Level 1: Setup and Main Results

The celebrated Chang–Wilson–Wolff theorem [**38**] states that if the square function of function f is uniformly bounded, then $e^{a|f|^2}$ is (locally) integrable for some positive a, which, in turn, implies that the distribution tails $|\{f \ge x\}|$ decay like e^{-ax^2}. This theorem holds true for both the discrete and the continuous versions of the square function. The main aim of this section is to present some steps in an attempt to get the *sharp* bounds for the distribution tails in a dyadic setting.

So, let $I = [0,1]$. As usual, symbol $\mathcal{D} = \mathcal{D}(I)$ stands for the collection of all dyadic subintervals of interval I. Recall that with each dyadic interval $J \in \mathcal{D}$, we associate the corresponding Haar function H_J (see formula on page xvii).

Now let $f: I \to \mathbb{R}$ be any integrable function on I, such that $\langle f \rangle_I = 0$. Then $f = \sum_{J \in \mathcal{D}} a_J H_J$, where coefficients a_J can be found from the formula $a_J = |J|^{-1} \int_I f H_J$ and the series converges in L^1 as well as almost everywhere. The dyadic square function Sf of function f is then defined by the following formula:

$$Sf = \sqrt{\sum_{J \in \mathcal{D}} a_J^2 \chi_J},$$

where $\chi_J = H_J^2$ is the characteristic function of the dyadic interval J. The quantity we want to investigate is

$$\mathbf{B}(x) = \sup\{|\{f \geq x\}| \colon \|Sf\|_{L^\infty} \leq 1\}, \qquad x \in \mathbb{R}.$$

Here is the summary of what we know about function $\mathbf{B}(x)$:

- $\mathbf{B}$ is a continuous nonincreasing function on $\mathbb{R}$;
- $\mathbf{B}(x) = 1$ for all $x \leq 0$ and $\mathbf{B}$ is strictly decreasing on $[0, +\infty)$;
- $\mathbf{B}(x) = 1 - \mathbf{A}^{-1}(x)$ for all $x \in [0, 1]$ where $\mathbf{A} \colon [0, \frac{1}{2}] \to [0, 1]$ is the "dyadic suspension bridge function" constructed in the beginning of Level 2;
- If $x \in [0, 1]$ and $\mathbf{B}(x)$ is a binary rational number (i.e., a number of the kind $\frac{k}{2^n}$ with some nonnegative integer k and n), then we can explicitly construct a finite linear combination f of Haar functions, for which $|\{f \geq x\}| = \mathbf{B}(x)$;
- There exists a positive constant c (whose exact value remains unknown to us) such that $\mathbf{B}(x) = c\Phi(x)$ for all $x \geq \sqrt{3}$, where Φ is the Gaussian "error function," i.e., $\Phi(x) = \displaystyle\int_x^\infty e^{-y^2/2}\, dy$.

Briefly, this means that we know $\mathbf{B}$ exactly for $x \leq 1$, we know it up to an absolute constant factor for $x \geq \sqrt{3}$, but we do not have any clear idea about what $\mathbf{B}$ may be equal to between 1 and $\sqrt{3}$.

1.9.3 Level 2: Definitions, Auxiliary Results, and Ideas of Proofs

***1.9.3.1 Construction of the Dyadic Suspension Bridge Function* A.**
• *Nonlinear mean M.* For any two real numbers a, b, we define their nonlinear mean $M[a, b]$ by

$$M[a, b] = \frac{a + b}{\sqrt{4 + (a - b)^2}}.$$

The nonlinear mean $M[a, b]$ has the following properties:

(1) $M[a, a] = a$;

(2) $M[a,b] = M[b,a]$;

(3) $M[a,b] \le \dfrac{a+b}{2}$ for all $a, b \ge 0$;

(4)

$$\frac{\partial}{\partial a} M[a,b] = \frac{4 + 2b^2 - 2ab}{[4 + (a-b)^2]^{3/2}}.$$

When $a, b \in [0,1]$, the right-hand side is strictly positive and does not exceed $\frac{3}{4}$ (the numerator is at most 6 and the denominator is at least 8). It follows immediately from here that

(5) $M[a,b]$ is strictly increasing in each variable in the square $[0,1]^2$ and $M[a,b]$ lies strictly between $M[a,a] = a$ and $M[b,b] = b$ if $a, b \in [0,1]$ and $a \ne b$;

(6) $|M[a,b] - a| \le \frac{3}{4}|a-b|$ for all $a, b \in [0,1]$.

• *Definition of* **A**. Let

$$D_n = \left\{ \frac{k}{2^n} : k = 0, 1, \ldots, 2^{n-1} \right\} \qquad n = 1, 2, 3, \ldots.$$

For any $t \in D_n \backslash D_{n-1}$ with $n \ge 2$, we define $t^{\pm} = t \pm 2^{-n} \in D_{n-1}$. Let $D = \bigcup_{n \ge 1} D_n$ be the set of all binary rational numbers on the interval $[0, \frac{1}{2}]$. We shall define function $\mathbf{A}\colon D \to [0,1]$, as follows. Put $\mathbf{A}(0) = 0$, $\mathbf{A}(\frac{1}{2}) = 1$. This completely defines $\mathbf{A}$ on D_1. Assume now that we already know the values of $\mathbf{A}$ on D_{n-1}. For each $t \in D_n \setminus D_{n-1}$, we put

$$\mathbf{A}(t) = M[\mathbf{A}(t^-), \mathbf{A}(t^+)].$$

This defines $\mathbf{A}$ inductively on the entire D. The first few steps of this construction are shown on Figure 1.20.

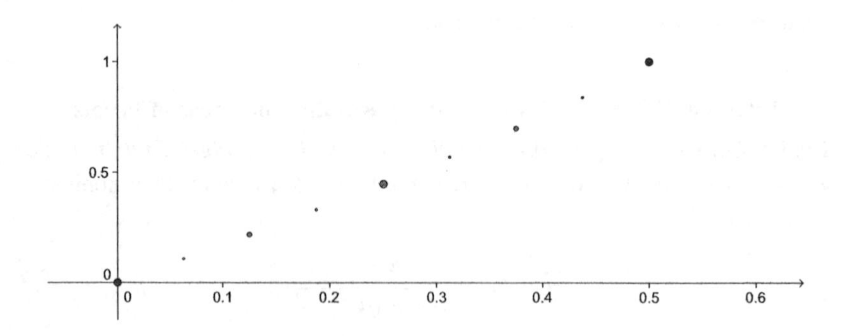

Figure 1.20 First steps in definition of **A**.

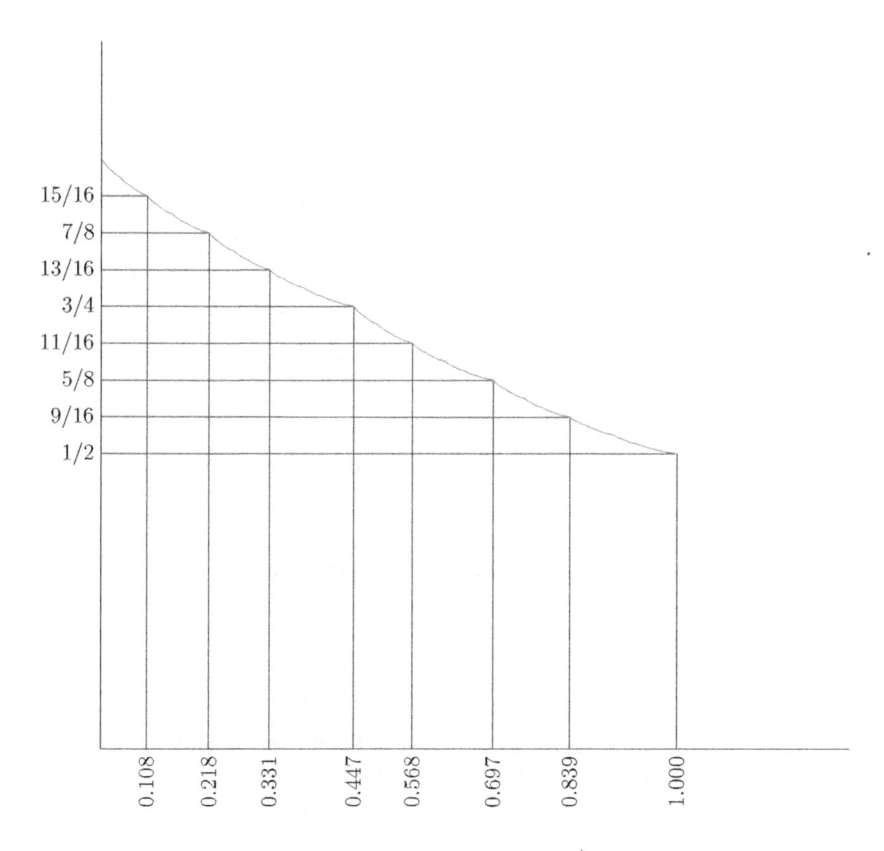

Figure 1.21 The graph of the function $\mathbf{B} = 1 - \mathbf{A}^{-1}$ on the interval $[0, 1]$.

After completing this procedure, our function $\mathbf{B}(x) = 1 - \mathbf{A}^{-1}(x)$ will look as shown in Figure 1.21.

Property (1.9.3.1) of the nonlinear mean M implies that the difference of values of $\mathbf{A}$ at any two neighboring points of D_n does not exceed $\left(\frac{3}{4}\right)^{n-1}$. It is not hard to derive from here that $\mathbf{A}$ is uniformly continuous on D and, moreover, $\mathbf{A} \in \mathrm{Lip}_\alpha$ with $\alpha = \log_2 \frac{4}{3}$. Thus, $\mathbf{A}$ can be extended continuously to the entire interval $[0, \frac{1}{2}]$. Property (1.9.3.1) implies that $\mathbf{A}$ is strictly increasing on D and, thereby, on $[0, \frac{1}{2}]$. Thus, the inverse function $\mathbf{A}^{-1}\colon [0,1] \to [0, \frac{1}{2}]$ is well defined and strictly increasing.

• *Properties of* $\mathbf{A}$. The main properties of $\mathbf{A}$ we shall need are as follows:

$$\mathbf{A}(t) \le 2t \qquad \text{for all } t \in [0, \tfrac{1}{2}],$$

the inequality

$$\mathbf{A}\left(\frac{s+t}{2}\right) \geq M[\mathbf{A}(s), \mathbf{A}(t)] \quad \text{for all } s, t \in [0, \tfrac{1}{2}],$$

and the fact that function $\dfrac{\mathbf{A}(t)}{t}$ is nondecreasing on $(0, \frac{1}{2}]$. The first assertion immediately follows from Property (1.9.3.1) of the nonlinear mean $M[a, b]$ by induction: at points $t = 0$ and $t = 1$, we have $\mathbf{A}(t) = 2t$, and if the inequality holds on D_{n-1}, then for $t \in D_n \setminus D_{n-1}$, we can estimate

$$\mathbf{A}(t) = M[\mathbf{A}(t^-), \mathbf{A}(t^+)] \leq \frac{\mathbf{A}(t^-) + \mathbf{A}(t^+)}{2} \leq t^- + t^+ = 2t.$$

Thus, the assertion is true on D and, by continuity, on the whole $[0, \frac{1}{2}]$. The proofs of the other two statements can be found on Level 3 in Sections 1.9.4.2 and 1.9.4.3.

***1.9.3.2 Continuity of* B.** By definition, $\mathbf{B}$ is nonincreasing positive on $\mathbb{R}$. It is easy to see that $\mathbf{B}(x) = 1$ for $x \leq 0$ (just consider the identically zero test function f). Now let $x \geq 0$. Take any test function f satisfying $\langle f \rangle_I = 0$ and $\|Sf\|_{L^\infty} \leq 1$. Construct a new function $g = g_{m,\delta}$ in the following way. Take an integer $m \geq 1$. Choose some $\delta \in (0, 2^{-3m})$. Let $I_j = [0, 2^{-j}]$, $J_j = I_j^+ = [2^{-(j+1)}, 2^{-j}]$ ($j = 0, 1, 2, \dots$). Let T_j be the linear mapping that maps J_j onto I (so, $T_0(x) = 2x - 1$, $T_1(x) = 4x - 1$, $T_2(x) = 8x - 1$, and so on). Put $f_j = f \circ T_j$ on J_j and $f_j = 0$ on $I \setminus J_j$. Now, let

$$g(x) = \delta \sum_{j=0}^{m-1} 2^j H_{I_j} + \sqrt{1 - 2^{2m}\delta^2} \sum_{j=0}^{m} f_j.$$

The first sum may look a bit strange as written, but it is just the Haar decomposition of the function $\begin{cases} 1, & 2^{-m} \leq x \leq 1; \\ 1 - 2^m, & 0 \leq x < 2^{-m} \end{cases}$ multiplied by δ (cf. Figure 1.22).

Then, clearly, $\langle g \rangle_I = 0$. Since f_j have a mean of 0, are supported by disjoint dyadic intervals, and none of the functions f_j from the second sum contains any of the function H_{I_k} from the first sum in its Haar decomposition, we have

$$(Sg)^2 \leq 1 - 2^{2m}\delta^2 + \delta^2 \sum_{j=0}^{m-1} 2^{2j} \leq 1$$

on I. Finally, for each $j = 0, 1, \dots, m-1$, we have

$$|\{g \geq \delta + \sqrt{1 - 2^{2m}\delta^2}\, x\} \cap J_j| \geq |\{f_j \geq x\} \cap J_j| = 2^{-(j+1)} |\{f \geq x\}|$$

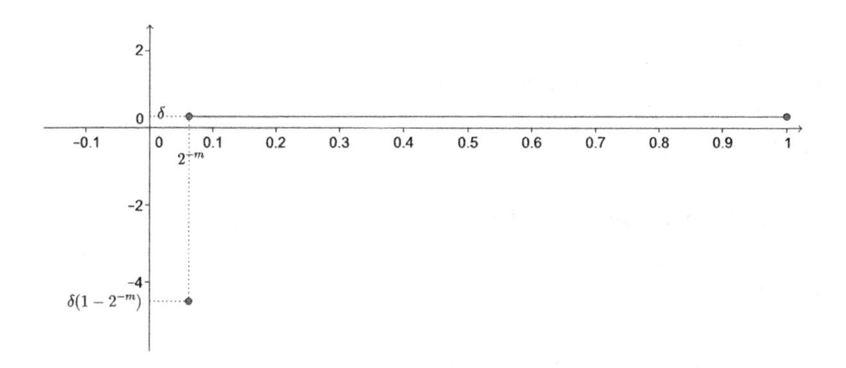

Figure 1.22 First summand of g.

and, thereby, for the entire interval I, we have the inequality

$$|\{g \ge \delta + \sqrt{1-2^{2m}\delta^2}\, x\}| \ge (1-2^{-m})|\{f \ge x\}|.$$

Now, let us fix an integer $m > 0$, then for every $x \in [0, 2^{m-1}]$ and $\delta \in [0, 2^{-3m}]$, we have

$$\delta + \sqrt{1-2^{2m}\delta^2}\, x \ge \delta + (1-2^{2m}\delta^2)x = x + \delta(1-2^{2m}\delta x) \ge x + \frac{\delta}{2}.$$

Hence, by definition of $\mathbf{B}$, we can write down the following estimate:

$$\mathbf{B}\left(x+\tfrac{\delta}{2}\right) \ge |\{g \ge x + \tfrac{\delta}{2}\}| \ge (1-2^{-m})|\{f \ge x\}|.$$

Taking the supremum over all test functions f on the right-hand side, we get

$$\mathbf{B}\left(x+\tfrac{\delta}{2}\right) \ge (1-2^{-m})\mathbf{B}(x).$$

Recalling that $\mathbf{B}$ is nonincreasing and $0 \le \mathbf{B} \le 1$, we conclude that

$$0 \le \mathbf{B}(x) - \mathbf{B}\left(x+\tfrac{\delta}{2}\right) \le 2^{-m},$$

which immediately implies the uniform continuity of $\mathbf{B}$ on any compact subset of $\mathbb{R}$.

One useful corollary of this continuity result is the possibility to restrict ourselves to the functions f that are *finite* linear combinations of the Haar functions in the definition of $\mathbf{B}$. Indeed, let $x \in \mathbb{R}$. Take any $\varepsilon > 0$. Choose $x' > x$ in such a way that $\mathbf{B}(x') \ge \mathbf{B}(x) - \varepsilon$. Choose a function f satisfying $\langle f \rangle_I = 0$ and $\|Sf\|_{L^\infty} \le 1$ such that $|\{f \ge x'\}| \ge \mathbf{B}(x') - \varepsilon$. Let f_n be the partial sums of the Haar series for f. Clearly, $\langle f_n \rangle_I = 0$ and $Sf_n \le Sf$ everywhere on I. Since f_n converge to f almost everywhere on I, we can choose n such that $|\{f_n \geq x\}| \geq |\{f \geq x'\}| - \varepsilon$. But then $|\{f_n \geq x\}| \geq \mathbf{B}(x) - 3\varepsilon$. Moreover, considering the functions $g_n = (1 - \frac{1}{n})f_n$ instead of f_n, we see that the supremum can be taken over finite linear combinations f, satisfying the strict inequality $\|Sf\|_{L^\infty} < 1$.

1.9.3.3 Main Inequality. Take any $\tau \in (-1,1)$ and any two functions $f_-, f_+ \colon I \to \mathbb{R}$ satisfying $\langle f_\pm \rangle_I = 0$ and $\|Sf_\pm\|_{L^\infty} \leq 1$. Consider function f defined by

$$f(x) = \tau H_I + \sqrt{1-\tau^2} \begin{cases} f_-(2x), & 0 \leq x < \frac{1}{2}; \\ f_+(2x-1), & \frac{1}{2} \leq x \leq 1. \end{cases}$$

It is easy to see that $\langle f \rangle_I = 0$. Also, we have

$$((Sf)(x))^2 = \tau^2 + (1-\tau^2) \begin{cases} ((Sf_-)(2x))^2, & 0 \leq x < \frac{1}{2}; \\ ((Sf_+)(2x-1))^2, & \frac{1}{2} \leq x \leq 1, \end{cases}$$

whence $\|Sf\|_{L^\infty} \leq 1$. Now, it immediately follows from our definition of f that, for every $x \in \mathbb{R}$,

$$|\{f \geq x\}| = \frac{1}{2}\left[\left|\left\{f_- \geq \frac{x+\tau}{\sqrt{1-\tau^2}}\right\}\right| + \left|\left\{f_+ \geq \frac{x-\tau}{\sqrt{1-\tau^2}}\right\}\right|\right].$$

But, according to the definition of $\mathbf{B}$, the right-hand side can be made as close to $\frac{1}{2}\left[\mathbf{B}\left(\frac{x+\tau}{\sqrt{1-\tau^2}}\right) + \mathbf{B}\left(\frac{x-\tau}{\sqrt{1-\tau^2}}\right)\right]$ as we wish by choosing the appropriate $f_\pm$. Since our function f belongs to the class of functions over which the supremum in the definition of $\mathbf{B}(x)$ is taken, we conclude that

$$\mathbf{B}(x) \geq \frac{1}{2}\left[\mathbf{B}\left(\frac{x+\tau}{\sqrt{1-\tau^2}}\right) + \mathbf{B}\left(\frac{x-\tau}{\sqrt{1-\tau^2}}\right)\right]. \tag{1.9.1}$$

From now on, we shall use the notation $X(x,\tau)$ for $\frac{x+\tau}{\sqrt{1-\tau^2}}$. And the inequality (1.9.1) will be referred to as *the main inequality*.

We shall call every nonincreasing, nonnegative continuous function B, satisfying the main inequality and the condition $B(x) = 1$ for $x \leq 0$, a supersolution. Our next claim is that $\mathbf{B}$ is just the *least* supersolution. Since $\mathbf{B}$ is a supersolution, it suffices to show that $\mathbf{B}(x) \leq B(x)$ for any other supersolution B. It is sufficient to check that for any finite linear combination f of the Haar functions satisfying $\|Sf\|_{L^\infty} < 1$, we have $|\{f \geq x\}| \leq B(x)$ for all $x \in \mathbb{R}$. We shall prove this statement by induction on the highest level of the Haar functions in the decomposition of f (the level of the Haar function H_J is just the number n, such that $|J| = 2^{-n}$). If f is identically 0, then the desired inequality immediately follows from the definition of supersolution. Assume that our inequality is proved for all linear combinations containing only Haar functions up to level $n-1$ and that f contains only Haar functions up to level n. Let τ be the coefficient at H_I in the decomposition of f. Note that we must have $|\tau| < 1$ (otherwise $Sf \geq 1$ on I). Let $T_\pm$ be the linear mappings that map I onto $I^\pm$. Put $f_\pm = (f \circ T_\pm \mp \tau)/\sqrt{1-\tau^2}$. The functions $f_\pm$ are also finite linear combinations of Haar functions, but they contain only Haar functions up to level $n-1$ (if $n = 0$, it means that $f_\pm$ are identically 0). Also, it is not hard to check that $\|Sf_\pm\|_{L^\infty} < 1$. Now, clearly,

$$|\{f \geq x\}| \leq \tfrac{1}{2}[|\{f_- \geq X(x,-\tau)\}| + |\{f_+ \geq X(x,\tau)\}|]$$
$$\leq \tfrac{1}{2}[B(X(x,-\tau)) + B(X(x,\tau))] \leq B(x)$$

by the induction assumption and the main inequality. We are done.

Now we shall characterize all triples (x_-, x, x_+) of real numbers, such that $x_\pm = X(x, \pm\tau)$ for some $\tau \in (-1,1)$. A straightforward computation shows that in such cases, we must have $x = M[x_-, x_+]$. Conversely, if $x = M[x_-, x_+]$, then we can take $\tau = \dfrac{x_+ - x_-}{\sqrt{4 + (x_+ - x_-)^2}}$ and check that $x_\pm = X(x, \pm\tau)$ for this particular τ. Thus, the main inequality can be restated in the following form:

$$B(x) \geq \tfrac{1}{2}[B(x_-) + B(x_+)]$$

for all triples x_-, x, x_+ satisfying the relation $x = M[x_-, x_+]$.

In the last part of this section, we show that it suffices to check the main inequality only in the case when all three numbers x_-, x, x_+ are nonnegative. Indeed, let $B(x)$ be a nonincreasing function such that $B(x) = 1$ for all $x \leq 0$. If $x \leq 0$, then the main inequality becomes trivial: $1 \geq \frac{1}{2}[B(x_-) + B(x_+)]$ for any two points $x_\pm$. If $x > 0$ and, say, $x_- < 0$ (note that the roles of x_- and x_+ are completely symmetric), we must have $x_- = X(x, -\tau)$ with $\tau > x > 0$. But then $X(x,\tau) > 0$ and the main inequality becomes stronger if we replace $\tau > x$ by $\tau = x$. Indeed, $B(X(x,-\tau))$ and $B(x)$ will stay the same while $B(X(x,\tau))$ will not decrease because B is nonincreasing. This remark allows us to forget about the negative semi-axis altogether and to define a supersolution as a nonnegative, nonincreasing, continuous function defined on $[0, +\infty)$ and satisfying the main inequality as well as the condition $B(0) = 1$.

1.9.3.4 Smooth Supersolutions and the Differential Main Inequality. Suppose now that a supersolution B is twice continuously differentiable on

$(0, +\infty)$, then we have the Taylor expansion

$$B(X(x, \pm\tau)) = B(x) \pm B'(x)\tau + \tfrac{1}{2}(xB'(x) + B''(x))\tau^2 + o(\tau^2) \quad \text{as } \tau \to 0.$$

Plugging this expansion into the main inequality, we see that we must have

$$xB'(x) + B''(x) \leq 0$$

for all $x > 0$. It is not hard to solve the corresponding linear differential equation: One possible solution is

$$\Phi(x) = \int_x^\infty e^{-y^2/2}\, dy$$

and the general solution is $C_1\Phi + C_2$, where C_1, C_2 are arbitrary constants.

Let $\Psi\colon (0, \Phi(-\infty)) \to (-\infty, +\infty)$ be the inverse function to Φ. By inverse function theorem, we have

$$\Psi' = \frac{1}{\Phi' \circ \Psi} = -e^{\Psi^2/2}.$$

Hence,

$$(B \circ \Psi)'' = e^{\Psi^2}((B' \circ \Psi) \cdot \Psi + B'' \circ \Psi).$$

Therefore, the differential main inequality is equivalent to concavity of $B \circ \Psi$ on $(0, \Psi(-\infty))$. Since for any nonnegative concave function G on $(0, \Phi(-\infty))$, the ratio $G(t)/t$ is nonincreasing, we conclude that the ratio $\dfrac{B(\Psi(t))}{t}$ is nonincreasing and, thereby, the ratio $\dfrac{B(x)}{\Phi(x)}$ is nondecreasing on $(-\infty, +\infty)$.

The last two conditions (the concavity of $B \circ \Psi$ and the nondecreasing property of the ratio $\dfrac{B(x)}{\Phi(x)}$) make perfect sense for all supersolutions, whether smooth or not. So, it would be nice to show that every supersolution can be approximated by a C^2-smooth one with arbitrary precision.

To do that, note that for every $x_-, x_+ \in \mathbb{R}$ and every $y \geq 0$, we have

$$M[x_- - y, x_+ - y] = M[x_-, x_+] - \frac{2y}{\sqrt{4 + (x_+ - x_-)^2}} \geq M[x_-, x_+] - y.$$

This allows us to conclude that if B is a supersolution, then so is $B(\cdot - y)$ for all $y \geq 0$. Also note that any convex combination of supersolutions is a supersolution as well. Now just take any nonnegative C^2 function η supported by $[0, 1]$ with total integral 1, for $\delta > 0$, define $\eta_\delta(x) = \delta^{-1}\eta(\delta^{-1}x)$, and consider the convolutions $B_\delta = B * \eta_\delta$. On the one hand, each B_δ is a supersolution. On the other hand, $B_\delta \to B$ pointwise as $\delta \to \infty$.

1.9.3.5 B *is strictly decreasing on* $[0, +\infty)$. Let us start with showing that $B(x) < 1$ for all $x > 0$. For this, it suffices to note that the inequality $\|Sf\|_{L^\infty} \le 1$ implies

$$\int_I f^2 = \int_I (Sf)^2 \le 1.$$

Now, if we consider the problem of maximizing $|\{f \ge x\}|$ under the restrictions $\langle f \rangle_I = 0$ and $\langle f^2 \rangle_I \le 1$, we shall get another function, $\mathcal{B}(x)$ on $[0, +\infty)$. Since we relaxed our restrictions, we must have $\mathbf{B} \le \mathcal{B}$ everywhere. But, unlike our original problem of finding $\mathbf{B}$, to find $\mathcal{B}$ exactly is a piece of cake: we have

$$\mathcal{B}(x) = \frac{1}{1+x^2} \qquad \text{for all } x \ge 0.$$

The reader can try to prove this statement herself/himself or to look up the proof on Level 3. At this point, we shall only mention that $\mathcal{B}(t)$ satisfies the condition $\mathcal{B}(0) = 1$ and the same main inequality. The derivation of that is almost exactly the same as before. Actually, the only result in this section that is impossible to repeat for $\mathcal{B}$ in place of $\mathbf{B}$ is to show that it is the *least* supersolution.

Now, when we know that $\mathbf{B}(x) \le \frac{1}{1+x^2} < 1$ for $x > 0$, the strict monotonicity becomes relatively easy. Indeed, assume that $\mathbf{B}(x) = \mathbf{B}(y) = a$ for some $0 < x < y$. Then $a < 1$. Due to the continuity of $\mathbf{B}$, we can choose the least $x \ge 0$ satisfying $\mathbf{B}(x) = a$. Inequality $x \ne 0$ holds because $\mathbf{B}(0) = 1 > a$, so we must have $x > 0$. Also, we still have $x < y$. Now take $\tau > 0$ so small that $X(x, -\tau) < x$ and $X(x, \tau) < y$. Then the Bellman inequality immediately implies that $\mathbf{B}(X(x, -\tau)) \le 2\mathbf{B}(x) - \mathbf{B}(X(x, \tau)) \le 2\mathbf{B}(x) - \mathbf{B}(y) = a$. Since we must also have $\mathbf{B}(X(x, -\tau)) \ge \mathbf{B}(x) = a$, we obtain $\mathbf{B}(X(x, -\tau)) = a$, which contradicts the minimality of x. It is worth mentioning that a similar argument can be used to derive continuity directly from the main inequality. We leave the details to the reader.

The strict monotonicity property implies that $\mathbf{B}^{-1}$ is well defined. Also, since $\mathbf{B}(x) \le \mathcal{B}(x)$, we must have $\mathbf{B}(x) \to 0+$ as $x \to \infty$. Thus, $\mathbf{B}^{-1}$ continuously maps the interval $(0, 1]$ onto $[0, +\infty)$. The main inequality is equivalent to the statement that

$$x = \mathbf{B}^{-1}(\mathbf{B}(x)) \le \mathbf{B}^{-1}\left(\frac{\mathbf{B}(x_-) + \mathbf{B}(x_+)}{2}\right)$$

for all triples x_-, x, x_+ of nonnegative numbers such that $x = M[x_-, x_+]$. Denoting $\mathbf{B}(x_-) = s$, $\mathbf{B}(x_+) = t$, we see that the last inequality is equivalent to

$$\mathbf{B}^{-1}\left(\frac{s+t}{2}\right) \geq M[\mathbf{B}^{-1}(s), \mathbf{B}^{-1}(t)]. \tag{1.9.2}$$

***1.9.3.6* B $= c\Phi$ *beyond* $\sqrt{3}$.** Our first task here will be to show that the function Φ satisfies the main inequality 1.9.1 if $x \geq \sqrt{3}$. Note that the inequality is an identity when $\tau = 0$. So it suffices to show that

$$\frac{\partial}{\partial \tau}[\Phi(X(x,-\tau) + \Phi(X(x,\tau))] \leq 0 \qquad \text{for all } \tau \in [0,1),$$

which, after a few simple algebraic manipulations, reduces to the inequality

$$(1 + x\tau)e^{-x\tau/(1-\tau^2)} \geq (1 - x\tau)e^{x\tau/(1-\tau^2)}.$$

If $x\tau \geq 1$, the left-hand side is nonnegative and the right-hand side is nonpositive. If $x\tau < 1$, we can rewrite the inequality we are proving in the form

$$\tfrac{1}{2}\log\frac{1 + x\tau}{1 - x\tau} - \frac{x\tau}{1 - \tau^2} \geq 0.$$

Expanding the left-hand side into a Taylor series with respect to τ, we obtain the inequality

$$\sum_{k \geq 0} x\left(\frac{x^{2k}}{2k+1} - 1\right)\tau^{2k+1} \geq 0$$

to prove. Observe that the coefficient in front of τ is always 0 and the coefficient at τ^3 is negative if $0 \leq x < \sqrt{3}$. That means that our inequality holds with the opposite sign for all sufficiently small τ if $0 \leq x < \sqrt{3}$ and, thereby, the main inequality fails for such x and τ as well. On the other hand, if $x \geq \sqrt{3}$, then *all* the coefficients on the left-hand side are nonnegative and the inequality holds.

Now let $c = \mathbf{B}(\sqrt{3})/\Phi(\sqrt{3})$. Consider function $B(x)$, defined by

$$B(x) = \begin{cases} \mathbf{B}(x), & x \leq \sqrt{3}; \\ c\Phi(x), & x \geq \sqrt{3}. \end{cases}$$

Note that, since the ratio $\dfrac{\mathbf{B}(x)}{\Phi(x)}$ is nondecreasing, we actually have $B(x) = \min\{\mathbf{B}(x), c\Phi(x)\}$ everywhere on $\mathbb{R}$. Indeed, $\mathbf{B}(\sqrt{3}) = c\Phi(\sqrt{3})$ by our choice of c, whence $\mathbf{B} \geq c\Phi$ on $[\sqrt{3}, +\infty)$ and $\mathbf{B} \leq c\Phi$ on $(-\infty, \sqrt{3}]$. Clearly, $B(x) = \mathbf{B}(x) = 1$ for $x \leq 0$, B is nonnegative, continuous, and nonincreasing. Let us check the main inequality for B. Take any triple x_-, x, x_+ with $x = M[x_-, x_+]$. If $x \le \sqrt{3}$, we have

$$B(x) = \mathbf{B}(x) \ge \tfrac{1}{2}[\mathbf{B}(x_-) + \mathbf{B}(x_+)] \ge \tfrac{1}{2}[B(x_-) + B(x_+)].$$

If $x \ge \sqrt{3}$, we have

$$B(x) = c\Phi(x) \ge \tfrac{1}{2}[c\Phi(x_-) + c\Phi(x_+)] \ge \tfrac{1}{2}[B(x_-) + B(x_+)].$$

Thus, B is a supersolution and, therefore, $\mathbf{B} \le B$ everywhere. But we also know that $\mathbf{B} \ge B$ everywhere. Thus, $\mathbf{B} = B$, i.e., $\mathbf{B} = c\Phi$ on $[\sqrt{3}, \infty)$.

1.9.3.7 $\mathbf{B} = 1 - \mathbf{A}^{-1}$ ***on*** $[0,1]$. The first observation to make here is that we know the value $\mathbf{B}(1)$ exactly: $\mathbf{B}(1) = \frac{1}{2}$. Indeed, the inequality $\mathbf{B}(1) \le \frac{1}{2}$ follows from the estimate $\mathbf{B}(x) \le \mathcal{B}(x) = \frac{1}{1+x^2}$ and the inequality $\mathbf{B}(1) \ge \frac{1}{2}$ follows from the consideration of the test-function $f = H_I$. Consider now the function $G(t) = \mathbf{B}^{-1}(1-t)$. It is continuous, increasing and maps $[0, \frac{1}{2}]$ onto $[0,1]$. According to the main inequality in the form (1.9.2), we must have

$$G\left(\frac{s+t}{2}\right) \ge M[G(s), G(t)] \qquad \text{for all } s,t \in [0, \tfrac{1}{2}], .$$

Also $G(0) = 0 = \mathbf{A}(0)$ and $G(\frac{1}{2}) = 1 = \mathbf{A}(\frac{1}{2})$. Since M is a monotone in each variable on $[0,1]^2$, we can easily prove by induction that $G \ge \mathbf{A}$ on D and, therefore, by continuity, on $[0, \frac{1}{2}]$. Applying $\mathbf{B}$ to both sides of this inequality, we conclude that $1 - t \le \mathbf{B}(\mathbf{A}(t))$ on $[0, \frac{1}{2}]$. Taking $t = \mathbf{A}^{-1}(x)$ ($x \in [0,1]$), we finally get

$$\mathbf{B}(x) \ge 1 - \mathbf{A}^{-1}(x) \qquad \text{for all } x \in [0,1].$$

It remains only to prove the reverse inequality. To this end, it would suffice to show that function

$$B(x) = \begin{cases} 1, & x \le 0; \\ 1 - \mathbf{A}^{-1}(x), & 0 \le x \le 1; \\ \frac{1}{1+x^2}, & x \ge 1. \end{cases}$$

is a supersolution. The only nontrivial property to check is the main inequality. It has been already mentioned earlier that we may restrict ourselves to the case when all three numbers x_-, x, x_+ are nonnegative. There are several possible cases:

1.9.3.8 Case 1: all three numbers are on $[\mathbf{0}, \mathbf{1}]$. In this case, we can just check the main inequality in the form (1.9.2), which reduces to the already mentioned inequality

$$\mathbf{A}\left(\frac{s+t}{2}\right) \geq M[\mathbf{A}(s), \mathbf{A}(t)] \quad \text{for all } s, t \in [0, \tfrac{1}{2}]$$

whose proof can be found on Level 3.

1.9.3.9 Case 2: $\boldsymbol{x > 1}$. Here all we need is to note that, since $\mathbf{A}(t) \leq 2t$, we have

$$1 - \mathbf{A}^{-1}(x) \leq 1 - \frac{x}{2} \leq \frac{1}{1+x^2} = \mathcal{B}(x)$$

on $[0, 1]$. Therefore, we can use the fact that the main inequality is true for $\mathcal{B}$ and write

$$B(x) = \mathcal{B}(x) \geq \tfrac{1}{2}[\mathcal{B}(x_-) + \mathcal{B}(x_+)] \geq \tfrac{1}{2}[B(x_-) + B(x_+)].$$

1.9.3.10 Case 3: $\boldsymbol{0 < x < 1,\ \ x_+ \geq 1}$. We can always assume that it is x_+ that is greater than 1, because the roles of x_+ and x_- in the main inequality are completely symmetric. Note that when $0 < x < 1$, we have

$$\frac{\partial}{\partial \tau} X(x, \tau) = \frac{1 + x\tau}{(1 - \tau^2)^{3/2}} > 0$$

for all $\tau \in (-1, 1)$. Thus, if $x_+ > x$, we must have $\tau > 0$ and $x_- = X(x, -\tau) < x$. The condition $x_- \geq 0$ implies that $\tau \leq x$.

First we consider the boundary case when $x_- = 0$. Then $x_+ = X(x, x) = \frac{2x}{\sqrt{1-x^2}}$, which is greater than or equal to 1 if and only if $x \geq \frac{1}{\sqrt{5}}$. Then the inequality we need to prove reduces to

$$B(x) \geq \tfrac{1}{2}\big[B(X(x, x)) + 1\big] = \frac{1 + x^2}{1 + 3x^2}.$$

Denote function $\frac{1+x^2}{1+3x^2}$ on the right-hand side by $F(x)$ and note that at the endpoints of this interval, we have identities $B(\frac{1}{\sqrt{5}}) = F(\frac{1}{\sqrt{5}}) = \frac{3}{4}$ and $B(1) = F(1) = \frac{1}{2}$. Recall also that $B \circ \Psi$ is concave on $\Psi^{-1}([\frac{1}{\sqrt{5}}, 1])$. Formally we proved this only for supersolutions, but since only arbitrarily small values of τ were used in the proof, we can conclude that this concavity result also holds for any nonnegative, nonincreasing, continuous function B, satisfying the main inequality just for the triples x_-, x, x_+ contained in $[\frac{1}{\sqrt{5}}, 1]$. So, it would suffice to show that function $F \circ \Psi$ is *convex* on the same interval, which is equivalent to the assertion that $xF'(x) + F''(x) \geq 0$ on $[\frac{1}{\sqrt{5}}, 1]$. A direct computation yields

$$xF'(x) + F''(x) = 4\frac{8x^2 - 3x^4 - 1}{(1 + 3x^2)^3}.$$

But

$$8x^2 - 3x^4 - 1 = 3x^2(1 - x^2) + (5x^2 - 1) \geq 0$$

on $[\frac{1}{\sqrt{5}}, 1]$ and we are done.

Now we are ready to handle the remaining case $0 < x_- < x < 1 < x_+$. Let $\widetilde{x}_+ = X(x, x)$ and let $\widetilde{x}_- = X(x, -\tau)$ where $\tau \in (0, 1)$ is chosen in such a way that $X(x, \tau) = 1$. Then $0 < x_- < \widetilde{x}_- < x < 1 < x_+ < \widetilde{x}_+$ and we have the main inequality for the triples 0, x, $\widetilde{x}_+$ and $\widetilde{x}_-$, x, 1. If

$$B(x_+) - B(\widetilde{x}_+) \leq B(0) - B(x_-) \quad \text{or} \quad B(x_-) - B(\widetilde{x}_-) \leq B(1) - B(x_+),$$

we can prove the desired main inequality for the triples x_-, x, x_+ by comparing it to the known main inequality for the triples 0, x, $\widetilde{x}_+$ or $\widetilde{x}_-, x, 1$, respectively. So, the only situation that is bad for us is the one when the strict inequalities

$$B(x+) - B(\widetilde{x}_+) > B(0) - B(x_-) \text{ and } B(x_-) - B(\widetilde{x}_-) > B(1) - B(x_+)$$

hold simultaneously. Now observe that, if four positive numbers a, b, c, d satisfy $a > c$ and $b > d$, then we also have $\dfrac{c}{c+b} < \dfrac{a}{a+d}$. Thus, in the bad situation, we must have

$$\frac{B(0) - B(x_-)}{B(0) - B(\widetilde{x}_-)} < \frac{B(x_+) - B(\widetilde{x}_+)}{B(1) - B(\widetilde{x}_+)}.$$

Since $\frac{\mathbf{A}(t)}{t}$ is nondecreasing on $[0, \frac{1}{2}]$, we can say that

$$\frac{B(0) - B(x_-)}{B(0) - B(\widetilde{x}_-)} \geq \frac{x_-}{\widetilde{x}_-}.$$

So, in the bad situation we must have the inequality

$$\frac{x_-}{\widetilde{x}_-} < \frac{B(x_+) - B(\widetilde{x}_+)}{B(1) - B(\widetilde{x}_+)}.$$

Note that everywhere in this inequality function $B(x)$ coincides with $\mathcal{B}(x) = \frac{1}{1+x^2}$. So, this is an elementary inequality (it contains fractions and square roots, of course, but still it is a closed form inequality involving functions given by explicit algebraic formulae). It turns out that exactly the opposite inequality is always true (the proof can be found on Level 4), so we are done with this case too.

***1.9.3.11 Optimal Functions for Binary Rational Values of* B.** By construction of the dyadic suspension bridge $\mathbf{A}$, for every point $t \in D\backslash\{0, \frac{1}{2}\}$, we have $\mathbf{A}(t) = M[\mathbf{A}(t_-) + \mathbf{A}(t_+)]$. Now let $x = \mathbf{A}(t)$ for some $t \in D$ and let $x_- = \mathbf{A}(t_-)$, $x_+ = \mathbf{A}(t_+)$. Then for the triples x_-, x, x_+, the main inequality becomes an identity. If we have a pair $f_\pm$ of finite linear combinations of Haar functions, such that $\|Sf_\pm\|_{L^\infty} \le 1$ and $|\{f_\mp \ge x_\pm\}| = \mathbf{B}(x_\pm)$, then, if we take $\tau \in (0,1)$, such that $x_\pm = X(x, \pm\tau)$ and define f by

$$f = \tau H_I + \sqrt{1-\tau^2}\begin{cases} f_-(2x), & 0 \le x < \frac{1}{2};\\ f_+(2x-1), & \frac{1}{2} \le x \le 1, \end{cases}$$

we shall get a finite linear combination of Haar functions satisfying $\|Sf\|_{L^\infty} \le 1$ and $|\{f \ge x\}| = \mathbf{B}(x)$. Since we, indeed, have such extremal linear combinations for $x = 0$ and $x = 1$ (the identically 0 function and the function H_I, respectively), we can now recursively construct an extremal linear combination for any $x = \mathbf{A}(t)$ with $t \in D$. Take, for instance, $\mathbf{A}(\frac{3}{8})$. The construction of the extremal function for this value reduces to finding the coefficient $\tau = \dfrac{\mathbf{A}(\frac{1}{2}) - \mathbf{A}(\frac{1}{4})}{\sqrt{4 + (\mathbf{A}(\frac{1}{2}) - \mathbf{A}(\frac{1}{4}))^2}}$ and two extremal functions: one for $\mathbf{A}(\frac{1}{4})$ and the other for $\mathbf{A}(\frac{1}{2})$. The construction of the extremal function for $\mathbf{A}(\frac{1}{4})$ reduces to finding the coefficient $\tau = \dfrac{\mathbf{A}(\frac{1}{2}) - \mathbf{A}(0)}{\sqrt{4 + (\mathbf{A}(\frac{1}{2}) - \mathbf{A}(0))^2}}$ and two more extremal functions: one for $\mathbf{A}(0)$ and the other for $\mathbf{A}(\frac{1}{2})$. But we know that the extremal function for $\mathbf{A}(0) = 0$ is 0 and the extremal function for $\mathbf{A}(\frac{1}{2}) = 1$ is H_I. So, we can put everything together and get a linear combination of 4 Haar functions that is extremal for $\mathbf{A}(\frac{3}{8})$. This construction is shown on Figure 1.23.

The resulting linear combination is

$$\frac{1}{\sqrt{26 - 2\sqrt{5}}}\Big[(\sqrt{5}-1)H_I + 2\sqrt{5}H_{I^-} + 2H_{I^+} + 4H_{I^{+-}}\Big],$$

which, indeed, equals $\mathbf{A}(\frac{3}{8}) = \frac{\sqrt{5}+1}{\sqrt{26-2\sqrt{5}}}$ on the union $I^{-+} \cup I^{++} \cup I^{+-+}$, whose measure is exactly $\frac{5}{8}$. The square function, in its turn, equals 1 on $I^- \cup I^{+-}$ and is strictly less than 1 on I^{++}.

The simplest picture is obtained when we construct an extremal function for $\mathbf{A}(2^{-n})$. What we get is function $\sqrt{\frac{3}{4^n-1}}(1 - 2^n\chi_{[0,2^{-n}]})$ that takes just two different values: one small positive value on a big set and one large negative value on a small set. The interested reader may amuse herself/himself with drawing more pictures, trying to figure out how many Haar functions are

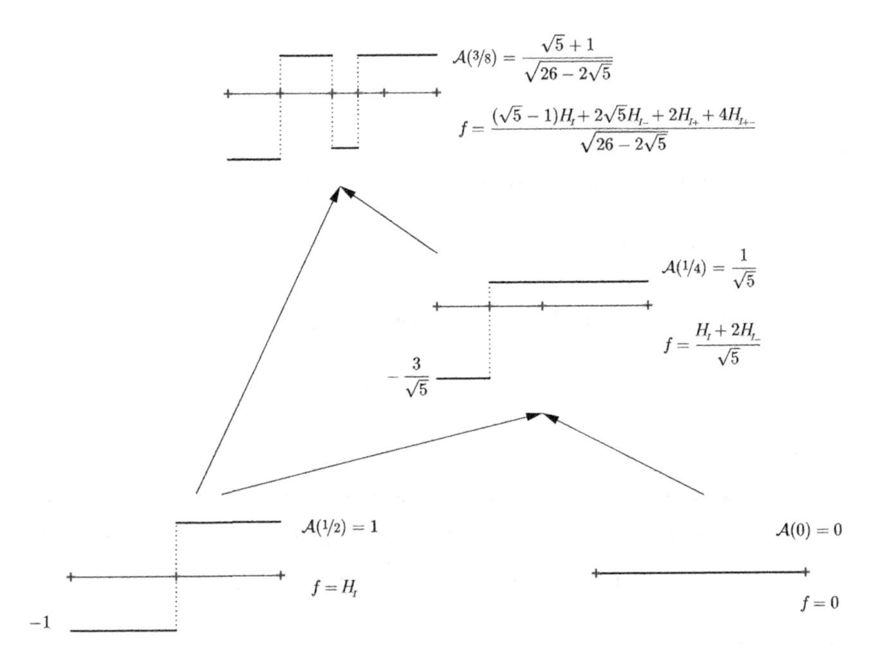

Figure 1.23 The construction of the extremal function for $\mathbf{A}(\frac{3}{8})$.

needed to construct an extremal function for any particular "good" value of x, or proving that for all other values of $x \in [0, 1]$ there are no extremal functions at all, but we shall stop here.

1.9.4 Level 3: Reductions to Elementary Inequalities

1.9.4.1 $\boldsymbol{\mathcal{B}(x) = \frac{1}{1+x^2}}$ ***for*** $\boldsymbol{x \geq 0}$**.** Recall that

$$\mathcal{B}(x) \stackrel{\text{def}}{=} \sup\Big\{|\{f \geq x\}|\colon \int_I f = 0,\ \int_I f^2 \leq 1\Big\}.$$

Considering the identically zero test-function f, we see that $\mathcal{B}(x) = 1$ for all $x \leq 0$. Now let $x > 0$. Putting

$$f(y) = \begin{cases} x, & 0 \leq y \leq \frac{1}{1+x^2}; \\ -\frac{1}{x}, & \frac{1}{1+x^2} < y \leq 1, \end{cases}$$

we see that $\mathcal{B}(x) \geq \frac{1}{1+x^2}$.

Now, take any test function f. Let $E = \{f \geq x\}$ and let $m = |E|$. Then

$$\int_{I \setminus E} f = -\int_E f \leq -mx$$

and

$$\int_{I\setminus E} f^2 \geq \frac{1}{|I\setminus E|}\left|\int_{I\setminus E} f\right|^2 \geq \frac{m^2x^2}{1-m}$$

by Cauchy–Schwartz inequality. Thus,

$$\int_I f^2 = \int_E f^2 + \int_{I\setminus E} f^2 \geq mx^2 + \frac{m^2x^2}{1-m} = \frac{m}{1-m}x^2.$$

Since this integral is bounded by 1, we get the inequality

$$\frac{m}{1-m}x^2 \leq 1,$$

whence $m \leq \dfrac{1}{1+x^2}$.

One more thing we want to do in this section is to show directly that $\mathcal{B}$ is a supersolution. If $x, X(x, \pm\tau) \geq 0$, the main inequality

$$\mathcal{B}(x) \geq \tfrac{1}{2}[\mathcal{B}(X(x,-\tau) + \mathcal{B}(X(x,\tau))]$$

reduces to

$$\frac{1}{1+x^2} \geq \frac{1}{2}\left[\frac{1-\tau^2}{1-2x\tau+x^2} + \frac{1-\tau^2}{1+2x\tau+x^2}\right] = \frac{(1-\tau^2)(1+x^2)}{(1+x^2)^2-4x^2\tau^2},$$

which is equivalent to

$$(1-\tau^2)(1+x^2)^2 \leq (1+x^2)^2 - 4x^2\tau^2,$$

or

$$(1+x^2)^2 \geq 4x^2,$$

which is obviously true.

1.9.4.2 The Inequality $\mathbf{A}(\frac{s+t}{2}) \geq M[A(s), A(t)]$. Since $\mathbf{A}$ is continuous, it suffices to check this inequality for $s, t \in D$. If $s, t \in D_1$, then our inequality turns into an identity. Suppose now that we already know that our inequality holds for all $s, t \in D_{n-1}$. To check its validity on D_n, we have to consider two cases:

• *Case 1:* $s \in D_n \setminus D_{n-1}, t \in D_{n-1}$. Let $s^\pm = s \pm 2^{-n}$. Note that s^- and s^+ are two neighboring points in D_{n-1}, whence they must lie on the same side of t (it is possible that one of them coincides with t). Denote $y = \mathbf{A}(s^-)$, $z = \mathbf{A}(s^+)$. By definition of the dyadic suspension bridge function $\mathbf{A}$, we then have

$$\mathbf{A}(s) = M[y, z].$$

Denote $x = \mathbf{A}(t)$. Then

$$M[\mathbf{A}(s), \mathbf{A}(t)] = M[M[y, z], x].$$

Note that $\frac{s^-+t}{2}$ and $\frac{s^++t}{2}$ are two neighboring points of D_n and point $\frac{s+t}{2} \in D_{n+1}$ lies between them in the middle. Hence,

$$\mathbf{A}\left(\frac{s+t}{2}\right) = M\left[\mathbf{A}\left(\frac{s^- + t}{2}\right), \mathbf{A}\left(\frac{s^+ + t}{2}\right)\right]$$

But since our inequality holds on D_{n-1}, we have

$$\mathbf{A}\left(\frac{s^- + t}{2}\right) \geq M\left[\mathbf{A}\left(s^-\right), \mathbf{A}\left(t\right)\right] = M[y, x]$$

and

$$\mathbf{A}\left(\frac{s^+ + t}{2}\right) \geq M\left[\mathbf{A}\left(s^+\right), \mathbf{A}\left(t\right)\right] = M[z, x].$$

Using the monotonicity of M in each argument on $[0, 1]^2$, we conclude that

$$\mathbf{A}\left(\frac{s+t}{2}\right) \geq M\left[M[z, x], M[y, x]\right].$$

Therefore, it would suffice to prove that

$$M\left[M[z, x], M[y, x]\right] \geq M[M[y, z], x]$$

for all numbers $x, y, z \in [0, 1]$, such that y and z lie on the same side of x. This will be left to Level 4, where the general scheme of the proof is presented. The complete description of all technical details can be found in [**135**].

• *Case 2:* $s, t \in D_n \setminus D_{n-1}$. Without loss of generality, we may assume that $s < t$. Again, let $s^\pm = s \pm 2^{-n}$, $t^\pm = t \pm 2^{-n} \in D_{n-1}$. Clearly, $s^- < s^+ \leq t^- < t^+$. Denote $x = \mathbf{A}(s^-)$, $y = \mathbf{A}(s^+)$, $z = \mathbf{A}(t^-)$, $w = \mathbf{A}(t^+)$. Then $x < y \leq z < w$.

By definition of the dyadic suspension bridge function $\mathbf{A}$, we have

$$\mathbf{A}(s) = M[x, y], \qquad \mathbf{A}(t) = M[z, w].$$

Now, note that $\frac{s+t}{2} \in D_n$ is also a middle point for pairs s^-, t^+ and s^+, t^- of points in D_{n-1}. Hence, by our assumption, we have

$$\mathbf{A}\left(\frac{s+t}{2}\right) \geq \max\{M[x, w], M[y, z]\}$$

and to prove the desired inequality for $\mathbf{A}$ in this case, it would suffice to show that

$$M[M[x, y], M[z, w]] \leq \max\{M[x, w], M[y, z]\},$$

provided that $0 \le x \le y \le z \le w \le 1$. This will be left to Level 4, where the general scheme of the proof is presented. The complete description of all technical details can be found in [**135**].

1.9.4.3 The ratio $\mathbf{A}(t)/t$ increases. Since $\mathbf{A}$ is continuous, it suffices to check this property for $t \in D$. We shall show by induction on m that for every $t_0 \in D_n \setminus \{\frac{1}{2}\}$, the ratio $\frac{A(t)-A(t_0)}{t-t_0}$ is nondecreasing on $D_{n+m} \cap (t_0, t_0+2^{-n}]$. The property to prove coincides with this statement for $n = 1,\ t_0 = 0$.

The base of induction $m = 1$ is fairly simple. The interval $(t_0, t_0 + 2^{-n}]$ contains just two points of D_{n+1}: $t_1 = t_0 + 2^{-(n+1)} \in D_{n+1} \setminus D_n$ and $t_2 = t_0 + 2^{-n} \in D_n$. By definition of $\mathbf{A}$ and property (1.9.3.1) of M on page 116, we have

$$A(t_1) = M[A(t_0), A(t_2)] \le \frac{A(t_0) + A(t_2)}{2},$$

whence

$$\frac{A(t_1) - A(t_0)}{t_1 - t_0} \le \frac{A(t_2) - A(t_0)}{2(t_1 - t_0)} = \frac{A(t_2) - A(t_0)}{t_2 - t_0}.$$

Now, assume that the statement is already proved for $m-1 \ge 1$. Let $t_0 \in D_n$ and, again, let $t_1 = t_0 + 2^{-(n+1)} \in D_{n+1} \setminus D_n$ and $t_2 = t_0 + 2^{-n} \in D_n \subset D_{n+1}$. By the induction assumption applied to $n+1$ and $m-1$ instead of n and m, we see that the ratio $\frac{A(t)-A(t_0)}{t-t_0}$ is nondecreasing on $D_{n+m} \cap (t_0, t_1]$ and the ratio $\frac{A(t)-A(t_1)}{t-t_1}$ is non-decreasing on $D_{n+m} \cap (t_1, t_2]$. Note also that, for $t \in (t_1, t_2]$, we have the identity

$$\frac{A(t) - A(t_0)}{t - t_0} = \frac{A(t_1) - A(t_0)}{t_1 - t_0} + \frac{t - t_1}{t - t_0}\left[\frac{A(t) - A(t_1)}{t - t_1} - \frac{A(t_1) - A(t_0)}{t_1 - t_0}\right].$$

Since $t \mapsto \frac{t-t_1}{t-t_0}$ is a positive increasing function on $(t_1, t_2]$, checking the nondecreasing property of the ratio $\frac{A(t)-A(t_0)}{t-t_0}$ can be reduced to showing that the factor $\frac{A(t)-A(t_1)}{t-t_1} - \frac{A(t_1)-A(t_0)}{t_1-t_0}$ is nonnegative and nondecreasing on $D_{n+m} \cap (t_1, t_2]$. We know that it is nondecreasing by the induction assumption and, therefore, it suffices to check its nonnegativity at the least element of $D_{n+m} \cup (t_1, t_2]$, which is $t' = t_1 + 2^{-(n+m)}$.

Let $x = \mathbf{A}(t_1)$. By construction of function $\mathbf{A}$, we have $\mathbf{A}(t') = y_m$, where the sequence y_j is defined recursively by $y_1 = \mathbf{A}(t_2),\ y_j = M[x, y_{j-1}]$ for all $j \ge 2$. We shall also consider the auxiliary sequence z_j defined recursively by $z_1 = \mathbf{A}(t_0),\ z_j = \frac{z_{j-1}+x}{2}$ for all $j \ge 2$.

Note that

$$\frac{\mathbf{A}(t_1) - A(t_0)}{t_1 - t_0} = \frac{x - z_1}{t_1 - t_0} = 2^{m-1}\frac{x - z_m}{t_1 - t_0}.$$

Also,

$$\frac{\mathbf{A}(t') - A(t_1)}{t' - t_1} = 2^{m-1}\frac{y_m - x}{t_2 - t_1}.$$

Since $t_2 - t_1 = t_1 - t_0 = 2^{-(n+1)}$, our task reduces to proving that $y_m - x \geq x - z_m$ or, equivalently, $x \leq \frac{z_m + y_m}{2}$. We shall show by induction on j that even a stronger inequality $x \leq M[z_j, y_j]$ holds for all $j \geq 1$.

For the base of induction we have the identity $x = M[z_1, y_1]$, following right from the definition of $\mathbf{A}$. Recall that t_0 and t_2 are two neighboring points of D_n and $t_1 \in D_{n+1}$ lies in the middle between them.

To make the induction step, it would suffice to show that for every triple $0 \leq z \leq x \leq y \leq 1$ satisfying $x \leq M[z, y]$, we also have

$$x \leq M\left[\frac{x+z}{2}, M[x, y]\right].$$

Unfortunately, we have managed to prove it only under the additional restriction $y - z \leq \frac{3}{4}$. Fortunately, this restriction holds automatically almost always. If $n \geq 2$, then using property (1.9.3.1) of the nonlinear mean (see page 116), we get

$$y_j - z_j \leq y_1 - z_1 = A(t_2) - A(t_0) \leq \left(\tfrac{3}{4}\right)^{n-1} \leq \tfrac{3}{4}$$

for all $j \geq 1$. Also, if $j \geq 2$, we have

$$\begin{aligned} y_j - z_j &= M[x, y_{j-1}] - \frac{z_{j-1} + x}{2} \leq \frac{x + y_{j-1}}{2} - \frac{z_{j-1} + x}{2} \\ &= \frac{y_{j-1} - z_{j-1}}{2} \leq \tfrac{1}{2} \end{aligned}$$

for all $n \geq 1$.

Thus, the only case we cannot cover by our induction step is $n = 1$, $j = 2$. We will have to add it to the base of induction. It is just the numerical inequality

$$\frac{1}{\sqrt{5}} \leq M\left[\frac{1}{2\sqrt{5}}, M\left[\frac{1}{\sqrt{5}}, 1\right]\right],$$

which we leave for the reader.

The last observation we want to make in this section is that instead of checking the inequality $x \leq M[\frac{x+z}{2}, M[x, y]]$ for all triples $0 \leq z \leq x \leq y \leq 1$ satisfying $y - z \leq \frac{3}{4}$, $x \leq M[z, y]$, we can check it only for the case $0 \leq y - z \leq \frac{3}{4}$, $x = M[z, y]$.

Indeed, since $M[z,y] \geq x$, $M[z,x] \leq M[x,x] = x$, and M is continuous, we can use the intermediate value theorem and find $y' \in [x,y]$, such that $M[z,y'] = x$. Obviously, $y' - z \leq y - z \leq \frac{3}{4}$ too. Now, if we know that $x \leq M[\frac{x+z}{2}, M[x,y']]$, we can just use the monotonicity of M twice and conclude that $x \leq M[\frac{x+z}{2}, M[x,y]]$ as well. This observation allows us to eliminate x altogether from the inequality to prove. All we need to show is that

$$M[z,y] \leq M\left[\frac{z+M[z,y]}{2}, M[M[z,y],y]\right],$$

whenever $0 \leq z \leq y \leq 1$ and $y - z \leq \frac{3}{4}$. This will be left to Level 4, where the general scheme of the proof is presented. The complete description of all technical details can be found in [**135**].

1.9.5 Level 4: Proofs of Elementary Inequalities

1.9.5.1 General Idea. We shall reduce all our elementary inequalities to checking nonnegativity of some polynomials of two or three variables with rational coefficients on the unit square $[0,1]^2$ or on the unit cube $[0,1]^3$. Since the polynomials that will arise this way are quite large (typically, they can be presented on one or two pages, but one of them, if written down in full, would occupy more than six pages), to check their nonnegativity by hand would be quite a tedious task, to say the very least. So, we will need some simple and easy program to test for nonnegativity that would allow us to delegate the actual work to a computer.

1.9.5.2 Nonnegativity Test. We shall start with polynomials of one variable. Suppose that we want to check that $P(x) = a_0 + a_1 x + a_2 x^2 + \cdots + a_n x^n \geq 0$ on $[0,1]$. Then, of course, we should at least check that $a_0 = P(0) \geq 0$. Suppose it is so. Write our polynomial in the form

$$P(x) = a_0 + x(a_1 + a_2 x + \cdots + a_n x^{n-1})$$

and replace the first factor x by x_1. We shall get a polynomial of two variables

$$Q(x, x_1) = a_0 + x_1(a_1 + a_2 x + \cdots + a_n x^{n-1}).$$

Clearly, if Q is nonnegative on $[0,1]^2$, then P is nonnegative on $[0,1]$. But Q is linear in x_1, so it suffices to check its nonnegativity at the endpoints $x_1 = 0$ and $x_1 = 1$. The first case reduces to checking that $a_0 \geq 0$, which has been done already, and the second case reduces to checking the nonnegativity of the polynomial

$$(a_0 + a_1) + a_2 x + \cdots + a_n x^{n-1} = P(0) + \frac{P(x) - P(0)}{x},$$

which is a polynomial of a smaller degree.

This observation leads us to the following informal algorithm:

(1) Is $P(0) \geq 0$? If not, stop and report failure. If yes, proceed.
(2) Is P constant? If yes, stop and report success. If no, proceed.
(3) Replace P by $P(0) + \frac{P(x)-P(0)}{x}$ and go back to step (1).

Of course, since we know the number of steps needed to reduce the polynomial to a constant exactly (it is just the degree of the polynomial), the "go to" operation will be actually replaced by a "for" loop in the real program. Otherwise, the algorithm we shall use is exactly as written. Here is the formal program for Mathematica, which the reader may want to play with a bit before proceeding, just to make sure it works as promised.

```
P[x_]=...;
flag=False;
n=Exponent[P[x],x];
For[k=0, k<n+1, k++,
    If[P[0]<0, flag=True; Break[] ];
    P[x_]=Expand[P[0]+(P[x]-P[0])/x]
   ];
If[flag, Print["Test failed"], Print["Test successful"]];
```

Of course, when running this program, instead of ellipsis, one needs to plug in the polynomial one wants to test.

If one thinks a bit about what this test really does, one would realize that what is actually being checked is the nonnegativity of the poly-affine form

$$Q(x_1, x_2, \ldots, x_n) = a_0 + a_1 x_1 + a_2 x_1 x_2 + \cdots + a_n x_1 x_2 \ldots x_n$$

on $[0,1]^n$ and the test really reduces to checking that all partial sums of the coefficients starting with a_0 are nonnegative. In this form, the test is well-known to any analyst as the statement that nonnegativity of Cesàro partial sums implies nonnegativity of Abel–Poisson ones. What is surprising here is not the test itself, but its uncanny effectiveness.

The test can easily be generalized to polynomials of more than one variable. All we need to do is treat a polynomial of two or more variables as a polynomial of one fixed variable with coefficients that are polynomials of other variables. In this way, checking the nonnegativity of one polynomial of, say, three variables is reduced to checking nonnegativity of several polynomials of two variables, to each of which we can apply our test again. It seems that the best way to program such a test is to write a recursive subroutine, but since the number of variables in all our applications does not exceed three, we just wrote the test for three variables, as follows:

```
LinearTest=Function[
  flag=False;
  nz=Exponent[R[x,y,z],z];
  For[kz=0, kz<nz+1, kz++,
      S[x_,y_]=R[x,y,0];
      ny=Exponent[S[x,y],y];
      For[ky=0, ky<ny+1, ky++,
          T[x_]=S[x,0];
          nx=Exponent[T[x],x];
          For[kx=0, kx<nx+1, kx++,
              If[T[0]<0, flag=True; Break[] ];
              T[x_]=Expand[T[0]+(T[x]-T[0])/x]
             ]
          If[flag, Break[] ];
          S[x_,y_]=Expand[S[x,0]+(S[x,y]-S[x,0])/y]
         ];
     If[flag, Break[] ];
     R[x_,y_,z_]=Expand[R[x,y,0]+(R[x,y,z]-R[x,y,0])/z]
     ];
  If[flag, Print["Test failed"], Print["Test succeeded"] ];
]
```

The way to apply the test to an actual polynomial is to execute the sequence of commands

```
R[x,y,z]=...;
LinearTest[];
```

where, again, ellipsis should be replaced with the actual polynomial that one wants to test. Note that we can interpret a polynomial of fewer than three variables as a polynomial of three variables, so this three-variable test can be applied verbatim with the same syntax to polynomials of two variables as well. Again, what is actually checked is the nonnegativity of a poly-affine form and the test reduces to checking that all the rectangular partial sums of the coefficients are nonnegative. The last observation implies, in particular, that the order in which the variables are used in the test is of no importance. On the other hand, it is quite possible that the test fails for $P(x)$, but succeeds for $P(1-x)$: just consider $4x - 6x^2 + 4x^3 - x^4 = 1 - (1-x)^4$. So, some clever fiddling with variables may occasionally help.

We omit the details of calculations of polynomials and of checking their nonnegativity in all specific cases we need. The reader can find all the details in [**135**]. Here we only give the list of inequalities we have to check:

- $M[M[y,x],M[z,x]] \geq M[M[y,z],x]$;
- $M[M[x,y],M[z,w]] \leq \max\{M[x,w],M[y,z]\}$;
- $\frac{x_-}{\tilde{x}_-} \geq \frac{\mathcal{B}(x_+)-\mathcal{B}(\tilde{x}_+)}{\mathcal{B}(1)-\mathcal{B}(\tilde{x}_+)}$;
- $M[z,y] \leq M\left[\frac{z+M[z,y]}{2}, M[M[z,y],y]\right]$.

1.10 More about Buckley's Inequality

Now we continue the investigation of the Bellman function related to Buckley's inequality, started in Section 1.2. In that section, we found the Bellman candidate $B(x) = 8(\log x_1 - x_2)$ and proved that $\mathbf{B}(x) \leq B(x)$. To prove the reverse inequality we need to construct an optimizer for every point $x \in \Omega_r$. In this case, there exists no weight supplying us with the value of the Bellman function and we need to construct a family of weights for each point $x = (x_1, x_2)$ from Ω_r. We denote such family of weights as w_n and we will build them recursively.

Fix a point $\tilde{x}$ with

$$\tilde{x}_1 e^{-\tilde{x}_2} = x_1 e^{-x_2}, \qquad \tilde{x}_1 > x_1, \tag{1.10.1}$$

and draw a straight line through these two points. Let us choose point $\tilde{x}$ in such a way that the distance between x and $\tilde{x}$ is exactly n times less than the distance between x and the point of intersection of our line with the boundary $x_2 = \log x_1$. This intersection point we denote by $(\beta_n, \log \beta_n)$. Put

$$x^{(n,k)} \stackrel{\text{def}}{=} \Big(x_1 - k\frac{x_1-\beta_n}{n},\ x_2 - k\frac{x_2-\log\beta_n}{n}\Big), \qquad k = -1, 0, 1, \ldots n,$$

where $x^{(n,0)} = x$, $x^{(n,-1)} = \tilde{x}$, and $x^{(n,n)}$ is our intersection point on the boundary. The neighboring points are equidistant.

Now we start to build the family of test weights $w_n^{(k)}$ for all points $x^{(n,k)}$ simultaneously, using the following rule: weight $w_n^{(k)}$ corresponding to point $x^{(n,k)}$ coincides on the right half I_+ with the scaled copy of $w_n^{(k-1)}$ and on the left half I_- it is the scaled copy of $w_n^{(k+1)}$ for $0 \leq k \leq n-1$. For exceptional endpoints, we have special rules. With point $x^{(n,n)}$ everything is clear, because on that boundary there are the Bellman points of the constant functions only.

Therefore, we must take $w_n^{(n)} = \beta_n$. Test function $w_n^{(-1)}$, corresponding to point $\tilde{x}$, can be taken as a multiple of $w_n = w_n^{(0)}$ due to our choice (1.10.1) and Lemma 1.2.3, i.e.,

$$w_n^{(-1)} = \Big(1 + \frac{x_1 - \beta_n}{n}\Big) w_n.$$

So, in n steps, we can express all $w_n^{(k)}$ in terms of w_n and β_n. Function w_n itself will be recursively expressed as a series of scaled copies of itself and a series of constant functions. So, function w_n is correctly defined on a set of positive measure (where it is piecewise constant). After a series of iterations of this definition, the measure of the set where the function remains undefined, tends to zero as a geometrical progression. Therefore, this recursive definition determines w_n correctly almost everywhere.

Now let $B_n^{(k)}$ be the sum in the definition of the Bellman function for the weight $w = w_n^{(k)}$, $0 \le k \le n-1$. Note that $w_n^{(n)} = \beta_n$ and $B_n^{(n)} = 0$. On the other end of the scale we have $B_n^{(-1)} = B_n^{(0)}$, since $w_n^{(-1)}$ and $w_n^{(0)}$ differ by a multiplicative constant. We have the following identity:

$$B_n^{(k)} = \frac{B_n^{(k+1)} + B_n^{(k-1)}}{2} + \Delta_{n,k},$$

where

$$\Delta_{n,k} = \frac{4(x_1 - \beta_n)^2}{[(n-k)x_1 + k\beta_n]^2},$$

because $\langle w_n^{(k)} \rangle = x_1 - \frac{k}{n}(x_1 - \beta_n)$ and $\langle w_n^{(k+1)} \rangle - \langle w_n^{(k-1)} \rangle = \frac{2}{n}(x_1 - \beta_n)$. Rewrite this equality in the form

$$B_n^{(k)} - B_n^{(k-1)} = B_n^{(k+1)} - B_n^{(k)} + 2\Delta_{n,k}$$

and sum it from $k = m$ to $k = n-1$:

$$B_n^{(m)} - B_n^{(m-1)} = B_n^{(n)} - B_n^{(n-1)} + 2\sum_{k=m}^{n-1} \Delta_{n,k}.$$

If we put $m = 0$ in this identity, we get $B_n^{(n-1)} = 2\sum_{k=0}^{n-1} \Delta_{n,k}$, therefore,

$$B_n^{(m)} - B_n^{(m-1)} = -2\sum_{k=0}^{m-1} \Delta_{n,k}.$$

Now we multiply this identity by -1 and take the sum from $m = 1$ to $m = n$:

$$B_n^{(0)} = 2\sum_{m=1}^{n}\sum_{k=0}^{m-1} \Delta_{n,k} = 8\sum_{m=1}^{n}\sum_{k=0}^{m-1} \frac{\left(\frac{x_1 - \beta_n}{n}\right)^2}{\left[x_1 - \frac{k}{n}(x_1 - \beta_n)\right]^2}.$$

This expression is the Riemannian sum of the integral

$$8\int_{\beta_n}^{x_1} d\tau \int_{\tau}^{x_1} \frac{dt}{t^2} = 8\Big[\log\frac{x_1}{\beta_n} - \frac{x_1-\beta_n}{\beta_n}\Big].$$

It is easy to see that $\beta_n \to \beta$, where β satisfies the following equation:

$$x_2 - \log\beta = \frac{x_1-\beta}{x_1}, \tag{1.10.2}$$

because β is the first coordinate of the point where the tangent line to the curve $x_2 = \log x_1 - \text{const}$ intersects the boundary $x_2 = \log x_1$. Therefore,

$$\lim_{n\to\infty} B_n^{(0)} = 8\Big[\log\frac{x_1}{\beta} - \frac{x_1-\beta}{\beta}\Big] = 8[\log x_1 - x_2].$$

So, we have proved the following theorem:

Theorem 1.10.1 *Let* $\mathbf{B}$ *be defined by* (1.2.2). *Then*

$$\mathbf{B}(x) = 8(\log x_1 - x_2).$$

We see that this function does not depend on A_∞-"norm" of the considered class of weights. It is no longer the case, if we consider the lower Bellman function, i.e., if we take infimum rather than supremum in the definition (1.2.2).

From now on we deal with the lower Bellman function

$$\mathbf{B}(x) = \mathbf{B}_{\min}(x) = \mathbf{B}(x_1, x_2; r)$$

$$\stackrel{\text{def}}{=} \inf_{w\in A_\infty^d(I,r)} \left\{ \frac{1}{|I|}\sum_{J\in\mathcal{D}(I)} |J| \left(\frac{\langle w\rangle_{J_+} - \langle w\rangle_{J_-}}{\langle w\rangle_J}\right)^2 : \right. \tag{1.10.3}$$

$$\left. \langle w\rangle_I = x_1,\ \langle \log w\rangle_I = x_2 \right\}.$$

Our aim is to prove the following theorem:

Theorem 1.10.2

$$\mathbf{B}_{\min}(x; r) = \frac{4(r^2-1)}{r^2\log r}(\log x_1 - x_2).$$

Our new function $\mathbf{B}$ is defined on the same domain Ω_r. It is clear that it satisfies the same boundary condition (Lemma 1.2.2) and the same homogeneity property (Lemma 1.2.3). The main inequality now has the opposite sign:

Lemma 1.10.3 (Main Inequality) *For every pair of points* $x^\pm$ *from* Ω_r, *such that their mean* $x = (x^+ + x^-)/2$ *is also in* Ω_r, *the following inequality holds*:

$$\mathbf{B}(x) \le \frac{\mathbf{B}(x^+) + \mathbf{B}(x^-)}{2} + \left(\frac{x_1^+ - x_1^-}{x_1}\right)^2. \tag{1.10.4}$$

The proof of this Lemma is absolutely the same as that for $\mathbf{B}_{\max}$. We need only replace "sup" by "inf" and "$\ge$" by "$\le$" in the corresponding formulas. However, Bellman induction requires some additional arguments. In the case of $\mathbf{B}_{\max}$, we were able to simply omit a positive term, while now we need to prove that this term is negligible.

LEMMA 1.10.4 (Bellman Induction) *Let g be a nonnegative function on $[1,r]$, $g(1)=0$, such that function $B(x)\stackrel{\text{def}}{=}g(x_1e^{x_2})$ satisfies inequality* (1.10.4) *in Ω_r. Then $\mathbf{B}(x)\ge B(x)$.*

PROOF The procedure of induction starts as before: we fix an interval I and a weight $w\in A_\infty^{d}(I,r)$. Then we repeatedly use the main inequality in the form

$$|J|\,B(\mathfrak{b}_J)\le|J^+|\,B(\mathfrak{b}_{J^+})+|J^-|\,B(\mathfrak{b}_{J^-})+|J|\Big(\frac{\langle w\rangle_{J^+}-\langle w\rangle_{J^-}}{\langle w\rangle_J}\Big)^2,$$

applying it first to I, then to the intervals of the first generation (that is, $I^\pm$) and so on until $\mathcal{D}_n(J)$:

$$\begin{aligned}|I|\,B(\mathfrak{b}_I)&\le|I^+|\,B(\mathfrak{b}_{I^+})+|I^-|\,B(\mathfrak{b}_{I^-})+|I|\Big(\frac{\langle w\rangle_{I^+}-\langle w\rangle_{I^-}}{\langle w\rangle_I}\Big)^2\\&\le\sum_{J\in\mathcal{D}_n(I)}|J|\,B(\mathfrak{b}_J)+\sum_{k=0}^{n-1}\sum_{J\in\mathcal{D}_k(I)}|J|\Big(\frac{\langle w\rangle_{J^+}-\langle w\rangle_{J^-}}{\langle w\rangle_J}\Big)^2.\end{aligned}$$

Therefore,

$$\sum_{k=0}^{n-1}\sum_{J\in\mathcal{D}_k(I)}|J|\Big(\frac{\langle w\rangle_{J^+}-\langle w\rangle_{J^-}}{\langle w\rangle_J}\Big)^2\ge|I|\,B(\mathfrak{b}_I)-\sum_{J\in\mathcal{D}_n(I)}|J|\,B(\mathfrak{b}_J),$$

and, passing to the limit as $n\to\infty$, we get

$$\sum_{J\in\mathcal{D}(J)}|J|\Big(\frac{\langle w\rangle_{J^+}-\langle w\rangle_{J^-}}{\langle w\rangle_J}\Big)^2\le|I|\,B(\mathfrak{b}_I)-\lim_{n\to\infty}\sum_{J\in\mathcal{D}_n(I)}|J|\,B(\mathfrak{b}_J).$$

To get the desired assertion we need to check that the limit in this formula is zero. To this end we rewrite this sum as an integral:

$$\sum_{J\in\mathcal{D}_n(I)}|J|\,B(\mathfrak{b}_J)=\int_I B(x^{(n)}(s))\,ds,$$

where $x^{(n)}(s)=\mathfrak{b}_J$, when $s\in J$, $J\in\mathcal{D}_n(I)$. As $n\to\infty$, we see that $|J|\to0$. Therefore, $x^{(n)}(s)\to(w(s),\log w(s))$. Since g is continuous and $g(1)=0$, we have $B(x^{(n)})=g(x_1^{(n)}e^{-x_2^{(n)}})\to g(1)=0$ for a. e. $s\in I$.

Now we have to justify the possibility of passing to the limit under the integral sign. First, we assume that the weight w is bounded and separated from zero. Then points $x^{(n)}(s)$ run over a bounded set. Continuous function B is bounded on this set and to pass to the limit we can use the Lebesgue dominated convergence theorem. Then, approximating an arbitrary weight w by truncated ones (with the same norm bound, see [**163**]) and using monotone convergence theorem, we get that this limit is zero for any w. □

Now we start our search for Bellman candidate. In the same way as we did for $\mathbf{B}_{\max}$, we come to the same matrix, as in (1.2.5), but it has to be positive, rather than negative, i.e., instead of (1.2.6)–(1.2.7) we have

$$g'' + \frac{8}{s^2} \geq 0, \tag{1.10.5}$$

$$(sg')' \geq 0. \tag{1.10.6}$$

As before, from the requirement for the matrix to be degenerate we come to the same equation and, therefore, to the same general solution:

$$g(s) = c \log s + c_1.$$

Boundary condition $g(1) = 0$ yields again $c_1 = 0$, condition (1.10.6) is fulfilled automatically, but (1.10.5) yields $c \leq 8$. Since we are interested in the maximal possible function g, it seems natural to take $c = 8$. However, that would be a wrong choice, of course, because $c = 8$ gives us $\mathbf{B}_{\max}$ satisfying (1.2.3), while we now need (1.10.4) to be true. Therefore, we have to chose c to be maximal constant, such that (1.10.4) holds.

Lemma 1.10.5 *Function*

$$B(x_1, x_2) = c(\log x_1 - x_2)$$

satisfies the main inequality (1.10.4) *for all pairs* $x^\pm \in \Omega_r$, *such that* $x = \frac{x^+ + x^-}{2}$ *is in* Ω_r *as well, if and only if*

$$c \leq c_{\max}(r) = \frac{4(r^2 - 1)}{r^2 \log r}.$$

Proof As before, put $\Delta = \frac{1}{2}(x^+ - x^-)$, so $x^\pm = x \pm \Delta$. Then

$$\begin{aligned} B(x) &- \frac{B(x^+) + B(x^-)}{2} - \Big(\frac{x_1^+ - x_1^-}{x_1}\Big)^2 \\ &= c \log x_1 - c x_2 - \frac{c}{2} \log(x_1^+ x_1^-) + \frac{c}{2}(x_2^+ + x_2^-) - \Big(\frac{x_1^+ - x_1^-}{x_1}\Big)^2 \end{aligned}$$

$$= \frac{c}{2} \log \frac{x_1^2}{(x_1 + \Delta_1)(x_1 - \Delta_1)} - 4\Big(\frac{\Delta_1}{x_1}\Big)^2$$

$$= -4\left[\frac{c}{8} \log\left(1 - \Big(\frac{\Delta_1}{x_1}\Big)^2\right) + \Big(\frac{\Delta_1}{x_1}\Big)^2\right] \le 0.$$

Therefore, the maximal acceptable constant c is

$$c_{\max}(r) = \inf\left\{\frac{8\big(\frac{\Delta_1}{x_1}\big)^2}{-\log\left(1 - \big(\frac{\Delta_1}{x_1}\big)^2\right)} : x \in \Omega_r,\ x \pm \Delta \in \Omega_r\right\}$$

For a fixed point x the maximal possible Δ_1 occurs for the points $x^\pm$ on the upper boundary $x_2^\pm = \log x_1^\pm$. This occurs if

$$\frac{\log(x_1 + \Delta_1) + \log(x_1 - \Delta_1)}{2} = x_2,$$

i.e., for

$$\Delta_1^2 = x_1^2 - e^{2x_2}.$$

Therefore, for all acceptable points, we have

$$\frac{\Delta_1^2}{x_1^2} \le 1 - \frac{e^{2x_2}}{x_1^2} \le 1 - \frac{1}{r^2}.$$

Since the function $t \mapsto \frac{t}{-\log(1-t)}$ decreases on $(0, 1)$, it attains its minimal value for maximal possible t, i.e., for $t = \big(\frac{\Delta_1}{x_1}\big)^2 = 1 - \frac{1}{r^2}$. Thus, we obtain the announced value of $c_{\max}$. □

So, we have found our Bellman candidate:

$$B(x; r) = \frac{4(r^2 - 1)}{r^2 \log r}(\log x_1 - x_2)$$

and we are ready to check that this candidate is the true Bellman function.

THEOREM 1.10.6

$$\mathbf{B}_{\min}(x; r) = \frac{4(r^2 - 1)}{r^2 \log r}(\log x_1 - x_2).$$

PROOF The inequality $\mathbf{B}(x) \ge B(x)$ follows from Lemmas 1.10.4 and 1.10.5. To get the opposite inequality we construct extremizers. First we find extremizers for points on the lower boundary, i.e., for $x_2 = \log \frac{x_1}{r}$. In this case, extremizers are very easy to find — all of them are proportional to the first

Haar function. That is clear from the fact that for the extremal value we need to take maximal possible Δ_1 for a given x. Namely, we take

$$w = x_1 + \Delta_1 H_I = x_1 + \sqrt{x_1^2 - e^{2x_2}}\, H_I = x_1\Big(1 + \frac{1}{r}\sqrt{r^2 - 1}\Big) H_I .$$

It is clear that $\langle w \rangle_I = x_1$,

$$\langle \log w \rangle_I = \frac{1}{2}\big(\log(x_1 + \Delta_1) + \log(x_1 - \Delta_1)\big) = x_2,$$

and

$$\frac{1}{|I|}\sum_{J\in\mathcal{D}(I)} |J| \Big(\frac{\langle w \rangle_{J_+} - \langle w \rangle_{J_-}}{\langle w \rangle_J}\Big)^2 = \Big(\frac{2\Delta_1}{x_1}\Big)^2 = 4\frac{x_1^2 - e^{2x_2}}{x_1^2} = 4\Big(1 - \frac{1}{r^2}\Big),$$

while

$$\mathbf{B}(x_1, \log\frac{x_1}{r}) = \frac{4(r^2 - 1)}{r^2 \log r}\log r = 4\Big(1 - \frac{1}{r^2}\Big).$$

To construct an extremizer for an arbitrary point x, we note that our candidate is linear with respect to x_2 and in this direction (i.e., for $\Delta_1 = 0$) the main inequality turns into equality. So, for the first splitting we take $x_1^\pm = x_1$ and $x_2^+ = \log x_1$, $x_2^- = \log\frac{x_1}{r}$. The only problem is in the fact that it is not a dyadic splitting, but this problem can be solved by the standard "dyadization": we take the nearest boundary point (in the vertical direction) as, say, x^-, and repeat the procedure, applying it to x^+. Since for each boundary point the extremal weight is known, in the first step, we determine the desired extremizer on one half of the interval, in the second step – on half of the rest, and so on. As a result, the extremizer is well defined almost everywhere. By construction, its Bellman point is just x, and the sum in the definition of the Bellman function is $B(x)$, because in each step we had equality in the Main inequality.

It is easy to formalize this procedure. Take binary decomposition of the ratio in which point x splits the segment of a vertical line between the boundaries, i.e.,

$$\frac{\log x_1 - x_2}{\log r} = \sum_{k=1}^{\infty} \alpha_k 2^{-k}, \qquad \alpha_k \in \{0, 1\}.$$

Then the desired extremizer is

$$w = x_1\left(1 + \frac{\sqrt{r^2 - 1}}{r}\sum_{k=1}^{\infty}\alpha_k H_{[2^{-k},\, 2^{-k+1}]}\right).$$

First, we check that $\mathfrak{b}_{[0,1]} = x$. Clearly, $\langle w \rangle_{[0,1]} = x_1$,

$$\begin{aligned}\langle \log w \rangle_{[0,1]} &= \log x_1 \\ &+ \frac{1}{2}\sum_{k=1}^{\infty} \alpha_k 2^{-k}\left[\log\left(1+\frac{\sqrt{r^2-1}}{r}\right)+\log\left(1-\frac{\sqrt{r^2-1}}{r}\right)\right] \\ &= \log x_1 + \frac{1}{2}\frac{\log x_1 - x_2}{\log r}\log\frac{1}{r^2} = x_2.\end{aligned}$$

And finally,

$$\begin{aligned}\sum_{J\in\mathcal{D}([0,1])} |J|\left(\frac{\langle w \rangle_{J_+} - \langle w \rangle_{J_-}}{\langle w \rangle_J}\right)^2 &= \sum_{k=1}^{\infty} \alpha_k 2^{-k}\frac{4(r^2-1)}{r^2} \\ &= \frac{\log x_1 - x_2}{\log r}\cdot\frac{4(r^2-1)}{r^2} = B(x;r).\end{aligned}$$

□

1.11 Hints and Answers

1.11.1 Problem 1.1.1

1.11.1.1 Answer.

$$\mathbf{B}(x) = 1 - |x_1 - x_2|.$$

1.11.1.2 Hint. Now, the domain of $\mathbf{B}$ is

$$\Omega = \{x = (x_1, x_2) \colon \; -1 \le x_i \le 1\}.$$

The function is concave, as before, and satisfies the following boundary and symmetry conditions:

$$\mathbf{B}(\pm 1, x_2) = \pm x_2, \qquad \mathbf{B}(x_1, \pm 1) = \pm x_1; \tag{1.11.1}$$

$$\mathbf{B}(x_1, x_2) = \mathbf{B}(x_2, x_1), \qquad \mathbf{B}(x) = \mathbf{B}(-x). \tag{1.11.2}$$

The obstacle condition is the same (1.1.7). The proof of Bellman induction (Proposition 1.1.8) is literally the same.

The following functions on $I = [0,1]$ can be taken as extremizers:

$$f_i = \chi_{[0,\frac{1+x_i}{2}]} - \chi_{[\frac{1+x_i}{2},1]}.$$

1.11.2 Problem 1.1.2

1.11.2.1 Answer.

$$\mathbf{B}^{\min}(x) = [x_1 + x_2 - 1]_+ \stackrel{\text{def}}{=} \begin{cases} 0, & x_1 + x_2 - 1 \le 0, \\ x_1 + x_2 - 1, & x_1 + x_2 - 1 > 0. \end{cases}$$

1.11.2.2 Hint. Function $\mathbf{B}^{\min}$ is convex, rather than concave (in the Main inequality we have the opposite sign). It satisfies the same boundary (1.1.3) and symmetry (1.1.4) conditions. The obstacle condition (1.1.7) holds with the opposite sign. The proof of Bellman induction literally repeats the proof of Proposition 1.1.8), with the only exception that we have to replace all inequalities by the opposite ones.

The following functions on $I = [0, 1]$ can be taken as extremizers:

$$f_1 = \chi_{[0,x_1]}, \qquad f_2 = \chi_{[1-x_2,1]}.$$

1.11.3 Problem 1.1.3

1.11.3.1 Answer.

$$\mathbf{B}^{\min}(x) = |x_1 + x_2| - 1.$$

1.11.3.2 Hint. This function $\mathbf{B}^{\min}$ is convex as well. It satisfies the same boundary (1.11.1) and symmetry (1.11.2) conditions as in Problem 1.1.1, but in the obstacle condition (1.1.7) we have to change the inequality sign. The proof of the Bellman induction literally repeats the proof of Proposition 1.1.8), with the only exception that we have to replace all inequalities by the opposite ones.

The following functions on $I = [0, 1]$ can be taken as extremizers:

$$f_1 = \chi_{[0,\frac{1+x_1}{2}]} - \chi_{[\frac{1+x_1}{2},1]}, \qquad f_2 = -\chi_{[0,\frac{1-x_2}{2}]} + \chi_{[\frac{1-x_2}{2},1]}.$$

1.11.4 Problem 1.2.1

For any given point $x^0 \in \Omega_r$, we can find two points $x^\pm$ on the boundary $x_2 = \log x_1$ such that $x^0 = \frac{x^- + x^+}{2}$. Then $w = \frac{1}{2}(x_1^- \mathbf{1}_{[0,\frac{1}{2}]} + x_1^+ \mathbf{1}_{[\frac{1}{2},1]})$ will be the desired weight from $A_\infty^d([0,1], r)$.

However, we can make more, we can construct a weight from $A_\infty([0,1], r)$. Consider the following straight line:

$$x_2 = x_2^0 + \frac{x_1}{x_1^0} - 1$$

passing through the point x^0. We can assume that $x_2^0 < \log x_1^0$, because otherwise the constant function $w = x_1^0$ is an appropriate weight. Since the boundary function $x_2 = \log x_1$ is strictly concave, our straight line has two points of intersection with this boundary, say, $x^\pm$. Note that the whole segment $[x^-, x^+]$ is inside the domain Ω_r because our line is in the strip between two tangent lines to the upper and lower boundaries of Ω_r:

$$x_2 = \log x_1^0 + \frac{x_1}{x_1^0} - 1 \qquad \text{and} \qquad x_2 = \log \frac{x_1^0}{r} + \frac{x_1}{x_1^0} - 1.$$

Since x^0 is a convex combination of the points $x^\pm$, there exists a number $\alpha \in (0,1)$ such that $x^0 = \alpha x^- + (1-\alpha)x^+$, and we can take the desired weight as a concatenation of two constant functions $w = x_1^- \mathbf{1}_{[0,\alpha]} + x_1^+ \mathbf{1}_{[\alpha,1]}$. It is clear that $\mathfrak{b}_{[0,1]}(w) = x^0$ and the property $w \in A_\infty([0,1], r)$ follows from the fact that the entire segment $[x^-, x^+]$ is inside domain Ω_r.

1.11.5 Problem 1.2.2

Complete answer to this problem is given in Section 1.10.

1.11.6 Problem 1.2.3

1.11.6.1 Answer.

$$\mathbf{B}(x; m, M) = 4M\sqrt{x_1 x_2} - x_1 x_2 + m^2 - 4mM.$$

1.11.6.2 Hint. The domain of $\mathbf{B}$ is

$$\Omega = \big\{x = (x_1, x_2) \colon m^2 \le x_1 x_2 \le M^2\big\}.$$

Evident properties:

- The function $\mathbf{B}$ does not depend on I.
- Homogeneity: $\mathbf{B}(x_1, x_2) = \mathbf{B}(x_1 x_2, 1) \stackrel{\text{def}}{=} g(x_1 x_2)$.
- Boundary condition: $\mathbf{B}|_{x_1 x_2 = m^2} = g(m^2) = 0$.

Main inequality:
For every pair $x^\pm \in \Omega$ such that $x = \frac{1}{2}x^+ + \frac{1}{2}x^- \in \Omega$, we have

$$\mathbf{B}(x) \ge \frac{\mathbf{B}(x^+) + \mathbf{B}(x^-)}{2} + \frac{|x_1^+ - x_1^-|\,|x_2^+ - x_2^-|}{4}.$$

In differential form,

$$\begin{pmatrix} \dfrac{\partial^2 B}{\partial x_1^2} & \dfrac{\partial^2 B}{\partial x_1 \partial x_2} \pm 1 \\[2ex] \dfrac{\partial^2 B}{\partial x_1 \partial x_2} \pm 1 & \dfrac{\partial^2 B}{\partial x_2^2} \end{pmatrix} \le 0,$$

or, in terms of g,

$$\begin{pmatrix} x_2^2 g'' & g' + x_1 x_2 g'' + \sigma \\[2ex] g' + x_1 x_2 g'' + \sigma & x_1^2 g'' \end{pmatrix} \le 0,$$

where $\sigma = \pm 1$.

The condition that this matrix has to be degenerate gives us a differential equation, whose general solution is $g(s) = 2c\sqrt{s} - s + c_1$ (this is quite a bit of work). Constant c_1 can be found from the boundary condition: $c_1 = m^2 - 2cm$, and constant c has to be chosen as small as possible to obtain the best estimate: $c = 2M$.

In this way, we have constructed the following Bellman candidate:

$$B(x; m, M) = 4M\sqrt{x_1 x_2} - x_1 x_2 + m^2 - 4mM.$$

Using Bellman induction, we check that $\mathbf{B}(x) \le B(x)$. The proof of the stated theorem follows from the inequality

$$\mathbf{B}(x; 0, M) \le 4M\sqrt{x_1 x_2}.$$

All details of the proof that this Bellman candidate is, in fact, the true Bellman function, i.e.,

$$\mathbf{B}(x; m, M) = 4M\sqrt{x_1 x_2} - x_1 x_2 + m^2 - 4mM,$$

can be found in [**187**].

1.11.7 Problem 1.3.1

Direct calculations yield

$$\sum_{i,j=1}^{2} \frac{\partial^2 B_\pm}{\partial x_i \partial x_j} \Delta_i \Delta_j$$

$$= \mp \frac{\left((x_1 \pm \sqrt{\delta^2 - x_2 + x_1^2}) \Delta_1 - \frac{1}{2}\Delta_2 \right)^2}{\sqrt{\delta^2 - x_2 + x_1^2}(1 \mp \delta)} \exp\left\{ x_1 \pm \sqrt{\delta^2 - x_2 + x_1^2} \mp \delta \right\}.$$

1.11.8 Problem 1.3.2

According to the preceding formula, the quadratic form of the Hessian is zero along the direction

$$\left(x_1 \pm \sqrt{\delta^2 - x_2 + x_1^2}\right)\Delta_1 - \frac{1}{2}\Delta_2 = 0.$$

Let y be a point running along the extremal, i.e., $\Delta_i = y_i - x_i$. Then the equation of the extremal passing through point x is

$$y_2 = x_2 + 2\left(x_1 \pm \sqrt{\delta^2 - x_2 + x_1^2}\right)(y_1 - x_1).$$

Let us first consider the sign "+". For this choice the Hessian is nonpositive (see Problem 1.3.1) and it corresponds to our Bellman function. This line is tangent to the upper boundary at point $(u, u^2 + \delta^2)$, where $u \stackrel{\text{def}}{=} x_1 + \sqrt{\delta^2 - x_2 + x_1^2}$, and intersects the lower boundary at points $(u_\pm, u_\pm^2)$, where $u_\pm = u \pm \delta$. However, only half of this segment is an extremal trajectory. Indeed, if we rewrite the equation of the extremal line in terms of u, we get

$$y_2 = 2uy_1 + \delta^2 - u^2.$$

Therefore,

$$\delta^2 - y_2 + y_1^2 = (u - y_1)^2$$

and looking at Theorem 1.3.10, we can see

$$B(y) = \frac{1 - |u - y_1|}{1 - \delta} \exp\{y_1 + |u - y_1| - \delta\}.$$

So, we see that if $y_1 < u$, then this function is linear, but it is not linear for $y_1 > u$. Therefore, the extremal line is a one-sided tangent segment with $y_1 \in [u - \delta, u]$. We will call such a segment *left tangent line*, because it is to the left of the tangency point. The whole domain Ω_δ is foliated by the left tangent lines which are the extremals of our function B.

For function B_- the picture is symmetrical: The domain Ω_δ is foliated by the right tangent lines, along which B_- is linear.

1.11.9 Problem 1.3.3

After multiplying the desired inequality by $(1 - \delta)\exp\left(\frac{\delta}{\sqrt{2}} - u\right)$, we get

$$2e^{-(1-\frac{1}{\sqrt{2}})\delta} \le 1 - \delta + e^{(\sqrt{2}-1)\delta}.$$

So, we need to check that the function

$$\omega(\delta) \stackrel{\text{def}}{=} 1 - \delta + e^{(\sqrt{2}-1)\delta} - 2e^{-(1-\frac{1}{\sqrt{2}})\delta}$$

is nonnegative for $\delta \in (0,1)$. Since $\omega(0) = \omega'(0) = 0$ and

$$\omega''(\delta) = (\sqrt{2}-1)^2 \left(e^{(\sqrt{2}-1)\delta} - e^{-(1-\frac{1}{\sqrt{2}})\delta}\right) \geq 0,$$

we have the required property.

1.11.10 Problem 1.3.4

The first three assertions could be verified by direct calculation, but in the fourth one it is useful to apply the cut-off lemma (Lemma 1.3.4): it is sufficient to check that $\log t \in \mathrm{BMO}_1$, which follows from $\langle \log^2 t \rangle_{[c,d]} - \langle \log t \rangle^2_{[c,d]} = 1 - \frac{cd}{(d-c)^2}\left(\log \frac{d}{c}\right)^2$.

1.11.11 Problem 1.3.5

The proof repeats the consideration of $\mathbf{B} = \mathbf{B}_{\max}$ in the text of this section. We only need to replace the word "concave" by the word "convex" and, therefore, reverse the sign in the inequalities. As to the extremizer, it can be obtained by reversing the the signs of x_1, u, and φ in the extremizer for $\mathbf{B}_{\max}$, given by (1.3.12) on page 26:

$$\varphi(t) = \begin{cases} \varepsilon \log \frac{\varepsilon t}{u - x_1} + u & 0 \leq t \leq \frac{u-x_1}{\varepsilon} \\ u & \frac{u-x_1}{\varepsilon} \leq t \leq 1 \end{cases},$$

where

$$u = x_1 - \sqrt{\varepsilon^2 - x_2 + x_1^2} + \varepsilon.$$

2

What You Always Wanted to Know about Stochastic Optimal Control, but Were Afraid to Ask

2.1 Disclaimer

The problems we were concerned with in Chapter 1 and the ones we will be concerned with in the chapters that follow have distinctive flavor of problems considered in a different branch of mathematics. Namely, they look very much like the problems of stochastic optimal control. Only the objects we are optimizing are usually related to singular integrals. To make this analogy more conspicuous and practically unavoidable, we need the language pertinent to stochastic optimal control, namely, the language of stochastic integrals and Itô's formula.

From the very beginning, we would like to warn the reader that we are not the experts either in stochastic integrals or in stochastic optimal control. So this chapter is filled more with reasonings (that we feel believable) than with rigorous proofs, which the reader can find in the standard textbooks on the subject cited in this chapter.

In our defense, we would like to draw the analogy of our situation with that of a person who might not be a zoologist at all, but still can describe the features common for elephant and giraffe.

2.2 Stochastic Integrals Are Not That Simple

The definition of stochastic integral is quite different from the Riemann sum definition. It should not be such a surprise, because we have to integrate a random process with respect to another random process. In this case, the Riemann sum definition does not lead to correctly defined object. Let us show why.

Let $w(s)$ denote Brownian motion started at 0, that is, $w(0) = 0$. For all $0 \leq t_1 < t_2$, random variable $w_{t_2} - w_{t_1}$ is a Gaussian variable with zero average and variance $\sqrt{t_2 - t_1}$, and for all $0 \leq t_1 < t_2 \leq t_3 < t_4$ random variables $w(t_2) - w(t_1)$, $w(t_4) - w(t_3)$ are independent.

We want to understand what this integral means:

$$\int_{s_1}^{s_2} \xi(t)\, dw(t).$$

As we will see later, the definition of stochastic integral starts with considering the class of step functions having finitely many steps. Each such function is associated with the usual partition of interval $[a, b]$ into finitely many subintervals. At first glance, this reminds us of the Riemann integral definition.

We have warned the reader that this superficial analogy with Riemann integral is deceiving. Stochastic integration is not the Riemann sum definition, as the following example shows.

EXAMPLE Consider two simplest Riemann sums, built on a partition of the interval $[s_1, s_2]$:

$$\Sigma_1 \stackrel{\text{def}}{=} \sum_{i=1}^{m} w(t_{i-1})(w(t_i) - w(t_{i-1})),$$

$$\Sigma_2 \stackrel{\text{def}}{=} \sum_{i=1}^{m} w(t_i)(w(t_i) - w(t_{i-1})).$$

If such refinement were small and if stochastic integral were a Riemann sum thing, these two random variables, Σ_1 and Σ_2, would have been close. Let us see, whether that is the case.

Notice that they are in $L^2(\Omega, \mathcal{F}, \mathbb{P})$ uniformly when the partition changes. Here $(\Omega, \mathcal{F}, \mathbb{P})$ is the probability space on which Brownian motion is given. Using the property of independence we can write:

$$\begin{aligned}
\mathbb{E}(\Sigma_1^2) &= 2\sum_{i<j} \mathbb{E}\left[w(t_{i-1})\big(w(t_i) - w(t_{i-1})\big)w(t_{j-1}) \cdot \big(w(t_j) - w(t_{j-1})\big)\right] \\
&\quad + \sum_i \mathbb{E}\left[w(t_{i-1})^2\big(w(t_i) - w(t_{i-1})\big)^2\right] \\
&= 2\sum_{i<j} \mathbb{E}\left[w(t_{i-1})\big(w(t_i) - w(t_{i-1})\big)w(t_{j-1})\right] \cdot \mathbb{E}\left[w(t_j) - w(t_{j-1})\right] \\
&\quad + \sum_i t_i(t_i - t_{i-1}) \leq s_2(s_2 - s_1).
\end{aligned}$$

Then,

$$\mathbb{E}(\Sigma_2^2) \le 2\mathbb{E}(\Sigma_1^2) + 2\mathbb{E}\left[\sum_i (w(t_i) - w(t_{i-1}))^2\right]^2$$
$$\le 2s_2(s_2 - s_1) + 2\mathbb{E}\left[\sum \xi_i^2\right]^2,$$

where $\xi_i \overset{\text{def}}{=} w(t_i) - w(t_{i-1})$ are Gaussian independent with zero average and with variance $|t_i - t_{i-1}|$. Then,

$$\mathbb{E}\left(\sum \xi_i^2\right)^2 = 2\sum_{i<j} \mathbb{E}\xi_i^2\, \mathbb{E}\xi_j^2 + \mathbb{E}\xi_i^4$$
$$= 2\sum_{i<j} |t_i - t_{i-1}||t_j - t_{j-1}| + 3\sum_i |t_i - t_{i-1}|^2 \le 5(s_2 - s_1)^2.$$

If the correct definition of integral were given by Riemann sum approach, then Σ_1, Σ_2 would have been close, in some sense, for large m. They are definitely not close in $L^1(\Omega, \mathbb{P})$. In fact,

$$\mathbb{E}(\Sigma_2 - \Sigma_1) = \mathbb{E}\sum_i (w(t_i) - w(t_{i-1}))^2 = \sum_i (t_i - t_{i-1}) = s_2 - s_1 \ne 0. \tag{2.2.1}$$

One of the most natural notions of closeness for random variables is the closeness in probability. It is also the weakest sense of closeness. Suppose Σ_1 and Σ_2 are close in probability. Then we use the following simple claim.

LEMMA 2.2.1 *Let $(\Omega, \mathbb{P})$ be a probability space and f_n be measurable functions, such that $f_n \Rightarrow 0$ in measure, and such that $\|f_n\|_{L^2(\Omega,\mathbb{P})} \le C < \infty$ with C independent of n. Then there is a subsequence f_{n_k}, such that $\|f_{n_k}\|_{L^1(\Omega,\mathbb{P})} \to 0$.*

PROOF Fix a decreasing to zero sequence of positive numbers ε_k and choose n_k, such that $\mathbb{P}\{|f_m| > \varepsilon_k\} \le 2^{-k}$ for all $m \ge n_k$. Denote

$$E_p \overset{\text{def}}{=} \bigcap_{k=p}^{\infty} \{\omega \in \Omega : |f_{n_k}| \le \varepsilon_k\}.$$

Then it is clear that

$$\mathbb{P}(\Omega \setminus E_p) \le 2^{-(p-1)}.$$

Now for any $j \ge p$, we can write the estimate

$$\int_\Omega |f_{n_j}|\, d\mathbb{P} = \int_{\Omega\setminus E_p} |f_{n_j}|\, d\mathbb{P} + \int_{E_p} |f_{n_j}|\, d\mathbb{P}.$$

The first integral is at most $\|f_{n_j}\|_{L^2(\Omega,\mathbb{P})}\sqrt{\mathbb{P}(\Omega\setminus E_p)} \leq 2^{-\frac{(p-1)}{2}}C$. The second integral is at most ε_p by the definition of E_p and the fact that $j \geq p$. □

In our case, $\{f_m\}$ is a sequence $\Sigma_1 - \Sigma_2$. Remember that both of these sums depend on the number m of points in the partition. We saw that the $L^2(\Omega, \mathbb{P})$ norms of $\Sigma_1 - \Sigma_2$ are uniformly bounded. But the $L^1(\Omega, \mathbb{P})$-norm of $\Sigma_1 - \Sigma_2$ does not go to zero, see (2.2.1).

We understand now that stochastic integral $\int_{s_1}^{s_2} \xi(t)\, dw(t)$ is a much more subtle thing, than Riemann sum integral. Stochastic integrals were understood by Kioshi Itô. The book **[199]** has a good explanation of this notion.

2.3 Itô's Definition of Stochastic Integral

Let $B\mathcal{F}$ be a sigma algebra of sets $A \subset \mathbb{R} \times \Omega$, such that for every $t \in [s_1, s_2]$, we have $A \cap ((-\infty, t] \times \Omega)$ in $B_t \times \mathcal{F}_t$, where B_t is Borel sigma algebra on $(-\infty, t]$ and $\mathcal{F}_t$ is a sigma algebra generated by $\{w(s)\}_{s\leq t}$. Let $M_2[s_1, s_2]$ be the set of real valued functions of $(t, \omega) \in \mathbb{R} \times \Omega$, measurable with respect to $B\mathcal{F}$, and such that:

(a) $f(t, \cdot)$ is measurable with respect to $\mathcal{F}_t$ for each t,
(b) $\int_{s_1}^{s_2} |f(t, \cdot)|^2\, dt < \infty$ with probability 1.

We often write $f(t)$ instead of $f(t, \omega)$, keeping in mind that for every t we deal with random variable $f(t)$. For all random functions f (random processes) Itô defines

$$\int_{s_1}^{s_2} f(t)\, dw(t). \tag{2.3.1}$$

Definition 2.3.1 $f \in M_2[s_1, s_2]$ is called a step function if there exists a finite partition of $[s_1, s_2]$, such that $f(t) = f(t_i, \omega)$, for $t \in [t_i, t_{i+1})$.

We introduce the stochastic integral for step functions in a natural way:

$$\int_{s_1}^{s_2} f\, dw(t) \stackrel{\text{def}}{=} \sum_i f(t_i) \cdot (w(t_{i+1}) - w(t_i))\,.$$

Proposition 2.3.2 *For every $f \in M_2[s_1, s_2]$, there exits a sequence of step functions, as above, such that with probability 1 we have*

$$\lim_{n\to\infty} \int_{s_1}^{s_2} |f(t) - f_n(t)|^2\, dt = 0. \tag{2.3.2}$$

Moreover, if in addition

$$\mathbb{E}\int_{s_1}^{s_2} |f(t)|^2\,dt < \infty,$$

then step functions can be chosen in such a way that

$$\lim_{n\to\infty} \mathbb{E}\int_{s_1}^{s_2} |f(t) - f_n(t)|^2\,dt = 0.$$

We need the following proposition.

PROPOSITION 2.3.3 *Let φ be a step function, as in Definition 2.3.1. Let $\delta, \varepsilon > 0$. Then*

$$\mathbb{P}\left[\left|\int_a^b \varphi(t)\,dw(t)\right| > \varepsilon\right] \le \frac{\delta}{\varepsilon^2} + \mathbb{P}\left[\int_a^b |\varphi(t)|^2\,dt > \delta\right].$$

This proposition immediately gives us the following: suppose that $f \in M_2[s_1, s_2]$ and f_n are step functions from Proposition 2.3.2. Then

$$\mathbb{P} - \lim_{n\to\infty} \int_{s_1}^{s_2} |f(t) - f_n(t)|^2\,dt = 0.$$

So,

$$\mathbb{P} - \lim_{n,m\to\infty} \int_{s_1}^{s_2} |f_m(t) - f_n(t)|^2\,dt = 0.$$

By definition,

$$\forall\varepsilon > 0\,,\, \mathbb{P}\left[\int_{s_1}^{s_2} \left|f_m(t) - f_n(t)\right|^2 dt > \varepsilon\right] \to 0\,, \quad m, n \to \infty.$$

Now we use Proposition 2.3.3 to see that

$$\limsup_{m,n\to\infty} \mathbb{P}\left[\left|\int_{s_1}^{s_2} f_n(t)\,dw(t) - \int_a^b f_m(t)\,dw(t)\right| > \varepsilon\right] \le \frac{\delta}{\varepsilon^2}$$

for any $\delta > 0$. So the sequence of random variables $\xi_n \overset{\text{def}}{=} \int_{s_1}^{s_2} f_n(t)\,dw(t)$ is Cauchy convergent in measure (in probability). So in probability it converges to a certain random variable ξ. It is easy to see that any other sequence of step functions, satisfying (2.3.2), converges to the same random variable ξ.

DEFINITION 2.3.4 This ξ is by definition $\int_{s_1}^{s_2} f\,dw(t)$.

Itô's stochastic integral is therefore constructed. Of course, the properties of this integral are somewhat more useful than its definition.

Itô's integral has several interesting properties that we now list.

If $\mathbb{E}\int_{s_1}^{s_2} |f(t)|^2\,dt < \infty$, then

$$\mathbb{E}\int_{s_1}^{s_2} f(t)\,dw(t) = 0. \tag{2.3.3}$$

Also

$$\mathbb{E}\left[\int_{s_1}^{s_2} f(t)\,dw(t)\right]^2 = \mathbb{E}\int_{s_1}^{s_2} |f(t)|^2\,dt. \tag{2.3.4}$$

If in addition $\mathbb{E}\int_{s_1}^{s_2} |g(t)|^2\,dt < \infty$, then

$$\mathbb{E}\left[\int_{s_1}^{s_2} f(t)\,dw(t)\cdot\int_{s_1}^{s_2} g(t)\,dw(t)\right] = \mathbb{E}\int_{s_1}^{s_2} f(t)\cdot g(t)\,dt. \tag{2.3.5}$$

Meaning that integral of the product is the product of integrals.

2.4 Stochastic Differential and Itô's Formula

Let $a(t) \in M_2[s_1, s_2]$. Let $b(t)$ be measurable with respect to $\mathcal{F}_t$ for every t, and be such that

$$\int_{s_1}^{s_2} |b(t)|\,dt < \infty$$

with probability 1.

Suppose $\zeta(t)$ is a random process, such that for all t_1 and t_2 from $[s_1, s_2]$ we have

$$\zeta(t_2) - \zeta(t_1) = \int_{t_1}^{t_2} b(t)\,dt + \int_{t_1}^{t_2} a(t)\,dw(t). \tag{2.4.1}$$

DEFINITION 2.4.1 Then the formula above defines stochastic differential:

$$d\zeta(t) = b(t)dt + a(t)dw(t).$$

This equality is just an abbreviation for equation (2.4.1).

REMARK 2.4.2 If $b = 0$, this integral is obviously a martingale on the filtration $\{\mathcal{F}_t\}_{t>0}$ of sigma algebras generated by Brownian motion.

2.4.1 Itô's Formula

THEOREM 2.4.3 *Let* ζ *be a process defined by* (2.4.1) *with parameters* a, b, *and let* $u(t, x)$ *be a sufficiently smooth function. Consider a new process*

$$\eta(t) \stackrel{\text{def}}{=} u(t, \zeta(t)).$$

Then the process η has stochastic differential, and

$$d\eta(t) = \left[u_t'(t,\zeta(t)) + u_x'(t,\zeta(t))b(t) + \frac{1}{2}u_{xx}''(t,\zeta(t)) \cdot a^2(t)\right] dt \\ + u_x'(t,\zeta(t)) \cdot a(t)dw(t). \tag{2.4.2}$$

PROOF The proof is quite subtle. See [**63, 199**]. □

2.4.2 Matrix Itô's Formula

Matrix Itô's formula also exists and will be used in Section 2.5.1. Let b be $d \times 1$ column of processes and let a be a $d \times d_1$ matrix of processes with entries in $M_2[s_1, s_2]$. Let $W(t)$ be a column of d_1 independent Brownian motions. Let $\zeta(t)$ be a $d \times 1$ process with stochastic differential

$$d\zeta(t) = b(t)\,dt + a\,dW(t).$$

Let $u(t,x)$ be a sufficiently smooth function, where $x \in \mathbb{R}^d$. Let $\eta(t) = u(t,\zeta(t))$. Then η also has stochastic differential, and matrix Itô's formula gives us the following:

$$d\eta(t) = \left[\frac{\partial u}{\partial t} + \nabla_x u(t,\zeta) \cdot b(t) + \tfrac{1}{2}\,\mathrm{trace}(a^* H_u(t,\zeta)a)\right] dt + \nabla_x u \cdot a\,dW(t). \tag{2.4.3}$$

Here $\cdot$ is the scalar product in $\mathbb{R}^d$, matrix H_u is an $d \times d$ matrix of second derivatives of u with respect to x-variables (Hessian matrix of u).

2.5 Bellman Functions of Stochastic Optimal Control Problems and Bellman PDEs

Let W consist of d_1 independent Brownian motions, it will be called d_1 dimensional Brownian motion. We call (vector) process α adapted if $\alpha(s)$ is measurable with respect to sigma algebra $\mathcal{F}_s$ generated by $W(t), 0 \le t \le s$. Let $a = a(\alpha, x)$ be $d \times d_1$ matrix. Let $\zeta(t)$ is a d-dimensional process given by

$$\zeta(t) = x + \int_0^t b(\alpha(s),\zeta(s))\,ds + \int_0^t a\big(\alpha(s),\zeta(s)\big)\,dW(s), \tag{2.5.1}$$

in other words the process starts at $x = \zeta(0) \in \mathbb{R}^d$ and satisfies a stochastic differential equation

$$d\zeta(t) = b(\alpha(t),\zeta(t))\,dt + a\big(\alpha(t),\zeta(t)\big)\,dW(t),$$

where α is a d_2-dimensional control process, we can choose it ourselves, but it must be adapted. Also values of the process α are often restricted: $\alpha(s, \omega) \in A \subset \mathbb{R}^{d_2}$.

DEFINITION 2.5.1 The expression $b(\alpha, x)$ is called a drift term.

Matrix function a is smooth and $d \times d_1$-dimensional, and b is a smooth column function of size d. Everything happens in domain $\Omega \subset \mathbb{R}^d$ (often but not always $\Omega = \mathbb{R}^d$).

The choice of adapted process $\alpha(s)$ gives us different motions, all started at the same initial $x \in \mathbb{R}^d$.

We can imagine this as a "broom" of motions, each hidden elementary event ω gives us "one stem of a broom."

Suppose we are given the (usually nonnegative) profit function $f(\alpha, x)$, meaning that on a trajectory of $\zeta(t)$, for the time interval $[t, t + \Delta t]$, the profit is $f(\alpha(t), \zeta(t))\Delta t + o(\Delta t)$. So on the whole trajectory, we earn

$$\int_0^\infty f(\alpha(t), \zeta(t))\, dt.$$

We are also given the (nonnegative) bonus function F. This is how much one is given (in a lump sum) at one's retirement.

DEFINITION 2.5.2 For a fixed process $\alpha = \alpha(s)$ we introduce the average profit:

$$v^\alpha(x) \stackrel{\text{def}}{=} \mathbb{E} \int_0^\infty f(\alpha(t), \zeta(t))\, dt + \mathbb{E} \limsup_{t \to \infty} F(\zeta(t)). \tag{2.5.2}$$

We want to choose a control process α to maximize the average profit.

DEFINITION 2.5.3 The optimal average gain is the quantity

$$v(x) \stackrel{\text{def}}{=} \sup_\alpha v^\alpha(x), \tag{2.5.3}$$

where supremum is taken over all adapted processes α.

DEFINITION 2.5.4 Function $v(x)$ is called the Bellman function of stochastic optimal control problem (2.5.1), (2.5.2).

For $\Omega \neq \mathbb{R}^d$ the following modification is needed. Let

$$\tau(\omega) \stackrel{\text{def}}{=} \inf\{t \geq 0 \colon \zeta(t) \notin \Omega\}$$

Then (2.5.2) changes to

$$v^\alpha(x) \stackrel{\text{def}}{=} \mathbb{E} \int_0^\tau f(\alpha(t), \zeta(t))\, dt + \mathbb{E} \limsup_{t \to \tau} F(\zeta(t)). \tag{2.5.4}$$

Usually the analysis consists of

a) writing Bellman PDE on v in Ω;
b) writing boundary conditions on (part of) the boundary of Ω if $\Omega \neq \mathbb{R}^d$;
c) solving it;
d) using "verification theorem," which says that under certain conditions on data a, b, F, f, Ω, A the classical solution of Bellman PDE is exactly v from (2.5.3).

2.5.1 Writing Bellman PDE

This part consists of (a) matrix Itô's formula from Section 2.4.2 and (b) Bellman's principle of dynamic programming. We will show the reader the idea of how these tools are used, but we will not pursue the mathematical rigor here, referring the interested reader to [**95**].

Our first assumption (the reader understands that unfortunately it is not justified) is that Bellman function v from (2.5.3) is sufficiently smooth. Then we can use Itô's formula (2.4.3) and get

$$\begin{aligned} dv(\zeta(s)) = &\sum_{k=1}^{d} \frac{\partial v}{\partial x_k}(\zeta(s)) \sum_{j=1}^{d_1} a_{kj}(\alpha(s), \zeta(s))\, dw(s)^j \\ &+ \sum_{k=1}^{d} \frac{\partial v}{\partial x_k}(\zeta(s)) b_k(\alpha(s), \zeta(s))\, ds \\ &+ \frac{1}{2} \sum_{i,j=1}^{d} \frac{\partial^2 v}{\partial x_i \partial x_j}(\zeta(s)) a^{ij}(\alpha(s), \zeta(s))\, ds, \end{aligned}$$

where

$$a^{ij}(\alpha, x) \stackrel{\text{def}}{=} \sum_{k=1}^{d_1} a_{ik}(\alpha, x) a_{kj}(\alpha, x)$$

is i, j matrix element of $d \times d$ matrix aa^*.

Introduce two linear differential operators with nonconstant coefficients:

$$\mathcal{L}_1(\alpha, x) \stackrel{\text{def}}{=} \sum_{k=1}^{d} b_k(\alpha, x) \frac{\partial}{\partial x_k},$$

$$\mathcal{L}_2(\alpha, x) \stackrel{\text{def}}{=} \sum_{i,j=1}^{d} a^{ij}(\alpha, x) \frac{\partial^2}{\partial x_i \partial x_j},$$

and

$$\mathcal{L}(\alpha, x) \stackrel{\text{def}}{=} \mathcal{L}_1(\alpha, x) + \mathcal{L}_2(\alpha, x).$$

In the first line of the formula for $dv(\zeta(t))$, all $\alpha(s)$, $x(s)$ and functions of $\alpha(s)$, $x(s)$ are $\mathcal{F}_s$-measurable, but $\mathbb{E}[dW(s)|\mathcal{F}_s] = 0$. Therefore, if we hit our formula for $dv(\zeta(t))$ by the expectation $\mathbb{E}$, then the first line becomes 0, and we get

$$\mathbb{E}\left[\frac{d}{dt}v(\zeta(t))\right] = \mathbb{E}[\mathcal{L}(\alpha(t), \zeta(t))v(\zeta(t))].$$

Or

$$\mathbb{E}v(\zeta(t)) = v(x) + \mathbb{E}\int_0^t [\mathcal{L}_1(\alpha(s), \zeta(s)) + \mathcal{L}_2(\alpha(s), \zeta(s))]v(\zeta(s))\, ds. \tag{2.5.5}$$

Now we need the second ingredient to write the Bellman equation: the Bellman principle or dynamic programming principle. It is in this next equality:

$$v(x) = \sup_\alpha \mathbb{E}\left[\int_0^\tau f(\alpha(s), \zeta(s))\, ds + \limsup_{t\to\tau} F(\zeta(s))\right] \tag{2.5.6}$$

$$= \sup_\alpha \mathbb{E}\left[\int_0^{\min(t,\tau)} f(\alpha(s), \zeta(s))\, ds + v(\zeta(\min(t,\tau)))\right], \quad \forall t > 0. \tag{2.5.7}$$

After a minute's thought one sees that this equality reflects the stationarity of Brownian motion and the fact that to be perfect one has to be perfect every moment.

Now plug $\mathbb{E}v(x(\min(t,\tau)))$ from (2.5.5) into (2.5.7). We get

$$0 = \sup_\alpha \mathbb{E}\int_0^{\min(t,\tau)} [f(\alpha(s), \zeta(s)) + \mathcal{L}(\alpha(s), \zeta(s))v(\zeta(s))]\, ds, \quad \forall t > 0.$$

Divide by t and tend t to zero. We "obtain" Bellman equation:

$$\sup_{\alpha\in A}\{\mathcal{L}(\alpha, x)v(x) + f(\alpha, x)\} = 0. \tag{2.5.8}$$

See details of obtaining (2.5.8) in the book of N. Krylov [**95**].

2.5.2 Obstacle Condition

Let f, F be nonnegative (this is not a big assumption as this is usually the case), and also let F be convex (this is often the case). Let us also assume that

$$\lim_{t \to \tau} \zeta(t) = \zeta(\tau), \quad \mathbb{P} - a.e. \tag{2.5.9}$$

This gives the following

$$\mathbb{E} \liminf_{t \to \tau} F(\zeta(t)) = \mathbb{E}F(\zeta(\tau)) \,. \tag{2.5.10}$$

Now let us also assume that the drift does not exist in (2.5.1):

$$b(\alpha, x) = 0. \tag{2.5.11}$$

Then the process $\{\zeta(t)\}$ is a martingale, and the convexity of F implies $\mathbb{E}F(\zeta(t)) \geq F(\mathbb{E}\zeta(t)) = F(x)$. Therefore,

$$v(x) \geq F(x), \ \forall x \in \Omega. \tag{2.5.12}$$

Another situation when the obstacle condition (2.5.12) is obviously satisfied is the following. We again assume (2.5.11), and also assume that $a(\alpha, \zeta)$ has a very simple form (see Section 2.5.5 in the following pages):

$$a(\alpha, x) = \begin{bmatrix} \alpha_1 \\ \vdots \\ \alpha_d \end{bmatrix} \stackrel{\text{def}}{=} \alpha.$$

Then, plugging $\alpha = 0$ into the definition (2.5.6) of $v(x)$, we immediately get $\zeta(s) = x$ identically, and so $v(x) \geq \mathbb{E}F(\zeta(\tau)) = F(x)$, if we use the fact that $f \geq 0$.

2.5.3 Boundary Conditions

We see that assumption (2.5.9) implies $\zeta(\tau) \in \partial\Omega$ if $\tau < \infty$. However, two things may happen. First of all, $\zeta(\infty)$ can lie inside the interior Ω°. Secondly, the landing points of the trajectories do not fill in the whole boundary. Here is the boundary condition:

$$v(x) = F(x), \quad \forall x \in \partial'\Omega. \tag{2.5.13}$$

Here $\partial'\Omega$ is the subset of $\partial\Omega$ accessible in finite time by the trajectories $\zeta(t)$.

2.5.4 Supersolutions

In applications to harmonic analysis one is also interested in supersolutions of the Bellman equation (2.5.8):

$$\begin{cases} \sup\limits_{\alpha\in A} [\mathcal{L}(\alpha,x)V(x)+f(\alpha,x)] \le 0, \ x\in\Omega, \\ V(x)\ge F(x), \ x\in\Omega. \end{cases} \tag{2.5.14}$$

Lemma 2.5.5 *Consider a problem of stochastic optimal control as at the beginning of Section 2.5. Let F be a continuous function on $\bar{\Omega}$, and let* (2.5.9) *be satisfied. Let a sufficiently smooth function V solve* (2.5.14), *and let v be the Bellman function, then $V\ge v$ in Ω.*

Proof Equation (2.5.8) states that $-\mathcal{L}(\alpha,x)V(x)\ge f(\alpha,x)$. Using (2.5.5) for V and then (2.5.14), one gets

$$\begin{aligned} V(x) &= \mathbb{E}V(\zeta(t)) - \mathbb{E}\int_0^t (\mathcal{L}(\alpha(s),\zeta(s))V)(\zeta(s))ds \\ &\ge \mathbb{E}F(\zeta(t)) + \mathbb{E}\int_0^t f(\alpha(s),\zeta(s))ds. \end{aligned}$$

Write $\liminf_{t\to\infty}$ of both parts, and notice that actually we can replace $\liminf \mathbb{E}F(\zeta(t)) \ge \mathbb{E}\lim F(\zeta(t))$ by (2.5.10) and Fatou's lemma. Then we get $V(x)\ge v^\alpha(x)$. It rests to take the supremum over the control process α. □

2.5.5 Special Matrices a Bring Us to Harmonic Analysis

Let us consider a very simple matrix a not depending on x:

$$d_1 = 1, \ a(\alpha,x) = \begin{bmatrix}\alpha_1\\ \vdots \\ \alpha_d\end{bmatrix} \stackrel{\text{def}}{=} \alpha. \tag{2.5.15}$$

If on top of that $b=0$, then operator $\mathcal{L}$ just involves Hessian matrix H_v of function v:

$$\mathcal{L}(\alpha)v(x) = \frac{1}{2}\sum_{i,j=1}^d \frac{\partial^2 v}{\partial x_i \partial x_j}\alpha_i\alpha_j = \frac{1}{2}(H_v(x)\alpha,\alpha).$$

We claim that this is the generic case of harmonic analysis problems in $\mathbb{R}^1$. Equation (2.5.8) becomes

$$\begin{cases} \sup_{\alpha\in A} \left[\frac{1}{2}(H_v(x)\alpha,\alpha) + f(\alpha,x)\right] = 0, \quad x\in\Omega, \\ \qquad\qquad\qquad\qquad v(x) \geq F(x),\ x\in\Omega. \end{cases} \tag{2.5.16}$$

If $b \neq 0$, then we just add the first-order differential operator (called a drift):

$$\begin{cases} \sup_{\alpha\in A} \left[\frac{1}{2}(H_v(x)\alpha,\alpha) + \sum_{k=1}^d b_k(\alpha,x)\frac{\partial}{\partial x_k}v(x) + f(\alpha,x)\right] = 0,\ x\in\Omega, \\ v(x) \geq F(x),\ x\in\Omega, \text{ or } v(x) = F(x),\ x\in\partial'\Omega. \end{cases} \tag{2.5.17}$$

REMARK 2.5.6 This is a nonlinear equation of the second order. In fact, let us fix x and let us find the values of parameters α that maximize the left-hand side for these fixed variables x. Then we plug these values $\alpha(x)$ back into the left-hand side. We will get a nonlinear differential expression involving the partial derivatives of v of the second and of the first (if the drift terms are present) order. The class of nonlinear equation obtained in this manner belongs to the realm of fully nonlinear PDEs. A vast literature exists on the existence and the regularity of solutions of fully nonlinear PDE. A good source can be [**120**].

2.5.6 Analysis on the Complex Plane. Conformal Restrictions

Complex analysis problems on $\mathbb{R}^2$ sometimes can be reduced to the analysis of the following Bellman equation:

$$d_1 = 2,\ a(\alpha,x) = \begin{bmatrix} \alpha_{1,1} & \alpha_{1,2} \\ \vdots & \vdots \\ \alpha_{d,1} & \alpha_{d,2} \end{bmatrix} =: \alpha. \tag{2.5.18}$$

$$\begin{cases} \sup_{\alpha\in A} \left[\frac{1}{2}\operatorname{trace}(\alpha^* H_v(x)\alpha) + \sum_{k=1}^d b_k(\alpha,x)\frac{\partial}{\partial x_k}v(x) + f(\alpha,x)\right] = 0,\ x\in\Omega, \\ v(x) \geq F(x),\ x\in\Omega \text{ or } v(x) = F(x),\ x\in\partial'\Omega. \end{cases} \tag{2.5.19}$$

REMARK 2.5.7 This equation is much more difficult to analyze than (2.5.17). On the other hand, we can easily notice that certain restrictions on α makes clear that Hessian of v should be replaced by Laplacian of v. This is explained in the following example.

Example Let $d_1 = d = 2$ and let matrix α have restrictions $\alpha \in A$ of the type that the first row is orthogonal to the second row and that the norms of the rows are equal. In other words,

$$\alpha_{1,1} = \alpha_{2,2},\ \alpha_{1,2} = -\alpha_{2,1}. \tag{2.5.20}$$

We can call such conditions on α's Cauchy–Riemann conditions. This is not just a superficial analogy with Cauchy–Riemann equations from complex analysis. The analogy is deeper. As we assumed that $d_1 = 2$, we have two independent degrees of freedom to control our random process. Our stochastic differential equation generates the process (x_1, x_2) with values in $\mathbb{R}^2$. The meaning of this $\mathbb{R}^2$-valued random process is that it "mirrors" two functions (call them (g_1, g_2)). The just mentioned two independent degrees of freedom mirror the fact that these functions depend each on two variables (naturally call them (x, y)). Then $\alpha_{i,j}$ is a close analog of the derivative of functions g_1 and g_2, where the second index j may be interpreted as a partial derivative. If $j = 1$ we can think about $\alpha_{i,1}$ as standing for a partial derivative of g_i with respect to x, and if $j = 2$, $\alpha_{i,2}$ is standing for a partial derivative of g_i with respect to y.

In Sections 2.6, 2.7, and in Chapter 3, we use the analogy of complex analysis with stochastic optimal control problems. For example, given a problem involving complex conjugate functions g_1, g_2, it seems only natural to "imbed" such a problem into stochastic optimal control environment. To take into account Cauchy–Riemann equations we use assumption (2.5.20) on control process, which will be used in (2.7.5). Similar situation occurs while we investigate martingale transform and use the equality of martingale differences in (2.6.41).

Notice that

$$\begin{aligned}\mathrm{trace}(\alpha^* H_v(x)\alpha) = \frac{\partial^2 v}{\partial x^2}(\alpha_{1,1}^2 + \alpha_{1,2}^2) + 2\frac{\partial^2}{\partial x \partial y}(\alpha_{1,1}\alpha_{2,1} + \alpha_{1,2}\alpha_{2,2})\\ + \frac{\partial^2 v}{\partial y^2}(\alpha_{2,1}^2 + \alpha_{2,2}^2)\end{aligned}$$

We see that if the Cauchy–Riemann conditions (2.5.20) are satisfied, then the mixed derivative $\frac{\partial v}{\partial x \partial y}$ disappears and we get

$$\mathrm{trace}(\alpha^* H_v(x)\alpha) = (\alpha_{1,1}^2 + \alpha_{1,2}^2)\left(\frac{\partial^2 v}{\partial x^2} + \frac{\partial^2 v}{\partial y^2}\right). \tag{2.5.21}$$

It is shown in Section 2.7 how this observation brings an approach to finding sharp constants in several interrelated complex analysis problems. This includes the problem of finding the norm of the Hilbert transform in

L^p (Pichorides constants). The reader will see soon how observation (2.5.21) brings about an obstacle problem for the Laplacian.

2.6 Almost Perfect Analogy between Stochastic Optimal Control and Harmonic Analysis: Functions of One Variable

In this section, we provide the selection of harmonic analysis problems on the real line that can be easily emulated as stochastic optimal control problems. The next section deals with harmonic or complex analysis problems on the plane that also can be mirrored by stochastic optimal control. The examples will be real, not artificial. Still we chose them for their (almost) perfectness and their simplicity for our illustrative goal. The next chapters contains several harmonic analysis problems solved by stochastic optimal control approach. But they will be far from being very simple.

As soon as we set up a stochastic optimal control problem that mirrors our original one we do the following:

- write down Bellman PDE,
- solve it (if we can),
- prove verification propositions.

REMARK 2.6.1 Verification means proving that the solution of a formally obtained Bellman PDE is in fact the Bellman function of the original harmonic analysis problem. Verification is needed because our Bellman PDE is always obtained in a rather formal fashion. There is no justification why the process of emulating a concrete harmonic analysis problem by a (formally) similar stochastic optimal control problem will give us the solution of the initial problem.

REMARK 2.6.2 Verification theorems are common inside the proper stochastic optimal control theory, because even without any harmonic analysis agenda whatsoever, the Bellman PDE for a stochastic optimal control problem is a relatively formal object, whose solution is not automatically equal to the Bellman function of a stochastic optimal control problem. An interesting example of the use of verification can be found e.g., in [**135**].

In the rest of this section, we collect several examples where we make the translation from the harmonic analysis language to stochastic optimal control language, get formal Bellman equation, solve it, and this allows us to solve the original problem from harmonic analysis. This is all illustrative, the hard calculations have already been done in Chapter 1.

2.6.1 A_∞^d Weights and Associated Carleson Measures: Buckley's Inequality

Recall that $\mathcal{D}$ is a dyadic lattice on the interval $I = [0,1]$. We recall also that to each interval $J \in \mathcal{D}$ one assigns its Haar function h_J normalized in L^2.

For a function $w \in L^2(I)$ we can write its Haar decomposition

$$w(x) = \langle w \rangle_I + \sum_{J \in \mathcal{D}} (w, h_J) h_J(x), \quad x \in I. \tag{2.6.1}$$

The series converges in $L^2(I)$, and we can write a Haar approximation to w by aborting the series like that

$$w_n(x) = \langle w \rangle_I + \sum_{k=1}^{n} \sum_{J \in \mathcal{D}_k} (w, h_J) h_J(x), \quad x \in I. \tag{2.6.2}$$

We recall that a nonnegative function on I a dyadic $A_\infty^d(I, r)$ weight if

$$\langle w \rangle_J \le r\, e^{\langle \log w \rangle_J}, \quad \forall J \in \mathcal{D}(I). \tag{2.6.3}$$

The family of such functions is denoted by A_∞^d. Here $\mathcal{D}$ is a dyadic lattice on $\mathbb{R}$, $\langle \cdot \rangle_J$ is the averaging over J. The result of Buckley stated in Section 1.2 says that the following inequality holds for $w \in A_\infty^d$:

$$\sum_{J \in \mathcal{D}(I)} \left(\frac{\langle w \rangle_{J_+} - \langle w \rangle_{J_-}}{\langle w \rangle_J} \right)^2 |J| \le C|I|, \tag{2.6.4}$$

Here C depends only on r in (2.6.3), but does not depend on I.

2.6.1.1 Emulating a Function by a Controlled Discrete Stochastic Process. Let $\Phi\colon [0,1] \to \mathbb{R}^d$ be a summable vector-function. It is useful to think about function Φ on interval $I = [0,1]$ as a random walk in $\mathbb{R}^d$. Here is a way to do that. We start at the initial point $x = \zeta(0) = \langle \Phi \rangle_I \in \mathbb{R}^d$. Put $J \stackrel{\text{def}}{=} I$ and introduce $\alpha(1) = \langle \Phi \rangle_{J_+} - \langle \Phi \rangle_J = \frac{1}{2}(\langle \Phi \rangle_{J_+} - \langle \Phi \rangle_{J_-}) = \langle \Phi \rangle_J - \langle \Phi \rangle_{J_-}$. Now let $\xi_1, \ldots, \xi_n, \ldots$ be Bernoulli independent random variables assuming values ± 1 with probability $\frac{1}{2}$. If $\xi_1 = 1$ we walk from $\zeta(0)$ to $\zeta(1) = \zeta(0) + \alpha(1)$, if $\xi_1 = -1$, we walk from $\zeta(0)$ to $\zeta(1) = \zeta(0) - \alpha(1)$. In other words, $\zeta(1) = \zeta(0) + \alpha(1)\xi_1$. We consider vector random variable $\alpha(2)$ built as follows: if $\xi_1 = 1$ then put $J \stackrel{\text{def}}{=} (I)_+$ and $\alpha(2) = \langle \Phi \rangle_{J_+} - \langle \Phi \rangle_J$. If $\xi_1 = -1$ then put $J \stackrel{\text{def}}{=} (I)_-$ and $\alpha(2) = \langle \Phi \rangle_{J_+} - \langle \Phi \rangle_J$.

Clearly $\alpha(2)$ is a random variable measurable with respect to σ-algebra generated by ξ_1.

Now our random walker jumps from $\zeta(1)$ to $\zeta(2) = \zeta(1) + \alpha(2)\xi_2 = x + \alpha(1)\xi_1 + \alpha(2)\xi_2$. At time n, we are at a random place

$$\zeta(n) = x + \alpha(1)\xi_1 + \alpha(2)\xi_2 + \cdots + \alpha(n)\xi_n. \tag{2.6.5}$$

It is absolutely clear that this formula is a fancy way to write the Haar expansion of function Φ already written in (2.6.2).

Notice that random variables $\alpha(1), \dots, \alpha(n+1)$ are measurable with respect to σ-algebra generated by $\zeta(0), \dots, \zeta(n)$. In our terminology, the discrete process $\{\alpha(k)\}_{k=1}^{\infty}$ is adapted to discrete process $\{\zeta(k)\}_{k=0}^{\infty}$. We call the discrete process $\{\alpha(k)\}_{k=1}^{\infty}$ the control process for the process $\zeta \stackrel{\text{def}}{=} \{\zeta(0), \dots, \zeta(n), \dots\}$.

Vector random process $\zeta(k) \stackrel{\text{def}}{=} \{(x_1(k), \dots, x_d(k))\}_{k=1}^{\infty}$ is supposed to mirror the behavior of Haar approximations to Φ.

Now we specify Φ as follows. Let $w \in A_\infty^d$, put $\Phi(t) \stackrel{\text{def}}{=} (w(t), \log w(t))$, $t \in I = [0, 1]$.

The just constructed random walk is in $\mathbb{R}^2$, and to incorporate the main property (2.6.3) we require that the random walk would stay in the domain

$$\Omega \stackrel{\text{def}}{=} \{(x_1, x_2) \in \mathbb{R}^2 : 1 \le x_1 e^{-x_2} \le r\}. \tag{2.6.6}$$

Notice that for our choice of Φ the left inequality holds automatically because by Jensen's inequality for every $J \in D$, we have

$$e^{\langle \log w \rangle_J} \le \langle w \rangle_J.$$

However, the right inequality is not automatic at all, it is present here because of the fact that $w \in A_\infty^d(I, r)$.

Remark 2.6.3 The reader may become suspicious as to why we do not say anything about the relation between two coordinates of vector process $\alpha(k)$? By construction, quantities $\alpha_1(k)$ and $\alpha_2(k)$ are of course related. Moreover, for a given a function Φ they are not at all random. But we consider here the whole class of test functions. By ignoring the dependence of $\alpha_1(k)$ and $\alpha_2(k)$ we obtain the whole class of test functions for the problem under consideration.

In fact, we will make several outrageous steps. We will make the time continuous. We will think about $\{\xi_k\}_{k=1}^{\infty}$ as a one-dimensional Brownian motion. We will not put any restriction on the values of vector control process $\alpha(s) = (\alpha_1(s), \alpha_2(s))$. Then we will formally write the problem of estimating the right-hand side of (2.6.4) as a problem of optimizing a profit function (which we still need to detect). The solution of the corresponding Bellman PDE will bring the constant in (2.6.4). But to prove that one will need verification propositions.

So we use the stochastic optimal control machinery to guess the answer, and then we should independently verify why our guess is correct. We illustrate the fact that it is correct in many cases in the rest of this chapter. Moreover, in the next chapters, we consider some rather involved problems of estimating singular integral operators, where this method is effective as well.

2.6.1.2 We Start Our Journey to a Formal Bellman PDE by Asking: "Who Moves?". We already saw the answer. The random movement will be in $\mathbb{R}^2$ (rather in Ω from (2.6.6)), and here are processes and control processes:

$$(x_1, x_2) = (\langle w \rangle_J\,,\ \langle \log w \rangle_J) \tag{2.6.7}$$
$$\alpha_1 = \langle w \rangle_{J_+} - \langle w \rangle_J,\ \alpha_2 = \langle \log w \rangle_{J_+} - \langle \log w \rangle_J.$$

Notice that

$$|\alpha_1| = \frac{1}{2} |\langle w \rangle_{J_-} - \langle w \rangle_{J_+}|. \tag{2.6.8}$$

We can interpret (2.6.5) as a very simple stochastic differential equation of the type (2.5.1) if we choose $b = 0$ and we choose matrix a exactly as in (2.5.15):

$$a(\alpha, x) = \begin{bmatrix} \alpha_1 \\ \alpha_2 \end{bmatrix}. \tag{2.6.9}$$

2.6.1.3 Profit and Bonus Functions. Function of profit can be read off (2.6.4) if one compares

$$\frac{1}{|I|} \sum_{J \in \mathcal{D}} \left(\frac{\langle w \rangle_{J_+} - \langle w \rangle_{J_-}}{\langle w \rangle_J} \right)^2 |J|$$

from (2.6.4) with (2.5.2) written as the average over random trajectories (lines of life).

Each line of life has as a starting point $J = I$ and then proceeds to J_{ξ_1}, then to $(J_{\xi_1})_{\xi_2}$, etc.

Thus, $\frac{1}{|I|} \sum_{\ell \subseteq I,\, \ell \in \mathcal{D}} \cdots$ plays the role of $\mathbb{E} \int_0^\infty \cdots$. This allows us to choose the plausible profit function if we take (2.6.7) and (2.6.8) into consideration:

$$f(\alpha, x) = \frac{4\alpha_1^2}{x_1^2}.$$

Bonus function $F \equiv 0$ here.

2.6.1.4 Formal Bellman PDE. Bellman equation reads now

$$\sup_{\alpha=(\alpha_1,\alpha_2)} \left[(H_v\alpha, \alpha) + \frac{8\alpha_1^2}{x_1^2} \right] = 0. \tag{2.6.10}$$

to be solved in Ω from (2.6.6) with the obstacle condition

$$v(x) \geq 0 \quad \forall x \in \Omega. \tag{2.6.11}$$

2.6.1.5 Homogeneity. Notice that the domain Ω is invariant under the transformation $T_\lambda(x_1, x_2) = (\lambda x_1, x_2 + \log\lambda)$, $\lambda > 0$. We can easily see that the random walk generated by $w \in A^d_\infty$ that started at (x_1, x_2) is transformed by T_λ to a corresponding random walk starting at $T_\lambda(x_1, x_2)$ (generated by λw). Notice that profit function (and the bonus function too) is invariant under this transformation. Therefore, we will be searching for a homogeneous solution of formal Bellman equation written below. Namely, v satisfies

$$v \circ T_\lambda = v. \tag{2.6.12}$$

2.6.1.6 Solution to Bellman PDE. It is easy to see that equation (2.6.10) can be rewritten in the following form. Consider the matrix

$$M_v \stackrel{\text{def}}{=} \begin{bmatrix} \frac{\partial^2 v}{\partial x_1^2} + \frac{8}{x_1^2}, & \frac{\partial^2 v}{\partial x_1 \partial x_2} \\ \frac{\partial^2 v}{\partial x_1 \partial x_2}, & \frac{\partial^2 v}{\partial x_2^2} \end{bmatrix}$$

Then Bellman equation (2.6.10) becomes

$$\sup_{\alpha=(\alpha_1,\alpha_2)} (M_v\alpha, \alpha) = 0. \tag{2.6.13}$$

Let us look carefully at equation (2.6.13). Notice that vector (α_1, α_2) runs over the whole $\mathbb{R}^2$ including zero vector. The conclusion is somewhat unpleasant. Equation (2.6.13) is actually an inequality only. It is equivalent to the requirement for the matrix $M_v(x_1, x_2)$ to be nonpositively defined for all $(x_1, x_2) \in \Omega$. So (2.6.13) is nothing else but

$$M_v(x_1, x_2) \leq 0, \ \forall (x_1, x_2) \in \Omega. \tag{2.6.14}$$

To make up for the missed information we should remember that v is supposed to be the optimal nonnegative function satisfying (2.6.14). Optimality should be expressed as a certain equality, and we have only inequality so far, something is missing. One may suspect that optimality can be expressed as the fact that for each $x = (x_1, x_2)$ there exists nonzero vector $\alpha = (\alpha_1, \alpha_2)$ depending on x such that $(M_v(x)\alpha, \alpha) = 0$. In other words, the optimality feels like the saturation of condition (2.6.14), the fact that this condition is barely satisfied, it degenerates at every x. This is all the more plausible because at the level of control this means the existence of the optimal control (optimal jump) at every step of our random walk.

Let us look at the problem of optimality from two different points of view; it may mean any of the following:

- either v is the smallest nonnegative function satisfying (2.6.14);
- or v saturates either inequality (2.6.14) or obstacle condition (2.6.11) at every point of Ω meaning that at every point either the matrix stays nonpositive but has a degenerate direction on which it vanishes or v itself vanishes.

The second choice will be our choice, and it will finally brings us a nonlinear PDE of Monge–Ampère type if we write the saturation of inequality (2.6.14) as follows:

$$\det M_v(x_1, x_2) = \left(\frac{\partial^2 v}{\partial x_1^2} + \frac{8}{x_1^2}\right)\frac{\partial^2 v}{\partial x_2^2} - \left(\frac{\partial^2 v}{\partial x_1 \partial x_2}\right)^2 = 0,\ \forall (x_1, x_2) \in \Omega. \tag{2.6.15}$$

Remark 2.6.4 In the paper [**174**], the authors show that under certain assumptions the Bellman function v defined in (2.5.7), and the smallest function solving Bellman equation and obstacle condition (2.5.17) coincide. We do not use it here, because in our examples we are not always under the assumptions when this result is proved. Instead, we adopt the second notion of optimality above, formally find the family of solutions of a corresponding Bellman PDE that saturates the first line of (2.5.17), and then try to verify that one of these solutions coincides with Bellman function v.

We need to solve the system (2.6.14), (2.6.15) subject to obstacle condition $v \geq 0$ in Ω and subject to homogeneity (2.6.12).

Here is the formula from Section 1.2 for the solution of this problem: $v(x_1, x_2) = 8(\log x_1 - x_2)$. Notice that this formula for v was obtained by solving an absolutely formal equation (2.6.15). So we need a verification that this solution of formal Bellman equation has anything to do with the estimate (2.6.4).

2.6.1.7 Verification Proposition. The Bellman function of our dyadic problem of estimating $\frac{1}{|I|}\sum_{J\in\mathcal{D}}\left(\frac{\langle w\rangle_{J_+} - \langle w\rangle_{J_-}}{\langle w\rangle_J}\right)^2 |J|$ was introduced in (1.2.2) of

Section 1.2. It has been proved in Lemma 1.2.4 of this section that we have the following inequality:

$$\mathbf{B}(x_1, x_2) \le 8(\log x_1 - x_2) = v(x_1, x_2). \tag{2.6.16}$$

After all, it turns out that the solution of formal Bellman PDE does have a lot to do with the real Bellman function of the problem defined in (1.2.2). In fact, in Section 1.10, we proved that $8(\log x_1 - x_2) = v(x_1, x_2)$ is exactly the Bellman function of Buckley's inequality.

In particular, this implies that the constant in equation (2.6.4) is bounded in a following way:

$$C(r) \le 8 \log r. \tag{2.6.17}$$

2.6.1.8 Saturation as the Way to Satisfy the Minimality Condition. We wish to find the minimal solution of (1.2.3). Our first step was to replace (1.2.3) by differential inequality (1.2.5), i.e., (2.6.10) or (2.6.14). Let us show that minimality implies saturation of conditions $M_v(x_1, x_2) \le 0, \;\; v(x_1, x_2) \ge 0$ at every point of Ω.

Before doing that let us introduce the following definition.

DEFINITION 2.6.5 We say that continuous v satisfies Monge–Ampère equation (2.6.15) in Ω if $v(x_1, x_2) - 8 \log x_1$ is a concave function, and at each point of Ω there is a direction passing through this point along which $v(x_1, x_2) - 8 \log x_1$ is linear.

REMARK 2.6.6 For smooth functions v this definition coincides with the properties: 1) v satisfies (2.6.14) 2) v satisfies (2.6.15).

THEOREM 2.6.7 *Let* $v_0 = \inf V$, *where the infimum is taken over all smooth nonnegative* V *satisfying* (2.6.14) *in* Ω. *Then* v_0 *is continuous and satisfies Monge–Ampère equation* (2.6.15) *in the sense of the preceding definition at every point* $(x_1, x_2) \in \Omega$, *where* $v_0 > 0$.

PROOF Let V be a smooth nonnegative function in Ω satisfying (2.6.14). Notice that for each such V we have that $V - 8 \log x_1$ is concave. Therefore, $v_0 - 8 \log x_1 = \inf(V - 8 \log x_1)$ is also concave. In particular, v_0 is continuous.

Now assign to V function $U = V - 8 \log x_1$. It is smooth, concave, and satisfies $U \ge 8 \log \frac{1}{x_1}$ in Ω. Consider u_0 that is the infimum of all smooth, concave functions in Ω majorizing $8 \log \frac{1}{x_1}$. It is again concave, continuous, and $u_0 \ge 8 \log \frac{1}{x_1}$. Let us prove that u_0 is degenerate concave in every point $(y_1, y_2) \in \Omega$, where $u_0(y_1, y_2) > 8 \log \frac{1}{y_1}$. Suppose this is not the case and u_0 is strictly concave at such point $y = (y_1, y_2)$. Then "cut off" a very small cap centered at the point $(y, u_0(y))$, making the new function u flat around $(y, u(y))$ and with $u(y) < u_0(y)$. This is possible in such a way that still $u(y_1, y_2) > 8 \log \frac{1}{y_1}$ and the same is true in a small neighborhood of y, where u_0 was flattened to u. Outside this neighborhood $u = u_0$. But obviously this means that u_0 is not the infimum claimed above. We come to contradiction.

Now consider $v_0 \stackrel{\text{def}}{=} u_0 + 8 \log x_1$. Obviously it is nonnegative, and we just proved that at any y where it is strictly positive there is a direction through y such that $v_0 - 8 \log x_1$ is linear. So v_0 satisfies equation (2.6.15) in the sense of our definition above. In particular, if this infimum happens to be smooth it satisfies (2.6.15) in the normal sense as soon as it is strictly positive. □

2.6.2 A Toy Two-Weight Inequality

In the article [**187**], it is proved that the following inequality always holds.

THEOREM 2.6.8 *Suppose that for all $J \in \mathcal{D}$*

$$\langle u \rangle_J \langle v \rangle_J \le 1. \tag{2.6.18}$$

Then for all $J \in \mathcal{D}$

$$\begin{aligned} \frac{1}{|J|} \sum_{K \in \mathcal{D}(J)} & |\langle u \rangle_{K+} - \langle u \rangle_{K-}| |\langle v \rangle_{K+} - \langle v \rangle_{K-}| |K| \\ & \le C \langle u \rangle_J^{1/2} \langle v \rangle_J^{1/2}. \end{aligned} \tag{2.6.19}$$

Let us apply our approach of (formally) writing Bellman equation for a Stochastic Optimal control problem emulating this estimate. It will be again a Monge–Ampère equation. We solve it and use the verification proposition to show that solution v of this formal equation actually proves Theorem 2.6.8 with a sharp constant C.

Using the pattern of the previous Section (2.6.1), we will move faster than before.

2.6.2.1 Who Moves? And Where?

$$(x_1, x_2) = (\langle u \rangle_J, \langle v \rangle_J).$$

Control processes are

$$\alpha_1 = \langle u \rangle_{J_+} - \langle u \rangle_J, \; \alpha_2 = \langle v \rangle_{J_+} - \langle v \rangle_J.$$

Hence,

$$|\alpha_1| = \frac{1}{2}|\langle u \rangle_{J_-} - \langle u \rangle_{J_+}|, \; |\alpha_2| = \frac{1}{2}|\langle v \rangle_{J_-} - \langle v \rangle_{J_+}| \tag{2.6.20}$$

The domain of definition (the domain where the random walk happens) is the domain in $\mathbb{R}^2$ filled in by averages of pairs of nonnegative functions, satisfying (2.6.18). So

$$\Omega \stackrel{\text{def}}{=} \{x = (x_1, x_2) : 0 \le x_1, 0 \le x_2, x_1 x_2 \le 1\}. \tag{2.6.21}$$

2.6.2.2 Profit and Bonus Functions. The bonus function $F = 0$ here. As in the previous problem $f^\alpha(x)$ is easy to find if we take into account (2.6.20), which would mean that $|\langle u \rangle_{J+} - \langle u \rangle_{J-}||\langle v \rangle_{J+} - \langle v \rangle_{J-}| = 4|\alpha_1||\alpha_2|$, and therefore

$$f(\alpha, x) = 4|\alpha_1|\,|\alpha_2|.$$

2.6.2.3 Formal Bellman PDE. Bellman equation (2.5.16) reads now as follows:

$$\sup_{\alpha=(\alpha_1,\alpha_2)\in\mathbb{R}^2} [(H_v \alpha, \alpha) + 8|\alpha_1||\alpha_2|] = 0. \tag{2.6.22}$$

to be solved in Ω from (2.6.21) with the obstacle condition

$$v(x) \ge 0 \quad \forall x \in \Omega. \tag{2.6.23}$$

2.6.2.4 Solution to Bellman PDE. It is easy to see that equation (2.6.22) can be rewritten in the following form. Consider two matrices

$$M_v^{\pm} \stackrel{\text{def}}{=} \begin{bmatrix} \frac{\partial^2 v}{\partial x_1^2}, & \frac{\partial^2 v}{\partial x_1 \partial x_2} \pm 4 \\ \frac{\partial^2 v}{\partial x_1 \partial x_2} \pm 4, & \frac{\partial^2 v}{\partial x_2^2} \end{bmatrix}$$

Then Bellman equation (2.6.22) becomes

$$\sup_{\alpha=(\alpha_1,\alpha_2)} (M_v^{\pm} \alpha, \alpha) = 0. \tag{2.6.24}$$

Exactly as in Section 2.6.1, we look at (2.6.24) and we see that it is the system of inequalities. In fact, notice that vector (α_1, α_2) runs over the whole $\mathbb{R}^2$ including zero vector. Hence, equations (2.6.24) are actually inequalities only. The system of these inequalities is equivalent to the requirement for the matrices $M_v^{\pm}(x_1, x_2)$ to be nonpositively defined for all $(x_1, x_2) \in \Omega$. So (2.6.24) is nothing else but

$$M_v^{+}(x_1, x_2) \le 0, \; M_v^{-}(x_1, x_2) \le 0, \; \forall (x_1, x_2) \in \Omega. \tag{2.6.25}$$

Exactly as before, we make up for the missed information that v is supposed to be the best nonnegative solution of (2.6.25) by guessing that we should add the saturation condition. Namely, we require that at every point $y = (y_1, y_2)$ of Ω either M^+ or M^- degenerate or $v = 0$. Here is the Bellman equation with which we are complementing requirement (2.6.25):

$$\det M_v^+(x_1, x_2) \cdot \det M_v^-(x_1, x_2) = 0\,, \ \forall (x_1, x_2) \in \Omega. \tag{2.6.26}$$

2.6.2.5 Homogeneity. Notice that the domain Ω is invariant under the transformation $T_\lambda(x_1, x_2) = (\lambda x_1, x_2/\lambda)\,, \ \lambda > 0$. We can easily see that the random walk generated by the pair of functions (u, v) that started at (x_1, x_2) is transformed by T_λ to a corresponding random walk starting at $T_\lambda(x_1, x_2)$ (generated by $(\lambda u, v/\lambda)$). Notice that profit function (and the bonus function too) is invariant under this transformation. Therefore, we will be searching for a homogeneous solution of formal Bellman equation written below. Namely, v satisfies

$$v \circ T_\lambda = v. \tag{2.6.27}$$

2.6.2.6 Solution of Formal Equation. In Section 1.11.6, we presented a nonnegative solution of (2.6.26) satisfying both concavities (2.6.25). We want it to satisfy also the homogeneity (2.6.27). This function is

$$v(x_1, x_2) = 16\sqrt{x_1 x_2} - 4x_1 x_2. \tag{2.6.28}$$

All of Ω except the upper boundary $\Gamma \stackrel{\text{def}}{=} \{(x_1, x_2) : x_1 \cdot x_2 = 1\}$ consists of points of degeneration of M_v^+, that is, we have $\det M_v^+(x_1, x_2) = 0$ in $\Omega \setminus \Gamma$. On the other hand, the reader can check that on Γ another degeneration condition holds: $\det M_v^-(x_1, x_2) = 0$, $(x_1, x_2) \in \Gamma$.

2.6.2.7 Verification. The dyadic Bellman function of Theorem 2.6.8 $\mathbf{B}(x_1, x_2)$ should be compared with v from (2.6.28) above. In article [**187**], the verification proposition shows that they are equal. In particular, the fact that

$$\mathbf{B}(x_1, x_2) = v(x_1, x_2) \tag{2.6.29}$$

proves that the best constant in Theorem 2.6.8 is equal to 16.

2.6.3 John–Nirenberg Inequality in Integral Form

We consider functions $\varphi \in L^2(I)$ belonging to the space $\mathrm{BMO}^2(I)$, meaning that

$$\sup_J \langle |\varphi(s) - \langle\varphi\rangle_J|^2\rangle_I^{\frac{1}{2}} < \infty$$

for all subintervals $J \subset I$. If this condition holds only for the dyadic subintervals $J \in \mathcal{D}(I)$, we will write $\varphi \in \mathrm{BMO}^{2,d}(I)$. We already mentioned in Section 1.4 that functions from BMO class belong to $L^p(J)$ for all $p < \infty$; moreover, an equivalent semi-norm can be obtained by replacing exponent 2 by p and $\frac{1}{2}$ by $\frac{1}{p}$ in the definition above.

The BMO^2 ball of radius ε centered at 0 will be denoted by $\mathrm{BMO}^2_\varepsilon$. The corresponding ball in $\mathrm{BMO}^{2,d}$ will be called $\mathrm{BMO}^{2,d}_\varepsilon$. Using the Haar decomposition

$$\varphi(s) = \langle\varphi\rangle_I + \sum_{J\in\mathcal{D}(I)} (\varphi, h_J)h_J(s),$$

we can write down the expression for the semi-norm in the following way:

$$\|\varphi\|^2_{\mathrm{BMO}^2(I)} = \sup_{J\subset I} \left(\langle\varphi^2\rangle_J - \langle\varphi\rangle_J^2\right). \tag{2.6.30}$$

We know that the reason for the indifference of BMO functions to exponent p lies in the following famous result:

THEOREM 2.6.9 (John–Nirenberg [**88**]) *There exist absolute constants c_1 and c_2 such that*

$$|\{s \in I\colon |\varphi(s) - \langle\varphi\rangle_I| \geq \lambda\}| \leq c_1 e^{-c_2\frac{\lambda}{\|\varphi\|}}|I|$$

for all $\varphi \in \mathrm{BMO}^2(I)$.

An equivalent, integral form of the same assertion is the following:

THEOREM 2.6.10 *There exists an absolute constant ε_0 such that for any $\varphi \in \mathrm{BMO}^2_\varepsilon(I)$ with $\varepsilon < \varepsilon_0$ the inequality*

$$\langle e^\varphi\rangle_I \leq c\, e^{\langle\varphi\rangle_I}$$

holds with a constant $c = c(\varepsilon)$ not depending on φ.

We will apply our formal approach to finding sharp ε_0 and $c(\varepsilon)$ in Theorem 2.6.10. Also the same formal approach helped to find the sharp constants in the dyadic version, see [**171**].

THEOREM 2.6.11 *There exists an absolute constant* ε_d *such that for any* $\varphi \in \mathrm{BMO}_\varepsilon^{2,d}(I)$ *with* $\varepsilon < \varepsilon_d$ *the inequality*

$$\langle e^\varphi \rangle_I \le c\, e^{\langle \varphi \rangle_I}$$

holds with a constant $c = c(\varepsilon)$ *not depending on* φ.

2.6.3.1 Formal Approach Gives an Unexpected Result. We saw in Chapter 1 that $\varepsilon_d < \varepsilon_0$. To find both constants one uses two different solutions of the formal Bellman equation. We are going to write this equation now in exactly the same way as before. However, the formal approach of previous Sections 2.6.1, 2.6.2 differs from the one in the present section.

Before, the minimal solution of the formal Bellman equation gave the solution of dyadic problem right away. The minimal solution of the formal Bellman equation coincided exactly with dyadic Bellman function. John–Nirenberg Theorems 2.6.10, 2.6.11 in the present section are not like that.

In our case now, we will have again a family of solutions of Bellman equation, there will be a minimal one, but it does not give the dyadic Bellman function as it was the case in Sections 2.6.1 and 2.6.2.

Below we start again with the dyadic problem. Write down the corresponding discrete time random walk in a certain appropriate domain Ω. Replace formally the discrete time by the continuous time and find the formal Bellman equation and boundary conditions. The profit function will be zero.

And then verification propositions of Section 1.3 will show that one solution (but not the minimal in the usual sense) is the dyadic Bellman function, which proves Theorem 2.6.11, and another solution is yet another Bellman function that proves Theorem 2.6.10. Notice that formality of approach is emphasized here by the fact that even though we will obtain correct constants for both dyadic and classical John–Nirenberg integral inequalities of Theorems 2.6.11 and 2.6.10 correspondingly, we achieve this by starting with dyadic problem and formally writing its discrete time random walk as a continuous time motion.

2.6.3.2 Who Moves? And Where? Random walk is undertaken by

$$(x_1, x_2) = (\langle \varphi \rangle_J, \langle \varphi^2 \rangle_J), \; J \in \mathcal{D}(I)$$

with control processes

$$\alpha_1 = \langle \varphi \rangle_{J_+} - \langle \varphi \rangle_J, \; \alpha_2 = \langle \varphi^2 \rangle_{J_+} - \langle \varphi^2 \rangle_J.$$

Domain of the random walk of our (x_1, x_2) is prescribed by

$$\Omega \stackrel{\text{def}}{=} \{(x_1, x_2) : 0 \le x_2 - x_1^2 \le \varepsilon^2\}. \tag{2.6.31}$$

The domain is bounded by two parabolas. We will call them upper and lower parabolas by the obvious reason.

2.6.3.3 Profit Function: Boundary Condition. The profit function is zero here.

$$f(\alpha, x) = 0. \tag{2.6.32}$$

This is because we are not optimizing any sum over all dyadic intervals unlike the previous examples of Buckley's inequality and our simple two-weight inequality in Sections 2.6.1, 2.6.2.

Instead of considering the obstacle condition we prefer to consider the boundary condition. There is an explanation for that: The constant test functions are allowed in our problem here, but they correspond only to some boundary points of the domain Ω from (2.6.31). Compare this with Section 2.6.2, where the constant test function generated the whole domain.

Notice that by the Lebesgue differentiation theorem for almost every point $x \in J$ with respect to Lebesgue measure we have for any $\varphi \in L^2(J)$ that

$$\lim_{|J|\to 0, J\in\mathcal{D}, x\in J} \langle\varphi\rangle_J^2 = \lim_{|J|\to 0, J\in\mathcal{D}, x\in J} \langle\varphi^2\rangle_J .$$

This tells us that any random walk generated by $\varphi \in \mathrm{BMO}^2(J) \subset L^2(J)$ almost surely lands on the lower parabola. Also, clearly any point of the lower parabola can be a landing point. Therefore, according to the scheme of Section 2.5.3, we have that $\partial'\Omega$ is the lower parabola $\{(x_1, x_2) : x_2 = x_1^2\}$.

Now let us introduce the following boundary condition on $\partial'\Omega$:

$$v(x_1, x_1^2) = F(x_1, x_1^2) \stackrel{\text{def}}{=} e^{x_1}. \tag{2.6.33}$$

We guess this boundary F by looking at the functional $\langle e^{\varphi}\rangle_J$ to be optimized. We can write this functional in the language of our random walk as $\mathbb{E}e^{x_1(\tau)}$, where τ is the exit time of the random walk from Ω. Looking at $\mathbb{E}e^{x_1(\tau)}$ we see that the boundary condition (2.6.33) is correct.

2.6.3.4 Formal Bellman Equation. From (2.6.32) we see that our formal scheme (2.5.16) makes us write the following relationship, where H_v is, as always, Hessian matrix of v.

$$\sup\{(H_v\alpha, \alpha)\colon \alpha = (\alpha_1, \alpha_2) \in \mathbb{R}^2\} = 0. \tag{2.6.34}$$

We should find v like that satisfying the boundary condition (2.6.33).

Again as before we see that this is just an inequality, saying just that we are looking for a concave function with boundary assumptions (2.6.33):

$$H_v \le 0. \tag{2.6.35}$$

To express the fact that we are looking for the best solution of (2.6.35), we again formally saturate it, transforming it into a Monge–Ampère equation:

$$\det H_v = 0. \tag{2.6.36}$$

Now we want to write a solution of (2.6.36) with boundary condition (2.6.33). As before, here we have many solutions.

They form a one parameter family

$$v_\delta(x) = \frac{1 - \sqrt{\delta^2 - x_2 + x_1^2}}{1-\delta} \exp\left\{x_1 + \sqrt{\delta^2 - x_2 + x_1^2} - \delta\right\}, \quad \delta \ge \varepsilon \tag{2.6.37}$$

2.6.3.5 Verification Propositions. Looking at this family we see that δ must be strictly less than 1. And we see that the width of the parabolic strip Ω satisfies $\varepsilon \le \delta$. This makes us think by the analogy with Sections 2.6.1 and 2.6.2 that $\varepsilon_d = 1$ and for any width $\varepsilon < 1$ we can nicely estimate $\langle e^\varphi \rangle_I$ for any $\varphi \in \mathrm{BMO}_\varepsilon^{2,d}(I)$. We start with the random dyadic walk, and thus it seems only natural that we should have obtained sharp constants in dyadic Theorem 2.6.11. This is what happened in two previous examples.

But here this is not the case. Actually, we start with dyadic random walk, obtained the solutions (2.6.37) by solving a formal Bellman equation. But we do not get ε_d yet. Instead, we get ε_0 from Theorem 2.6.10. verification propositions of Section 1.3 shows that $\varepsilon_0 = 1$, and that for any $\varepsilon < \varepsilon_0$, functions from $\mathrm{BMO}_\varepsilon^2$ (notice the disappearance of d) will have summable exponent and

$$\langle e^\varphi \rangle_I \le c(\varepsilon) e^{\langle \varphi \rangle_I}, \quad c(\varepsilon) \stackrel{\text{def}}{=} \frac{e^{-\varepsilon}}{1-\varepsilon}.$$

Verification propositions of that section also show that the constants $\varepsilon_0, c(\varepsilon)$ are sharp.

Paper [**171**] contains the calculation of ε_d and $c_d(\varepsilon), \varepsilon < \varepsilon_d$, by choosing the right parameter δ_0 and v_{δ_0} from the family (2.6.37).

This example reflects a very important difference between the finite difference inequality to which Bellman function should satisfy and its differential form.

As we have seen, our formal method brought us some surprises, but still found the best constants in both dyadic and classical John–Nirenberg inequalities.

2.6.4 Burkholder–Bellman Function

Burkholder proved the following harmonic analysis theorem that answers the question, which was open for quite a while. Let I be an arbitrary interval, $\mathcal{D}(I)$ be its dyadic lattice, and let f, g be two summable functions on I.

DEFINITION 2.6.12 We remind the reader of some notions and facts from Sections 1.7, 1.8. Function g is a a martingale transform of f if

$$|\langle g\rangle_{J+} - \langle g\rangle_{J-}| = |\langle f\rangle_{J+} - \langle f\rangle_{J-}|, \qquad \forall J \in \mathcal{D}(I).$$

THEOREM 2.6.13 *Let $p \in (1,\infty)$ and let $p^* = \max(p, \frac{p}{p-1})$. Let g be a martingale transform of $f \in L^p(I)$, and let also*

$$|\langle g\rangle_{I}| \le |\langle f\rangle_{I}|. \tag{2.6.38}$$

Then

$$\langle |g|^p\rangle_I \le (p^*-1)^p \langle |f|^p\rangle_I\,. \tag{2.6.39}$$

The constant $(p^*-1)^p$ is sharp in the sense that there exists a sequence $\{f_n\}$ and their martingale transforms $\{g_n\}$ such that

$$\lim_{n\to\infty} \frac{\|g_n\|_p}{\|f_n\|_p} = p^* - 1. \tag{2.6.40}$$

Again, we would like to show that Theorem 2.6.13 lies nicely in the realm of our formal approach through stochastic optimal control.

So exactly as in the previous three examples we are heading to Bellman equation (2.5.16). The only details to clarify are: what are profit and bonus function and one more thing, which was not present in the previous three examples. Namely, before we had no restrictions on vector α's in (2.5.16). This time our 3D vectors α will be forced to lie in a subset of $\mathbb{R}^3$, which will be called A. The restriction $\alpha \in A$ will be very essential for the solution of corresponding formal Bellman equation. As before, the solution will provide us with the proof of Theorem 2.6.13.

2.6.4.1 Who Moves? And Where? Quite naturally the dyadic random walk is generated by function f, its martingale transform g, and by $|f|^p$. So, unlike in the previous examples, the walk will be in 3D.

$$x = (x_1, x_2, x_3) = (\langle f\rangle_J, \langle g\rangle_J, \langle |f|^p\rangle_J), \; J \in \mathcal{D}(I).$$

The vector control process is

$$\alpha = (\alpha_1, \alpha_2, \alpha_3) = (\langle f\rangle_{J_+} - \langle f\rangle_J,\ \langle g\rangle_{J_+} - \langle g\rangle_J,\ \langle |f|^p\rangle_{J_+} - \langle |f|^p\rangle_J).$$

We use very much below the fact that controls α_1 and α_2 are related. The fact that g is a martingale transform of f means exactly that

$$\alpha \in A \stackrel{\text{def}}{=} \{\alpha \in \mathbb{R}^3 \colon \alpha_1 = \pm\alpha_2\}. \tag{2.6.41}$$

2.6.4.2 Profit and Bonus Function. No sum over all dyadic intervals $I \in \mathcal{D}(J)$ is involved, so the profit function vanishes, i.e., $f(\alpha, x) = 0$. Now we need to understand what is the bonus function in Burkholder's problem.

Notice that random walk stays in the domain

$$\Omega \stackrel{\text{def}}{=} \{x = (x_1, x_2, x_3) \in \mathbb{R}^3 \colon |x_1|^p \le x_3\}. \tag{2.6.42}$$

To detect bonus function F, let us work only with $1 < p \le 2$ and with test functions $f \in L^2(J)$. By simple density argument this is enough. For such f, our dyadic random walk will land almost surely on $\partial\Omega$ by the Lebesgue differentiation theorem. Therefore, denoting by τ the first moment of exit from Ω, $\lim_{t\to\tau} |x_2(t)|^p$ exists almost certainly.

Notice that we want to maximize $\langle |g|^p\rangle_J$, which is the same in the language of random walk as to maximize $\mathbb{E}\lim_{t\to\tau} |x_2(t)|^p$. Introduce

$$F(x) = F(x_1, x_2, x_3) \stackrel{\text{def}}{=} |x_2|^p,\ x \in \Omega. \tag{2.6.43}$$

By the classical maximal theorems, we can notice that

$$\mathbb{E}\lim_{t\to\tau} |x_2(t)|^p = \mathbb{E}\lim_{t\to\tau} F(\zeta(t)) = \lim_{t\to\tau} \mathbb{E}F(\zeta(t)).$$

By the convexity of F, we see that the last expression is at least $\lim_{t\to\tau} F(\mathbb{E}\zeta(t)) = |x_2(0)|^p = F(x)$.

We see that maximizing $\langle |g|^p\rangle_J = \mathbb{E}\lim_{t\to\tau} |x_2(t)|^p$ over all random walks starting at $x \in \Omega$ we always stay above $F(x(0))$, where F is from (2.6.43). And this is true for any starting point $x(0) \in \Omega$.

We found that our bonus function is in fact given by (2.6.43).

2.6.4.3 Formal Bellman Equation. So we have the Bellman equation

$$\sup\{(H_v\alpha, \alpha) \colon \alpha = (\alpha_1, \alpha_2, \alpha_3),\ |\alpha_1| = |\alpha_2|\} = 0 \tag{2.6.44}$$

in $\Omega = \{x = (x_1, x_2, x_3) \colon |x_1|^p \le x_3\}$ (this is a convex domain by the way), with obstacle condition

$$v(x_1, x_2, x_3) \ge |x_2|^p. \tag{2.6.45}$$

2.6.4.4 Saturation. The next logical step toward the proof of Burkholder's Theorem 2.6.13 is to act as before, namely, to impose the saturation condition on (2.6.44). Again, (2.6.44) only pretends to be an equation. In fact, it means the following. Consider any point $a \in \Omega$, and let $\Pi^+(a)$ stand for 2D hyperplane passing through $a = (a_1, a_2, a_3)$ given by condition $x_1 + x_2 = \text{const} = a_1 + a_2$. Similarly, $\Pi^-(a)$ stands for 2D affine hyperplane passing through $a = (a_1, a_2, a_3)$ given by condition $x_1 - x_2 = \text{const} = a_1 - a_2$. It is easy to see that condition (2.6.44) is equivalent to

$$H_v|\Pi^+(a) \leq 0, \quad H_v|\Pi^-(a) \leq 0, \qquad \forall a \in \Omega. \tag{2.6.46}$$

In its turn, this is equivalent to the following pair of restrictive concavities of v:

$$H_{v|\Pi^+(a)} \leq 0, \quad H_{v|\Pi^-(a)} \leq 0, \qquad \forall a \in \Omega. \tag{2.6.47}$$

We now require that at every point of Ω one of these restrictive concavities is saturated.

$$\det H_{v|\Pi^+(a)} \cdot \det H_{v|\Pi^-(a)} = 0, \qquad \forall a \in \Omega. \tag{2.6.48}$$

The latter condition plus (2.6.46) means that at every point $a \in \Omega$ function v must be concave along two hyperplanes and for each point in one of these hyperplanes, it should be degenerated, meaning that Hessian matrix $H_v(a)$ will have an eigenvector with eigenvalue 0 in one of these $\Pi^\pm(a)$.

A priori, the choice of $\pm$ for each a can be extremely complicated. But extra information helps.

2.6.4.5 Symmetry. In Section 1.8, it was observed that

$$v(\pm a_1, \pm a_2, a_3) = v(a_1, a_2, a_3). \tag{2.6.49}$$

This will help us to choose between $\Pi^\pm(a)$ along which hyperplane to degenerate. But we also need homogeneity.

2.6.4.6 Homogeneity. The following easy property was shown in Section 1.8:

$$v \circ T_\lambda = \lambda^p v. \tag{2.6.50}$$

2.6.4.7 How to Find a Candidate for the Solution? Here is a heuristic reasoning that brings (as we will see) the right solution. By symmetry (2.6.49), we need to consider only one quarter of Ω, namely,

$$O \stackrel{\text{def}}{=} \Omega \cap \{a \in \mathbb{R}^3 \colon a_1 > 0, \ a_2 > 0\}.$$

In O, it is a bad idea to degenerate along Π^-. In fact, the degeneration direction is that along which v is linear, but by homogeneity, at infinity v grows much faster than a linear function. This is bad for the choice of Π^- because the intersection of Π^- with O extends to infinity.

So let us try to impose

$$\det H_{v|\Pi^+(a)} = 0, \quad \forall a \in O.$$

By symmetry, this degeneration is switching to $\det H_{v|\Pi^-(a)} = 0$ when we cross the hyperplanes $a_1 = 0$ or $a_2 = 0$. Then it becomes $\det H_{v|\Pi^+(a)} = 0$ for $a \in \Omega$ such that $a_1 < 0, a_2 < 0$. We have to solve all these Monge–Ampère equations effectively with obstacle condition $v(x) \geq |x_2|^p$ in the whole Ω.

Section 1.8 does that by presenting a formula for v. It satisfies concavities (2.6.47), homogeneity (2.6.50), obstacle condition and degeneracy condition (2.6.48). Here we repeat this formula for Burkholder's Bellman function.

Define the following function on $\mathbb{R}^2_+ = \{z = (z_1, z_2) \colon z_i > 0\}$:

$$F_p(z_1, z_2) = \begin{cases} [z_1^p - (p^* - 1)^p z_2^p], \text{ if } z_1 \leq (p^* - 1)z_2, \\ p(1 - \frac{1}{p^*})^{p-1}(z_1 + z_2)^{p-1}[z_1 - (p^* - 1)z_2], \text{ if } z_1 \geq (p^* - 1)z_2. \end{cases} \tag{2.6.51}$$

Note, for $p = 2$ the expressions above are reduced to $F_2(z_1, z_2) = z_1^2 - z_2^2$.

The equation $F_p(|x_2|, |x_1|) = F_p(v^{\frac{1}{p}}, x_3^{\frac{1}{p}})$ determines implicitly the Burkholder's Bellman function v.

We obtain it in a quite formal way.

2.6.4.8 Verification Proposition. However, the verification propositions of Section 1.8 shows that v coincides with the dyadic Bellman function $\mathbf{B}$ of Burkholder's inequality (2.6.39). In particular, we see that for $x_2 \leq x_1$ the following holds:

$$v(x) \leq (p^* - 1)^p x_3,$$

which proves the main inequality (2.6.39) of Burkholder's Theorem 2.6.13. Also, verification procedure in Section 1.8 shows that the constant is sharp.

REMARK 2.6.14 Burkholder's function often made an impression of being obtained by an ad hoc approach. We hope that we have shown here that being considered as a stochastic optimal control problem, its nature became more lucid, and each step in finding it becomes logical.

We suggest the reader to return to Chapter 1 and to try to repeat what we just did by determining what kind of stochastic control problems underline each problem considered in Chapter 1. This includes the assignment of variables, the determination of control processes, the allotment of the profit and bonus functions.

In the next section, we will show how some other classical constants (like e.g., Pichorides constants) of harmonic analysis can be explained from the formal stochastic optimal control approach.

So far, we considered only dyadic problems for function of one real variable. For these problems, the mother equation was Bellman equation (2.5.16).

In the following section, the same scheme will lead us to yet another Bellman equation.

2.7 Almost Perfect Analogy between Stochastic Optimal Control and Harmonic Analysis: Functions on the Complex Plane

In this section, the problems will concern functions of complex variable and the problems will stop being dyadic. We will see that (2.5.19) will start to play the major part.

Recall that given real-valued integrable function f on $[-\pi, \pi)$, we can define a new function in the unit disc as follows:

$$h(z) \stackrel{\text{def}}{=} \frac{1}{2\pi} \int_{-\pi}^{\pi} \frac{e^{it} + z}{e^{it} - z} f(t)\, dt.$$

The real part $f(z)$ of $h(z)$ is harmonic in the unit disc, and it is called the harmonic extension of f, the imaginary $g(z)$ part of $h(z)$ is of course also harmonic and it is called the harmonic conjugate of $f(z)$, and it is often denoted by $g \stackrel{\text{def}}{=} \tilde{f}$. Notice that always $\tilde{f}(0) = 0$. Function $g(z)$ has non-tangential boundary values $g(t)$ when $z \to e^{it}$ in any cone inside the unit disc, this boundary function is the same as the Hilbert transform of f: $g(t) = \mathcal{H}f(t)$ for a. e. $t \in [-\pi, \pi]$. All this can be found in any textbook about Hardy spaces, e.g., [**61, 68**].

Let us consider Pichorides problem of defining the norm in $L^p(\mathbb{T})$ of the operator of harmonic conjugation. We denote the Hilbert transform (harmonic conjugation) by $\mathcal{H}$ and the problem is to give a sharp constant in the inequality

$$\|c_1 f + c_2 \mathcal{H} f\|_p \le C(p, c_1, c_2) \|f\|_p. \tag{2.7.1}$$

THEOREM 2.7.1

$$C(p, 0, 1) = \begin{cases} \tan \frac{\pi}{2p}, & 1 < p \le 2, \\ \cot \frac{\pi}{2p}, & 2 \le p < \infty. \end{cases}$$

Here f is **real-valued** function on the circle $\mathbb{T}$.

2.7.1 Stochastic Optimal Control and $C(p, 0, 1)$

2.7.1.1 Random Walk in $\mathbb{R}^3$ Related to Pichorides Problem. We follow the approach of Section 2.6. However, the problem now is not dyadic and the functions depend on two real variables. Denote the harmonic extension of $|f|^p$ by h and keep symbols f, g to denote the harmonic extension of functions f, g from $\mathbb{T}$ to $\mathbb{D}$.

For all future goals, one can think that f, g are smooth functions on the circle. It is important to keep in mind that

$$f + ig \quad \text{is holomorphic in } \mathbb{D}. \tag{2.7.2}$$

In application to Theorem 2.7.1, this is the case as $g = \mathcal{H}f$. Moreover, $g(0) = 0$ in this case. In what follows, the condition $g(0) = 0$ does not play any role. But in a crucial place in the future, we will need another, slightly relaxed restriction on initial values $f(0), g(0)$, namely,

$$|g(0)| \leq |f(0)|. \tag{2.7.3}$$

REMARK 2.7.2 It is clear that some restriction of that type is needed if one wants to estimate $\|g\|_p$ from above in terms of $\|f\|_p$.

Consider the two-dimensional Brownian motion $W = (w_1, w_2)$ starting at $0 \in \mathbb{C}$, and introduce the random walk

$$\zeta(t) = (x_1(t), x_2(t), x_3(t)) \stackrel{\text{def}}{=} (f \circ W(t), g \circ W(t), h \circ W(t)).$$

The control processes are written in the following matrix (where $z = r + is$):

$$a(\alpha) \stackrel{\text{def}}{=} \begin{bmatrix} \alpha_{1,1}, & \alpha_{1,2} \\ \alpha_{2,1}, & \alpha_{2,2} \\ \alpha_{3,1}, & \alpha_{3,2} \end{bmatrix} = \begin{bmatrix} \frac{\partial f}{\partial r} \circ W(t), & \frac{\partial f}{\partial s} \circ W(t) \\ \frac{\partial g}{\partial r} \circ W(t), & \frac{\partial g}{\partial s} \circ W(t) \\ \frac{\partial h}{\partial r} \circ W(t), & \frac{\partial h}{\partial s} \circ W(t) \end{bmatrix}. \tag{2.7.4}$$

In particular, the Cauchy–Riemann equations imply the following form of matrix a:

$$a(\alpha) = \begin{bmatrix} \alpha_{1,1}, & \alpha_{1,2} \\ -\alpha_{1,2}, & \alpha_{1,1} \\ \alpha_{3,1}, & \alpha_{3,2} \end{bmatrix}. \tag{2.7.5}$$

Now we can write Itô's formula (2.4.3) for each component of random walk $\zeta(t) = (f \circ W(t), g \circ W(t), h \circ W(t))$.

The harmonicity of f, g, h immediately implies that the stochastic differential $d\zeta(t)$ can be written in the form (2.5.1) without drift, i.e., $b = 0$, and with matrix a from (2.7.5).

2.7.2 Who Moves? And Where?

We already saw that

$$\zeta(t) = x + \int_0^t a(\alpha)\, dW(t),$$

where a is given in (2.7.5), (2.7.4).

It is immediate by definition that the random walk happens in domain

$$\Omega \stackrel{\text{def}}{=} \{x = (x_1, x_2, x_3) \in \mathbb{R}^3 \colon |x_1|^p \le x_3\}. \tag{2.7.6}$$

2.7.2.1 Profit Function and Bonus Function. Here profit function vanishes:

$$f(\alpha, x) = 0.$$

Bonus function is obviously exactly the same as in Section 2.6.13,

$$F(x) = F(x_1, x_2, x_3) = |x_2|^p.$$

2.7.2.2 Restriction on Controls. As in Section 2.6.13, it is essential to write down that matrix α belongs to a specific class. Let A stand for family of matrices like in (2.7.5).

$$\alpha_{1,1} = \alpha_{2,2}, \qquad \alpha_{2,1} = -\alpha_{1,2}. \tag{2.7.7}$$

In other words $\alpha \in A$ if and only if it has the form as the right-hand side of (2.7.5).

2.7.2.3 Formal Bellman Equation. We now use equation (2.5.19) as a pattern. This is the difference with Section 2.6, where the pattern was equation (2.5.16).

Let us use the following notations:

$$\alpha_i = (\alpha_{i,1}, \alpha_{i,2}), \qquad i = 1, 2, 3.$$

The scalar product of two vectors has a standard notation $(\cdot, \cdot)$.

We write down (2.5.19) using these Cauchy–Riemann relationships (2.7.7), which means that $(\alpha_1, \alpha_2) = 0$.

Then we get

$$\begin{aligned}\sup_{\alpha \in A} \Bigg[(\alpha_1, \alpha_1) \left(\frac{\partial^2 v}{\partial x_1^2} + \frac{\partial^2 v}{\partial x_2^2} \right) + 2(\alpha_1, \alpha_3) \frac{\partial^2 v}{\partial x_1 \partial x_3} \\ + 2(\alpha_2, \alpha_3) \frac{\partial^2 v}{\partial x_2 \partial x_3} + (\alpha_3, \alpha_3) \frac{\partial^2 v}{\partial x_3^2} \Bigg] = 0.\end{aligned} \tag{2.7.8}$$

We need to solve this "equation" with obstacle condition

$$v(x_1, x_2, x_3) \geq |x_2|^p, \quad x = (x_1, x_2, x_3) \in \Omega, \tag{2.7.9}$$

where Ω is given in (2.7.6).

As in all cases considered in Section 2.6, this is not an equation, but rather an inequality. Also unlike those cases, this inequality states some very involved type of "concavity." The main idea is to saturate it of course. But now this becomes quite a complicated task. We need a simplification.

2.7.2.4 Homogeneity. Notice that the domain Ω is invariant under the transformation $T_\lambda(x_1, x_2, x_3) = (\lambda x_1, \lambda x_2, \lambda^p x_3)$, $\lambda > 0$. We can easily see that the random walk generated by $f, g, |f|^p$ that started at (x_1, x_2, x_3) is transformed by T_λ to a corresponding random walk starting at $T_\lambda(x_1, x_2, x_3)$ (generated by $\lambda f, \lambda g, \lambda^p|f|^p$). Notice that profit function (and the bonus function too) is homogeneous under this transformation in the sense that

$$v \circ T_\lambda = \lambda^p v. \tag{2.7.10}$$

2.7.2.5 Reduction of the Number of Variables. The idea that such problems sometimes allow us to reduce the number of variables belongs to Burkholder. In our case, we formulate this idea in two lemmas. The first one is just for warming up. The second one is a convenient variant of the first one, and we will use it.

LEMMA 2.7.3 *Let $u(x, t)$ be a concave function in convex domain*

$$O = \{(x, t) \in \mathbb{R}^{n+1} : t \geq \rho(x)\}.$$

Consider

$$U(x) \stackrel{\text{def}}{=} \sup_t\{u(x, t) : t \geq \rho(x)\}.$$

Then U is concave on $\mathbb{R}^n$.

REMARK 2.7.4 This lemma was used several times in the problems of Bellman type related to the Monge–Ampère equation. Our condition (2.7.8) is much more involved than the mere concavity condition. This is why we need a more sophisticated Lemma 2.7.6 below.

REMARK 2.7.5 Notice that the supremum of the family of concave function is not concave generically. It is the infimum of concave function that is concave. However, lemma shows an important situation when we get the supremum of concave functions again concave.

LEMMA 2.7.6 *Let $v(x_1, x_2, x_3)$ be a sufficiently smooth function in convex domain $O = \{x \in \mathbb{R}^3 : x_3 \geq \rho(x_1, x_2)\}$ satisfying (2.7.8). Consider*

$$V(x_1, x_2) \stackrel{\text{def}}{=} \sup_{x_3}\{v(x_1, x_2, x_3) \colon x_3 \geq \rho(x_1, x_2)\}.$$

Suppose that for every $(x_1, x_2) \in \mathbb{R}^2$ the supremum is attained inside O. Then V is superharmonic in $\mathbb{R}^2$.

PROOF Given $x' = (x_1, x_2)$, let $x_3 = t(x')$ be the point where the supremum is attained. Then $V(x') = v(x', t(x'))$. Direct calculation shows that

$$\Delta V(x') = \Delta v(x', t(x')) + 2\frac{\partial^2 v}{\partial x_1 \partial x_3} \cdot \frac{\partial t}{\partial x_1} + 2\frac{\partial^2 v}{\partial x_2 \partial x_3} \cdot \frac{\partial t}{\partial x_2}$$
$$+ \frac{\partial^2 v}{\partial x_3^2} \cdot \left[\left(\frac{\partial t}{\partial x_1}\right)^2 + \left(\frac{\partial t}{\partial x_2}\right)^2\right] + \frac{\partial v}{\partial x_3} \cdot \left[\frac{\partial^2 t}{\partial x_1^2} + \frac{\partial^2 t}{\partial x_2^2}\right].$$

The last term vanishes as $\frac{\partial v}{\partial x_3}(x', t(x')) = 0$ by the fact that we attain maximum at $x_3 = t(x')$. Hence,

$$\Delta V(x') = \Delta v(x', t(x')) + 2\frac{\partial^2 v}{\partial x_1 \partial x_3} \cdot \frac{\partial t}{\partial x_1} + 2\frac{\partial^2 v}{\partial x_2 \partial x_3} \cdot \frac{\partial t}{\partial x_2}$$
$$+ \frac{\partial^2 v}{\partial x_3^2} \cdot \left[\left(\frac{\partial t}{\partial x_1}\right)^2 + \left(\frac{\partial t}{\partial x_2}\right)^2\right]. \tag{2.7.11}$$

Now plug $\alpha_1 = (1, 0)$, $\alpha_2 = (0, 1)$, $\alpha_3 = \left(\frac{\partial t}{\partial x_1}, \frac{\partial t}{\partial x_2}\right)$ into (2.7.8). We get the right-hand side of (2.7.11). And by (2.7.8), it is nonpositive. Lemma is proved. □

We expect that minimal (or otherwise saturated) solution of (2.7.8) satisfying homogeneity condition (2.7.10) and obstacle condition (2.7.9) will give us the proof of Theorem 2.7.1.

It is easy to saturate (2.7.8), one just requires that for every point of domain Ω there exists $\alpha \in A, \alpha \neq 0$, such that it makes the left-hand side of (2.7.8) to vanish. We already confessed that we do not know how to solve saturated (2.7.8). Let β be a positive number, and let us introduce

$$v_\beta(x) = v(x) - \beta x_3, \qquad V_\beta(x_1, x_2) \stackrel{\text{def}}{=} \sup_{x_3}\{v_\beta(x) \colon x \in \Omega\}.$$

Obviously v_β satisfies (2.7.8) as soon as v does. Hence, by Lemma 2.7.6 function V_β satisfies

$$\Delta V_\beta(x_1, x_2) \leq 0, \qquad (x_1, x_2) \in \mathbb{R}^2. \tag{2.7.12}$$

Also the obstacle condition (2.7.9) together with the inequality $x_3 \geq |x_1|^p$ valid in Ω show that V_β satisfies the new obstacle condition:

$$V_\beta(x_1, x_2) \geq F_\beta(x_1, x_2) \stackrel{\text{def}}{=} |x_2|^p - \beta |x_1|^p. \tag{2.7.13}$$

Homogeneity for v listed in (2.7.10) becomes now

$$V_\beta(\lambda x_1, \lambda x_2) = \lambda^p V_\beta(x_1, x_2), \quad \lambda > 0. \tag{2.7.14}$$

2.7.2.6 Obstacle Problems for Superharmonic Functions. We formally reduced everything to this obstacle problem: to find the smallest β such that function $F_\beta = |x_2|^p - \beta|x_1|^p$ has a superharmonic majorant V_β satisfying homogeneity (2.7.14).

We consider cases $1 < p \leq 2$ first. We will solve our obstacle problem by producing a very simple superharmonic function bigger than our obstacle with an optimal β. The verification proposition will show that β is optimal.

Consider for $1 < p \leq 2$ the following function

$$u(z) \stackrel{\text{def}}{=} u_p(z) \stackrel{\text{def}}{=} \Re\, z^p. \tag{2.7.15}$$

We understand this formula as follows. Function z^p is well-defined analytic function in the right half-plane. We consider its real part in this half-plane and then mirror reflect it into another half-plane. It is easy to see that the resulting function is subharmonic everywhere except the origin. See Exercise 2.8.4 at the end of this chapter.

To check the subharmonicity at the origin, we need to check only that

$$\int_{-\pi/2}^{\pi/2} \cos p\theta \, d\theta \geq 0,$$

which is true for $1 < p \leq 2$.

Now choose constants A_p and B_p in such a way that

$$|\sin\theta|^p \leq A_p \cos p\theta + B_p \cos^p \theta, \quad \forall \theta \in [-\pi/2, \pi/2]. \tag{2.7.16}$$

Inequality (2.7.16) means exactly what we need: that homogeneous superharmonic function

$$A_p u(z) = A_p \Re\,(x_1 + i x_2)^p$$

is bigger than

$$F_{B_p}(x_1, x_2) = |x_2|^p - B_p |x_1|^p$$

in the right half-plane (then its symmetric extension will have majorization property also in the left half-plane). Let us find the smallest $B_p = \beta$ for which

A_p exists such that (2.7.16) holds. We will see soon that this is the following constant: $B_p = \left(\tan\frac{\pi}{2p}\right)^p$.

Plug $\theta = \pm\pi/2$ into (2.7.16), we see that A_p must be negative. Plugging $\theta = 0$, we see that B_p must be positive.

Plugging $\theta = \pm\frac{\pi}{2p}$, we see that necessarily

$$B_p \geq \left(\tan\frac{\pi}{2p}\right)^p.$$

It is reasonable to check whether $B_p = \left(\tan\frac{\pi}{2p}\right)^p$ is in fact the smallest possible constant for which inequality (2.7.16) holds.

LEMMA 2.7.7 *Inequality* (2.7.16) *holds with* $B_p = \left(\tan\frac{\pi}{2p}\right)^p$ *and* $A_p = -\left(\sin^p\frac{\pi}{2p}\right)\left(1+\tan\frac{\pi}{2p}\right)$.

2.7.2.7 Verification Propositions. We use the notations of Theorem 2.7.1. Our approach is consistently the same as before. We found the constant by solving a formal Bellman equation in its form reduced to obstacle condition for superharmonic functions. As the approach was completely formal, we need to verify that this educated formal guessing gives in fact the right constant.

PROPOSITION 2.7.8 *Let* $1 < p \leq 2$. *Then* $C(p,0,1)^p \leq B_p = \left(\tan\frac{\pi}{2p}\right)^p$. *Moreover,* $\|g\|_p^p \leq B_p\|f\|_p^p$ *for any two real-valued functions such that* (2.7.2) *and* (2.7.3) *are satisfied.*

PROOF Let f be a smooth test function on the circle and g be another real-valued harmonic function such that $f + ig$ is analytic and such that $|g(0)| \leq |f(0)|$. We proved a pointwise inequality

$$|g(z)|^p - B_p|f(z)|^p \leq -A_p u(f(z) + ig(z)), \quad \forall z \in \mathbb{D}. \tag{2.7.17}$$

This is just a superposition of (2.7.16) with the map $(f,g) : \mathbb{D} \to \mathbb{C}$, $z \mapsto f(z) + ig(z)$. Plug $z = e^{it}$ and integrate over the circle. Function u is superharmonic. We substitute inside it a holomorphic function $f + ig$. So $u(f(z) + ig(z))$ is superharmonic on the disc and continuous up to the boundary. We can integrate (2.7.17) over the circle. Subharmonicity of u from (2.7.15) gives us

$$\frac{1}{2\pi}\int_0^{2\pi} u(f(e^{it}) + ig(e^{it}))\,dt \geq u(f(0) + ig(0)) \geq 0. \tag{2.7.18}$$

For the second inequality above, see Problem 2.8.6. Therefore,

$$\frac{1}{2\pi}\int_0^{2\pi} |g(e^{it}|^p\,dt - B_p\frac{1}{2\pi}\int_0^{2\pi} |f(e^{it}|^p\,dt \leq 0.$$

This verifies one of our inequality $C(p, 0, 1) \le B_p$ if $1 < p \le 2$. □

REMARK 2.7.9 If one guesses (2.7.17), then the whole machinery above is not needed. Our goal was to show a formal but logical way to come to this inequality from "the first principles." And the way through stochastic optimal control is such a logical path.

PROPOSITION 2.7.10 *If* $1 < p \le 2$, *then* $C(p, 0, 1) \ge B_p = \left(\tan \frac{\pi}{2p}\right)^p$.

PROOF Let us fix $a > p$ and consider function in the unit disc given by the formula

$$\Phi_a(z) = \left(\frac{1+z}{1-z}\right)^{\frac{1}{a}}.$$

It is very easy to see that it maps the unit disc conformally onto the cone domain $\{w : |\arg w| < \frac{\pi}{a}\}$. Boundary goes into the boundary. Therefore, if we put $f \stackrel{\text{def}}{=} \Re\Phi_a, g \stackrel{\text{def}}{=} \Im\Phi_a$ (of course, $f(0) = 1, g(0) = 0$) we automatically get the equality

$$\|g\|_p^p = \left(\tan \frac{\pi}{2a}\right)^p \|f\|_p^p.$$

Tending a to p, we prove the proposition. □

2.7.2.8 The Case $p > 2$.

REMARK 2.7.11 The sharp constant of Pichorides inequality for $p > 2$ is of course exactly the same as for the conjugate exponent $p' \in (1, 2)$. This follows from duality, and the fact that the adjoint operator to the Hilbert transform is minus the Hilbert transform: $\mathcal{H}^* = -\mathcal{H}$. However, we wish to follow the same path as before to emphasize the difficulties that would arise in using our previous method but for $p > 2$. We will succeed, and we will build the correct function U_p that plays the role of Bellman function. But it will be not as simple as in (2.7.15).

Function u given by (2.7.15) in the right half-plane and extended by symmetry into the left half-plane is not subharmonic on the plane anymore if $p > 2$. We need to take a closer look at the superharmonic obstacle problem with the obstacle function $F_\beta = |x_2|^p - \beta|x_1|^p$. The homogeneity of order p reduces one variable and the obstacle now is

$$\psi_\beta(\theta) = |\sin\theta|^p - \beta|\cos\theta|^p.$$

The homogeneity of the majorant makes $V_\beta = r^p\varphi(\theta)$, where $r = \sqrt{x_1^2 + x_2^2}$, and θ is the argument of $x_1 + ix_2$, and its superharmonicity becomes

$$L\varphi(\theta) \stackrel{\text{def}}{=} \varphi''(\theta) + p^2\varphi(\theta) \le 0. \tag{2.7.19}$$

To find the smallest β for which ψ_β has a majorant satisfying (2.7.19), we now look at the range $0 \le \theta \le \pi$ and notice that $L\psi$ is itself negative near two end points of $[0,\pi]$. In fact, for $\theta \in [0,\pi/2]$, $L\psi = p(p-1)[\sin^{p-2}\theta - \beta\cos^{p-2}\theta]$, and

$$L\psi(\theta) \le 0, \quad \theta \in [0,\theta'] \cup [\pi - \theta', \pi], \quad \theta' = \theta'(\beta, p). \tag{2.7.20}$$

Now we take function $u(\theta) \stackrel{\text{def}}{=} A\cos p(\theta - \frac{\pi}{2})$, $A > 1$, (for which $Lu = 0$ on the whole interval $[0,\pi]$), and notice that for very large positive β, $\psi_\beta = \sin^p\theta - \beta|\cos\theta|^p$ is smaller than $u(\theta)$on the whole interval $[0,\pi]$.

Let us notice that two zeros of ψ_β will start to move toward the end of the interval $[0,\pi]$, when we start to diminish β. However, they cannot become closer to the endpoints than zeros of u. These zeros are

$$\theta_0 \stackrel{\text{def}}{=} \frac{\pi}{2p'}, \quad \text{and} \quad \pi - \theta_0, \tag{2.7.21}$$

where $p' = \frac{p}{p-1}$ is a conjugate exponent of p.

Both ψ_β and u are symmetric with respect to $\theta = \pi/2$, so we will look only at the approach of the zero of ψ_β to θ_0 from the right. We choose two constants A, β in the definition of u and β_p in such a way that two conditions happen simultaneously:

$$\psi_\beta(\theta_0) = u(\Theta_0),$$

and

$$\psi'_\beta(\theta_0) = u'(\theta_0).$$

One can easily see that the solutions $\beta = \beta_p$ and $A = A_p$ are

$$(\beta_p)^{\frac{1}{p}} = \tan\theta_0 = \tan\frac{\pi}{2p'} = \cot\frac{\pi}{2p}, \quad A_p = \frac{\cos^{p-1}\left(\frac{\pi}{2p}\right)}{\sin\frac{\pi}{2p}}. \tag{2.7.22}$$

Now consider

$$\varphi(\theta) \stackrel{\text{def}}{=} \begin{cases} \psi_{\beta_p}(\theta), & \theta \in [0,\theta_0] \cup [\pi - \theta_0, \pi], \\ A_p\cos p\left(\theta - \frac{\pi}{2}\right), & \theta \in [\theta_0, \pi - \theta_0]. \end{cases} \tag{2.7.23}$$

Let us compare θ' and θ_0: θ' is the root of $\tan^{p-2}\theta' = \beta_p$ while θ_0 is the root of $\tan^p \theta_0 = \beta_p$. So $\tan\theta_0 = \tan\frac{\pi}{2p'} > 1$ because $p > 2$. Therefore,

$$\tan^{p-2}\theta' = \beta_p = (\tan\theta_0)^p > (\tan\theta_0)^{p-2}.$$

We see that $\theta_0 < \theta'$. This fact and the tangential kissing of u and ψ_{β_p} at θ_0 that is strictly less than θ' show together that

$$L\varphi \le 0 \tag{2.7.24}$$

on $[0, \pi]$ in the sense of distribution.

2.7.2.9 Majorization. We are left to show that $\varphi \ge \psi_{\beta_p}$. By the definition of φ, we need to do this only when $\varphi = A_p \cos p(\theta - \frac{\pi}{2})$, that is on $[\theta_0, \pi - \theta_0]$.

For $\theta \in [\theta_0, \pi - \theta_0] = [\frac{\pi}{2p'}, \pi - \frac{\pi}{2p'}]$, put $t \stackrel{\text{def}}{=} \frac{\pi}{2} - \theta$. By symmetry, the only thing we now need to check is that for $t \in [0, \frac{\pi}{2p}]$ and for $p > 2$, the following trigonometric inequality holds

$$\cos^p t - \frac{\cos^{p-1}(\pi/2p)}{\sin(\pi/2p)} \cos pt \le \cot^p(\pi/2p)\sin^p t. \tag{2.7.25}$$

To prove it, let us denote $a \stackrel{\text{def}}{=} \frac{\cos^{p-1}(\pi/2p)}{\sin(\pi/2p)}$ and consider

$$f(t) \stackrel{\text{def}}{=} \frac{\cos^p t - a\cos pt}{\sin^p t}, \qquad t \in (0, \pi/2p].$$

We get

$$f'(t) = p\frac{a\cos(p-1)t - \cos^{p-1} t}{\sin^{p+1} t},$$

$$f(\pi/2p) = \cot^p(\pi/2p), \qquad f'(\pi/2p) = 0.$$

We want to check that $f'(t) \le 0$ for $t \in (0, \pi/2p]$. This is the same as to prove that on this interval, we have

$$f_1(t) \stackrel{\text{def}}{=} \frac{\cos^{p-1} t}{\cos(p-1)t} \le a = \frac{\cos^{p-1}(\pi/2p)}{\sin(\pi/2p)} = f_1(\pi/2p). \tag{2.7.26}$$

Equality (2.7.26) is valid because

$$f_1'(t) = (p-1)\frac{\cos^{p-2} t\,\sin(p-2)t}{\cos^2(p-1)t} \ge 0.$$

2.7.2.10 Superharmonic Majorant of Obstacle Function F_{β_p}, $p > 2$. Given function φ from (2.7.23), we can consider function of $x_1 + ix_2 = re^{i\theta}$ given by

$$V \stackrel{\text{def}}{=} V_{\beta_p} \stackrel{\text{def}}{=} r^p \varphi(\theta), \quad \theta \in [0, \pi], \tag{2.7.27}$$

and we continue it by reflection in the real line to the lower half-plane. It was built to be superharmonic in all points except may be the real axis. At all points of the real axis except for the origin, it is superharmonic as well. See Problem 2.8.8. Direct calculation shows that also

$$\int_0^\pi \varphi(\theta)\, d\theta \le 0.$$

This implies that V satisfies the mean value inequality at the origin as well. Therefore, V is superharmonic on the whole plane. We have just proved that

$$F_{\beta_p}(x_1, x_2) \le V(x_1, x_2). \tag{2.7.28}$$

2.7.2.11 Verification Proposition. We are going to prove that the constant $\beta_p =$ gives exactly $C(p, 0, 1)$ for $p > 2$.

PROPOSITION 2.7.12 *Let $p > 2$. Then $C(p, 0, 1) \le (\beta_p)^{\frac{1}{p}} = \cot \frac{\pi}{2p}$. Moreover, $\|g\|_p^p \le \beta_p \|f\|_p^p$ for any two real-valued functions such that* (2.7.2) *and* (2.7.3) *are satisfied.*

We already remarked that just duality will suffice to get this proposition from the analogous proposition for $1 < p \le 2$ proved above. However, we can also show this proposition by using the special function V constructed in (2.7.27).

The proof repeats verbatim the proof of Proposition 2.7.8 if we take into consideration that V_β is superharmonic, that obstacle inequality (2.7.28) holds, and that $V(f(0), g(0)) \le 0$ (see Exercise 2.8.9).

PROPOSITION 2.7.13 *Let $p > 2$. Then $C(p, 0, 1) \ge (\beta_p)^{\frac{1}{p}} = \cot \frac{\pi}{2p}$.*

PROOF Consider $a > p$, $b = a' = \frac{a}{a-1}$, and $F_a(z) \stackrel{\text{def}}{=} i\left(\frac{1+z}{1-z}\right)^{\frac{1}{a}}$. Then $F(0) = i$. We write $F_a = f_a + ig_a$, $g_a(0) = 1$. This is a conformal map of the unit disc onto the cone of angular size $\frac{\pi}{a}$. In particular, we have a pointwise equality $|\Im F_a(e^{ix})| = \tan \frac{\pi}{2b} |\Re F_a(e^{ix})|$. This implies $\|g_a\|_p = \tan \frac{\pi}{2b} \|f_a\|_p$. Now let us estimate $\|g_a - 1\|_p^p$ from below. Clearly by triangle inequality

$$\left(\tan \frac{\pi}{2b} \|f_a\|_p - 1\right)^p \le \|g_a - 1\|_p^p.$$

Also by definition of $C(p,0,1)$ we have

$$\|g_a - 1\|_p^p \le C(p,0,1)^p \|f_a\|_p^p,$$

just because $g_a - 1 = \mathcal{H} f_a$. Combine these two inequalities:

$$C(p,0,1)\|f_a\|_p \ge \left(\tan\frac{\pi}{2b}\|f_a\|_p - 1\right)$$

and divide it by $\|f_a\|_p$ to get

$$C(p,0,1) \ge \tan\frac{\pi}{2b} - 1/\|f_a\|_p.$$

By construction $\lim_{a\to p}\|f_a\|_p = \infty$, and this proves the proposition by tending a to p and b to p'. □

REMARK 2.7.14 Of course the second part ($p > 2$) of Theorem 2.7.1 follows from the first just because $\mathcal{H}^* = -\mathcal{H}$ and the norms of an operator and its dual are the same. However, we wanted to illustrate how nontrivial a solution of the obstacle problem can be. In our example, the solution for $p > 2$ was more involved than for $1 < p < 2$. We also wanted to show that stochastic optimal control approach that brought us in both cases to superharmonic obstacle problem was fruitful and brought the right answer in both cases. Also in Proposition 2.7.12, we proved a bit more than just the sharp estimate of the Hilbert transform.

2.7.3 Stochastic Optimal Control and $C(p, 1, i)$

THEOREM 2.7.15

$$C(p,1,i) = \begin{cases} \frac{1}{\cos\frac{\pi}{2p}}, & 1 < p \le 2, \\ \frac{1}{\cos\frac{\pi}{2p'}}, & 2 < p < \infty. \end{cases}$$

This can be obtained immediately from Theorem 2.7.1 just by the following two elementary applications of Minkowski's inequality:

$$\|(f^2+g^2)^{\frac{1}{2}}\|_p \ge (\|f\|_p^2 + \|g\|_p^2)^{\frac{1}{2}}, \quad 1 < p \le 2, \tag{2.7.29}$$

$$\|(f^2+g^2)^{\frac{1}{2}}\|_p \le (\|f\|_p^2 + \|g\|_p^2)^{\frac{1}{2}}, \quad 2 \le p < \infty. \tag{2.7.30}$$

However, we again can use our stochastic optimal control pattern, and we immediately come to the same type of superharmonic obstacle problem, but only the obstacle function changes. Now instead of $|x_2|^p - \beta|x_1|^p$, it becomes

$$F_\beta \stackrel{\text{def}}{=} |x_1 + ix_2|^p - \beta|x_1|^p = (x_1^2 + x_2^2)^{\frac{p}{2}} - \beta|x_1|^p.$$

Again we need to find the smallest $\beta = \beta_p$, for which F_β possesses a superharmonic majorant V with $V(x,y) \le 0$ if $|y| \le |x|$.

2.7.4 The Majorant for $1 < p \leq 2$

Here as always $x_1 + ix_2 = re^{i\theta}$, $r = \sqrt{x_1^2 + x_2^2}$. Then

$$V(x_1, x_2) = -\left(\tan \frac{\pi}{2p}\right) r^p \cos p\theta, \quad \theta \in [-\pi/2, \pi/2] \tag{2.7.31}$$

extended to the left half-plane by symmetry with respect to the imaginary axis is a superharmonic function in the whole plane, and it is a majorant of $F_{\beta_p} = r^p - \beta_p r^p |\cos \theta|^p$, where

$$\beta_p = \left(\frac{1}{\cos \frac{\pi}{2p'}}\right)^p. \tag{2.7.32}$$

2.7.5 Verification Propositions

Now we repeat verbatim the verification arguments from Section 2.7.2.11.

2.7.6 The Majorant for $2 < p < \infty$

Set ($r = \sqrt{x_1^2 + x_2^2}$)

$$V(x_1, x_2) = \begin{cases} r^p - \left(\frac{1}{\cos \frac{\pi}{2p'}}\right)^p |x_1|^p, & \theta \in \left[0, \frac{\pi}{2p'}\right] \cup \left[\pi - \frac{\pi}{2p'}, \pi\right], \\ \left(\tan \frac{\pi}{2p'}\right) r^p \cos p\left(\theta - \frac{\pi}{2}\right), & \theta \in \left[\frac{\pi}{2p'}, \pi - \frac{\pi}{2p'}\right]. \end{cases} \tag{2.7.33}$$

Then we extend V into the lower half-plane by the symmetry with respect to the real axis.

The reader can compare this formula with (2.7.23) and (2.7.27).

Of course one also needs to prove that the majorization $V \geq F_{\beta_p}$ holds (where β_p is from (2.7.32)). Exactly as in 2.7.2.9, this reduces to a trigonometric inequality:

$$1 - \left(\frac{1}{\cos \frac{\pi}{2p'}}\right)^p \cos^p \theta \leq \left(\tan \frac{\pi}{2p}\right) \cos p\left(\theta - \frac{\pi}{2}\right), \quad \theta \in \left[\frac{\pi}{2p'}, \frac{\pi}{2}\right]. \tag{2.7.34}$$

2.8 A Problem of Gohberg–Krupnik from the Point of View of Stochastic Optimal Control

We showed in Section 2.7 that

$$C(p,1/2,i/2) = \max\left(\frac{1}{2\cos\frac{\pi}{2p}}, \frac{1}{2\sin\frac{\pi}{2p}}\right).$$

Actually, we showed a bit more, namely, that for all real-valued f, we have

$$\|cf(0) + f/2 + i\mathcal{H}f/2\|_p \le C(p,1/2,i/2)\|f\|_p, \qquad \forall a \in \mathbb{R}, \qquad (2.8.1)$$

where the constant is sharp. This is because $|\mathcal{H}f(0)| = 0 \le |c+1/2|$ for any $c \in \mathbb{R}$. Let us consider a very special case $c = \frac{1}{2}$.

DEFINITION 2.8.1 Operator $f \mapsto \frac{1}{2}(f(0) + f + i\mathcal{H}f)$, $f \in L^p(\mathbb{T})$, is called the Riesz projection, and is denoted by $\mathbb{P}_+$.

The sharp inequality (2.8.1) gives the norm of $\mathbb{P}_+$ in L^p on real functions. Notice that for $p = 2$, this norm is $C(2,1/2,i/2) = \frac{1}{\sqrt{2}} < 1$. But $\mathbb{P}_+$ on complex-valued functions obviously has the same norm as the orthogonal projection of $L^2(\mathbb{T})$ onto the Hardy space $\mathcal{H}^2(\mathbb{T})$, and as such must have norm 1. So even for $p = 2$, the effect of passing from real-valued functions to complex-valued functions is very noticeable.

We naturally conclude that it is interesting to find $\|\mathbb{P}_+\|_p$, or, in other words, to find the sharp constant in

$$\|f(0)/2 + f/2 + i\mathcal{H}f/2\|_p \le C_{\mathbb{C}}(p,1/2,i/2)\|f\|_p \qquad (2.8.2)$$

for complex-valued f. Obviously, $C_{\mathbb{C}}(p,1/2,i/2) > C(p,1/2,i/2)$ at least for p close to 2 (we will see that for all p, $1 < p < \infty$, this is the case).

Put $\mathbb{P}_- \stackrel{\text{def}}{=} I - \mathbb{P}_+$ and keep in mind the formula $\mathbb{P}_- = \frac{1}{2}(I - i\mathcal{H} - K_0)$, where $K_0 f = f(0)$. Gohberg and Krupnik posed in 1968 the problem of finding the norms $\|\mathbb{P}_\pm\|_p$ on $L^p_{\mathbb{C}}(\mathbb{T})$.

It was solved in 2000 by Hollenbeck and Verbitsky [**69**]. We will show how their result can be interpreted in the language of stochastic optimal control. This explains some formulas in their proof that may seem ad hoc, but that are quite natural in our framework. The constant turns out to be

$$C_{\mathbb{C}}(p,1/2,i/2) = \frac{1}{\sin\frac{\pi}{p}}. \qquad (2.8.3)$$

Hollenbeck and Verbitsky proved even a more interesting inequality with the sharp constant

THEOREM 2.8.2

$$\|\max(|\mathbb{P}_+ f|, |\mathbb{P}_- f|)\|_p \leq \frac{1}{\sin\frac{\pi}{p}} \|f\|_p, \quad 1 < p \leq 2. \tag{2.8.4}$$

REMARK 2.8.3 The corresponding constant for $p > 2$ is unknown. We will discuss how one can try to solve this problem.

2.8.1 Random Walk in $\mathbb{R}^5$ Related to Gohberg–Krupnik Problem

Given a smooth function f on $\mathbb{T}$, we put

$$f_+(z) \stackrel{\text{def}}{=} \mathbb{P}_+ f(z), \quad f_-(z) \stackrel{\text{def}}{=} \overline{\mathbb{P}_- f(1/\bar{z})}, \qquad z = \xi + i\eta \in \mathbb{D}.$$

These are two analytic functions in the disc $\mathbb{D}$. Notice that on the circle $\mathbb{T}$

$$f_+(z) + \overline{f_-(z)} = \mathbb{P}_+ f(z) + \mathbb{P}_- f(z) = f(z), \quad z \in \mathbb{T}. \tag{2.8.5}$$

In what follows, we need that the initial value $f_-(0)$ satisfies

$$f_-(0) = 0. \tag{2.8.6}$$

Now we use the following notations for harmonic functions $\Re f_\pm, \Im f_\pm$:

$$f_+ \stackrel{\text{def}}{=} f_1 + i f_2, \quad f_- \stackrel{\text{def}}{=} f_3 + i f_4.$$

Consider the two-dimensional Brownian motion $W = (w_1, w_2)$ starting at $0 \in \mathbb{C}$, and introduce the random walk

$$\zeta(t) = (\zeta_i(t))_{i=1}^5 \stackrel{\text{def}}{=} (f_i \circ W(t))_{i=1}^5,$$

where $f_5(\zeta)$ is the harmonic extension of $|f|^p$ into the unit disc.

The control processes are written in the following matrix (where $\zeta = \xi + i\eta$):

$$a(\alpha) \stackrel{\text{def}}{=} \begin{bmatrix} \alpha_{1,1}, \ \alpha_{1,2} \\ \alpha_{2,1}, \ \alpha_{2,2} \\ \alpha_{3,1}, \ \alpha_{3,2} \\ \alpha_{4,1}, \ \alpha_{4,2} \\ \alpha_{5,1}, \ \alpha_{5,2} \end{bmatrix} = \begin{bmatrix} \frac{\partial f_1}{\partial \xi} \circ W(t), \ \frac{\partial f_1}{\partial \eta} \circ W(t) \\ \dots\dots\dots\dots\dots \\ \dots\dots\dots\dots\dots \\ \dots\dots\dots\dots\dots \\ \frac{\partial f_5}{\partial \xi} \circ W(t), \ \frac{\partial f_5}{\partial \eta} \circ W(t) \end{bmatrix}. \tag{2.8.7}$$

In particular, the Cauchy–Riemann equations imply the following form of matrix a:

$$a(\alpha) = \begin{bmatrix} \alpha_{1,1}, \ \alpha_{1,2} \\ -\alpha_{1,2}, \ \alpha_{1,1} \\ \alpha_{3,1}, \ \alpha_{3,2} \\ -\alpha_{3,2}, \ \alpha_{3,1} \\ \alpha_{5,1}, \ \alpha_{5,2} \end{bmatrix}. \tag{2.8.8}$$

Now we can write formula Itô (2.4.3) for each component of random walk $\zeta_i(t) = f_i \circ W(t),\ i = 1, \ldots, 5$.

The harmonicity of f_i immediately implies that the stochastic differential $d\zeta(t)$ can be written in the form (2.5.1) without any drift and with matrix a from (2.8.8).

2.8.2 Who Moves? And Where?

We already saw that

$$\zeta(t) = x + \int_0^t a(\alpha)\, dW(t),$$

where a is given in (2.8.8), (2.8.7).

Let us see that the random walk happens in domain

$$\Omega \stackrel{\text{def}}{=} \{x \in \mathbb{R}^5 \colon \big((x_1 + x_3)^2 + (x_2 - x_4)^2\big)^{\frac{p}{2}} \le x_5\}.$$

We want to use from now on some complex notations in our $\mathbb{R}^5$. Let $\mathfrak{w} = x_1 + ix_2, \mathfrak{z} = x_3 + ix_4$. Slightly abusing notations, we use two different ways to write down vector x: 1) $x = (x_1, \ldots, x_5)$, and 2) $x = (\mathfrak{w}, \mathfrak{z}, x_5)$.

$$\Omega \stackrel{\text{def}}{=} \{x \in \mathbb{C} \times \mathbb{C} \times \mathbb{R} \colon |\mathfrak{w} + \bar{\mathfrak{z}}|^p \le x_5\}|,. \tag{2.8.9}$$

In fact, (2.8.5) shows that the harmonic extension of $f_+ + \bar{f_-}$ from $\mathbb{T}$ to $\mathbb{D}$ is the harmonic extension of f. Using the symbol $P\star$ for the operator of harmonic extension, we can write the following chain of inequalities:

$$\begin{aligned} f_5 &= P \star |f|^p \ge |P \star f|^p = |P \star (f_+ + \bar{f_-})|^p \\ &= |P \star (f_+) + \overline{P \star f_-})|^p = |f_+ + \bar{f_-}|^p = |f_1 + if_2 + f_3 - if_4|^p, \end{aligned}$$

which implies (2.8.9).

2.8.2.1 Profit Function and Bonus Function. Here profit function vanishes:

$$f(\alpha, x) = 0.$$

Bonus function is obvious from the need to estimate L^p-norm of $\max(|f_+|, |f_-|)$: $F(x) = \max(|\mathfrak{w}|, |\mathfrak{z}|)^p$.

2.8.2.2 Restriction on Controls. As in Sections 2.6.13 and 2.7, it is essential to write down that matrix α belongs to a specific class. We denote the family of matrices as in (2.8.8) by symbol A.

2.8.2.3 Formal Bellman Equation. We now use equation (2.5.19) as a pattern. We did this already in Section 2.7. This is the difference with the Section 2.6, where the pattern was equation (2.5.16).

Let us use the following notation:

$$\alpha_i = (\alpha_{i,1}, \alpha_{i,2}), \qquad i = 1, \ldots, 5.$$

The scalar product of two such 2D vectors has a standard notation $(\cdot, \cdot)$. We write down (2.5.19) using these Cauchy–Riemann relationships (2.8.8). Then we get

$$\begin{aligned}\sup_{\alpha\in A}\Bigg[&(\alpha_1,\alpha_1)\left(\frac{\partial^2 v}{\partial x_1^2}+\frac{\partial^2 v}{\partial x_2^2}\right)+(\alpha_3,\alpha_3)\left(\frac{\partial^2 v}{\partial x_3^2}+\frac{\partial^2 v}{\partial x_4^2}\right)\\ &+2\sum_{i=1,2,5;\, j=3,4,5}(\alpha_i,\alpha_j)\frac{\partial^2 v}{\partial x_i \partial x_j}-(\alpha_5,\alpha_5)\frac{\partial^2 v}{\partial x_5^2}\Bigg]=0.\end{aligned} \tag{2.8.10}$$

We need to solve this "equation" with obstacle condition

$$v(x) \geq F(x) = \max(|x_1 + ix_2|, |x_3 - ix_4|)^p, \quad x \in \Omega, \tag{2.8.11}$$

where Ω is given in (2.8.9).

As in all cases considered in Section 2.7.2.3 this is not an equation, but rather an inequality. Unlike the cases of the previous section, this inequality states some very involved type of "concavity." The main idea is to saturate it of course. But now this becomes quite a complicated task. We need a simplification.

2.8.2.4 Homogeneity. Notice that the domain Ω is invariant under the transformation

$$T_\lambda x = (\lambda \mathfrak{w}, \bar{\lambda}\mathfrak{z}, |\lambda|^p x_5),\ \lambda \in \mathbb{C}.$$

We can easily see that the random walk generated by $f_+, f_-, |f|^p$ that started at $x \in \Omega$ is transformed by T_λ to a corresponding random walk starting at $T_\lambda(x)$ (generated by $\lambda f_+, \bar{\lambda} f_-, |\lambda|^p |f|^p$). Notice that profit function (and the bonus function too) is homogeneous under this transformation in the sense that

$$v \circ T_\lambda = |\lambda|^p v. \tag{2.8.12}$$

2.8.2.5 Saturation and Reducing the Number of Variables. Saturating (2.8.10) seems to be a very nasty problem because there are too many variables. Let us first get rid of x_5 exactly as we did it in Section 2.7.

LEMMA 2.8.4 *Let $u(\mathfrak{w}, \mathfrak{z}, x_5)$ be a sufficiently smooth function in convex domain $O = \{x \colon x_5 \geq \varphi(\mathfrak{w}, \mathfrak{z})\}$ satisfying* (2.8.10). *Consider*

$$U(\mathfrak{w}, \mathfrak{z}) \stackrel{\text{def}}{=} \sup_{x_5}\{u(\mathfrak{w}, \mathfrak{z}, x_5) \colon x_5 \geq \varphi(\mathfrak{w}, \mathfrak{z})\}.$$

Suppose that for every $(\mathfrak{w}, \mathfrak{z}) \in \mathbb{C}^2$, the supremum is attained inside O. Then U is pluri-superharmonic in $\mathbb{C}^2$.

PROOF Exactly as in the proof of Lemma 2.7.6, we will obtain that

$$\Delta_{\mathfrak{w}} U(\mathfrak{w}, \mathfrak{z}) \leq 0, \quad \Delta_{\mathfrak{z}} U(\mathfrak{w}, \mathfrak{z}) \leq 0. \tag{2.8.13}$$

Let us notice the invariance of (2.8.10) under certain transformations. We consider $R_c(\mathfrak{w}, \mathfrak{z}, x_5) = (\mathfrak{w} + c\mathfrak{z}, \mathfrak{z}, x_5)$ with $c \in \mathbb{R}$, and look at $u \circ R_c$. This function also satisfies (2.8.10) if we transform matrix α to matrix α', where to obtain α' one needs to add to the first two rows of α the next two rows multiplied by c. Replacing u by $u \circ R_c$ and α by α' and noticing that $\alpha' \in A$ as soon as $\alpha \in A$, we can see that $u \circ R_c$ also satisfies (2.8.10).

Similarly, consider $R_{ic}(\mathfrak{w}, \mathfrak{z}, x_5) = (\mathfrak{w} + ic\mathfrak{z}, \mathfrak{z}, x_5)$ with $c \in \mathbb{R}$, and look at $u \circ R_{ic}$. To see that this function also satisfies (2.8.10) if u does, one needs to transfer α in the following way. First change the sign of the fourth row, then change the places of the third and the fourth row, and finally add to the first two rows the third and the fourth rows (new) multiplied by c.

All this is said to claim that if $U(\mathfrak{w}, \mathfrak{z}, x_5)$ satisfies (2.8.10), then $U(\mathfrak{w} + (c_1 + ic_2)\mathfrak{z}, \mathfrak{z}, x_5)$ satisfies (2.8.10). Similarly, $U(\mathfrak{w}, \mathfrak{z} + (c_1 + ic_2)\mathfrak{w}, x_5)$ satisfies (2.8.10). Now we see that (2.8.13) can be extended to the fact that

$$\Delta_{\lambda_1 \cdot \mathfrak{w} + \lambda_2 \cdot \mathfrak{z}} U \leq 0, \qquad \lambda_1, \lambda_2 \in \mathbb{C}.$$

Lemma is proved. □

We expect that minimal (or otherwise saturated) solution of (2.8.10) satisfying homogeneity condition (2.8.12) and obstacle condition (2.8.11) will give us the proof of Theorem 2.8.2.

Consider two new functions:

$$u_\beta(\mathfrak{w}, \mathfrak{z}, x_5) = v(\mathfrak{w}, \mathfrak{z}, x_5) - \beta x_5,$$

$$U(\mathfrak{w}, \mathfrak{z}) \stackrel{\text{def}}{=} U_\beta(\mathfrak{w}, \mathfrak{z}) \stackrel{\text{def}}{=} \sup_{x_5}\{u_\beta(\mathfrak{w}, \mathfrak{z}, x_5) \colon (\mathfrak{w}, \mathfrak{z}, x_5) \in \Omega\}.$$

Obviously, u_β satisfies (2.8.10) as soon as v does. Hence, by Lemma 2.8.4, function U is pluri-superharmonic.

Also the obstacle condition (2.8.11) together with the inequality $x_5 \geq |\mathfrak{w} + \bar{\mathfrak{z}}|^p$ valid in Ω show that U satisfies the new obstacle condition:

$$U(\mathfrak{w}, \mathfrak{z}) \geq F_\beta(\mathfrak{w}, \mathfrak{z}) \stackrel{\text{def}}{=} \max(|\mathfrak{w}|, |\mathfrak{z}|)^p - \beta|\mathfrak{w} + \bar{\mathfrak{z}}|^p. \qquad (2.8.14)$$

Homogeneity for u listed in (2.8.12) becomes now

$$U(\lambda\mathfrak{w}, \bar{\lambda}\mathfrak{z}) = |\lambda|^p U(\mathfrak{w}, \mathfrak{z}), \quad \lambda \in \mathbb{C}. \qquad (2.8.15)$$

2.8.2.6 Obstacle Problems for Pluri-Superharmonic Functions. We formally reduced everything to this obstacle problem: to find the smallest β such that function $F_\beta = \max(|\mathfrak{w}|, |\mathfrak{z}|)^p - \beta|\mathfrak{w} + \bar{\mathfrak{z}}|^p$ has a pluri-superharmonic majorant $U(\mathfrak{w}, \mathfrak{z})$ satisfying homogeneity (2.8.12). Notice that the obstacle problem in Section 2.7 is very much like that but of course simpler because it is in $\mathbb{C}$, and not in $\mathbb{C}^2$.

We consider case $1 < p \leq 2$ first. We will solve our obstacle problem by producing a very simple pluri-superharmonic function bigger than our obstacle with an optimal β. This function was found in [**69**]. The verification proposition will show that β is optimal.

Again (compare with (2.7.15)) for $1 < p \leq 2$, it is possible to take

$$U(\mathfrak{w}, \mathfrak{z}) \stackrel{\text{def}}{=} \Re\,(\mathfrak{w}\mathfrak{z})^{p/2}. \qquad (2.8.16)$$

We understand this formula as follows. Function $U(\mathfrak{w}, \mathfrak{z}) = \Re\,\phi(\mathfrak{w}\mathfrak{z})$. Function $\phi(\mathfrak{z}) = \mathfrak{z}^{p/2}$ is a well-defined analytic function in the whole plane slitted by the ray $(-\infty, 0]$. Now $\Re\phi(\mathfrak{z})$ on two sides of the cut equals $r^{p/2}\cos\frac{p\pi}{2}$ and $r^{p/2}\cos(-\frac{p\pi}{2})$. So $\Re\phi(\mathfrak{z})$ is continuous on the slit.

This allows us to define $\phi(\mathfrak{z})$ as a subharmonic function in the whole plane without the origin. See Exercise 2.8.11.

The sub-mean value property at the origin is just an elementary calculation:

$$\int_{-\pi}^{\pi} \cos\frac{pt}{2}\,dt = \frac{4}{p}\sin\frac{p\pi}{2} > 0, \quad 1 < p \leq 2.$$

Function $-B_p U(\mathfrak{w}, \mathfrak{z})$, $B_p > 0$ will turn out to be the right pluri-superharmonic majorant of the obstacle if we choose the constant correctly. We will show this soon.

Let us ask the following question (compare this with (2.7.16)): What is the smallest $A_p > 0$ such that there exists $B_p > 0$ for the following to hold:

$$\max(|\mathfrak{w}|^p, |\mathfrak{z}|^p) - A_p|\mathfrak{w} + \bar{\mathfrak{z}}|^p \leq -B_p U(\mathfrak{w}, \mathfrak{z}) = -B_p \Re\,(\mathfrak{w}\mathfrak{z})^{p/2}\,? \qquad (2.8.17)$$

This problem is symmetric and homogeneous. So it reduces to the equivalent obstacle problem in one complex variable. We need to find the smallest

$A_p > 0$ such that there exists $B_p > 0$ such that on the whole complex plane

$$1 - A_p|1+\mathfrak{z}|^p \le -B_p U(1,\bar{\mathfrak{z}}) = -B_p \Re\,(\bar{\mathfrak{z}})^{p/2}\,. \tag{2.8.18}$$

It is easy to find candidates for A_p, B_p. On the rays $\Gamma_\pm \stackrel{\text{def}}{=} \{\mathfrak{z}\colon \mathfrak{z} = re^{\pm i\frac{\pi}{p}}\}$ the right-hand side vanishes. Therefore, it is easy to see that putting $r = \cos\frac{\pi}{p}, \mathfrak{z} = re^{\frac{i\pi}{p}}$, we obtain

$$A_p \ge \left(\frac{1}{\sin\frac{\pi}{p}}\right)^p. \tag{2.8.19}$$

It is natural to try to take the minimal value of $A_p = \left(\frac{1}{\sin\frac{\pi}{p}}\right)^p$. Now one can find B_p by requiring that left- and right-hand side of (2.8.18) are not only equal (and equal to 0) at points $P\pm \stackrel{\text{def}}{=} \cos\frac{\pi}{p}e^{\pm\frac{i\pi}{p}}$ but that also they are tangent to each other at these points. Then one can find

$$B_p = \frac{2\left|\cos\frac{\pi}{p}\right|}{\sin\frac{\pi}{p}}. \tag{2.8.20}$$

See Exercise 2.8.12.

2.8.2.7 Why (2.8.18) Holds with These A_p, B_p? One considers in [**69**] the function

$$H(t,r) \stackrel{\text{def}}{=} A_p(1+2r\cos t + r^2)^{p/2} - B_p r^{p/2}\cos\frac{pt}{2} - 1$$

in the rectangle $S \stackrel{\text{def}}{=} \{(t,r)\colon |t| \le \pi, 0 \le r \le 1$. One needs to prove that $H \ge 0$ in the strip S, if A_p, B_p are given by (2.8.19), (2.8.20) correspondingly. This is a statement of the sort (2.7.25) but much more involved, because (2.7.25) is a one-dimensional statement. However, here we deal with functions on the plane. The scheme of [**69**] consists of four steps:

- $\min\{H(t,r)\colon (t,r)\in S\}$ is attained inside S;
- one finds three points inside S, where $\nabla H = 0$, at $(\pm\frac{\pi}{p}, |\cos\frac{\pi}{p}|)$ and at $(0, r_0)$, where r_0 is a solution of

$$2A_p(1+r_0)^{p-1} = B_p r_0^{p/2-1}.$$

- one shows that r_0 is a unique solution of this equation, and that $(0, r_0)$ is a saddle point of H, and hence it is not a minimum point; for that, one evaluates the Hessian of H at the point $(0, r_0)$ and shows that its determinant is negative for $1 < p < 2$;

- one shows that there are no other critical points of H in S. This claim is reduced in [**69**] to showing that the following equation has only one solution $t = \frac{\pi}{p}$ on interval $(0, \pi)$, A_p, B_p are given by (2.8.19), (2.8.20) correspondingly.

$$\sin^{p-1} t \left(\sin \frac{(2-p)t}{2} \right)^{\frac{2-p}{2}} = \frac{B_p}{2A_p} \left(\sin \frac{p}{2} t \right)^{\frac{p}{2}}. \tag{2.8.21}$$

REMARK 2.8.5 The third and the fourth items are technically rather long and complicated, the reader can find them in [**69**]. Our goal was rather to show why the pluri-superharmonic obstacle problem appears naturally from the stochastic optimal control considerations. To solve the obstacle problem is an entirely different story. We saw that it is an interesting and challenging problem. The example of (2.7.25) and especially the one we are dealing with now show quite conspicuously the difficulties arising in the solution of the corresponding obstacle problem.

2.8.2.8 Verification Propositions. Inequality (2.8.17) shows that for $1 < p \leq 2$ we have

$$\int_{-\pi}^{\pi} \max(|f_+(e^{it})|, |f_-(e^{it})|)^p \, dt \leq \left(\frac{1}{\sin \frac{\pi}{p}} \right)^p \int_{-\pi}^{\pi} |f_+(e^{it})|^p \, dt. \tag{2.8.22}$$

In fact, let us plug $w = f_+(\zeta)$, $z = f_-(\zeta)$, $\zeta \in \mathbb{D}$, into (2.8.17). The result is continuous up to the boundary $\mathbb{T}$ (the test function f was chosen to be smooth), the composition is pointwise smaller than the superharmonic function $-B_p U(f_+(\zeta), f_-(\zeta))$. Now we use the mean value property of subharmonic function $U(f_+(\zeta), f_-(\zeta))$:

$$\int_{-\pi}^{\pi} U(f_+(e^{it}), f_-(e^{it})) \, dt \geq U(f_+(0), f_-(0)) = U(f_+(0), 0) = 0. \tag{2.8.23}$$

The equality $U(f_+(0), f_-(0)) = U(f_+(0), 0)$ follows from the facts that $f_-(0) = 0$ (see (2.8.6)) and $U(f_+(0), f_-(0)) = \Re\, \phi(f_+(0) \cdot f_-(0)) = \Re\, \phi(0) = 0$, where $\Re\, \phi(z) = \Re\, z^{p/2}$ is a subharmonic function in the whole plane defined after (2.8.16).

To prove the sharpness of the constant in (2.8.22), we follow the scheme of Proposition 2.7.13.

PROPOSITION 2.8.6 *The constant* $C_{\mathbb{C}}(p, 1/2, i/2)$ *from* (2.8.2) *has the estimate from below* $C_{\mathbb{C}}(p, 1/2, i/2) \geq \frac{1}{\sin \frac{1}{p}}$.

PROOF We start as in Proposition 2.7.13. Consider $\Psi_a(z) \stackrel{\text{def}}{=} \left(\frac{1+z}{1-z}\right)^{\frac{1}{a}}$, where $a > p$ and in the future $a \to p$. Consider now

$$\Phi_a \stackrel{\text{def}}{=} \alpha\,\Re\,\Psi_a + i\beta\,\Im\,\Psi_a,$$

where α, β will be used momentarily. Observe that

$$\Phi_a + i\mathcal{H}(\Phi_a) = (\alpha+\beta)(\Re\,\Psi_a + i\Im\,\Psi_a).$$

Hence,

$$\frac{\|\Phi_a + i\mathcal{H}(\Phi_a)\|_p}{\|\Phi_a\|_p} = \frac{\left(\int_{-\pi}^{\pi} |\alpha+\beta|^p |\Re\,\Psi_a + i\Im\,\Psi_a|^p\,dt\right)^{1/p}}{\left(\int_{-\pi}^{\pi} |\alpha\Re\,\Psi_a + i\beta\Im\,\Psi_a|^p\,dt\right)^{1/p}}.$$

Now notice that pointwise on one half of the circle $\mathbb{T}$, we have $\Im\,\Psi_a = \left(\tan\frac{\pi}{2a}\right)\Re\,\Psi_a$, and on another half we have $\Im\,\Psi_a = -\left(\tan\frac{\pi}{2a}\right)\Re\,\Psi_a$. Using this observation, it is easy to compute the ratio in the right-hand side:

$$\frac{\|\Phi_a + i\mathcal{H}(\Phi_a)\|_p}{\|\Phi_a\|_p} = \frac{2^{1/p}|\alpha+\beta|}{|\alpha\cos\frac{\pi}{2a} + i\beta\sin\frac{\pi}{2a}|^p + |\alpha\cos\frac{\pi}{2a} - i\beta\sin\frac{\pi}{2a}|^p}.$$

Now choose $\alpha = \sin^2\frac{\pi}{2a}$, $\beta = \cos^2\frac{\pi}{2a}$. Then we obtain

$$\frac{\|\Phi_a + i\mathcal{H}(\Phi_a)\|_p}{\|\Phi_a\|_p} = \frac{2}{\sin\frac{\pi}{a}}. \tag{2.8.24}$$

We can also write

$$\frac{\|2\mathbb{P}_+ f\|_p}{\|\Phi_a\|_p} = \frac{\|\Phi_a + i\mathcal{H}(\Phi_a) - 1\|_p}{\|\Phi_a\|_p} \geq \frac{\|\Phi_a + i\mathcal{H}(\Phi_a)\|_p}{\|\Phi_a\|_p} - \frac{1}{\|\Phi_a\|_p}.$$

Plug (2.8.24) into the right-hand side of the above equation. Then make a tend to p from the right. Then $\frac{2}{\sin\frac{\pi}{a}} \to \frac{2}{\sin\frac{\pi}{p}}$ of course, and simultaneously $\|\Phi_a\|_p$ tends to infinity. This completes the proof of the proposition. □

2.8.2.9 The Case $p > 2$. We do not know how to find the smallest constant A_p, for which there exists a pluri-superharmonic function $V(\mathfrak{w}, \mathfrak{z})$, $V(0,0) = 0$, such that

$$\max(|\mathfrak{w}|^p, |\mathfrak{z}|^p) - A_p|\mathfrak{w} + \bar{\mathfrak{z}}|^p \leq V(\mathfrak{w}, \mathfrak{z})\,. \tag{2.8.25}$$

As in Section 2.7.33, the majorant V is probably a compound function that is equal to $-B\,U(\mathfrak{w}, \mathfrak{z})$ in one part of $\mathbb{C}^2$ (call it O), and is equal to the obstacle $\max(|\mathfrak{w}|^p, |\mathfrak{z}|^p) - A_p|\mathfrak{w} + \bar{\mathfrak{z}}|^p$ on $\mathbb{C}^2 \setminus O$.

Notice also that we can use (1) homogeneity and (2) the saturation of pluri-superharmonicity. The latter means that at every point of O there is a complex line along which function V is harmonic.

Exercises

PROBLEM 2.8.1 Consider the two-dimensional Brownian motion $W = (w_1, w_2)$ starting at $0 \in \mathbb{C}$, let u be a harmonic function. Introduce a new process

$$\zeta(t) \stackrel{\text{def}}{=} u(w_1(t) + iw_2(t)).$$

Prove its stochastic differential $d\zeta(t)$ has no drift.

PROBLEM 2.8.2 Prove that if u is harmonic function in a neighborhood of the origin and W is two-dimensional Brownian motion starting at the origin, then stochastic differential $du \circ W(t)$ does not have drift.

PROBLEM 2.8.3 Prove Lemma 2.7.3 on page 185.

PROBLEM 2.8.4 Let $1 < p \leq 2$. Function z^p is well-defined analytic function in the right half-plane. Consider its real part in this half-plane and then mirror reflect it into the left half-plane. As a result of this construction, we now have the function u in the whole plane. Prove that it is subharmonic everywhere except for the origin. Hint: prove that in a neighborhood G of any point of the imaginary axis such that $0 \notin G$ u is the maximum of two harmonic functions. This is used on page 187.

PROBLEM 2.8.5 Prove Lemma 2.7.7 on page 188.

PROBLEM 2.8.6 Let function u be defined as in Problem 2.8.4. Let $|y| \leq x$. Prove that $u(x + iy) \geq 0$.

PROBLEM 2.8.7 Consider the function $\Phi_\alpha(z) \stackrel{\text{def}}{=} \left(\frac{1+z}{1-z}\right)^\alpha$, $0<\alpha<2$. Prove that it conformally maps the unit disc onto the domain (cone) $\{w\colon |\arg w| \leq \frac{\pi\alpha}{2}\}$. This was used in Proposition 2.7.10 on page 189.

PROBLEM 2.8.8 Let V be function from (2.7.27) on page 192. Show that at all points of the real axis except the origin, it is superharmonic by calculating its Laplacian at such points.

PROBLEM 2.8.9 Let V be function from (2.7.27) on page 192. Let $|y| \leq x$. Show that $V(x, y) \leq 0$.

PROBLEM 2.8.10 Prove that function defined in (2.7.33) on page 194 is subharmonic in the whole plane.

PROBLEM 2.8.11 Let $1 < p \leq 2$. Function $\phi(z) = z^{p/2}$ from page 200 is a well-defined analytic function in the whole plane slitted by the ray $(-\infty, 0]$. Now $\Re\phi(z)$ on two sides of the cut equals $r^{p/2}\cos\frac{p\pi}{2}$ and $r^{p/2}\cos(-\frac{p\pi}{2})$. So $\Re\phi(z)$ is continuous on the slit. Show that this allows us to define $\phi(z)$ as a subharmonic function in the whole plane without the origin.

PROBLEM 2.8.12 Prove Formula (2.8.20) on page 201.

3

Conformal Martingale Models: Stochastic and Classical Ahlfors–Beurling Operators

3.1 Estimates of Subordinated Martingales

In this section, we address the question of finding the best L^p-norm constant for differentially subordinated martingale with certain extra properties.

In fact, this goal is pursued for the sake of better understanding of one particular singular integral operator, the Ahlfors–Beurling transform B given by the following relationship:

$$B(\bar{\partial} f) = \partial f, \quad \forall\, f \in C_0^\infty(\mathbb{C}).$$

Another way to represent operator B is as a singular integral operator:

$$Bf(z) \stackrel{\text{def}}{=} p.v. \frac{1}{\pi} \int_{\mathbb{C}} \frac{f(\zeta)}{(\zeta - z)} dm_2(\zeta).$$

The first representation shows that B has norm 1 in $L^2(\mathbb{C})$ because clearly

$$\|\bar{\partial} f\|_{L^2(\mathbb{C})} = \|\partial f\|_{L^2(\mathbb{C})}. \tag{3.1.1}$$

It is an open problem to find its norm in $L^p(\mathbb{C})$. This problem has an interesting story and is related to many other open problems of analysis and especially of calculus of variations, see, e.g., [**3**]. The well-known conjecture is as follows:

$$\|B\|_p = \max\left(p - 1, \frac{1}{p - 1}\right).$$

It has been understood long ago that Ahlfors–Beurling operator has tight connection with martingale estimates. That will be demonstrated below in this chapter. In fact, our goal is to obtain several nontrivial estimates of the norm of B in $L^p(\mathbb{C})$.

In [**5**], Bañuelos and Janakiraman make the observation that the martingale associated with the Ahlfors–Beurling operator is, in fact, a conformal martingale (we will see that below). They show that Burkholder's proof in [**24**] naturally accommodates for the conformality, and leads to an improvement in the previously obtained estimates of $\|B\|_p$. In Bañuelos–Janakiraman paper, the Ahlfors–Beurling transform was associated with the left-hand-side conformality. Therefore, the constant in the first part of Theorem 3.2.1 below plays an important role in getting nice (but not optimal) estimates of norm $\|B\|_p$.

Also, as Ahlfors–Beurling transform B is associated with the left-hand-side conformality (see Section 3.5), three questions naturally arise:

(1) If $2 < p < \infty$, what is the best constant C_p in the left-hand-side conformality problem: $\|Y\|_p \le C_p \|X\|_p$, where Y is conformal and $d\langle Y\rangle \le d\langle X\rangle$? In other words, it would be interesting to improve the constant $D_p \stackrel{\text{def}}{=} \sqrt{\frac{p^2-p}{2}}$ in the first part of Theorem 3.2.1. We do not know how to do that.

(2) Similarly, since we know that $\|B\|_p = \|B\|_{p'}$, if $1 < p < 2$, what is the best constant C_p in the left-hand-side conformality problem? We will answer this question below. The sharp constant will be found in the first part of Theorem 3.2.4 below.

(3) Can we compare $D_{p'}$ and C_p, $1 < p < 2$? We will do that; the comparison will be not to our advantage, namely, we will see that $D_p < C_{p'}$. However, in this chapter, we will be able to prove that $\|B\|_p$ is strictly below $2D_p$ when $p \to \infty$.

Let $\mathcal{O} = (\Omega, \mathfrak{A}, P)$ be a probability space with filtration $\mathfrak{A}$, generated by a two-dimensional Brownian motion $W(t)$. Let $X(t) = \int_0^t \nabla X(s) \cdot dW(s)$ and $Y(t) = \int_0^t \nabla Y(s) \cdot dW(s)$ be two $\mathbb{R}^2$-valued martingales on this probability space. We spend a bit of time explaining the notation and reconciling it with that used in (2.5.1) of Chapter 2.

Thus,

$$X = \begin{bmatrix} X_1 \\ X_2 \end{bmatrix}, \; Y = \begin{bmatrix} Y_1 \\ Y_2 \end{bmatrix}.$$

In Chapter 2 we used matrix α, whose elements were processes $\alpha_{ij}(s)$ adapted to Brownian motion. Here we use

$$\nabla X_i(s) := (\alpha_{i1}(s), \alpha_{i2}(s)), \;\; i = 1, 2,$$

and the following definitions.

Definition 3.1.1 The processes $\langle X_i\rangle \stackrel{\text{def}}{=} \int_0^t |\nabla X_i|^2 \, dt$, where $|\nabla X_i|^2 \stackrel{\text{def}}{=} \alpha_{i1}^2 + \alpha_{i2}^2$, $i = 1, 2$, are called quadratic variations of X_i. The process $\langle X\rangle \stackrel{\text{def}}{=}$

$\int_0^t (|\nabla X_1|^2 + |\nabla X_i|^2)\,dt$ is called the quadratic variation of the vector process X. We denote $|\nabla X|^2 \stackrel{\text{def}}{=} |\nabla X_1|^2 + |\nabla X_i|^2$. Similarly, we define the quadratic variation of Y_1, Y_2, Y.

DEFINITION 3.1.2 Covariation of (X_1, X_2) is the process

$$\langle X_1, X_2 \rangle \stackrel{\text{def}}{=} \int_0^t (\alpha_1(s), \alpha_2(s))_{\mathbb{R}^2}\, ds = \int_0^t (\nabla X_1, \nabla X_2)\, dt,$$

where

$$\alpha_i \stackrel{\text{def}}{=} (\alpha_{i1}, \alpha_{i2}),\ i = 1, 2.$$

The same definition will be used for Y.

REMARK 3.1.3 The standard notation for derivatives of quadratic variation, namely, for $|\nabla X|^2, |\nabla Y|^2, |\nabla X_i|^2, |\nabla Y_i|^2$ in the literature, is $d\langle X\rangle$, $d\langle Y\rangle$, $d\langle X_i\rangle$, $d\langle Y_i\rangle$, and for the derivatives of mutual covariations – $d\langle X_1, X_2\rangle$, $d\langle Y_1, Y_2\rangle$. We will be using that notation too.

Let us consider two $\mathbb{R}^2$-valued martingales X and Y, such that the quadratic variation of Y runs slower than the quadratic variation of X, i.e., $d\langle Y\rangle(s) \le d\langle X\rangle(s)$, or, equivalently,

$$\begin{aligned} |\nabla Y(s)| &= \sqrt{|\nabla Y_1(s)|^2 + |\nabla Y_2(s)|^2} \\ &\le \sqrt{|\nabla X_1(s)|^2 + |\nabla X_2(s)|^2} = |\nabla X(s)|, \qquad s \ge 0, \end{aligned}$$

We use the notation $\|X\|_p = \sup_{t\ge 0} \|X(t)\|_p = \sup_{t\ge 0}(\mathbb{E}|X(t)|^p)^{1/p}$.

If the condition of subordination holds, then Y is said to be differentially subordinate to X. If for $1 < p < \infty$, we have $\mathbb{E}\|X\|_p < \infty$, then the Burkholder–Davis–Gundy and Doob inequalities imply that $\mathbb{E}\|Y\|_p < \infty$ (see [**164**]), and there exists a universal constant C_p, such that $\|Y\|_p \le C_p\|X\|_p$. An evident problem then is to find the best constant C_p.

3.2 Conformal Martingales and the Ahlfors–Beurling Transform

As we already saw in Chapter 1, Burkholder solved completely the problem of sharp estimate $\|Y\|_p \le C_p\|X\|_p$ if martingale Y is subordinated to martingale X.

It was explained in Chapter 1 that the sharp constant is the following:

$$C_p = p^* - 1, \qquad p^* = \max\{p, p'\}, \quad p' = \frac{p}{p-1}.$$

We wish to obtain analogous results when X, Y are $\mathbb{R}^2$-valued martingales, one being subordinated to the other, but they have some extra symmetry conditions, called conformality and defined below.

We identify $\mathbb{R}^2$-valued martingales $X = (X_1, X_2)$, $Y = (Y_1, Y_2)$ with complex valued martingales $X_1 + iX_2$, $Y_1 + iY_2$. A complex-valued martingale $Y = Y_1 + iY_2$ is said to be *conformal* if the quadratic variations of the coordinate martingales are equal and their mutual covariation is 0:

$$d\langle Y_1 \rangle = d\langle Y_2 \rangle, \quad d\langle Y_1, Y_2 \rangle = 0.$$

The first part of the following theorem was proved in [**5**]. The second part can be proven along the same lines; see [**16**].

THEOREM 3.2.1 (*One-sided conformality treated by Burkholder's method*)

(1) (*left-hand-side conformality*) *Suppose that* $2 \leq p < \infty$. *If* Y *is a conformal martingale and* X *is any martingale, such that* $d\langle Y \rangle \leq d\langle X \rangle$, *then*

$$\|Y\|_p \leq \sqrt{\frac{p^2 - p}{2}} \|X\|_p.$$

(2) (*right-hand-side conformality*) *Suppose that* $1 < p \leq 2$. *If* Y *is a conformal martingale and* X *is any martingale, such that* $d\langle X \rangle \leq d\langle Y \rangle$, *then*

$$\|X\|_p \leq \sqrt{\frac{2}{p^2 - p}} \|Y\|_p.$$

REMARK 3.2.2 The sharpness of these constants has not been established yet to the best of our knowledge.

Natural questions that arise are (1) what are the estimates for the left-side conformality case and $1 < p \leq 2$, (2) and what are the estimates for the right-side conformality case and $p \geq 2$? We answer these questions in this section. We follow [**17**]. We present the sharp constants for these situations and in Section 3.5 we show the application of the result to the estimates of Ahlfors–Beurling operator.

For the case when both complex-valued martingales X, Y are conformal, this was done in [**16**] for $p \geq 2$. The two-side conformality sharp estimate for $0 < p < 2$ is done in [**7**]. Moreover, this paper of Bañuelos–Osękowski extends the result to the conformal martingales in $\mathbb{R}^d$. The extension is in the language of Bessel processes, which allows for non-integer d as well!

To describe the answer, we need some notation. The Laguerre function L_p with index p solves the following equation:

$$sL_p''(s) + (1 - s)L_p'(s) + pL_p(s) = 0.$$

These functions are discussed below and their properties are reviewed in Section 3.3.1.2; see also [**16, 41, 42**].

Given $p > 1$, we denote by z_p the least positive root in $(0, 1)$ of the bounded Laguerre function L_p.

THEOREM 3.2.3 *Let* $Y = (Y_1, Y_2)$ *be a conformal martingale, and* $X = (X_1, X_2)$ *be an arbitrary martingale.*

(1) *(left-hand-side conformality) Let* $1 < p \leq 2$. *Suppose* $d\langle Y_1\rangle \leq d\langle X\rangle$. *Then the best constant in the inequality* $\|Y\|_p \leq A_p\|X\|_p$ *is*

$$A_p = \frac{z_p}{1 - z_p}. \tag{3.2.1}$$

(2) *(right-hand-side conformality) Let* $2 \leq p < \infty$. *Suppose* $d\langle X\rangle \leq d\langle Y_1\rangle$. *Then the best constant in the inequality* $\|X\|_p \leq A_p\|Y\|_p$ *is*

$$A_p = \frac{1 - z_p}{z_p}. \tag{3.2.2}$$

Obviously then we get the following claim, which should be compared with Theorem 3.2.1.

THEOREM 3.2.4 *Let* $Y = (Y_1, Y_2)$ *be a conformal martingale and* $X = (X_1, X_2)$ *be an arbitrary martingale.*

(1) *(left-hand-side conformality) Let* $1 < p \leq 2$. *Suppose* $d\langle Y\rangle \leq d\langle X\rangle$. *Then the best constant in the inequality* $\|Y\|_p \leq C_p\|X\|_p$ *is*

$$C_p = \frac{1}{\sqrt{2}} \cdot \frac{z_p}{1 - z_p}. \tag{3.2.3}$$

(2) *(right-hand-side conformality) Let* $2 \leq p < \infty$. *Suppose* $d\langle X\rangle \leq d\langle Y\rangle$. *Then the best constant in the inequality* $\|X\|_p \leq C_p\|Y\|_p$ *is*

$$C_p = \sqrt{2} \cdot \frac{1 - z_p}{z_p}. \tag{3.2.4}$$

For asymptotics of z_p, C_p, $C_{p'}$ as $p \to \infty$, see Section 3.4.3.

3.3 Proof of Theorem 3.2.3: Right-Hand Side Conformality, $2 < p < \infty$

Below $|\cdot|$ denotes Euclidean norm in $\mathbb{R}^2$.

In Sections 2.6, 2.7, our course of action was as follows:

- (1) Given a harmonic analysis problem, we searched for the way to embed it into stochastic optimal control scheme of Bellman.

- (2) For that, we first identified what can serve as the underlying controlled stochastic differential equation, and what can serve as controls.
- (3) We call this step "who moves."
- (4) Then we identified the data of the harmonic analysis problem that can serve as profit function and bonus function.
- (5) We always identified the domain where the trajectories of the stochastic differential equation will live.
- (6) We often identified the boundary conditions.
- (7) Then we *formally* replaced our discrete random walk by continuous one, by introducing Brownian motion.
- (8) Then we defined the Bellman function of the stochastic optimal control problem that was *formally* assigned to the initial harmonic analysis problem.
- (9) At last, we used Bellman's formalism to derive Hamilton–Jacobi–Bellman partial differential equation on the latter Bellman function.
- (10) Then we solved (hopefully) that nonlinear partial differential equation.
- (11) Lastly, we (with a huge surprise) verified that this *formal* Bellman function really often helps to completely solve the original harmonic analysis problem.

In short, without any rigorous base of doing so, we manufactured a problem in stochastic optimal control, considered its Hamilton–Jacobi–Bellman partial differential equation and Bellman function, found Bellman function of stochastic optimal control by solving Hamilton–Jacobi–Bellman partial differential equation. In conclusion, without any rigorous base, we believe that this Bellman function helps to solve the original harmonic analysis problem. That conclusion is dangerous, but we tried (and very often succeeded) to use verification arguments to show that our formal approach, in fact, works.

The problems we are considering in this section dictate to us a slightly different approach. In the items above, one should now delete the word "formal." In fact, the problems considered in Theorems 3.2.3, 3.2.4 are already the problems of stochastic optimal control. So they are subject to formalism of Chapter 2. It is clear who moves: just martingales X, Y; and it is clear what the controls are, see (3.3.2) below. In particular, finding sharp constants A_p, C_p in Theorems 3.2.3, 3.2.4 is equivalent to finding a certain solution to Hamilton–Jacobi–Bellman equation that the reader will soon see in (3.3.4). As was always the case in Chapter 2, this "equation" will be, in fact, a partial differential inequality. To find its "best" solution, we will, as before, try to saturate it. This scheme will succeed. We will find the Bellman function that will have the following, rather typical, behavior:

- the domain of definition is partitioned into two subdomains,
- in one of these subdomains, Bellman function coincides with obstacle function, and it just satisfies partial differential inequalities (3.3.4),
- on the second subdomain, Bellman function lies strictly above the obstacle, and there it saturates partial differential inequalities (3.3.4), which means that it satisfies some (rather involved) nonlinear partial differential equation, called Hamilton–Jacobi–Bellman equation, that, by the way (we do not use this), belongs to the class of so-called fully nonlinear equations,
- certain a priori symmetry properties of hypothetical solution allow us to reduce the just mentioned nonlinear partial differential equation to a nonlinear ordinary differential equation, which we will solve.

We start with writing down the obstacle function. It is $V(\vec{x},\vec{y}) = |\vec{x}|^p - c^p|\vec{y}|^p$. Using the material of Chapter 2, we can see that our problem of finding the best constant in (3.2.2) is equivalent to finding the smallest constant c, for which there exists a majorant $U(\vec{x},\vec{y}) \geq V(\vec{x},\vec{y})$, $U(0,0) \leq 0$, $\vec{x} \in \mathbb{R}^2$, $\vec{y} \in \mathbb{R}^2$, such that for X and Y, as above, the process $U(X,Y)$ is a supermartingale. We shall explain this below.

But first, when such U is obtained:

$$\mathbb{E}|X|^p - c^p\mathbb{E}|Y|^p = \mathbb{E}[V(X,Y)] \leq \mathbb{E}[U(X,Y)] \leq 0. \tag{3.3.1}$$

Let us introduce the Bellman function, as in (2.5.3) of Chapter 2. Now we can use our main "war horse" – the Hamilton–Jacobi–Bellman equation (2.5.8) of Section 2.5.1.

First of all, we have to use the matrix a that gives us processes X, Y. By the properties of Y, this matrix will have a special form, described below:

$$a \stackrel{\text{def}}{=} \begin{bmatrix} \alpha_{1,1}(s), & \alpha_{1,2}(s) \\ \alpha_{2,1}(s), & \alpha_{2,2}(s) \\ \beta_{1,1}(s), & \beta_{3,2}(s) \\ \beta_{2,1}(s), & \beta_{2,2}(s) \end{bmatrix} \tag{3.3.2}$$

with the property that the lower 2×2 matrix has orthogonal rows of the same norm for every s and ω in the underlying probability space. Using the notation $\alpha_1(s) = (\alpha_{1,1}(s),\ \alpha_{1,2}(s))$, $\alpha_2(s) = (\alpha_{2,1}(s),\ \alpha_{2,2}(s))$, $\beta_1(s) = (\beta_{1,1}(s),\ \beta_{1,2}(s))$, $\beta_2(s) = (\beta_{2,1}(s),\ \beta_{2,2}(s))$, we can write the orthogonality condition, as follows:

$$|\beta_1| = |\beta_2|,\ (\beta_1, \beta_2)_{\mathbb{R}^2} = \beta_{1,1}\beta_{2,1} + \beta_{1,2}\beta_{2,2} = 0, \tag{3.3.3}$$

which is just another way to express the conformality of $Y = (Y_1, Y_2)$, because $d\langle Y_1\rangle = |(\beta_{1,1}, \beta_{1,2})|^2$, $d\langle Y_2\rangle = |(\beta_{2,1}, \beta_{2,2})|^2$, and $d\langle Y_1, Y_2\rangle = (\beta_1, \beta_2) = 0$.

Here is our main partial differential inequality, where $\vec{x} = (x_1, x_2), \vec{y} = (y_1, y_2)$ and where α is an arbitrary 2×2 matrix, while matrix β is any matrix satisfying (3.3.3):

$$\begin{aligned} &\sum_{i,j=1}^{2} U_{x_i x_j}(\alpha_i, \alpha_j)_{\mathbb{R}^2} + \Delta_y U |\beta_1|^2 \\ &+ \sum_{i,j=1}^{2} 2U_{x_i y_j}(\alpha_i, \beta_j) \le 0 \end{aligned} \tag{3.3.4}$$

for all vectors $\alpha_1, \alpha_2, \beta_1, \beta_2 \in \mathbb{R}^2$, such that $|\alpha_1|^2 + |\alpha_2|^2 \le |\beta_1|^2 = |\beta_2|^2$ and such that $(\beta_1, \beta_2) = 0$.

REMARK 3.3.1 Property (3.3.4) implies (actually it is equivalent to) the property that $U(X, Y)$ is a supermartingale for all $\mathbb{R}^2$ martingales X and Y, such that Y is conformal. This can be obtained by direct use of Itô's formula and conformality property.

We want to emphasize to the reader that this inequality is just a special case of (2.5.8) of Section 2.5.1. In particular, there is no randomness whatsoever in (3.3.4). All α_i, β_i in (3.3.4) are usual $\mathbb{R}^2$ vectors, and not processes anymore.

We are going to use the following symmetry: the functions U, V depend only on the lengths $|\vec{x}|$ and $|\vec{y}|$. To see this for U, we can average U over independent rotations of (x_1, x_2)-planes and (y_1, y_2)-planes. It is easy to notice that (3.3.4) is invariant under such rotations. Since the obstacle V is, of course, invariant, we can see that the average of rotations satisfies (3.3.4) and simultaneously lies over V.

Alternatively, we could reason that after rotations of martingales X and Y, they will still be of the same class as before, and thus, $U(X, Y)$ will remain a supermartingale after this rotation, and can be averaged over the rotations and the average will still be a supermartingale and will still majorize $V(X, Y)$ pointwise.

Using the notation $x = \sqrt{x_1^2 + x_2^2}$, $y = \sqrt{y_1^2 + y_2^2}$, we can write:

$$\begin{aligned} u(x, y) &\stackrel{\text{def}}{=} U(x_1, x_2, y_1, y_2), \\ v(x, y) &\stackrel{\text{def}}{=} V(x_1, x_2, y_1, y_2) = |x|^p - c^p |y|^p. \end{aligned} \tag{3.3.5}$$

In what follows, we use the differential notation for the following vectors:

$$\nabla X_i = (\alpha_{i1}, \alpha_{i2})^T, \quad \nabla Y_i = (\beta_{i1}, \beta_{i2})^T, \quad i = 1, 2.$$

Let us introduce new vectors:

$$h_1 \stackrel{\text{def}}{=} \frac{(x_1, x_2) \cdot (\nabla X_1, \nabla X_2)^T}{|x|}, \qquad h_2 \stackrel{\text{def}}{=} \frac{(-x_2, x_1) \cdot (\nabla X_1, \nabla X_2)^T}{|x|},$$

$$k \stackrel{\text{def}}{=} \frac{(y_1, y_2) \cdot (\nabla Y_1, \nabla Y_2)^T}{|y|}.$$

Let us notice that we have two simple but important identities. Since $d\langle Y_1\rangle = d\langle Y_2\rangle$, we will have the first equality below:

$$|k|^2 = |\nabla Y_1|^2. \tag{3.3.6}$$

The second one follows from the latter definition of vectors $h_i, i = 1, 2$.

$$|h_1|^2 + |h_2|^2 = |\nabla X_1|^2 + |\nabla X_2|^2. \tag{3.3.7}$$

We also have the following identities:

$$U_{x_1x_2} = u_{xx} \cdot \frac{x_1x_2}{|\vec{x}|^2} - u_x \cdot \frac{x_1x_2}{|\vec{x}|^3},$$

$$U_{x_ix_i} = u_{xx} \cdot \frac{x_i^2}{|\vec{x}|^2} + u_x \cdot \frac{x_{3-i}^2}{|\vec{x}|^3}, \qquad 1 \le i \le 2,$$

$$U_{y_iy_i} = u_{yy} \cdot \frac{y_i^2}{|\vec{y}|^2} + u_y \cdot \frac{y_{3-i}^2}{|\vec{y}|^3}, \qquad 1 \le i \le 2,$$

$$U_{x_iy_j} = u_{xy} \cdot \frac{x_iy_j}{|\vec{x}| \cdot |\vec{y}|}, \qquad 1 \le i, j \le 2.$$

The properties (3.3.6), (3.3.7) allow us to rewrite (3.3.4) as follows:

$$u_{xx}|h_1|^2 + \frac{u_x}{x}|h_2|^2 + 2u_{xy}(h_1, k) + (u_{yy} + \frac{u_y}{y})|k|^2 \le 0.$$

This should be valid for all vectors h_1, h_2, and k satisfying the subordination condition $d\langle Y_1\rangle \ge d\langle X\rangle$, which, as (3.3.6), (3.3.7) show, assumes now the following form:

$$|h_1|^2 + |h_2|^2 \le |k|^2, \tag{3.3.8}$$

Henceforth, here is the Bellman "equation" on function u:

$$u_{xx}|h_1|^2 + \frac{u_x}{x}|h_2|^2 + 2u_{xy}(h_1, k) + \left(u_{yy} + \frac{u_y}{y}\right)|k|^2 \le 0. \tag{3.3.9}$$

So we need to clarify when this quadratic form is negative on the cone (3.3.8).

3.3.1 A Simplified Setting: $X = X_1$, $|h_1|^2 + |h_2|^2 = |k|^2$

In the previous section, we worked with the case when both $\nabla X_1, \nabla X_2 \in \mathbb{R}^2$ and $|h_1|^2 + |h_2|^2 \leq |k|^2$. Let us assume now that $\nabla X_2 = 0$, $|\nabla X_1| = |\nabla Y_1|$; we can restrict ourselves to the case $X_2 = 0$. Then $|h_1| = |k|$, $h_2 = 0$, and condition (3.3.9) reduces to

$$u_{xx} + 2u_{xy} + \left(u_{yy} + \frac{u_y}{y}\right) \leq 0, \tag{3.3.10}$$

$$u_{xx} - 2u_{xy} + \left(u_{yy} + \frac{u_y}{y}\right) \leq 0 \tag{3.3.11}$$

In many similar situations (see [**5, 16, 24**]), the best majorant in the simplified setting is also the best one in the general case. Hence, we may hope for the same effect in our problem and look first for functions u satisfying (3.3.10) and (3.3.11). We will proceed as follows:

1. We use the homogeneity of $u(x, y)$ to reduce the partial differential inequalities to ordinary differential inequalities for a function of one real variable $g(r)$.
2. We assume that the optimal u (and g) will solve (with equality) one of the two differential equations, wherever it is above the obstacle function v. Then solve the easier looking equation, which will be the one with $-u_{xy}$.
3. We will find the smallest constant c for which there exists a majorant satisfying (3.3.11). It will turn out that $u_{xy} \leq 0$ for this solution, and hence (3.3.10) holds as well.

3.3.1.1 Homogeneity and Reduction in Variables. The function u satisfies the same homogeneity condition as $v_c(x, y) = x^p - c^p y^p$. Namely, for all $t \in \mathbb{R}$,

$$u(tx, ty) = t^p u(x, y).$$

To see this, suppose that u is a suitable majorant of v. Then $u_t(x, y) = \frac{1}{t^p} u(tx, ty) \geq \frac{1}{t^p} v(tx, ty) = v(x, y)$ is also a majorant and as can be easily checked, satisfies (3.3.10) and (3.3.11). Therefore u_t is also a suitable majorant for each $t > 0$. Now take the infimum over all t to get a suitable homogeneous of degree p majorant satisfying the homogeneity condition.

Define

$$g(r) = u(1 - r, r), \qquad 0 \leq r \leq 1.$$

Then,

$$u(x,y) = (x+y)^p u\left(1 - \frac{y}{x+y}, \frac{y}{x+y}\right) = (x+y)^p g\left(\frac{y}{x+y}\right). \tag{3.3.12}$$

Set

$$\mathcal{L}_p g(r) \stackrel{\text{def}}{=} r g''(r) + (1-r) g'(r) + p g(r).$$
$$\mathcal{H}_p g(r) \stackrel{\text{def}}{=} -r(1-r) g''(r) + (p-1)(1-2r) g'(r) + p(p-1) g(r) \tag{3.3.13}$$

Substituting (3.3.12) into (3.3.10) and (3.3.11) gives us the following conditions on g:

$$\mathcal{L}_p g(r) + 4r \mathcal{H}_p g(r) \leq 0, \tag{3.3.14}$$
$$\mathcal{L}_p g(r) \leq 0. \tag{3.3.15}$$

The operator $\mathcal{L}_p$ is the Laguerre operator, the equation $\mathcal{L}_p f = 0$ is the Laguerre equation and its solutions are the Laguerre functions.

In Section 3.3.1.2, we gather all the properties of Laguerre functions that we need. We inherit the obstacle condition from function U. Here it is:

$$g(r) \geq v_c(r) = (1-r)^p - c^p r^p.$$

Finally, note that for $0 \leq r \leq 1$,

$$u_{xy}(1-r, r) = \mathcal{H}_p g(r). \tag{3.3.16}$$

Since $v_c(0) = 1$ for all c, we have $g(0) \geq 1$. As g is the minimal possible function, it is likely that it solves either $\mathcal{L}_p g = 0$ or $\mathcal{L}_p g(r) + 4r\mathcal{H}_p g(r) = 0$ wherever $g > v_c$. We consider first the simpler equation $\mathcal{L}_p g = 0$ and attempt to construct such majorant g with the help of its solutions.

3.3.1.2 The Laguerre Functions. Consider the Laguerre equation

$$sL'' + (1-s)L' + pL = 0. \tag{3.3.17}$$

It has two linearly independent solutions L_p and $\tilde{L}_p$.

$$L_p(s) = 1 - ps + \frac{p(p-1)}{4} s^2$$
$$+ \cdots + (-1)^n \frac{p(p-1)\cdots(p-n+1)}{n!^2} s^n + \cdots, \tag{3.3.18}$$

$$\tilde{L}_p(s) = L_p(s) \log \frac{1}{|s|} + H(s), \tag{3.3.19}$$

H is analytic in a neighborhood of 0. Evidently, $L_p(x)$ is a bounded analytic function in $[0, 1]$ and $\tilde{L}_p$ is unbounded near 0. Denote by z_p the smallest zero of L_p on the interval $[0, 1]$.

LEMMA 3.3.2 *For every solution to the Laguerre equation* (3.3.17), *its smallest zero in* $[0, 1]$ *is at most* z_p.

PROOF Notice that $\tilde{L}_p(0) = +\infty$. Consider the Wronskian $W(s) = \tilde{L}_p'(s)L_p(s) - L_p'(s)\tilde{L}_p(s)$. By (3.3.19), we have

$$W(s) = \frac{-L_p^2}{s} + H'L_p - L_p'H,$$

which is strictly negative for s close to 0, say for $0 \leq s \leq \varepsilon$. Since $W'(s) = -\frac{1-s}{s}W(s)$, we have $(\log(-W))'(s) \geq -\frac{1-\varepsilon}{\varepsilon}$ for $\varepsilon \leq x \leq 1$, and hence, W does not change the sign on $[0, 1]$ and is strictly negative. Since L_p changes sign at z_p from positive to negative, we have $L_p'(z_p) < 0$ (L_p and L_p' have no common zeros; otherwise, (3.3.17) would make L_p'' and hence, all the derivatives of L_p vanish at such a common zero) and, hence,

$$W(z_p) = -L_p'(z_p)\tilde{L}_p(z_p) = |L_p'(z_p)|\tilde{L}_p(z_p).$$

Since $W < 0$, it follows that $\tilde{L}_p(z_p) < 0$. Now consider $f = c_1 L_p + c_2 \tilde{L}_p$ for $c_2 > 0$. Then $f(z_p) < 0$ and $f(s) > 0$ for x close to 0. Therefore, f has a zero in $(0, z_p)$. The same arguments work for $c_2 < 0$. □

LEMMA 3.3.3 *The function* L_p *is strictly convex on* $(0, z_p]$ *for* $1 < p < \infty$*; it is strictly concave on* $(0, z_p]$ *for* $0 < p < 1$.

PROOF First consider the case $1 < p < \infty$. Starting with the Laguerre equation and then differentiating it, we get

$$sL_p'' + (1 - s)L_p' + pL_p = 0, \tag{3.3.20}$$

$$sL_p''' + (2 - s)L_p'' + (p - 1)L_p' = 0. \tag{3.3.21}$$

Then $L_p''(0) = \frac{p(p-1)}{2} > 0$. Let $x_1 > 0$ be the first positive point where $L_p''(x_1) = 0$. Suppose that $x_1 < z_p$. Then $L_p(x_1) > 0$ and (3.3.20) implies that $L_p'(x_1) < 0$. Then (3.3.21) yields $L_p'''(x_1) > 0$ and so L_p'' is strictly increasing at x_1, which is not possible. Therefore $x_1 > z_p$ and L_p is strictly convex on $(0, z_p]$. A similar argument shows that L_p is strictly concave on $(0, z_p]$ for $0 < p < 1$. □

LEMMA 3.3.4 *We have* $(sL_p')' = -pL_{p-1}$, $sL_p' = p(L_p - L_{p-1})$.

PROOF Differentiating the Laguerre equation

$$(sL_p')' - sL_p' + pL_p = 0 \tag{3.3.22}$$

gives us

$$(sL_p')'' - (sL_p')' + pL_p' = 0.$$

Multiply this by s and differentiate again to get

$$s(sL_p')''' + (1-s)(sL_p')'' + (p-1)(sL_p')' = 0.$$

Thus, $(sL_p')'$ solves the Laguerre equation with constant $p-1$ and hence is a multiple of L_{p-1}. It remains to use that $(sL_p')'(0) = -p$.

To get the second identity just apply the Laguerre equation. □

From now on, in this section, we assume that $p > 1$.

LEMMA 3.3.5 *We have* $L_p' < 0$ *on* $(0, z_p]$, $z_p < z_{p-1}$.

PROOF Since z_{p-1} is the root of L_{p-1}, by Lemma 3.3.4, we have $(sL_p')'(z_{p-1}) = 0$. Then by (3.3.22), we get $-z_{p-1}L_p'(z_{p-1}) + pL_p(z_{p-1}) = 0$. Since $L_p \geq 0$, $L_p'' > 0$ on $(0, z_p]$, we have $L_p' < 0$ on $(0, z_p]$, and it follows that $z_{p-1} > z_p$. □

The following result improves the assertion of Lemma 3.3.2.

LEMMA 3.3.6 *Let* $f \in C^2[0,1]$ *be a supersolution of the Laguerre equation,*

$$sf''(s) + (1-s)f'(s) + pf(s) \leq 0.$$

Then $f(z_p) \leq 0$.

PROOF Set

$$T(s) = sL_p'(s)f(s) - sf'(s)L_p(s).$$

Then $T(0) = 0$,

$$\begin{aligned} T'(s) &= L_p'(s)f(s) - f'(s)L_p(s) + sL_p''(s)f(s) - sf''(s)L_p(s) \\ &\geq L_p'(s)f(s) - f'(s)L_p(s) - (1-s)L_p'(s)f(s) - pL_p(s)f(s) \\ &\quad + (1-s)f'(s)L_p(s) + pf(s)L_p(s) = T(s). \end{aligned}$$

Therefore,

$$T(s) \geq 0, \qquad s \in [0,1],$$

and hence,

$$0 \leq T(z_p) = z_pL_p'(z_p)f(z_p),$$

and $f(z_p) \leq 0$. □

LEMMA 3.3.7 *We have $L_p < L_{p-1}$, $L_p' < 0$ on $(0, z_{p-1}]$, $L_p < 0$ on $(z_p, z_{p-1}]$. Furthermore, L_p has exactly one root in $[0, z_{p-1}]$.*

PROOF By Lemma 3.3.4, for $s > 0$ we have $L_p(s) = L_{p-1}(s)$ if and only if $L_p'(s) = 0$.

Suppose that $L_p(s) = L_{p-1}(s)$ for some $s \in (0, z_{p-1})$. First of all, $L_p' < 0$ on $(0, z_p]$. Therefore, $s > z_p$. Next $L_{p-1} > 0$ in (z_p, z_{p-1}) and $L_p < 0$ in some interval $(z_p, z_p + \epsilon)$. If L_p is positive at a point in $(z_p, z_{p-1}]$, then for some $y \in (z_p, z_{p-1}]$, we have $L_p(y) < 0$, $L_p'(y) = 0$, and then $L_{p-1}(y) < 0$, which is impossible. □

COROLLARY 3.3.8 *The function L_p is convex in $(0, z_{p-1})$.*

PROOF Lemma 3.3.3 gives this to us in $(0, z_p]$. Suppose that $L_p''(s) = 0$ for some $x \in (z_p, z_{p-1})$. By the previous lemma, $L_p(s) < 0$, and the Laguerre equation implies that $L_p'(s) > 0$. Since $L_p'(0) < 0$, there is a point $y \in (0, x)$ such that $L_p'(y) = 0$. This contradicts the previous lemma. □

LEMMA 3.3.9 *We have $0 < z_p < 1$.*

PROOF By Lemma 3.3.5, it suffices to verify that $L_p(1) < 0$, $1 < p \leq 2$. By (3.3.18), we have

$$\begin{aligned} L_p(1) &= 1 - p + \frac{p(p-1)}{4} + \cdots + (-1)^n \frac{p(p-1)\cdots(p-n+1)}{n!^2} + \cdots \\ &= (p-1)\left(-1 + \frac{p}{4} + \frac{p(2-p)}{3!^2} \cdots \right. \\ &\qquad \left. + \frac{p(2-p)(3-p)\cdots(n-1-p)}{n!^2} + \cdots \right). \end{aligned}$$

Since $p(2-p) \leq 1$, $1 < p \leq 2$, we get

$$\frac{L_p(1)}{p-1} \leq -1 + \frac{1}{2} + \frac{1}{3!^2} \cdots + \frac{(n-2)!}{n!^2} + \cdots < -1 + e - 2 < 0.$$

□

LEMMA 3.3.10 *We have $z_p \leq 2/(p+1)$.*

PROOF By (3.3.22), we have

$$\int_0^{z_p} (sL_p'(s))'\, ds = \int_0^{z_p} sL_p'(s)\, ds - p \int_0^{z_p} L_p(s)\, ds.$$

By convexity of L_p on $[0, z_p]$, we get

$$L'_p(s) \leq L'_p(z_p), \qquad 0 \leq s \leq z_p,$$
$$\int_0^{z_p} L_p(s)\, ds \geq -L'_p(z_p)\frac{z_p^2}{2},$$

and hence,

$$z_p L'_p(z_p) \leq \frac{z_p^2}{2} L'_p(z_p) + p\frac{z_p^2}{2} L'_p(z_p).$$

It remains to use that $L'_p(z_p) < 0$. □

3.3.1.3 The Function H_pL_p. If

$$u(x, y) = (x + y)^p L_p(\frac{y}{x + y}),$$

then by (3.3.16) we have

$$\mathcal{H}_p L_p(s) = u_{xy}(1 - s, s). \tag{3.3.23}$$

LEMMA 3.3.11 *We have $\mathcal{H}_p L_p(s) = s(sL'_p)'''$.*

PROOF We start with the identities

$$\begin{aligned}
(sL'_p)' &= sL''_p + L'_p = sL'_p - pL_p, \\
(sL'_p)'' &= (sL'_p - pL_p)' = sL''_p - (p - 1)L'_p \\
&= -[(p - s)L'_p + pL_p], \\
(sL'_p)''' &= -[(p - s)L''_p + (p - 1)L'_p].
\end{aligned} \tag{3.3.24}$$

Since

$$\mathcal{H}_p L_p(s) = s(s - 1)L''_p + (p - 1)(1 - 2s)L'_p + p(p - 1)L_p,$$
$$(p - 1)[sL''_p + (1 - s)L'_p + pL_p] = 0,$$

we get

$$\mathcal{H}_p L_p(s) = -s[(p - s)L''_p + (p - 1)L'_p].$$

It remains to use (3.3.24). □

LEMMA 3.3.12 *If $2 < p < \infty$, then $\mathcal{H}_p L_p < 0$ on $(0, z_{p-1}]$. If $1 < p < 2$, then $\mathcal{H}_p L_p > 0$ on $(0, z_{p-1}]$. In particular, those inequalities hold on $(0, z_p]$.*

PROOF Let $p > 2$. By Lemma 3.3.3, L_{p-1} is strictly convex on $[0, z_{p-1}]$. Therefore, by Lemma 3.3.4, $(sL_p')'$ is strictly concave in $[0, z_{p-1}]$, and by Lemma 3.3.11, $\mathcal{H}_p L_p < 0$ in $(0, z_{p-1}]$. Similarly, the case $1 < p < 2$ follows from the fact that L_{p-1} is strictly concave in $[0, z_{p-1}]$.

The last claim follows from Lemma 3.3.3. □

Thus, for $2 < p < \infty$, L_p satisfies both (3.3.14) and (3.3.15) on $[0, z_p]$. Therefore, by (3.3.23), $\mathcal{U}$ satisfies (3.3.10) and (3.3.11) on $\{(x, y) : \frac{y}{x+y} \in [0, z_p]\}$.

3.3.1.4 The Obstacle Function v_{c_p}. Set

$$c_p = \frac{1 - z_p}{z_p},$$

$$v_{c_p}(s) = (1 - s)^p - \left(\frac{1 - z_p}{z_p}\right)^p s^p.$$

Then $v_{c_p}(z_p) = 0$, and by the repetition of (3.3.23) and the fact that $v_{xy} = 0$ we get

$$\mathcal{H}_p v_{c_p} = 0,$$

$$\mathcal{L}_p v_{c_p}(s) = sp\left[(p-1)(1-s)^{p-2} - p\left(\frac{1 - z_p}{z_p}\right)^p s^{p-2}\right],$$

and

$$\mathcal{L}_p v_{c_p}(s) > 0, \qquad 0 \le s < s_p,$$
$$\mathcal{L}_p v_{c_p}(s) < 0, \qquad s_p < s \le 1,$$

where number s_p is defined by

$$\left(\frac{1 - s_p}{s_p}\right)^{p-2} = \frac{p}{p-1}\left(\frac{1 - z_p}{z_p}\right)^p.$$

It is just the root of $\mathcal{L}_p v_{c_p}$.

LEMMA 3.3.13 *If $p > 2$, then $s_p < z_p$.*

PROOF It suffices to verify that

$$\frac{p-1}{p} < \left(\frac{1 - z_p}{z_p}\right)^2. \tag{3.3.25}$$

First, suppose that $2 < p < 3$. Estimate (3.3.25) is equivalent to

$$z_p < \frac{p}{p + \sqrt{p^2 - p}}.$$

By (3.3.18), it suffices to check that

$$\begin{aligned}0 > L_p\left(\frac{p}{p+\sqrt{p^2-p}}\right) &= 1 - \frac{p^2}{p+\sqrt{p^2-p}} + \frac{p(p-1)}{4}\cdot\frac{p^2}{(p+\sqrt{p^2-p})^2}\\ &- \sum_{n\geq 3}\frac{p(p-1)(p-2)(3-p)\cdots(n-1-p)}{n!^2}\cdot\frac{p^n}{(p+\sqrt{p^2-p})^n}.\end{aligned}$$

This follows from the estimate

$$0 > 1 - \frac{p^2}{p+\sqrt{p^2-p}} + \frac{p^3(p-1)}{4(p+\sqrt{p^2-p})^2}$$

or equivalently, for $2 < p < 3$,

$$\begin{aligned}&(p^2-p-\sqrt{p^2-p})(p+\sqrt{p^2-p}) > \frac{p^3(p-1)}{4},\\ &\iff 4(p^2-2p)\sqrt{p^2-p} > p^4-5p^3+8p^2-4p,\\ &\iff 4\sqrt{p^2-p} > (p-1)(p-2).\end{aligned}$$

Second, if $p \geq 3$, then we use that by Lemma 3.3.10, $z_p \leq 2/(p+1)$. Therefore,

$$\left(\frac{1-z_p}{z_p}\right)^2 \geq \left(\frac{p-1}{2}\right)^2 > \frac{p-1}{p}.$$

□

In a similar way, we have

LEMMA 3.3.14 *If $1 < p < 2$, then*

$$\frac{p-1}{p} > \left(\frac{1-z_p}{z_p}\right)^2.$$

3.3.1.5 The Touching Points. For large a, we have

$$aL_p(r) > v_{c_p}(r), \quad r\in[0,z_p), \qquad aL_p'(z_p) < v_{c_p}'(z_p).$$

Now we lower the value of a until either (i) the graph of aL_p on $[0,z_p)$ first touches the graph of v_{c_p} or (ii) $aL_p'(z_p) = v_{c_p}'(z_p)$.

Let us analyze the case (i). The touching point s satisfies the equalities

$$\begin{cases}(1-s)^p - c_p^p s^p = aL_p(s),\\ -p(1-s)^{p-1} - pc_p^p s^{p-1} = aL_p'(s),\end{cases} \tag{3.3.26}$$

or, equivalently,

$$\begin{cases} -pc_p^p s^{p-1} = apL_p(s) + a(1-s)L_p'(s), \\ -p(1-s)^{p-1} = -apL_p(s) + asL_p'(s). \end{cases}$$

Hence,

$$\frac{c_p^p s^{p-1}}{(1-s)^{p-1}} = \frac{pL_p(s) + (1-s)L_p'(s)}{-pL_p(s) + sL_p'(s)},$$

which implies that

$$c_p^p = \frac{(1-s)^p L_p'(s) + p(1-s)^{p-1} L_p(s)}{s^p L_p'(s) - ps^{p-1} L_p(s)} \stackrel{\text{def}}{=} F(s).$$

Next, we differentiate the function F and, by Lemma 3.3.12, obtain that

$$F'(s) = \frac{p(1-s)^{p-2}}{s^p} \frac{L_p(s)\mathcal{H}_p L_p(s)}{(sL_p'(s) - pL_p(s))^2} < 0, \qquad 0 < s < z_p.$$

Since

$$F(z_p) = \left(\frac{1-z_p}{z_p}\right)^p = c_p^p,$$

we see that function F on interval $[s, z_p]$ is strictly monotone and $F(s) = c_p^p = F(z_p)$, which makes the case (i) impossible.

Thus, we have

THEOREM 3.3.15 *For $c_p = \frac{1-z_p}{z_p}$ and for some $a_p > 1$, the function v_{c_p} touches $a_p L_p$ at z_p and $v_{c_p} < a_p L_p$ on $[0, z_p)$.*

Let us define

$$g(s) = \begin{cases} a_p L_p(s), & 0 < s \leq z_p, \\ v_{c_p}(s), & z_p < s \leq 1. \end{cases} \tag{3.3.27}$$

Then $g \in C^1[0,1]$. By Lemma 3.3.13, $\mathcal{L}_p g \leq 0$. Furthermore, by Proposition 3.3.12, we have $\mathcal{H}_p g \leq 0$. Therefore, g majorizes the obstacle function v_{c_p} and satisfies (3.3.14) and (3.3.15). Thus, the majorant $u(x,y) = (x+y)^p g(\frac{y}{x+y})$ satisfies (3.3.10) and (3.3.11).

3.3.1.6 Sharpness of the Constant. It remains to indicate that for any $c < \frac{1-z_p}{z_p}$, the function v_c has no majorant satisfying (3.3.14) and (3.3.15). Note that for $c < \frac{1-z_p}{z_p}$, $v_c(z_p) > 0$. So any possible supersolution f of the Laguerre equation, such that $f \geq v_c$ satisfies the inequality $f(z_p) > 0$. However, this contradicts Lemma 3.3.6.

3.3.2 The General Case, $2 < p < \infty$

Let $c_p = \frac{1-z_p}{z_p}$. In Section 3.3.1, we consider conformal martingales $Y = (Y_1, Y_2)$ and real martingales X satisfying $d\langle X\rangle = d\langle Y_i\rangle$, and established the sharp estimate

$$\|X\|_p \leq \left(\frac{1-z_p}{z_p}\right) \|Y\|_p,$$

where z_p is the smallest root of the bounded on $(0,1)$ the Laguerre function L_p. We started with $v(x,y) = x^p - c_p^p y^p$ and found a majorant $u(x,y)$ satisfying the required inequalities (3.3.10) and (3.3.11) for quadratic forms. Now we turn to the general case where X is a complex-valued martingale and $d\langle X\rangle \leq d\langle Y_i\rangle$. The function u should satisfy (3.3.9) and (3.3.8). We will show that the function u obtained in the simple setting in Section 3.3.1 works also in the general case. Henceforth, u will denote this function, and g will be its corresponding one-dimensional function defined in (3.3.27).

For the right-hand-side conformality with $p > 2$, the general quadratic form requirement is

$$u_{xx}|h_1|^2 + \frac{u_x}{x}|h_2|^2 + 2u_{xy}(h_1, k) + \left(u_{yy} + \frac{u_y}{y}\right)|k|^2 \leq 0 \qquad (3.3.28)$$

for all vectors h_1, h_2, and k satisfying

$$|h_1|^2 + |h_2|^2 \leq |k|^2. \qquad (3.3.29)$$

Let us set $|k| = 1$, $a = |h_1|$, and $\beta = (|h_1|^2 + |h_2|^2)^{1/2}$. Then for all β such that $a \leq \beta \leq 1$, we obtain an equivalent form of (3.3.28):

$$w(a) \stackrel{\text{def}}{=} -\left(\frac{u_x}{x} - u_{xx}\right) a^2 + 2|u_{xy}|a + \frac{u_x}{x}\beta^2 + \left(u_{yy} + \frac{u_y}{y}\right) \leq 0. \qquad (3.3.30)$$

LEMMA 3.3.16 *For $x > 0$, $u_x > 0$ and $-u_{xx} < \frac{u_x}{x} - u_{xx} < |u_{xy}|$.*

Let us first take this lemma for granted and finish the reduction of the general case, $2 < p < \infty$ to the special case.

The special case says that we have the following inequality for function u constructed above:

$$u_{xx}\beta^2 + 2|u_{xy}|\beta + \left(u_{yy} + \frac{u_y}{y}\right) \leq 0, \quad \text{for all } \beta \leq 1. \qquad (3.3.31)$$

We need to prove that (3.3.30) holds if (3.3.31) does.

If $\frac{u_x}{x} - u_{xx} \leq 0$ then the maximum of the expression in (3.3.30) on interval $0 \leq a \leq \beta$ is attained for $a = \beta$ obviously. This is the case of (3.3.31), and we are done.

So we assume $\frac{u_x}{x} - u_{xx} > 0$, then the maximum in the quadratic expression $w(a)$ considered on $[0, \infty)$ is attained at point $a_* \stackrel{\text{def}}{=} \frac{|u_{xy}|}{\frac{u_x}{x} - u_{xx}}$. Lemma 3.3.16 claims that $a_* \geq 1 \geq \beta$, but $0 \leq a \leq \beta$, so the maximum of the expression $w(a)$ is attained on the right endpoint $a = \beta$. We are again under the condition (3.3.31).

We proved that function u constructed above satisfies (3.3.30) just because it satisfies (3.3.31).

Proof of Lemma 3.3.16 When $u = v$, we have $v_x = px^{p-1} > 0$ and

$$\frac{v_x}{x} - v_{xx} = -p(p-2)x^{p-2} < 0 = |v_{xy}|.$$

From now on, we assume that u corresponds to the Laguerre function $g = a_p L_p$,

$$u(x, y) = (x + y)^p g\left(\frac{y}{x + y}\right), \qquad 0 \leq \frac{y}{x + y} < z_p.$$

A simple computation shows that $u_x(1 - s, s) = pg(s) - sg'(s) > 0$ since g and $-g'$ are strictly positive in $(0, z_p)$.

It remains to show that

$$\frac{u_x}{x} - u_{xx} < |u_{xy}|. \tag{3.3.32}$$

By (3.3.23) and Proposition 3.3.12, $u_{xy}(1 - s, s) = \mathcal{H}_p g(s) < 0$ in $(0, z_p)$, and hence $|u_{xy}| = -u_{xy}$. Therefore, (3.3.32) is equivalent to

$$\frac{u_x}{x} - u_{xx} + u_{xy} < 0. \tag{3.3.33}$$

Furthermore, by the Laguerre equation, we have

$$\left(\frac{u_x}{x}\right)(1 - s, s) = \frac{pg(s)}{1 - s} - \frac{s}{1 - s} g'(s), \tag{3.3.34}$$

$$\begin{aligned} u_{xx}(1 - s, s) &= s^2 g''(s) - 2(p-1)sg'(s) + p(p-1)g(s) \\ &= s^2 g''(s) - 2(p-1)sg'(s) - (p-1)sg''(s) \\ &\quad - (p-1)(1-s)g'(s) \\ &= -s(p-1-s)g''(s) - (p-1)(1+s)g'(s), \end{aligned} \tag{3.3.35}$$

and

$$u_{xy}(1 - s, s) = \mathcal{H}_p g(s) = -s(p - s)g''(s) - s(p-1)g'(s). \tag{3.3.36}$$

By (3.3.34), (3.3.35), and (3.3.36), condition (3.3.33) is equivalent to

$$\frac{pg(s) - sg'(s)}{1-s} - sg''(s) + (p-1)g'(s)$$
$$= \frac{1}{1-s}\left(\mathcal{H}_p g(s) + (p-2)(sg'(s) - pg(s))\right) < 0.$$

The latter inequality holds because $\mathcal{H}_p g$ and $sg'(s) - pg(s)$ are strictly negative on $(0, z_p)$. □

REMARK 3.3.17 Notice that (3.3.4) can be formally written down as a fully nonlinear PDE of Hamilton–Jacobi type:

$$\sup_{\alpha_i, \beta_i, i=1,2} \left(\sum_{i,j=1}^{2} U_{x_i x_j}(\alpha_i, \alpha_j) + \Delta_y U |\alpha_1|^2 + \sum_{i,j=1}^{2} 2U_{x_i y_j}(\alpha_i, \beta_j) \right) = 0. \tag{3.3.37}$$

In this section, we have solved this equation with a certain obstacle condition. This solution provided us with the sharp constants in the part of Theorem 3.2.4 dealing with $p \geq 2$.

3.4 Proof of Theorem 3.2.3: Left-Hand Side Conformality, $1 < p < 2$

In this section, we show that the same methods extend to the case of left-hand-side conformality when $1 < p < 2$. Notice that the main partial differential inequality (3.3.9) holds true again, exactly as in Section 3.3, but now we need to satisfy the nonpositivity of the same quadratic form on a different cone. See (3.4.3) and (3.4.2) in Section 3.4.2.

Let us begin with the special case $d\langle X\rangle = d\langle Y_i\rangle$. The obstacle functions are

$$v_c(x, y) = y^p - c^p x^p, \qquad v_c^*(s) = s^p - c^p(1-s)^p,$$

and the majorants

$$u(x, y) = (x+y)^p g\left(\frac{y}{x+y}\right), \qquad g(s) = U(1-s, s)$$

should satisfy the quadratic form inequalities (3.3.10), (3.3.11), (3.3.14), and (3.3.15). The function v_c^* takes the value $-c^p$ at $s = 0$ and increases to 1 at $s = 1$. We start with

$$c_p = \frac{z_p}{1 - z_p}$$

and the function $v^*_{c_p}$. Arguing as in Section 3.3.1.4 (see Lemma 3.3.14), we obtain that

$$\mathcal{L}_p v^*_{c_p}(s) < 0, \qquad z_p < s \le 1.$$

We have by definition of $v^*_{c_p}$

$$0 > v^*_{c_p}(r), \quad r \in [0, z_p), \qquad 0 < v^{*\prime}_{c_p}(z_p).$$

Hence, for $a = 0$ we have

$$aL_p(r) > v^*_{c_p}(r), \quad r \in [0, z_p), \qquad aL'_p(z_p) < v^{*\prime}_{c_p}(z_p).$$

Now we lower the value of a until either (i) the graph of aL_p on $[0, z_p)$ first touches the graph of $v^*_{c_p}$ or (ii) $aL'_p(z_p) = v^{*\prime}_{c_p}(z_p)$.

Let us analyze the case (i). The touching point s satisfies the equalities

$$\begin{cases} s^p - c_p^p(1-s)^p = aL_p(s), \\ ps^{p-1} + pc_p^p(1-s)^{p-1} = aL'_p(s). \end{cases} \tag{3.4.1}$$

Note that (3.4.1) is similar to (3.3.26). Arguing as in Section 3.3.1.5, we have

$$\frac{1}{c_p^p} = \frac{(1-s)^p L'_p(s) + p(1-s)^{p-1}L_p(s)}{s^p L'_p(s) - ps^{p-1}L_p(s)} \stackrel{\text{def}}{=} F(s),$$

$$F'(s) = \frac{p(1-s)^{p-2}}{s^p} \frac{L_p(s)\mathcal{H}_p L_p(s)}{(sL'_p(s) - pL_p(s))^2}.$$

By Lemma 3.3.12, $\mathcal{H}_p L_p > 0$, and hence, $F' > 0$ in $(0, z_p)$. Since

$$F(z_p) = \left(\frac{1-z_p}{z_p}\right)^p = c_p^{-p},$$

we see that function F on interval $[s, z_p]$ is strictly monotone and $F(s) = c_p^{-p} = F(z_p)$, which makes case (i) impossible.

Thus, in the case $1 < p < 2$, for $c_p = \frac{z_p}{1-z_p}$ and for some $a_p < 0$, the function $v^*_{c_p}$ touches a_pL_p at z_p and $v^*_{c_p} < a_pL_p$ on $[0, z_p)$.

The best majorant satisfying the required inequalities (3.3.10), (3.3.11) on quadratic forms is

$$u(x, y) = (x+y)^p g\left(\frac{y}{x+y}\right),$$

where

$$g(s) = \begin{cases} a_pL_p(s), & 0 < s \le z_p, \\ v^*_{c_p}(s), & z_p < s \le 1. \end{cases}$$

3.4.1 Sharpness of the Constant

Once again, for any $c < \frac{1-z_p}{z_p}$, the function v_c has no majorant satisfying (3.3.15). Indeed, $v_c(z_p) > 0$, and any supersolution f of the Laguerre equation such that $f \geq v_c$ satisfies the inequality $f(z_p) > 0$. However, this contradicts to Lemma 3.3.6. Since the Bellman function (which has the best constant) satisfies the corresponding quadratic form inequalities, it follows that our constant is sharp.

3.4.2 The General Case

For the left-hand-side conformality with $1 < p < 2$, the general quadratic form requirement is

$$u_{xx}|h_1|^2 + \frac{u_x}{x}|h_2|^2 + 2u_{xy}(h_1, k) + \left(u_{yy} + \frac{u_y}{y}\right)|k|^2 \leq 0 \qquad (3.4.2)$$

for all vectors h_1, h_2, and k satisfying

$$|k|^2 \leq |h_1|^2 + |h_2|^2. \qquad (3.4.3)$$

Setting $|k| = 1$, $a = |h_1|$, and $\beta = (|h_1|^2 + |h_2|^2)^{1/2}$, we obtain an equivalent form:

$$-\left(\frac{u_x}{x} - u_{xx}\right)a^2 + 2|u_{xy}|a + \frac{u_x}{x}\beta^2 + \left(u_{yy} + \frac{u_y}{y}\right) \leq 0,$$

$$\text{for all } \beta \geq \max(1, a). \qquad (3.4.4)$$

3.4.2.1 The Case When X Is Real-Valued and $h_2 = 0$. In this case, (3.4.2) becomes

$$u_{xx}\beta^2 + 2|u_{xy}|\beta + \left(u_{yy} + \frac{u_y}{y}\right) \leq 0, \qquad \beta \geq 1. \qquad (3.4.5)$$

When $u = v_c$, we have $u_{xx} = -c^p p(p-1)x^{p-2} < 0$, $u_{xy} = 0$, and therefore the maximum value in the left-hand side is attained for minimal $\beta = 1$. Thus, (3.4.5) follows from (3.3.10), (3.3.11) which hold for u as above.

Let us assume now that u corresponds to the Laguerre function $g = a_p L_p$. Then $u(x, y) = (x + y)^p g(\frac{y}{x+y})$, and as above,

$$u_{xx}(1-s, s) = s^2 g''(s) - 2(p-1)sg'(s) + p(p-1)g(s),$$
$$u_{xy}(1-s, s) = -s(1-s)g''(s) + (p-1)(1-2s)g'(s) + p(p-1)g(s).$$

Since $g \leq 0$, $g' > 0$, $g'' < 0$, and $\mathcal{H}_p g \leq 0$ in $[0, z_p]$, we have

$$sg''(s) - (p-1)g'(s) < 0, \qquad 0 \leq s \leq z_p,$$

and hence,

$$u_{xx} - u_{xy} \leq 0, \qquad u_{xx} \leq 0, \qquad u_{xy} \leq 0.$$

Therefore,

$$\frac{-|u_{xy}|}{u_{xx}} \leq 1,$$

and we obtain that the maximum value in the left-hand side of (3.4.5) is attained at $\beta = 1$, which is the special case considered in the Section 3.3, so (3.4.5) is satisfied in this case.

Thus, u always satisfies (3.4.5). This completes the argument for the case when X is real-valued, $d\langle Y_i\rangle \leq d\langle X\rangle$, and $h_2 = 0$.

3.4.2.2 The Case When X is Complex Valued. Now we deal with (3.4.4) in full generality.

If $\frac{u_x}{x} - u_{xx} \leq 0$, then the maximal value of the expression in the left-hand side of (3.4.4) for $a \leq \beta$ and for fixed β occurs when $a = \beta$, and we return to (3.4.5). Therefore, from now on we assume that $\frac{u_x}{x} - u_{xx} > 0$. This can happen only if $u > v = v_{c_p}$ since $\frac{v_x}{x} - v_{xx} = -c_p^p p(2-p)x^{p-2} < 0$. Therefore, we can assume that $u > v$. Since $u_{xy} \leq 0$, the maximal value of the expression in the left-hand side of (3.4.4) as a function of $a \in [0, \infty)$ occurs at

$$a_* = \frac{|u_{xy}|}{\frac{u_x}{x} - u_{xx}}.$$

As in the proof of Lemma 3.3.16,

$$\left(\frac{u_x}{x} - u_{xx} + u_{xy}\right)(1-s, s)$$
$$= \frac{1}{1-s}\big(\mathcal{H}_p g(s) - (2-p)(sg'(s) - pg(s))\big) \leq 0$$

on $[0, z_p]$, and hence, $a_* \geq 1$.

If $\beta \leq a_*$, then the maximal value in $a \in [0, \beta]$ in the left-hand side of (3.4.4) is attained at β and we return to (3.4.5). If $a_* < \beta$, then the maximal value is at $a = a_*$, and it remains to verify that

$$u_{xy}^2 + \left(\frac{u_x}{x}\beta^2 + u_{yy} + \frac{u_y}{y}\right)\left(\frac{u_x}{x} - u_{xx}\right) \leq 0. \tag{3.4.6}$$

Since

$$u_x(1-s, s) = pg(s) - sg'(s) < 0,$$

the maximal value in the left-hand side of (3.4.6) is attained for $\beta = a_* = a$ and we return once again to (3.4.5), which completes the proof for the case of complex X.

3.4.3 Estimates on z_p and Optimal Constants in Theorem 3.2.4

In the next section, we will need the asymptotic behavior of constants in the first and second parts of Theorem 3.2.4. For that we start with the known asymptotics of zeros of Laguerre functions.

Let J_0 be the Bessel function of order zero,

$$J_0(x) = \sum_{n\geq 0}(-1)^n \frac{x^{2n}}{2^{2n}(n!)^2}.$$

Denote its first positive zero by j_0. It is known (see, for example, [**198, section 15.51**]) that

$$j_0 \approx 2.404826.$$

Next, we use a Mehler–Heine type formula (see [**176, Theorem 8.1.3**]),

$$\lim_{n\to\infty,\, n\in\mathbb{N}} L_n(x/n) = J_0(2\sqrt{x}).$$

Then it is natural to assume that convergence holds also for zero. This is proved in [**176**].

$$\lim_{n\to\infty} nz_n = \frac{j_0^2}{4}.$$

Hence,

$$\lim_{n\to\infty} \frac{C_n}{n} = \frac{4\sqrt{2}}{j_0^2} \approx 0.97815.$$

Then simple interpolating argument shows

$$\lim_{p\to\infty} \frac{C_p}{p} = \frac{4\sqrt{2}}{j_0^2} \approx 0.97815. \tag{3.4.7}$$

This is the asymptotic of the constant in Theorem 3.2.4 (2).

For the asymptotic of the constant in Theorem 3.2.4 (1), we notice that by (3.3.18), for large p we have

$$0 = L_{p'}(z_{p'}) = 1 - (1+\varepsilon)(1-\delta) + \frac{(1+\varepsilon)\varepsilon}{4}(1-\delta)^2 + \cdots$$
$$+ \frac{(1+\varepsilon)\varepsilon(1-\varepsilon)\cdots(n-2-\varepsilon)}{n!^2}(1-\delta)^n + \cdots,$$

where $\varepsilon = 1/(p-1)$, $\delta = 1 - z_{p'}$, and hence

$$\delta - \varepsilon\left(1 - \sum_{n\geq 2}\frac{(n-2)!}{n!^2}\right) = O(\varepsilon^2 + \delta^2), \qquad \delta \to 0+,\ \varepsilon \to 0+.$$

Denote

$$q_0 = 1 - \sum_{n\geq 2}\frac{(n-2)!}{n!^2} \approx 0.718282. \tag{3.4.8}$$

Then,

$$\lim_{p\to\infty} p(1 - z_{p'}) = q_0, \tag{3.4.9}$$

and hence,

$$\lim_{p\to\infty}\frac{C_{p'}}{p} = \frac{1}{q_0\sqrt{2}} \approx 0.98444,$$

where $C_{p'}$ is the optimal constant in Theorem 3.2.4 (1).

Thus, we have

$$\lim_{p\to\infty}\frac{C_{p'}(LHC)}{C_p(RHC)} \approx 1.006, \tag{3.4.10}$$

where the letters LHC, RHC stand for the left-hand and right-hand conformality correspondingly.

Finally, by (3.3.18), for $p > 2$ we have

$$\begin{aligned}
L_{p'}\left(1 - \frac{q_0}{p}\right) &= 1 - \frac{p}{p-1}\left(1 - \frac{q_0}{p}\right) + \frac{p}{4(p-1)^2}\left(1 - \frac{q_0}{p}\right)^2 + \cdots \\
&\quad + \frac{p(p-2)\cdots((n-2)(p-1)-1)}{n!^2(p-1)^n}\left(1 - \frac{q_0}{p}\right)^n + \cdots \\
&< -\frac{1-q_0}{p-1} + \frac{p}{4(p-1)^2}\left(1 - \frac{q_0}{p}\right)^2 + \cdots + \frac{p(n-2)!}{n!^2(p-1)^2}\left(1 - \frac{q_0}{p}\right)^2 + \cdots \\
&= \frac{1-q_0}{p(p-1)^2}\left(-p^2 + p + (p-q_0)^2\right) < 0.
\end{aligned}$$

Hence, $L_{p'}(1 - \frac{q_0}{p}) < 0$, and thus

$$\frac{q_0}{p} < 1 - z_{p'}. \tag{3.4.11}$$

Then for $p > 2$

$$C_{p'}(LHC) < \frac{p}{\sqrt{2}q_0} - \frac{1}{\sqrt{2}}. \tag{3.4.12}$$

3.5 Burkholder, Bellman, and Ahlfors–Beurling Operator in L^p for Large p

One of the primary applications for Burkholder's theorem has come in Fourier analysis in estimating the L^p norm of the Ahlfors–Beurling transform. Let m denote plane Lebesgue measure.

The Ahlfors–Beurling transform is a singular integral operator acting on $L^p(\mathbb{C})$ and defined by

$$Bf(z) = -\frac{1}{\pi}\text{p.v.}\int_{\mathbb{C}} \frac{f(w)}{(z-w)^2}dm(w).$$

This operator arises naturally in the quasiconformal mapping theory due to the way it relates the complex derivative operators. If $\partial = \frac{\partial_x - i\partial_y}{2}$ and $\bar{\partial} = \frac{\partial_x + i\partial_y}{2}$, then

$$B = \frac{\partial}{\bar{\partial}} = \frac{\partial^2}{\Delta}.$$

Recall that the (singular) Riesz transforms (see e.g., [**173**]) are operators given by

$$R_1 = \frac{\partial_x}{\sqrt{\Delta}}, \qquad R_2 = \frac{\partial_y}{\sqrt{\Delta}}.$$

An alternative representation in terms of the second-order Riesz transforms [**56**] is particularly important for us: $B = R_2^2 - R_1^2 - i2R_1R_2$. One of the fundamental open problems for this operator is the computation of its L^p norm $\|B\|_p$. This question gains prominence due to the information it would yield regarding the Beltrami equation (see [**137**]) and for the proof of the former Gehring–Reich conjecture (and presently Astala's theorem) [**2**]. The problem of finding the norm $\|B\|_p$ attracts mathematicians in different areas of analysis and probability. It remains unsolved.

It is a conjecture by Iwaniec [**85**] that the norm constant is $\|B\|_p = p^* - 1$, the same constant as in Burkholder's theorem for martingales; by duality it is known that $\|B\|_p = \|B\|_{p'}$. The lower bound $\|B\|_p \geq p^* - 1$ was proved by Lehto [**104**]. Estimates of $\|B\|_p$ from above have been obtained and gradually improved, relying on critical theorems of Burkholder described in Section 1.8; see [**6, 9, 22, 23, 137**]. They are still considerably bigger than $p^* - 1$.

The first major breakthrough in finding the connection between martingale estimates and the Ahlfors–Beurling operator came in [**9**], where Bañuelos and Wang show that if a function $f \in L^p(\mathbb{R}^2)$ is extended harmonically as $U_f(x,t)$

to the upper half-space $\mathbb{R}^3_+$, then for the martingale $X = U_f(W(t))$ there exists a differentially subordinated martingale Y satisfying (essentially)

$$X_\tau \text{ have the same distribution as } f(x),$$

and

$$\mathbb{E}\,[Y_\tau | W(\tau) = x] = Bf(x), \quad d\langle Y\rangle \le 16\, d\langle X\rangle.$$

Here $W(t)$ is three-dimensional Brownian motion, τ is its exit time from $\mathbb{R}^3_+$, and the conditional expectation $\mathbb{E}\,[Y_\tau | W(\tau) = x]$ is the average value of Y_τ over paths that exit at x. This then implies (essentially)

$$\|Bf\|_p = \|\mathbb{E}\,[Y_\tau | B_\tau = x]\|_p \le \|Y_\tau\|_p \le 4(p^* - 1)\|X_\tau\|_p \le 4(p^* - 1)\|f\|_p.$$

The first inequality follows from Jensen's and the second one follows from Burkholder's theorem. Thus, we have $\|B\|_p \le 4(p^* - 1)$.

It is enticing to try to fit the Ahlfors–Beurling operator estimate within a general framework derived from stochastic control theory. This automatically asks for the construction of some special object, which we call Bellman function in this book. The search for such a martingale model and a certain Bellman function is what we are going to do below with partial success.

Notice Bañuelos–Wang's paper [**9**] was doing exactly this (without mentioning the stochastic optimal control). Namely, they used a pre-fabricated Bellman function, and this was just the Burkholder function from Section 1.8. They used a certain martingale model of Ahlfors–Beurling transform based on the composition of harmonic extensions with Brownian motion.

But it turns out that one can move further in this direction by using (a) different martingale models of the Ahlfors–Beurling operator and (b) different Bellman function. The examples of such an approach are [**137**] and [**152**]. For example in [**137**], a special function $\mathcal{B}$ called the the Bellman function is found in relation to the problem. Using the Bellman function approach, Nazarov and Volberg [**137**] obtain a better estimate $\|B\|_p \le 2(p^* - 1)$.

Paper [**137**] did not use any martingale models per se, it used the heat extension and a new Bellman function totally different from the one in Section 1.8. But in [**6**] the authors redo the work done in [**9**] by Bañuelos and Wang, but this time with heat extensions and space-time Brownian motion and also obtain $\|B\|_p \le 2(p^* - 1)$. So these two papers, [**6, 137**], can be viewed as giving a stochastic and non-stochastic approaches correspondingly.

The next step has been done in [**5**], where Bañuelos and Janakiraman make the observation that the martingale associated with the Ahlfors–Beurling transform is in fact a conformal one. They show that Burkholder's proof in

[**24**] naturally accommodates for this property and leads to an improvement in the estimate of $\|B\|_p$.

Using our knowledge from Section 3.2, we are going to give the martingale model of the Ahlfors–Beurling operator, which gives us probably the best estimate of this operator in L^p when p is large. This approach uses heavily the conformality.

For example, Theorem 3.2.1 (1) of [**5**] then gives

$$p > 2 \Rightarrow \|B\|_p \leq \sqrt{2p(p-1)}. \tag{3.5.1}$$

In particular, one obtains the best to date universal estimate (recall that $p^* = \max(p, p')$)

$$\|B\|_p \leq 1.535(p^* - 1). \tag{3.5.2}$$

Observe that (3.5.1) implies

$$\limsup_{p\to\infty} \frac{\|B\|_p}{p} \leq \sqrt{2}. \tag{3.5.3}$$

Notice a very simple fact about the Alhfors–Beurling operator:

$$\|B\|_p = \|B\|_{p'}. \tag{3.5.4}$$

This follows from a simple remark that the kernel of integral operators B and B^* look alike.

We are going to show in this section that Ahlfors–Beurling operator has $\|B\|_p$ at most twice the left-hand conformality constant. This will be shown later, now we notice that left-hand conformality results from Theorem 3.2.1, Theorem 3.2.4, (3.4.12) give us some estimates of the Ahlfors–Beurling operator.

But notice that the sharpness of the constant in Theorem 3.2.1 (1) is not known. On the other hand, constant $C_{p'}, p' < 2$, in Theorem 3.2.4 (1) is sharp, and, using (3.5.4) and this sharp constant, we can hopefully improve on (3.5.1). This is in fact what happens for large p, but it cannot be obtained naively, because, for example, (3.4.12) would give us only

$$p > 2 \Rightarrow \|B\|_p \leq \sqrt{2}\left(\frac{p}{q_0} - 1\right),$$

which is worse than (3.5.1). However, we are going to use Theorem 3.2.4 (1) in combination with some extra properties of B to prove

$$\limsup_{p\to\infty} \frac{\|B\|_p}{p} \leq 1.3922. \tag{3.5.5}$$

In particular, it breaches the previous asymptotic threshold

$$\limsup_{p\to\infty} \frac{\|B\|_p}{p} \le \sqrt{2},$$

which follows from (3.5.1) or, alternatively, from [**54**].

The following theorem is the goal of this section.

THEOREM 3.5.1 *Let $p > 2$. Then*

$$\|B\|_p < \left(\frac{p+3}{2}\pi\right)^{1/(2p)} \cdot \frac{p-q_0}{q_0},$$

where q_0 is defined in (3.4.8).

This estimate together with (3.4.9) gives us

$$\limsup_{p\to\infty} \frac{\|B\|_p}{p} \approx 1.3922 < \sqrt{2}.$$

Furthermore, since

$$\left(\frac{1003}{2}\pi\right)^{1/2000} < 1.004 \text{ and } 1.4\,q_0 > 1.005,$$

and function

$$p \to \left(\frac{p+3}{2}\pi\right)^{1/(2p)} \cdot \frac{p-q_0}{p}$$

is decreasing for sufficiently large p, we have

$$\|B\|_p < 1.4p, \qquad p \ge 1000. \tag{3.5.6}$$

To prove Theorem 3.5.1, we use the martingale theory as found in [**6, 9, 137**] and a symmetry lemma from [**54**]. We review this material in the following sections.

3.5.1 Martingales and the Ahlfors–Beurling Transform

Below u is a complex-valued function of two arguments: of a complex argument $z = x + iy$ and of a real positive argument t. The symbol ∇u always means gradient of u in x, y variables, written as a column vector in $\mathbb{C}^2$: $\nabla u \stackrel{\text{def}}{=} (\frac{\partial u}{\partial x}, \frac{\partial u}{\partial y})^T$. Also in the following, φ, f, g are always smooth functions with compact support on the plane $\mathbb{C}$. Given a function $\varphi \in L^p(\mathbb{C})$, we denote the heat extension to $\mathbb{R}^3_+$ by letter u^φ. Here by the heat extension, we understand the solution of

$$\frac{\partial u^\varphi}{\partial t} - \frac{1}{2}\Delta u^\varphi = 0.$$

Let

$$h_t(z) \stackrel{\text{def}}{=} \frac{1}{2\pi t} e^{-\frac{|z|^2}{2t}}.$$

It is a fundamental solution of $\frac{\partial}{\partial t} - \frac{1}{2}\Delta$.

By z, w possibly with indices, we denote point on the complex plane $\mathbb{C}$, by $dm(z)$ Lebesgue measure on $\mathbb{C}$.

DEFINITION 3.5.2 We denote below by $r \cdot c$ the scalar product of a row-vector $r \in \mathbb{R}^2$ and a column-vector $c \in \mathbb{C}^2$.

Let $W^w(t)$ be the two-dimensional Brownian motion $W^w(t) = (W_1(t), W_2(t))$ starting at w, then the stochastic process $u^\varphi(W^w(t), T-t)$ is a martingale. This follows from Itô's formula of Section 2.4 and is left as an exercise to the reader.

We denote by $\mathbb{E}^w$ the expectation on Brownian paths $W^w(t)$, see, e.g., [**63, 199**].

The reader can exercise the use of Itô's formula to prove that the composition of Brownian motion and the function solving heat equation $u^\varphi(W(t), T-t)$ is a martingale. Indeed, Itô's formula of Section 2.4 gives us the following relationship:

$$\varphi(W^w(T)) - u^\varphi(w,T) = \int_0^T dW^w(t) \cdot \nabla u^\varphi(W^w(t), T-t). \tag{3.5.7}$$

Moreover, a wonderful fact from Section 2.4 is that "the product of stochastic integrals is a (non-stochastic) integral of the product." Therefore,

$$\mathbb{E}^w|\varphi(W^w(T)) - u^\varphi(w,T)|^2 = \int_0^T \int_{\mathbb{C}} |\nabla u^\varphi(z, T-t)|^2 h_t(z-w) dm(z) dt. \tag{3.5.8}$$

We want to combine this equality with the following lemma.

LEMMA 3.5.3 *For all $p \in [1, \infty)$, we have*

$$\int_{\mathbb{C}} \mathbb{E}^w |\varphi(W^w(T)) - u^\varphi(w,T)|^p dm(w) \le C(p)\|\varphi\|_p^p.$$

PROOF Of course

$$\int_{\mathbb{C}} \mathbb{E}^w|\varphi(W^w(T)) - u^\varphi(w,T)|^p dm(w)$$

$$\le C(p)\left(\int_{\mathbb{C}} \mathbb{E}^w |\varphi(W^w(T))|^p dm(w) + \int_{\mathbb{C}} |u^\varphi(w,T)|^p dm(w)\right).$$

But

$$\int_{\mathbb{C}} \mathbb{E}^w |\varphi(W^w(T))|^p dm(w) = \int_{\mathbb{C}} \int_{\mathbb{C}} |\varphi(z)|^p h_T(z-w) dm(z) dm(w)$$
$$= \int_{\mathbb{C}} |\varphi(z)|^p dm(z) = \|\varphi\|_p^p.$$

Also pointwise, $|u^\varphi(w,T)|^p \le u^{|\varphi|^p}(w,T)$, and so

$$\int_{\mathbb{C}} |u^\varphi(w,T)|^p dm(w)) \le \int_{\mathbb{C}} u^{|\varphi|^p}(w,T) dm(w) = \int_{\mathbb{C}} |\varphi|^p dm = \|\varphi\|_p^p.$$

Lemma is proved. □

If we use this lemma and integrate (3.5.8) with respect to $dm(w)$, we get that uniformly in T:

$$\int_0^T \int_{\mathbb{C}} |\nabla u^\varphi(z, T-t)|^2 dt\, dm(z) \le C\|\varphi\|_2^2. \tag{3.5.9}$$

Let I be the identity 2×2 matrix and M be some real 2×2 matrix.

Definition 3.5.4 Let φ be a smooth function with compact support on the plane $\mathbb{C}$. The expression $I \star \varphi$ given below will be called heat martingale.

$$(I^w \star \varphi)(t) \stackrel{\text{def}}{=} \int_0^t dW^w(s) \cdot \nabla u^\varphi(W^w(s), t-s).$$

Given a 2×2 matrix M, we denote the martingale transform of $I \star \varphi$ by the matrix M as

$$(M^w \star \varphi)(t) \stackrel{\text{def}}{=} \int_0^t dW^w(s) \cdot M\, \nabla u^\varphi(W^w(s), t-s).$$

Definition 3.5.5 For a matrix

$$M \stackrel{\text{def}}{=} \begin{bmatrix} a, & b \\ c, & d \end{bmatrix}$$

consider

$$S_M^T \varphi(z) \stackrel{\text{def}}{=} \int_{\mathbb{C}} \mathbb{E}^w \left[(M^w \star \varphi)(T) | W^w(T) = z\right] h_T(z-w) dm(w).$$

Also for any matrix M, introduce the operator

$$B_M \stackrel{\text{def}}{=} aR_1^2 + dR_2^2 + (b+c)R_1R_2,$$

where $R_i, i = 1, 2$ are Riesz transforms.

For a special matrix A,

$$A \stackrel{\text{def}}{=} \begin{bmatrix} 1, & i \\ i, & -1 \end{bmatrix} \tag{3.5.10}$$

$$A^w \star f = \int_0^T dW^w(t) \cdot \begin{bmatrix} \partial_x + i\partial_y \\ i(\partial_x + i\partial_y) \end{bmatrix} u^f(W^w(t), T-t), \tag{3.5.11}$$

which is

$$2\int_0^T dW^w(t) \cdot \begin{bmatrix} \bar{\partial} \\ i\bar{\partial} \end{bmatrix} u^f(W^w(t), T-t). \tag{3.5.12}$$

The corresponding operator B_A is then $R_1^2 - R_2^2 + 2iR_1R_2$, that is, it is the Ahlfors–Beurling transform.

The following theorem holds.

THEOREM 3.5.6 *For any smooth function with compact support*

$$B_M(\varphi) = \lim_{T\to\infty} S_M^T\varphi, \tag{3.5.13}$$

where the convergence is in the weak sense.

PROOF By linearity and symmetry, it is enough to check the equality for the following matrix M:

$$M \stackrel{\text{def}}{=} \begin{bmatrix} 1, & 0 \\ 0, & 0 \end{bmatrix}$$

Moreover, we can think then that the test function φ is real-valued. Take another real-valued function ψ, which is smooth and with compact support on $\mathbb{C}$. We are going to prove that

$$(R_1^2\varphi, \psi) \stackrel{\text{def}}{=} (B_M\varphi, \psi) = \lim_{T\to\infty} (S_M^T\varphi, \psi)$$
$$= \lim_{T\to\infty} \int_{\mathbb{C}}\int_{\mathbb{C}} \mathbb{E}^w[(M^w\star\varphi)(T)|W^w(T) = z]\psi(z)h_T(z-w)dm(w)dm(z). \tag{3.5.14}$$

First we notice an obvious equality:

$$\int_{\mathbb{C}}\int_{\mathbb{C}} \mathbb{E}^w\left[(M^w\star\varphi)(T)|W^w(T) = z\right]\psi(z)h_T(z-w)dm(w)dm(z)$$
$$= \int_{\mathbb{C}} \mathbb{E}^w\left[\psi(W^w(T))(M^w\star\varphi)(T)\right]dm(w).$$

Now we use Definition 3.5.4 and (3.5.7) for function ψ. Then

$$\int_{\mathbb{C}} \mathbb{E}^w \left[\psi(W^w(T))(M^w \star \varphi)(T)\right] dm(w)$$

$$= \int_{\mathbb{C}} \mathbb{E}^w \left[\int_0^T dW^w(s) \cdot \nabla u^\psi(W^w(s), T-s) \cdot \right.$$

$$\left. \int_0^T dW_1^w(s) \frac{\partial u^\varphi}{\partial x}(W^w(s), T-s) \right] dm(w)$$

$$+ \int_{\mathbb{C}} u^\psi(w,T) \mathbb{E}^w \left[\int_0^T dW_1^w(s) \frac{\partial u^\varphi}{\partial x}(W^w(s), T-s) \right] dm(w) \stackrel{\text{def}}{=} I_1 + I_2.$$

To estimate I_2, we first use Cauchy inequality twice:

$$|I_2| \le \|u^\psi(\cdot,T)\|_{L^2(\mathbb{C})} \left(\int_{\mathbb{C}} \mathbb{E}^w \left| \int_0^T dW_1^w(s) \cdot \right. \right.$$

$$\left. \left. \frac{\partial u^\varphi}{\partial x}(W^w(s), T-s) \right|^2 dm(w) \right)^{1/2}.$$

But

$$\mathbb{E}^w \left| \int_0^T dW_1^w(s) \cdot \frac{\partial u^\varphi}{\partial x}(W^w(s), T-s) \right|^2$$

$$= \int_0^T \int_{\mathbb{C}} \left| \frac{\partial u^\varphi}{\partial x} \right|^2 (z, T-s) h_s(z-w) ds\, dm(z).$$

Hence,

$$|I_2| \le \|u^\psi(\cdot,T)\|_{L^2(\mathbb{C})} \left(\int_0^T \int_{\mathbb{C}} \left| \frac{\partial u^\varphi}{\partial x} \right|^2 (z, T-s) h_s \right.$$

$$\left. \times (z-w) ds\, dm(z) dm(w) \right)^{1/2}$$

$$= \|u^\psi(\cdot,T)\|_{L^2(\mathbb{C})} \left(\int_0^T \int_{\mathbb{C}} \left| \frac{\partial u^\varphi}{\partial x} \right|^2 (z, T-s) ds dm(z) \right)^{1/2}$$

$$\le C \|u^\psi(\cdot,T)\|_2 \|\varphi\|_2,$$

where the last inequality follows from (3.5.9). As to the term $\|u^\psi(\cdot,T)\|_2^2$, it is obviously at most $\|u^\psi(\cdot,T)\|_\infty \|u^\psi(\cdot,T)\|_1$. But

$$\|u^\psi(\cdot,T)\|_1 \le \|\psi\|_1, \quad \|u^\psi(\cdot,T)\|_\infty \le \frac{1}{2\pi T} \|\psi\|_\infty,$$

just because $u^\psi(z,T) = \int h_T(z-w)\psi(w)dm(w)$. Hence,

$$\lim_{T\to\infty} I_2 = 0. \tag{3.5.15}$$

To calculate I_1, we write it first as follows:

$$I_1 = \int_{\mathbb{C}} \mathbb{E}^w \left[\int_0^T dW_1^w(s) \frac{\partial u^\psi}{\partial x}(W^w(s), T-s) \right.$$
$$\left. \int_0^T dW_1^w(s) \frac{\partial u^\varphi}{\partial x}(W^w(s), T-s) \right] dm(w),$$

where we used the fact that W_1^w and W_2^w are independent. Now we use again the fact from Section 2.3: the product of two stochastic integrals is the integral (non-stochastic) of the product. Then we can rewrite

$$I_1 = \int_{\mathbb{C}} \mathbb{E}^w \int_0^T \left[\frac{\partial u^\psi}{\partial x}(W^w(s), T-s) \frac{\partial u^\varphi}{\partial x}(W^w(s), T-s)\, ds \right] dm(w).$$

In other words,

$$I_1 = \int_{\mathbb{C}} \int_{\mathbb{C}} \left[\int_0^T \frac{\partial u^\psi}{\partial x}(z, T-s) \frac{\partial u^\varphi}{\partial x}(z, T-s) h_s(z-w) ds \right] dm(z) dm(w)$$
$$= \int_{\mathbb{C}} \left[\int_0^T \frac{\partial u^\psi}{\partial x}(z,s) \frac{\partial u^\varphi}{\partial x}(z,s) \right] dm(z) ds.$$

We will prove in Lemma 3.5.7 below the following formula:

$$\int R_1^2 \varphi \cdot \psi\, dm = \lim_{T\to\infty} \int_0^T \int_{\mathbb{C}} \frac{\partial u^\varphi}{\partial x}(z,t)\, \frac{\partial u^\psi}{\partial x}(z,t)\, dm(z) dt. \tag{3.5.16}$$

Taking (3.5.16) temporarily for granted, we prove that

$$\lim_{T\to\infty} I_1 = (R_1^2\varphi, \psi). \tag{3.5.17}$$

Together (3.5.15) and (3.5.17) prove the theorem. We are left to show (3.5.16) for any two smooth functions with compact support. The theorem is fully proved modulo of the next lemma. □

Now we prove (3.5.16).

Lemma 3.5.7 *Let $\varphi, \psi \in C_0^\infty(\mathbb{C})$. Then the integral*

$$\int R_1^2 \varphi \cdot \psi\, dm = \lim_{T\to\infty} \int_0^T \int \frac{\partial u^\varphi}{\partial x}(z,t) \cdot \frac{\partial u^\psi}{\partial x}(z,t)\, dm(z) dt. \tag{3.5.18}$$

Proof The proof involves Parseval's formula. We can think that φ, ψ are real-valued. Consider $\varphi, \psi \in C_0^\infty(\mathbb{C})$ and now write

$$\begin{aligned}\int \psi R_1^2 \varphi dxdy &= \iint \frac{\xi_1^2}{|\xi|^2} \overline{\hat{\psi}(\xi)} \hat{\varphi}(\xi) d\xi \\ &= \iint_0^\infty e^{-t|\xi|^2} \xi_1^2 \hat{\varphi}(\xi) \hat{\psi}(-\xi) d\xi dt \\ &= \lim_{T\to\infty} \iint_0^T e^{-t|\xi|^2} \xi_1^2 \hat{\varphi}(\xi) \hat{\psi}(-\xi) d\xi dt,\end{aligned}$$

the last equality is by the Lebesgue dominant convergence theorem. We continue:

$$\begin{aligned}&\iint_0^T e^{-t|\xi|^2} \xi_1^2 \hat{\varphi}(\xi) \hat{\psi}(-\xi) d\xi dt \\ &\quad = -\int_0^T \int i\xi_1 \hat{\varphi}(\xi) e^{-\frac{1}{2}t|\xi|^2} \cdot i\xi \hat{\psi}(-\xi) e^{-\frac{1}{2}t|\xi|^2} d\xi dt \\ &\quad = \int_0^T \int i\xi_1 \hat{\varphi}(\xi) e^{-\frac{1}{2}t|\xi|^2} \cdot \overline{i\xi \hat{\psi}(\xi) e^{-\frac{1}{2}t|\xi|^2}} d\xi dt \\ &\quad = \int_0^T \int \frac{\partial u^\varphi}{\partial x}(z,t) \frac{\partial u^\psi}{\partial x}(z,t) dm(z) dt.\end{aligned}$$

Lemma is proved. □

3.5.2 How Martingale Representation of Ahlfors–Beurling Transform Can be Used

Let us write the following simple chain of equalities and inequalities with two smooth functions with compact support:

$$\begin{aligned}&(B_M \varphi, \psi) \qquad (3.5.19)\\ &= \lim_{T\to\infty} \int_{\mathbb{C}} \int_{\mathbb{C}} \mathbb{E}^w[(M^w \star \varphi)(T) | W^w(T) = z] \psi(z) h_T(z-w) dm(w) dm(z).\end{aligned}$$

It is sufficient to give the estimate for each fixed T of

$$\begin{aligned}&\int_{\mathbb{C}} \int_{\mathbb{C}} |\mathbb{E}^w\,[(M^w \star \varphi)(T) | W^w(T) = z]|\, \psi(z) | h_T(z-w) dm(w) dm(z) \\ &\quad = \int_{\mathbb{C}} |\mathbb{E}^w\,[|\psi(W^w(T)| |(M^w \star \varphi)(T)|]|\, dm(w)\end{aligned}$$

$$\le \int_{\mathbb{C}} \left(\mathbb{E}^w(|\psi(W^w(T)|^q)\right)^{1/q} \left(\mathbb{E}^w(|(M^w \star \varphi)(T)|^p)\right)^{1/p} dm(w)$$

$$\le \left(\int_{\mathbb{C}} \mathbb{E}^w(|\psi(W^w(T)|^q) dm(w)\right)^{1/q}$$

$$\left(\int_{\mathbb{C}} \mathbb{E}^w(|(M^w \star \varphi)(T)|^p) dm(w)\right)^{1/p}$$

The estimate of the first term is easy:

$$\int_{\mathbb{C}} \mathbb{E}^w(|\psi(W^w(T)|^q) dm(w) = \int_{\mathbb{C}}\int_{\mathbb{C}} |\psi(z)|^q h_T(z-w) dm(z) dm(w) = \|\psi\|_q^q. \tag{3.5.20}$$

Now we wish to estimate the second term.

THEOREM 3.5.8 *If* $\mathbb{E}^w|M^w \star \varphi|^p \le C(p)\mathbb{E}^w|I^w \star \varphi|^p$ *for all* T *then with the same constant* $C(p)$*, we have* $\|B_M\|_p \le C(p)$.

PROOF From the calculation above, we see that

$$\int_{\mathbb{C}} |\mathbb{E}^w\left[(M^w \star \varphi)(T)|W^w(T) = z\right]| \, |\psi(z)| h_T(z-w) dm(w) dm(z)$$

$$\le C(p)\|\psi\|_q \left(\int_{\mathbb{C}} \mathbb{E}^w(|I^w \star \varphi(T)|^p) dm(w)\right)^{1/p}$$

$$\le C(p)\|\psi\|_q \left(\int_{\mathbb{C}} \mathbb{E}^w(|\varphi(W^w(T) - u^\varphi(w,T)|^p) dm(w)\right)^{1/p}.$$

We applied (3.5.7) in the last line.

Now we use that $\int_{\mathbb{C}} |u^\varphi(w,T)|^p dm(w) \le \frac{C}{T^{p-1}}$. We are left with inequality

$$\int_{\mathbb{C}} |\mathbb{E}^w\left[(M^w \star \varphi)(T)|W^w(T) = z\right]| \, \psi(z)| dm(w)| dm(z)$$

$$\le C(p)\|\psi\|_q \left(\int_{\mathbb{C}} \mathbb{E}^w(|\varphi(W^w(T)|^p) dm(w)\right)^{1/p} + C(p)\|\psi\|_q \frac{C}{T^{1/q}}$$

$$\le C(p)\|\psi\|_q\|\varphi\|_p + C(p)\|\psi\|_q \frac{C}{T^{1/q}}, \tag{3.5.21}$$

where we used (3.5.20) in the last inequality.

Thus,

$$\limsup_{T\to\infty} \int_{\mathbb{C}} |\mathbb{E}^w [(M^w \star \varphi)(T)|W^w(T) = z]| \, |\psi(z)| h_T(z-w) dm(w) dm(z)$$
$$\leq C(p) \|\psi\|_q \|\varphi\|_p. \tag{3.5.22}$$

We have just reduced the problem of estimating operator B_M to (3.5.23) below. □

What can be the reason for the inequality

$$\mathbb{E}^w |M^w \star |\varphi^p \leq C(p) \mathbb{E}^w |I^w \star \varphi|^p \tag{3.5.23}$$

to be true for all T?

The reason is simple, we know that $M^w \star \varphi$ is a martingale transform of $I^w \star \varphi$. Clearly martingale $M^w \star \varphi$ is dominated by martingale $I^w \star \varphi$ if the operator norm of M has the following estimate: $\|M\| \leq 1$. Moreover, some extra symmetry properties, for example, the orthogonality of $M^w \star \varphi$ (for special M) also helps the estimate (3.5.23).

Therefore, we have for matrix $M = A$ from (3.5.10)

$$\|B\|_p \leq \|A\|(p^* - 1) = 2(p^* - 1), \tag{3.5.24}$$

where $\|A\|$ is the matrix norm of A, because $\|A\| = 2$.

Using $M = \begin{bmatrix} 1, & 0 \\ 0, & -1 \end{bmatrix}$ or $M = \begin{bmatrix} 0, & i \\ i, & 0 \end{bmatrix}$ whose norms are 1, we obtain the following estimates for the real and imaginary parts of operator B:

$$\|R_1^2 - R_2^2\|_p \leq (p^* - 1), \ \|2R_1R_2\|_p \leq (p^* - 1). \tag{3.5.25}$$

Remark 3.5.9 Estimate in (3.5.24) is far from being sharp. Estimates in (3.5.25) are actually equalities. This was proved in [**62**] and then in [**18**] by a different method that uses the language of laminates, which is an important notion from the calculus of variations.

Here is another important remark.

Remark 3.5.10 If we know extra properties of martingale $M \star g$, for example, that it is conformal, then the constant $(p^* - 1)$ immediately gets improved. See, e.g., Theorems 3.2.1, 3.2.4.

Theorem 3.5.11

$$p > 2 \Rightarrow \|B\varphi\|_p \leq \sqrt{2p(p-1)} \|\varphi\|_p, \tag{3.5.26}$$

PROOF Consider the complex-valued martingale $Y = A \star f = Y_1 + iY_2$ given in (3.5.11). Let $f = \varphi_1 + i\varphi_2$ be the splitting to the real and imaginary parts. Then we see immediately that

$$\nabla Y_1 = \Re \begin{bmatrix} (\partial_x + i\partial_y)(u^{\varphi_1} + iu^{\varphi_2}) \\ i(\partial_x + i\partial_y)(u^{\varphi_1} + iu^{\varphi_2}) \end{bmatrix} (W(t), T - t)$$

So at the space–time point $(W(t), T - t)$

$$\nabla Y_1 = \begin{bmatrix} u_x^{\varphi_1} - u_y^{\varphi_2} \\ -u_y^{\varphi_1} - u_x^{\varphi_2} \end{bmatrix}.$$

Similarly,

$$\nabla Y_2 = \Im \begin{bmatrix} (\partial_x + i\partial_y)(u^{\varphi_1} + u^{\varphi_2}) \\ i(\partial_x + i\partial_y)(u^{\varphi_1} + u^{\varphi_2}) \end{bmatrix} (W(t), T - t)$$

So

$$\nabla Y_2 = \begin{bmatrix} u_y^{\varphi_1} + u_x^{\varphi_2} \\ u_x^{\varphi_1} - u_y^{\varphi_2} \end{bmatrix}.$$

Obviously, we have pointwise equality $|\nabla Y_1| = |\nabla Y_2|$, and these vectors are orthogonal in $\mathbb{R}^2$. Thus, Y is a conformal martingale and theorem follows from Theorem 3.2.1. □

3.5.3 The Real Part of $B\varphi$

If $\varphi = \varphi_1 + i\varphi_2$, then the real part of $B\varphi$ is

$$\Re(B\varphi) = (R_1^2 - R_2^2)\varphi_1 - 2R_1R_2\varphi_2$$

We introduce

$$\tau_p := \left(\frac{1}{2\pi} \int_0^{2\pi} |\cos(\theta)|^p d\theta \right)^{-1/p}.$$

Next we give the following elementary lemma first proved in [**54**]. This lemma is also crucial for our proof of the new asymptotic estimate on $\|B\|_p$.

LEMMA 3.5.12 $\|B\|_p \le \tau_p \sup_{\|\varphi\|_p = 1} \|\Re(B\varphi)\|_p$.

PROOF Let B_θ denote the operator $e^{-i\theta}B$, $\theta \in [0, 2\pi)$. For any $z \in \mathbb{C}$, observe that

$$\begin{aligned} \Re(B_\theta\varphi)(z) &= \Re(B\varphi)(z)\cos\theta + \Im(B\varphi)(z)\sin\theta \\ &= |B\varphi(z)|\cos(\theta - \delta(z)), \end{aligned}$$

for some angle $\delta(z)$ depending on z. Taking the absolute value, raising to the p-th power, and averaging over θ gives us then the following:

$$\frac{1}{2\pi}\int_0^{2\pi} |\Re(B_\theta\varphi)(z)|^p d\theta = |B\varphi(z)|^p \tau_p^{-p}.$$

Now integrate both sides with respect to z to get

$$\tau_p^{-p}\|B\varphi\|_p^p = \frac{1}{2\pi}\int_0^{2\pi} \|\Re(B_\theta\varphi)\|_p^p \, d\theta.$$

It is clear that

$$\sup_{\|\varphi\|_p=1} \|\Re(B_\theta\varphi)\|_p = \sup_{\|\varphi\|_p=1} \|\Re(B\varphi)\cos\theta + \Im(B\varphi)\sin\theta\|_p$$

is constant in θ. Thus, we have

$$\begin{aligned}
\|B\|_p^p = \sup_{\|\varphi\|_p=1} \|B\varphi\|_p^p &= \frac{\tau_p^p}{2\pi} \sup_{\|\varphi\|_p=1} \int_0^{2\pi} \|\Re(B_\theta\varphi)\|_p^p d\theta \\
&= \frac{\tau_p^p}{2\pi} \sup_{\|\varphi\|_p=1} \int_0^{2\pi} \|\Re(B(e^{i\theta}\varphi)\|_p^p d\theta \\
&\leq \frac{\tau_p^p}{2\pi} \int_0^{2\pi} \sup_{\|\varphi\|_p=1} \|\Re(B(e^{i\theta}\varphi)\|_p^p d\theta \\
&= \tau_p^p \sup_{\|\varphi\|_p=1} \|\Re(B\varphi)\|_p^p.
\end{aligned}$$

□

3.5.4 Proof of Theorem 3.5.1

Here we use the notation $\|B\|_{L^{p'}(\mathbb{C},\mathbb{R})}$ for the norm of the operator B on the space of $L^{p'}$-integrable real-valued functions defined on $\mathbb{C}$.

LEMMA 3.5.13 *Let $1 < p < \infty$. For $\varphi \in L^p(\mathbb{C})$, we have*

$$\|\Re(B\varphi)\|_p \leq \|B\|_{L^{p'}(\mathbb{C},\mathbb{R})}\|\varphi\|_p.$$

PROOF In the following, ψ denotes a real-valued function. Let

$$T_1 \stackrel{\text{def}}{=} R_1^2 - R_2^2, \; T_2 \stackrel{\text{def}}{=} 2R_1R_2.$$

$$\begin{aligned}
\|\Re(B\varphi)\|_p &= \sup_{\|\psi\|_{p'}=1} \int (T_1\varphi_1 - T_2\varphi_2)\psi \\
&= \sup_{\|\psi\|_{p'}=1} \int (\varphi_1 T_1\psi - \varphi_2 T_2\psi) \\
&= \sup_{\|\psi\|_{p'}=1} \int (\varphi_1, \varphi_2) \cdot (T_1\psi, -T_2\psi) \\
&\leq \|\varphi\|_p \sup_{\|\psi\|_{p'}=1} \|B\psi\|_{p'} = \|\varphi\|_p \|B\|_{L^{p'}(\mathbb{C},\mathbb{R})}.
\end{aligned}$$

□

If ψ is real valued, then $A \star \psi$ is a conformal martingale (this is true even for complex-valued ψ), and on real vectors $\|A\| \leq \sqrt{2}$.

Hence, by Theorem 3.2.4, we have

$$\|B\|_{L^{p'}(\mathbb{C},\mathbb{R})} \leq \frac{z_{p'}}{1 - z_{p'}}.$$

By Lemmas 3.5.12 and 3.5.13, we get

$$\|B\|_p \leq \tau_p \|B\|_{L^{p'}(\mathbb{C},\mathbb{R})} \leq \tau_p \frac{z_{p'}}{1 - z_{p'}}. \tag{3.5.27}$$

It remains to use Wallis' formula:

$$\tau_{2n}^{-2n} = \frac{1}{2\pi} \int_0^{2\pi} |\cos(\theta)|^{2n} d\theta = \left(\frac{1 \cdot 3 \cdot \ldots \cdot (2n-1)}{2 \cdot 4 \cdot \ldots \cdot (2n)} \right)^2 > \frac{2}{(2n+1)\pi}.$$

By monotonicity of $\frac{1}{2\pi} \int_0^{2\pi} |\cos(\theta)|^p d\theta$, we obtain

$$\frac{1}{2\pi} \int_0^{2\pi} |\cos(\theta)|^p d\theta > \frac{2}{(p+3)\pi},$$

and hence

$$\tau_p < \left(\frac{p+3}{2} \pi \right)^{1/(2p)};$$

this together with (3.5.27) and (3.4.11) proves Theorem 3.5.1.

REMARK 3.5.14 It is interesting to note that Theorem 3.5.11 uses Theorem 3.2.1 in its left conformality part for $p > 2$, and Theorem 3.5.1 uses a complementary Theorem 3.2.4 also in its left conformality part but for $p < 2$.

4

Dyadic Models: Application of Bellman Technique to Upper Estimates of Singular Integrals

4.1 Dyadic Shifts and Calderón–Zygmund Operators

This chapter is devoted to several applications of the Bellman function technique to (weighted) estimates of Calderón–Zygmund operators. In particular, we have in mind the application to nonhomogenous weighted Calderón–Zygmund operators. Here the term "nonhomogeneous" is used to emphasize that the underlying measure, with respect to which we will integrate and with respect to which we consider the weights, can be very far from the nice Lebesgue measure. It might be non-doubling, for example.

The harmonic analysis of Calderón–Zygmund operators with respect to such measures allowed mathematicians to achieve several long-standing goals. For example, it allowed Tolsa [**179**] to solve Painlevé's problem.

This type of harmonic analysis got the name "nonhomogeneous harmonic analysis." It was developed for the purpose of solving Painlevé's problem mentioned above, in a sequence of papers by F. Nazarov. S. Treil, A. Volberg [**129–133**] and Tolsa [**179**], see also the books [**180, 194**].

Papers [**129–132**] were concerned with unweighted nonhomogeneous Calderón–Zygmund operators. But weighted theory represents an extra challenge.

We will look at certain weighted problems in homogeneous and nonhomogeneous situations from the perspective that Calderón–Zygmund operators are combinations of dyadic shifts. This point of view is due to T. Hytönen's paper [**72**], in which he proved the A_2 theorem, but some of its technical elements were already widely used in [**128–132**].

Another point of view is that Calderón–Zygmund operators can be dominated by the so-called sparse dyadic shifts. This approach was invented by Lerner [**105–109**] and was recently developed in a huge number of papers, of which we will mention only [**43**], [**44**], [**96**], and [**195**]. Sparse shifts are positive operators, and that is their big advantage over dyadic shifts. Dyadic shifts are "simple" operators, they have a nice dyadic structure, but they have signed kernels. Even being simple and dyadic, they are still essentially singular operators.

However, we will not use the sparse domination approach here. Firstly, it is not easy to "Bellmanize" it; secondly, the dyadic shifts approach gives us the decomposition of Calderón–Zygmund operators, while the sparse domination approach gives us the majorization result. Of course this majorization suffices for most estimates. But when one is concerned with the estimate of the Calderón–Zygmund operators from below, or when one is concerned with the estimate of the Calderón–Zygmund operators that is similar to various unweighted or weighted $T1$ theorems (see about $T1$ theorems, e.g., in [**48, 97, 100, 132, 194**]), the approach via dyadic shifts seems to be merited.

Also, although nonlinear methods of sparse domination appear to yield stronger results, at least in questions around the A_2 theorem, there is still independent interest toward linear representations by dyadic shifts, which are better suited, e.g., for iterative applications, as in [**47**], or multiparameter extensions, as in [**115, 145**]. It is also of some theoretical interest whether the nonlinear methods are fundamentally stronger, or whether the same results could be recovered via linear representations.

Besides, the dyadic shifts are quite beautiful objects, whose study seems to have an intrinsic interest.

4.1.1 Probability Space of Random Dyadic Lattices: Nonhomogeneous Dyadic Shifts

Consider a positive finite measure μ supported on a compact set E lying in $\frac{1}{2}Q_0$. By Q_0, we denote the unit cube centered at the origin. By $\frac{1}{2}Q_0$, we denote the cube of side-length $\frac{1}{2}$ centered at the origin of $\mathbb{R}^d$. We recall a construction of probability space of dyadic lattices made in [**128, 132**]. Suppose $\mathcal{D}_N$ denotes some dyadic grid of squares of size 2^{-N} of $\mathbb{R}^d$. Now for each such Q there are 2^d choices of its father. As soon as one father is chosen, all others are fixed as parallel translations of it. Consider all choices of grids of fathers as equally probable. Now having one grid of fathers fixed, consider 2^d choices of their fathers. Choose one independent of the previous choice and again let all choices be equally probable. Continuing in the same fashion, we build $\mathcal{D}_{N-1}(\omega)$, $\mathcal{D}_{N-2}(\omega)$, etc.

Notice that grids $\mathcal{D}_k$, $k > N$, are also built just by splitting each cube of the initial grid $\mathcal{D}_N$ to 2^d dyadic sons to obtain the grid $\mathcal{D}_{N+1}$, and then by repeated splitting, grids $\mathcal{D}_N, \mathcal{D}_{N+1}, \dots$ are uniquely defined. But grids $\mathcal{D}_0, \mathcal{D}_1, \dots, \mathcal{D}_{N-1}$ are defined by our random procedure above. Hence, we build a huge (but finite) number of dyadic lattices $\mathcal{D} = \cup_{k=0}^{\infty} \mathcal{D}_k$, and we consider them as a probability space by assigning the same probability to each lattice.

We consider the set of lattices above as a probability space, we write it as $(\Omega_0, \mathcal{P}_0)$, and we identify $\omega \in \Omega_0$ with a corresponding lattice. Sometimes we write it as $\mathcal{D}(\omega)$.

In the natural probability space of lattices $(\Omega_0, \mathcal{P}_0)$ just built, consider a subset Ω of lattices, such that our cube $\frac{1}{2}Q_0$ (containing $\operatorname{supp} \mu$) is contained in some $Q \in \mathcal{D}_0$. Put $\mathcal{P} = \frac{\mathcal{P}_0}{\mathcal{P}_0(\Omega)} \mathbf{1}_\Omega$. The probability space $(\Omega, \mathcal{P})$ is what we will be using now.

In what follows, all $\mathcal{D} = \mathcal{D}(\omega) = \cup_{k=0}^{\infty} \mathcal{D}_k$ are from $(\Omega, \mathcal{P})$. We recall that Ω has finitely many elements and $\mathcal{P}$ just assigns equal probability to each of them. So let Q be in such a $\mathcal{D}$ and let Q_i, $i = 1, \dots, 2^d$ be its children. For any $f \in L^1(\mu)$, we denote $\mathcal{E}_Q f = \langle f \rangle_{1,\mu} \mathbf{1}_Q$,

$$\mathcal{E}_k := \sum_{Q \in \mathcal{D}_k} \mathcal{E}_Q f,$$

and

$$\Delta_k f = (\mathcal{E}_{k+1} - \mathcal{E}_k) f, \ \Delta_Q f := \Delta_k f \cdot \mathbf{1}_Q, Q \in \mathcal{D}_k.$$

Now let $f \in L_0^2(\mu)$. Here subscript 0 means that $\int f \, d\mu = 0$. Then

$$f = \sum_{Q \in \mathcal{D}} \Delta_Q f,$$

and

$$\Delta_Q f = \sum_{i=1}^{2^d - 1} (f, h_Q^i)_\mu h_Q^i,$$

where h_Q^i are called μ-Haar functions and they have the following properties:

- $(h_Q^i, h_R^j)_\mu = 0, Q \neq R,$
- $(h_Q^i, h_Q^j)_\mu = 0, i \neq j,$
- $\|h_Q^i\|_\mu = 1,$
- $h_Q^i = \sum_{m=1}^{2^d - 1} c_{Q,m}^i \mathbf{1}_{Q_m},$
- $|c_{Q,m}^i| \leq 1/\sqrt{\mu(Q_m)}.$

In the abovementioned, Q_m are children of Q. In the following, we will start to skip index i. We already abbreviated $h_Q^{i,\mu}$, as h_Q^i. Now we will abbreviate it further, as h_Q, unless stated otherwise.

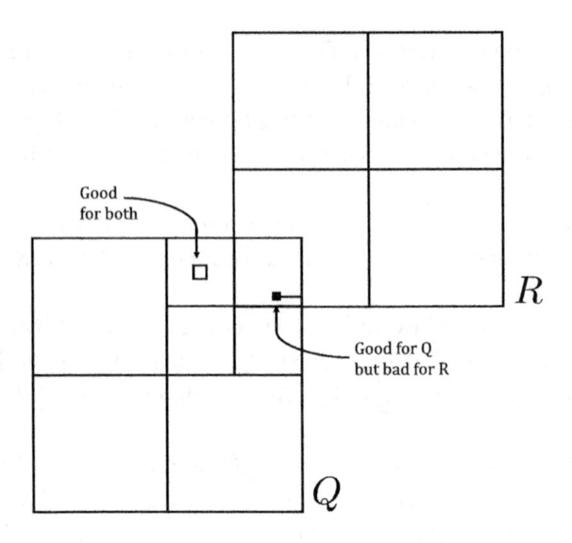

Figure 4.1 Good for one lattice and bad for another.

DEFINITION 4.1.1 Cube $Q \in \mathcal{D}(\omega)$ that has its side-length at least 2^{-N} and at most 1, is called **good** ((r,γ)-good) if for any R in the same $\mathcal{D}(\omega)$, but such that $\ell(R) \geq 2^r \ell(Q)$, one has

$$\operatorname{dist}(Q, \operatorname{sk}(R)) \geq \ell(R)^{1-\gamma}\ell(Q)^{\gamma}, \tag{4.1.1}$$

where $\operatorname{sk}(R) := \bigcup_{m=1}^{2^d-1} \partial Q_m$. Again, Q_m are all children of Q.

Given $Q \in \mathcal{D}_k$, we denote $g(Q) = k$, which is the generation of Q. Our main "tool" is going to be the famous "dyadic shifts." But they will be with respect to nonhomogenous measure. Their typical building blocks will be Haar projections with respect to nonhomogeneous measure μ. This is the only slight difference from [**72, 73, 76**].

DEFINITION 4.1.2 We denote by $\mathbb{S}_{m,n}$ (shift of complexity (m,n) or shift of complexity $\max(m,n)$), the operator given by the kernel

$$f \to \sum_{L \in \mathcal{D}} \int_L a_L(x,y) f(y)\, d\mu(y),$$

where

$$a_L(x,y) = \sum_{\substack{Q \subset L, R \subset L \\ g(Q)=g(L)+m,\, g(R)=g(L)+n}} c_{L,Q,R} h_Q^i(x) h_R^j(y), \tag{4.1.2}$$

where $h^i_Q := h^{\mu,i}_Q$, $h^j_R := h^{\mu,j}_R$ are Haar functions orthogonal and normalized in $L^2(d\mu)$, and $|c_{L,Q,R}|$ are such that

$$\sum_{Q,R} |c_{L,Q,R}|^2 \leq 1. \tag{4.1.3}$$

Often we will skip superscripts i, j. One will always skip superscript μ.

REMARK 4.1.3 In particular, it is easy to see that if a_L has the form (4.1.2) and satisfies

$$|a_L(x,y)| \leq \frac{1}{\mu(L)}, \tag{4.1.4}$$

then (4.1.3) is automatically satisfied, and we know that we are dealing with dyadic shift. In fact, the above sum of squares is the Hilbert–Schmidt norm of operator with kernel a_L. Inequality (4.1.4) allows us to see immediately that the Hilbert–Schmidt norm is bounded by 1.

REMARK 4.1.4 Sometimes in literature on dyadic shifts, the following is slightly different, but basically an equivalent definition is used: the operator, which sends $\Delta^n_L(L^2(\mu))$ to itself, and has the kernel $a_L(x,y)$ satisfying estimate (4.1.4), is called a local dyadic shift of order n. Here $\Delta^n_L(L^2(\mu))$ denotes the space of $L^2(\mu)$ functions supported on L and having constant values on children Q of L, such that $g(Q) = g(L) + n + 1$. Now dyadic shift of order n is an operator of the form $\mathbb{S}_n f := \sum_{L\in\mathcal{D}} \int_L a_L(x,y) f(y)dy$, where a_L corresponds to local shift of order n.

All these definitions bring us operators, which obviously satisfy the following:

$$\|\mathbb{S}_{m,n}\|_{L^2(\mu)\to L^2(\mu)} \leq 1, \tag{4.1.5}$$

$$\|\mathbb{S}_n\|_{L^2(\mu)\to L^2(\mu)} \leq n+1. \tag{4.1.6}$$

We also need to define generalized shifts, but only of complexity $(0, 1)$.

DEFINITION 4.1.5 (Generalized shifts, paraproducts, and their adjoint operators). Let the operator be given by

$$\Pi f := \sum_{L\in\mathcal{D}} \langle f\rangle_{L,\mu} \sqrt{\mu(L)} \sum_{\ell\subset L, |\ell|=2^{-s}|L|} \sum_j c_{L,\ell,j} \cdot h^j_\ell,$$

where $\{c_{L,\ell}\}$ satisfy *not just* the condition $\mu(L) \sum_{\ell\subset L, |\ell|=2^{-s}|L|} |c_{L,\ell}|^2 \leq 1$, that would be required for the usual $(0, s)$-shift normalization condition, but a rather stronger Carleson condition:

$$\sum_{L\subset R,\, L\in\mathcal{D}} \mu(L) \sum_{\ell\subset L, |\ell|=2^{-s}|L|} |c_{L,\ell}|^2 \le \mu(R). \tag{4.1.7}$$

Then Π is called a generalized shift of complexity $(0, s)$. Generalized shifts of complexity $(0, 1)$ will be called *paraproducts*. Operators adjoint to paraproducts Π will be denoted by Π^*.

4.1.2 Main Theorem

We are in a position to formulate our main results. Let us recall in what case an operator T is called an operator with a Calderón–Zygmund kernel of order m. First we give a more general definition, and for that we need a notion of *geometrically doubling metric space.*

DEFINITION 4.1.6 A metric space $(X, |\cdot|)$ is called geometrically doubling if every ball $B(x, r)$ can be covered by at most $N < \infty$ balls of radius $r/2$, and N does not depend on either x or r.

It is easy to see that if there exists a positive measure μ, $\operatorname{supp}\mu = X$, such that it has the doubling property $\mu(B(x, 2r)) \le M\,\mu(B(x, r))$ (with universal $M < \infty$), then that metric measure space $(X, |\cdot|, \mu)$ is geometrically doubling. Conversely, any geometrically doubling metric space carries such doubling measure μ. This is harder to prove; see **[90]**.

DEFINITION 4.1.7 Let X be a geometrically doubling metric space. Let $\lambda(x, r)$ be a positive function, increasing and doubling in r, i.e., $\lambda(x, 2r) \leqslant C\lambda(x, r)$, where C does not depend on x or r.

We call $K(x, y)\colon X \times X \to \mathbb{R}$ a Calderón–Zygmund kernel, associated with function λ, if

$$\begin{aligned}
|K(x,y)| &\leqslant \min\left(\frac{1}{\lambda(x,|xy|)}, \frac{1}{\lambda(y,|xy|)}\right),\\
|K(x,y)-K(x',y)| &\leqslant \frac{|xx'|^{\varepsilon}}{|xy|^{\varepsilon}\lambda(x,|xy|)}, \quad |xy| \geqslant C|xx'|, \\
|K(x,y)-K(x,y')| &\leqslant \frac{|yy'|^{\varepsilon}}{|xy|^{\varepsilon}\lambda(y,|xy|)}, \quad |xy| \geqslant C|yy'|.
\end{aligned} \tag{4.1.8}$$

By $B(x, r)$, we denote the ball in $|\cdot|$ metric, i.e., $B(x, r) = \{y \in X : |yx| < r\}$.

Let μ be a measure on X such that $\mu(B(x, r)) \leqslant C\lambda(x, r)$, where C does not depend on x or r. We say that T is an operator with a Calderón–Zygmund kernel K on our metric space X if

$$Tf(x) = \int K(x, y) f(y) d\mu(y), \ \forall x \notin \operatorname{supp}\mu, \ \forall f \in C_0. \tag{4.1.9}$$

We call T a Calderón-Zygmund operator, if it is an operator with a Calderón–Zygmund kernel, and also

$$T \text{ is bounded } L^2(\mu) \to L^2(\mu).$$

Let us notice that $\mathbb{R}^d$ is of course a geometrically doubling metric space with the usual Euclidean metric. Apply the above definition to

$$\lambda(x,r) = r^m, \ |xy| = |x-y|.$$

Then we have a Calderón–Zygmund kernel of order m in $\mathbb{R}^d$ and Calderón–Zygmund operators of order m in $\mathbb{R}^d$. Measure μ, satisfying $\mu(B(x,r)) \leq Cr^m$ (or more generally $\mu(B(x,r)) \leqslant C\lambda(x,r)$), but not satisfying any estimate from below is called *nonhomogeneous*. The corresponding Calderón–Zygmund operator then is also called *nonhomogeneous*.

THEOREM 4.1.8 *Let $\mu(B(x,r)) \leq r^m$. Let T be an operator with a Calderón–Zygmund kernel of order m in $\mathbb{R}^d$. Let it be defined on characteristic functions of cubes. Let the following conditions be satisfied:*

$$\|T1_Q\|_{2,\mu}^2 \leq C_0\mu(Q), \quad \text{for all cubes } Q, \tag{4.1.10}$$

$$\|T^*1_Q\|_{2,\mu}^2 \leq C_0\mu(Q), \quad \text{for all cubes } Q. \tag{4.1.11}$$

Then there exists a probability space of dyadic lattices $(\Omega, \mathcal{P})$ and constants $c_{1,T}, c_{2,T}, c_{3,T}, \varepsilon_T$, such that

$$\begin{aligned} T = c_{1,T}\int_\Omega \Pi(\omega)\, d\mathcal{P}(\omega) + c_{2,T}\int_\Omega \Pi^*(\omega)\, d\mathcal{P}(\omega) \\ + c_{3,T}\sum_{n=0}^\infty 2^{-n\varepsilon_T}\int_\Omega \mathbb{S}_n(\omega)\, d\mathcal{P}(\omega). \end{aligned} \tag{4.1.12}$$

Moreover, $\varepsilon_T > 0$. Constants $c_{1,T}, c_{2,T}, c_{3,T}$ are bounded from above by constants depending only on Calderón–Zygmund parameters of the kernel, on m and d, and on the best constants in the so-called $T1$ conditions (4.1.10), (4.1.11). *Moreover, $\varepsilon_T > 0$, and its bound from below depends only on the Calderón–Zygmund parameters of the kernel, on m and d, and on the best constants in conditions* (4.1.10), (4.1.11).

REMARK 4.1.9 The same thing holds for general geometrically doubling metric space X (not just $\mathbb{R}^d$) and any nonhomogeneous Calderón–Zygmund operator having a Calderón–Zygmund kernel in the generalized sense above. Of course, the measure should satisfy

$$\mu(B(x,r)) \leqslant C\lambda(x,r).$$

We prefer to prove the $\mathbb{R}^d$ version, just for the sake of avoiding some technicalities. For example, the construction of the suitable probability space of random dyadic lattices on X is a bit more involved than such construction in $\mathbb{R}^d$. See two different constructions of suitable probability spaces of dyadic lattices in [**74, 125**].

We immediately obtain the nonhomogeneous $T1$ theorem:

COROLLARY 4.1.10 *Let operator T be a Calderón–Zygmund operator of order m. Let μ be a measure of growth, which is at most m: $\mu(B(x,r)) \le r^m$. Then $\|T\|_{2,\mu} \le C(C_0, d, m, \varepsilon)$, where ε is from Definition 4.1.7, and C_0 is from $T1$ assumptions* (4.1.10), (4.1.11).

REMARK 4.1.11 The $T1$ theorem has a long history. For $\mu = m_d$, it was proved by G. David, J.-L. Journé [**48**]. For homogeneous (doubling) measures μ, it was proved by M. Christ [**39**]. In the case of nonhomogeneous μ, the proof is involved, but it has a relatively short exposition in [**128, 192**]. Here we follow the lines of [**128**] to prove the decomposition to random dyadic shift in Theorem 4.1.8. Just [**128**] is not quite enough for that goal, and we use a beautiful step of Hytönen as well. Then the nonhomogeneous $T1$ theorem is just a corollary of the decomposition result, because all shifts of order n, involved in (4.1.12) have norms at most $n+1$ (see the discussion above), but decomposition (4.1.12) has an exponentially decreasing factor.

4.1.3 Proof of (1.12)

Good cubes were introduced above in Definition 4.1.1. The cube is **bad** if it is not good.

LEMMA 4.1.12 $\mathcal{P}\{Q \text{ is bad}\} \le C_1 2^{-c_2 r}$.

PROOF By the definition of badness, it is easy to see that this probability can be estimated by the sum of volumes of $2^{-\gamma s}$-neighborhoods of the boundary of a unit cube, where s runs over all integers greater or equal to r. The reader can easily understand why the unit cube is involved by using the scaling invariance. □

4.1.3.1 We can Think That $\mathcal{P}\{Q$ is good$\}$ Is Independent of Q. This would be obvious if we did not pass from Ω_0 to Ω. In that case, no change would be needed.

Let us consider all cubes of all lattices of Ω that have side-length of at least 2^{-N} and at most 1 that lie at a distance of at most 1 from Q_0. We call the finite family of such cubes $\mathcal{F}$.

Now we will modify our probability space of dyadic lattices $(\Omega, \mathcal{P})$. The new space will be called $(\hat{\Omega}, \mathbb{P})$. The elementary event again will be a dyadic lattice $\hat{\mathcal{D}}$. And $\hat{\Omega}$ will be a finite set of such lattices. Probability $\mathbb{P}$ will be uniformly distributed. Let us consider all cubes of all such lattices that have side-length of at least 2^{-N} and at most 1, that lie at distance of at most 1 from Q_0. We will see that in our construction below, this finite family coincides with $\mathcal{F}$.

We will also slightly modify the notion of goodness. In this new space, the following property will be satisfied:

$$\mathbb{P}(Q \text{ is good}|\, Q \in \mathcal{F}) = \frac{1}{2}. \tag{4.1.13}$$

We already know that if r is large and fixed, and if Q is one of the cubes of one of the dyadic lattice of Ω whose side-length is at least 2^{-N} and at most 1, then

$$p_Q \stackrel{\text{def}}{=} \mathcal{P}(Q \text{ is good}) \geq 1 - C_1 2^{-c_2 r} \geq \frac{3}{4}. \tag{4.1.14}$$

Notice that p_Q is just the ratio, where in the numerator we have the number N_Q of lattices from Ω, in which Q is good, and in the denominator we have the number D_Q of all lattices in Ω that contain Q, as one of its cubes. Thus, $N_Q \leq D_Q \leq \frac{4}{3} N_Q$.

Now for any given $\mathcal{D}$ that contains Q, we consider $2N_Q$ copies of it. We choose (arbitrarily) D_Q among these copies. We call these chosen copies red, other copies are called blue. Now we change the meaning of "good." Cube Q will be called *really good* if it is a good cube of a red copy. If it is a bad cube or a good cube of a blue copy, we shall call it bad.

We need one more step. The previous step was just for one cube $Q \in \mathcal{F}$. In fact, given $\mathcal{D}$, we need to consider $2N_Q$ copies for *every* cube from $\mathcal{D}$. This means that we actually consider $N_{\mathcal{D}} \stackrel{\text{def}}{=} \Pi_{Q \in \mathcal{D}} 2N_Q$ copies for each $\mathcal{D} \in \Omega$.

We do this copying in a natural way: enumerate $Q \in \mathcal{D} \cap \mathcal{F}$, make copies for Q_1, as above (Q_1-copies). As before, we choose D_{Q_1} red copies, the rest are blue. Now copy each such copy $2N_{Q_2}$ times, whether red or blue, and for each Q_1-copy choose D_{Q_2} copies among those Q_1Q_2-copies, and call them Q_2-red, the rest are Q_2-blue, etc. The ratio of red Q_1-copies among all copies will still be $\frac{D_{Q1}}{2N_{Q_1}}$. The ratio of red Q_2-copies among all copies will still be $\frac{D_{Q2}}{2N_{Q_2}}$. We continue this process till we list all of the cubes $Q \in \mathcal{F}$.

Now it is clear, that given $\mathcal{D} \in \Omega$ and $Q \in \mathcal{D}$, we have

$$\frac{\text{number of copies such that they are } Q - \text{red and } Q \text{ was good}}{\text{number of all copies containing } Q}$$
$$= \frac{D_Q}{2N_Q} p_Q = \frac{1}{2}.$$

This precisely means (4.1.13).

If we would not be pursuing the finite number of elements property of our probability space, then the above construction could have been made easier.

For every $Q \in \mathcal{F}$ take a random variable $\xi_Q(\omega')$, which is equally distributed on $[0, 1]$. We know that

$$\mathcal{P}(Q \text{ is good}) \geq p_Q.$$

We call Q "really good" if

$$\xi_Q \in [0, \frac{1}{2p_Q}].$$

Otherwise, Q joins bad cubes. Then the probability space is $\hat{\Omega} = \{\omega, \{\xi_Q(\omega')\}_{Q \in \mathcal{D}(\omega)}\}$ provided with the natural probability measure $\mathbb{P}$.

$$\mathbb{P}(Q \text{ is really good}) = \frac{1}{2},$$

and we are done.

We can call random variable $\xi_Q(\omega')$ "the temperature" of Q. For $Q \in \mathcal{F}$, we can interpret the event of "real goodness" as the property of being good in the sense of Definition 4.1.1 plus having not very high temperature.

REMARK 4.1.13 We do not want to use the "really good" expression below. But everywhere below when we write "good" we mean "really good" in the above sense. We need this only to have the probability of being good the same for all cubes.

4.1.4 The Whole Bilinear Form of the Operator and Its Good Part

Let $f, g \in L^2_0(\mu)$, having a constant value on each cube from $\mathcal{D}_N$. We can write

$$f = \sum_Q \sum_j (f, h^j_Q) h^j_Q, \quad g = \sum_R \sum_i (g, h^i_R) h^i_R.$$

First, we state and prove the theorem that says that essential part of bilinear form of T can be expressed in terms of pair of cubes, where the smallest one is good. This is almost what has been done in [**128**]. The difference is that in [**128**] an error term (very small) appeared. To eliminate the error term, we follow the idea of Hytönen [72]. In fact, the work [72] beautifully improved on "good-bad" decomposition of [128, 129, 132] by replacing inequalities by the equality and getting rid of the error term.

THEOREM 4.1.14 *Let T be any linear operator. Then the following equality holds:*

$$\frac{1}{2}\mathbb{E}\sum_{\substack{Q,R,i,j\\ \ell(Q)\geqslant\ell(R)}}(Th^j_Q,h^i_R)(f,h^j_Q)(g,h^i_R) =$$

$$\mathbb{E}\sum_{\substack{Q,R,i,j\\ \ell(Q)\geqslant\ell(R),\ R\text{ is good}}}(Th^j_Q,h^i_R)(f,h^j_Q)(g,h^i_R).$$

The same is true if we replace $\geqslant$ by $>$.

PROOF We denote

$$\sigma(T) = \sum_{\ell(Q)\geqslant\ell(R)}(Th^j_Q,h^i_R)(f,h^j_Q)(g,h^i_R).$$

$$\sigma'(T) = \sum_{\substack{\ell(Q)\geqslant\ell(R)\\ R\text{ is good}}}(Th^j_Q,h^i_R)(f,h^j_Q)(g,h^i_R).$$

We would like to get a relationship between $\mathbb{E}\sigma(T)$ and $\mathbb{E}\sigma'(T)$. We fix g and put

$$g_{good} := \sum_{R\text{ is good}}(g,h^i_R)h^i_R.$$

Then,

$$\sum_Q\sum_{R\text{ is good}}(Th^j_Q,h^i_R)(f,h^j_Q)(g,h^i_R)$$
$$=\Big(T(f),\sum_{R\text{ is good}}(g,h^i_R)h^i_R\Big) = (T(f),g_{good})\,.$$

Taking expectations, we obtain

$$\begin{aligned}\mathbb{E}\sum_{Q,R}(Th^j_Q,h^i_R)(f,h^j_Q)(g,h^i_R)\mathbf{1}_{R\text{ is good}} &= \mathbb{E}(T(f),g_{good})\\ &=(T(f),\mathbb{E}\,g_{good}) = \frac{1}{2}(T(f),g)\\ &=\frac{1}{2}\mathbb{E}\sum_{Q,R}(Th^j_Q,h^i_R)(f,h^j_Q)(g,h^i_R).\end{aligned} \tag{4.1.15}$$

Next, fix Q, R such that $\ell(Q) < \ell(R)$. Then the goodness of R does not depend on Q, and so

$$\frac{1}{2}(Th_Q^j, h_R^i)(f, h_Q^j)(g, h_R^i) = \mathbb{E}\Big((Th_Q^j, h_R^i)(f, h_Q^j)(g, h_R^i)\mathbf{1}_{R\,\text{is good}}|Q, R\Big).$$

Let us explain this equality. The right-hand side is conditioned: meaning that the left-hand side involves the fraction of two numbers: 1) the number of all lattices containing Q, R in it and such that R (the one that is larger by size) is good and 2) the number of lattices containing Q, R in it. This fraction is exactly $\frac{1}{2}$. The equality has been explained.

Now we fix a pair of Q, R, $\ell(Q) < \ell(R)$, and multiply both sides by the probability that this pair is in the same dyadic lattice from our family. This probability is just the ratio of the number of dyadic lattices in our family containing elements Q and R to the number of all dyadic lattices in our family. After multiplication, by this ratio and the summation of all terms with $\ell(Q) < \ell(R)$ we get finally,

$$\begin{aligned}&\frac{1}{2}\mathbb{E}\sum_{\ell(Q)<\ell(R)}(Th_Q^j, h_R^i)(f, h_Q^j)(g, h_R^i)\\&= \mathbb{E}\sum_{\ell(Q)<\ell(R)}(Th_Q^j, h_R^i)(f, h_Q^j)(g, h_R^i)\mathbf{1}_{R\,\text{is good}}.\end{aligned} \tag{4.1.16}$$

Now we use first (4.1.15) and then (4.1.16):

$$\begin{aligned}&\frac{1}{2}\mathbb{E}\sum_{Q,R}(Th_Q^j, h_R^i)(f, h_Q^j)(g, h_R^i)\\&= \mathbb{E}\sum_{Q,R}(Th_Q^j, h_R^i)(f, h_Q^j)(g, h_R^i)\mathbf{1}_{R\text{ good}}\\&= \mathbb{E}\Bigg(\sum_{\ell(Q)<\ell(R)}(Th_Q^j, h_R^i)(f, h_Q^j)(g, h_R^i)\mathbf{1}_{R\text{ good}}\\&\quad + \sum_{\ell(Q)\geqslant\ell(R)}(Th_Q^j, h_R^i)(f, h_Q^j)(g, h_R^i)\mathbf{1}_{R\text{ good}}\Bigg)\\&= \frac{1}{2}\mathbb{E}\sum_{\ell(Q)<\ell(R)}(Th_Q^j, h_R^i)(f, h_Q^j)(g, h_R^i)\\&\quad + \mathbb{E}\sum_{\ell(Q)\geqslant\ell(R),R\text{ good}}(Th_Q^j, h_R^i)(f, h_Q^j)(g, h_R^i),\end{aligned}$$

and therefore,

$$\mathbb{E} \sum_{\ell(Q)\geqslant\ell(R),\ R \text{ is good}} (Th^j_Q, h^i_R)(f, h^j_Q)(g, h^i_R) \\ = \frac{1}{2}\mathbb{E} \sum_{\ell(Q)\geqslant\ell(R)} (Th^j_Q, h^i_R)(f, h^j_Q)(g, h^i_R), \tag{4.1.17}$$

which is the statement of our theorem. □

Now we skip i, j for the sake of brevity. We have just reduced the estimate of the bilinear form $\sum_{Q,R\in\mathcal{D}}(Th_Q, h_R)(f, h_Q)(g, h_R)$ to the estimate over *all* dyadic lattices in our family, but summing over pairs Q, R, where the smaller in size is always *good*:

$$\mathbb{E} \sum_{Q,R\in\mathcal{D},\text{ smaller is good}} (Th_Q, h_R)(f, h_Q).$$

Split it into two "triangular" sums:

$$\mathbb{E} \sum_{Q,R\in\mathcal{D},\ \ell(R)<\ell(Q),\ R \text{ is good}} (Th_Q, h_R)(f, h_Q)(g, h_R)$$

and

$$\mathbb{E} \sum_{Q,R\in\mathcal{D},\ \ell(Q)\leq\ell(R),\ Q \text{ is good}} (Th_Q, h_R)(f, h_Q)(g, h_R).$$

They are basically symmetric, so we will work only with the second sum.

First consider $\sigma_0 := \mathbb{E}\sum_{Q\in\mathcal{D},\ Q \text{ is good}}(Th_Q, h_Q)(f, h_Q)(g, h_Q)$. We do not care where Q is good or not and estimate the coefficient (Th_Q, h_Q) in the most simple way. Recall that $h_Q = \sum_{j=1}^{2^d} c_{Q,j}1_{Q_j}$, where Q_j are children of Q. We also remember that $|c_{Q,j}| \leq 1/\sqrt{\mu(Q_j)}$. Estimating

$$|c_{Q,j}||c_{Q,j'}||(T1_{Q_j}, 1_{Q_{j'}})| \leq \frac{1}{\sqrt{\mu(Q_j)}}\frac{1}{\sqrt{\mu(Q_j)}}C_0^2\sqrt{\mu(Q_j)}\sqrt{\mu(Q_j)} \leq C_0^2$$

by (4.1.10), we can conclude that σ_0/C_0^2 actually splits to at most 4^d shifts of order 0.

Similarly, we can work with

$$\sigma_s := \mathbb{E} \sum_{Q,R\in\mathcal{D},\ Q\subset R,\ \ell(Q)=2^{-s}\ell(R),\ Q \text{ is good}} (Th_Q, h_R)(f, h_Q)(g, h_R). \tag{4.1.18}$$

for $s = 1, \ldots, r-1$. We need r to be large, but not too big, it depends on d only, and is chosen in (4.1.14).

4.1.5 Decomposition of the Inner Sum

Now we start to work with σ_s, $s \geq r$, from (4.1.18). Fix a pair Q, R, and let R_1 be a descendant of R such that $\ell(R_1) = 2^r \ell(Q)$. Consider the son R_2 of R_1 that contains Q. We know that Q is good, in particular,

$$\begin{aligned}\operatorname{dist}(Q, \partial R_2) \geq \operatorname{dist}(Q, \operatorname{sk}(R_1)) \geq \ell(R_1)^{1-\gamma}\ell(Q)^{\gamma} &= \ell(R_1)2^{-r\gamma}\\ &= 2 \cdot 2^{-r\gamma}\ell(R_2).\end{aligned}$$

Number γ will be chosen to be less than 1, so

$$\operatorname{dist}(Q, \partial R_2) \geq 2^{-r}\ell(R_1) = \ell(Q). \tag{4.1.19}$$

We want to estimate (Th_Q, h_R).

Lemma 4.1.15 *Let $Q \subset R$, $S(R)$ be the son of R containing Q, and let $\operatorname{dist}(Q, \partial S(R)) \geq \ell(Q)$. Let T be a Calderón–Zygmund operator with parameter ε in (4.1.7). Then*

$$(Th_Q, h_R) = \langle h_R \rangle_{S(R)} (h_Q, \Delta_Q T^* 1)_\mu + t_{Q,R},$$

where

$$|t_{Q,R}| \leq \int_Q \int_{R\setminus S(R)} \frac{\ell(Q)^\varepsilon}{(\operatorname{dist}(t,Q) + \ell(Q))^{m+\varepsilon}} |h_Q(s)||h_R(t)|\, d\mu(s)|\, d\mu(t)$$

$$\int_Q \int_{\mathbb{R}^d\setminus S(R)} \frac{\ell(Q)^\varepsilon}{(\operatorname{dist}(t,Q) + \ell(Q))^{m+\varepsilon}} |h_Q(s)||\langle h_R \rangle_{S(R),\mu}|\, d\mu(s)|\, d\mu(t).$$

Proof We write $h_R = h_R \cdot 1_{R\setminus S(R)} + \langle h_R \rangle_{S(R),\mu} 1_{S(R)}$. Then we continue by

$$h_R = \langle h_R \rangle_{S(R),\mu} 1 + h_R \cdot 1_{R\setminus S(R)} - \langle h_R \rangle_{S(R),\mu} (1 - 1_{S(R)}).$$

Then (denoting by c_Q the center of Q) we write

$$\begin{aligned}(Th_Q, h_R) =& \langle h_R \rangle_{S(R)} (h_Q, \Delta_Q T^* 1)_\mu \\ &+ \int_{R\setminus S(R)} \int_Q [K(x,y) - K(x,c_Q)] h_Q(y) h_R(x)\, d\mu(y)\, d\mu(x) \\ &- \int_{\mathbb{R}^d\setminus S(R)} \int_Q [K(x,y) - K(x,c_Q)] h_Q(y) \langle h_R \rangle_{S(R),\mu} d\mu(y)\, d\mu(x).\end{aligned}$$

Now the usual Calderón–Zygmund estimate of the kernel finishes the lemma. In this estimate, we used (4.1.19). □

After proving this lemma, let us consider two integral terms above separately $t_1 := t_{1,Q,R} := \int_{R\setminus S(R)} \cdots$ and $t_2 := t_{2,Q,R} := \int_{\mathbb{R}^d\setminus S(R)} \cdots$.

In the second integral, we estimate h_Q in $L^1(\mu)$: $\|h_Q\|_{1,\mu} \le \sqrt{\mu(Q)}$, and we estimate $\langle h_R\rangle_{S(R),\mu}$ as follows:

$$|\langle h_R\rangle_{S(R),\mu}| \le 1/\sqrt{\mu(S(R))}.$$

Integral itself is at most (recall that $\mu(B(x,r) \le r^m)$)

$$\int_{\mathbb{R}^d\setminus S(R)} \frac{\ell(Q)^\varepsilon}{(\operatorname{dist}(t,Q)+\ell(Q))^{m+\varepsilon}}\,d\mu(t) \le \frac{\ell(Q)^\varepsilon}{\operatorname{dist}(Q,\operatorname{sk}(R))^\varepsilon}. \tag{4.1.20}$$

So if Q is good, meaning that $\operatorname{dist}(Q,\operatorname{sk}(R)) \ge \ell(R)^{1-\gamma}\ell(Q)^\gamma$ then (4.1.20) gives us

$$|t_{2,Q,R}| \le \left(\frac{\mu(Q)}{\mu(S(R))}\right)^{1/2}\frac{\ell(Q)^{1-\varepsilon\gamma}}{\ell(R)^{1-\varepsilon\gamma}}. \tag{4.1.21}$$

In the first integral we estimate h_Q in $L^1(\mu)$: $\|h_Q\|_{1,\mu} \le \sqrt{\mu(Q)}$, and we *cannot* estimate h_R in $L^\infty(\mu)$: $\|h_R\|_{L^\infty(R\setminus S(R))}$. The problem is that this supremum is bounded by $1/\sqrt{\mu(s(R))}$ for a sibling $s(R)$ of $S(R)$. But because doubling is missing, this can be an uncontrollably bad estimate. Therefore, we estimate here $\|h_R\|_{1,\mu} \le \sqrt{\mu(R)}$. As to the integral kernel itself, we are forced to estimate in L^∞ as $L^1(\mu)$ has been just spent. So we get the term $\frac{\ell(Q)^\varepsilon}{\operatorname{dist}(Q,\operatorname{sk}(R))^{m+\varepsilon}}$.

So if Q is good, meaning that $\operatorname{dist}(Q,\operatorname{sk}(R)) \ge \ell(R)^{1-\gamma}\ell(Q)^\gamma$ then (4.1.20) gives us

$$|t_{1,Q,R}| \le (\mu(Q)\mu(R))^{1/2}\frac{\ell(Q)^\varepsilon}{\ell(R)^{m+\varepsilon-(m+\varepsilon)\gamma}\ell(Q)^{(m+\varepsilon)\gamma}}.$$

Choose $\gamma := \frac{\varepsilon}{2(m+\varepsilon)}$. Then we get

$$|t_{1,Q,R}| \le \left(\frac{\ell(Q)}{\ell(R)}\right)^{\varepsilon/2}\frac{\sqrt{\mu(Q)}\sqrt{\mu(R)}}{\ell(R)^m} \le \left(\frac{\mu(Q)}{\mu(R)}\right)^{1/2}\left(\frac{\ell(Q)}{\ell(R)}\right)^{\varepsilon/2}. \tag{4.1.22}$$

We again used that $\mu(B(x,r) \le r^m$ in the last inequality. Compare now (4.1.21) and (4.1.22). We see that for small γ (and our γ is small) $\varepsilon/2 \le 1-\gamma\varepsilon$, and we can conclude that estimate (4.1.22) holds for both terms $t_{1,Q,R}, t_{2,Q,R}$.

Now notice that expressions ($k \geq 0$)

$$\sum_{Q\subset R,\, \ell(Q)=2^{-r-k}\ell(R),\, Q \text{ is good}} (Th_Q, h_R)(f, h_Q)(g, h_R)$$

can be split to three sums:

$$\sum_R \sum_{Q\subset R,\, \ell(Q)=2^{-r-k}\ell(R),\, Q \text{ is good}} t_{1,Q,R}(f, h_Q)(g, h_R),$$

$$\sum_R \sum_{Q\subset R,\, \ell(Q)=2^{-r-k}\ell(R),\, Q \text{ is good}} t_{2,Q,R}(f, h_Q)(g, h_R),$$

and

$$\sum_R \sum_{Q\subset R,\, \ell(Q)=2^{-r-k}\ell(R),\, Q \text{ is good}} \langle h_R\rangle_{S(R)} (h_Q, \Delta_Q T^* 1)_\mu (f, h_Q)(g, h_R).$$

Obviously, the first sum is the bilinear form of a shift of complexity $(0, r+k)$ having the coefficient $2^{-\frac{\varepsilon(r+k)}{2}}$ in front, just look at (4.1.22). The second sum is also the bilinear form of a shift of complexity $(0, r+k)$ having the coefficient $2^{-\frac{(1-\varepsilon\gamma)(r+k)}{2}}$ in front. We just look at (4.1.21) and notice that $\sum_{Q\subset S(R)} \left((\mu(Q)/\mu(S(R))^{1/2} \right)^2 \leq 1$.

This is exactly what we need, and the part of $\mathbb{E}\sum_{\ell(Q)\leq \ell(R),\, Q \text{ is good}} \cdots$, which is given by $Q \subset R$ is represented as the sum of shifts of complexity $(0, n)$, $n \geq r$, with exponential coefficients of the form $2^{-\delta n}$, $n > 0$. However, this is done up to the third sum.

We cannot take care of the third sum individually. Instead we sum the third sums in all $k \geq 0$ and all h^i_Q, $i = 1, \ldots, 2^d - 1$, h^j_R, $j = 1, \ldots, 2^d - 1$ (we recall that the index i was omitted, now we put it back). After summing over k and i, j we get ($F(L)$ denotes the dyadic father of L)

$$\sum_{i=1}^{2^d-1} \sum_{L\in\mathcal{D}} (g, h^i_{F(L)}) \langle h^i_{F(L)} \rangle_{L,\mu} \sum_{Q\subset L,\, \ell(Q)\leq 2^{-(r-1)}\ell(L),\, Q \text{ is good}} (\Delta_Q f, \Delta_Q(T^*1)).$$

We introduce the following operator:

$$\pi(g) := \sum_{i=1}^{2^d-1} \sum_{L\in\mathcal{D}} (g, h^i_{F(L)}) \langle h^i_{F(L)} \rangle_{L,\mu} \sum_{Q\subset L,\, \ell(Q)\leq 2^{-(r-1)}\ell(L),\, Q \text{ is good}} \Delta_Q(T^*1).$$

This is the same as

$$\pi(g) = \sum_{L\in\mathcal{D}} \langle \Delta_{F(L)} g \rangle_{L,\mu} \sum_{Q\subset L,\, \ell(Q)\leq 2^{-(r-1)}\ell(L),\, Q \text{ is good}} \Delta_Q(T^*1).$$

We can rewrite this formula by summing telescopically first over $L \in \mathcal{D}$ such that $\ell(L) \geq 2^{(r-1)}\ell(Q)$. Then we get (we assume $\int f\, d\mu = 0$ for simplicity)

$$\pi(g) = \sum_{R\in\mathcal{D}} \langle g\rangle_R \sum_{Q\subset R,\, \ell(Q)\leq 2^{-(r-1)}\ell(R),\, Q \text{ is good}} \Delta_Q(T^*1).$$

We check now by inspection the following equality:

$$\sum_{i,j=1}^{2^d-1}\sum_{k\geq 0}\sum_{Q\subset R,\, \ell(Q)=2^{-r-k}\ell(R),\, Q \text{ is good}} \langle h_R^i\rangle_{S(R)}(h_Q^i, \Delta_Q T^*1)_\mu (f, h_Q^i)(g, h_R^j)$$
$$= (f, \pi(g)) = (\pi^* f, g). \tag{4.1.23}$$

Operator $\pi(g)$ is a $(0, r-1)$ generalized dyadic shift, it is just

$$\pi(g) = \sum_{J=1}^{2^d-1}\sum_{R\in\mathcal{D}} \langle f\rangle_R \sum_{Q\subset R,\, \ell(Q)\leq 2^{-(r-1)}\ell(R),\, Q \text{ is good}} (h_Q^j, T^*1)h_Q^j.$$

To see that, we need just to check the Carleson condition. We fix $L \in \mathcal{D}$ and we can see by (4.1.19) that the estimate

$$\sum_{Q\subset L,\, \ell(Q)\leq 2^{-(r-1)}\ell(L),\, Q\cap\partial L=\emptyset} \|\Delta_Q(T^*1)\|_\mu^2 \leq C\mu(L) \tag{4.1.24}$$

is enough.

To prove (4.1.24), we need

Lemma 4.1.16 *Let T be a Calderón–Zygmund operator satisfying* (4.1.11). *Let S be any dyadic square, and let $S' = 1.1S$. Then*

$$\sum_{Q\subset S} \|\Delta_Q T^*1\|_\mu^2 \leq C_1\mu(S'),$$

where $C_1 = C(C_0, d, \varepsilon)$ with C_0 from (4.1.11).

Proof We write $\Delta_Q T^*1 = \Delta_Q T^*1_{S'} + \Delta_Q T^*(1 - 1_{S'})$. The first term is easy:

$$\sum_{Q\subset S} \|\Delta_Q T^*1_{S'}\|_\mu^2 \leq \|T^*1_{S'}\|_\mu^2 \leq C_0\mu(S')$$

just by (4.1.11).

To estimate $\sum_{Q\subset S}\|\Delta_Q T^*(1-1_{S'})\|_\mu^2$ we fix $f\in L^2(\mu)$, $\|f\|_\mu=1$, we write

$$
\begin{aligned}
(f,\Delta_Q T^*(1-1_{S'})) &= (T\Delta_Q f, 1-1_{S'}) \\
&= \int_{\mathbb{R}^d\setminus S'}\int_Q [K(x,y)-K(x,c_Q)]\,\Delta_Q f(y)\,d\mu(y)\,d\mu(x).
\end{aligned}
$$

Such an integral we already saw in (4.1.20), and we can estimate it by

$$
\|\Delta_Q f\|_\mu\sqrt{\mu(Q)}\cdot\int_{\mathbb{R}^d\setminus S'}\frac{\ell(Q)^\varepsilon}{(\operatorname{dist}(x,Q)+\ell(Q))^{m+\varepsilon}}\,d\mu(x)\le C\sqrt{\mu(Q)}\frac{\ell(Q)^\varepsilon}{\ell(S)^\varepsilon}.
$$

As f was arbitrary $f\in L^2(\mu)$, $\|f\|_\mu=1$, we get that

$$
\|\Delta_Q T^*(1-1_{S'})\|_\mu^2\le C\mu(Q)\left(\frac{\ell(Q)}{\ell(S)}\right)^{2\varepsilon}.
$$

Taking the sum over Q, we obtain

$$
\sum_{Q\subset S}\|\Delta_Q T^*(1-1_{S'})\|_\mu^2\le\|T^*1_{S'}\|_\mu^2\le C_0\mu(S').
$$

Lemma is proved. □

We proved (4.1.24), but it also proves that π is a bounded generalized shift.

4.1.6 The Decomposition of the Outer Sum

We are left to decompose

$$
\mathbb{E}\sum_{Q\cap R=\emptyset,\,\ell(Q)\le\ell(R),\,Q\text{ is good}}(Th_Q,h_R)(f,h_Q)(g,h_R)
$$

into the bilinear form of (s,t)-shifts with exponentially small in $\max(s,t)$ coefficients.

Denote

$$
D(Q,R):=\ell(Q)+\operatorname{dist}(Q,R)+\ell(R).
$$

Also let $L(Q,R)$ be a dyadic interval from the same lattice such that $\ell(L(Q,R))\in(2D(Q,R),4D(Q,R))$ that contains R.

Exactly as we did this before, we can estimate

$$
(Th_Q,h_R)=\int_R\int_Q[K(x,y)-K(x,c_Q)]h_Q(y)h_R(x)\,d\mu(y)\,d\mu(x)
$$

by estimating $\|h_Q\|_{1,\mu}\le\sqrt{\mu(Q)}$, $\|h_R\|_{1,\mu}\le\sqrt{\mu(R)}$, and $\frac{\ell(Q)^\varepsilon}{\operatorname{dist}(Q,R)^{m+\varepsilon}}\le\ell(Q)^{\varepsilon/2}/\ell(R)^{\varepsilon/2+m}$ if $\operatorname{dist}(Q,R)\le\ell(R)$. Otherwise, the estimate is

$\ell(Q)^{\varepsilon}/D(Q,R)^{m+\varepsilon}$. These two estimates are both united into the following one obviously:

$$|(Th_Q, h_R)| \leq C\left(\frac{\ell(Q)}{\ell(R)}\right)^{\varepsilon/2} \frac{\ell(R)^{\varepsilon/2}}{D(Q,R)^{m+\varepsilon/2}} \sqrt{\mu(Q)}\sqrt{\mu(R)}. \quad (4.1.25)$$

Of course, in this estimate we used not only that Q is good, but also that $\ell(Q) \leq 2^{-r}\ell(R)$. Only having this latter condition can we apply the estimate on $\operatorname{dist}(Q,R)$ from Definition 4.1.1 that was used in getting (4.1.25).

However, if $\ell(Q) \in [2^{-r-1}\ell(R), \ell(R)]$, we use just a trivial estimate of coefficient $(Th_Q, h_R)|$, namely,

$$|(Th_Q, h_R)| \leq C(C_0, d), \quad (4.1.26)$$

where C_0 is from (4.1.10). This is not dangerous at all because such pairs Q, R will be able to form below only shifts of complexity (s,t), where $0 \leq s \leq t \leq r$; the number of such shifts is at most $\frac{r(r+1)}{2}$, and let us recall that r is not a large number, it depends only on d (see (4.1.14)).

Now in a given $\mathcal{D} \in \Omega$, a pair of Q, R may or may not be inside $L(Q,R)$ ($R \subset L(Q,R)$ by definition). But the ratio of nice lattices (those when both Q, R are inside $L(Q,R)$) with respect to all lattices in which both Q, R are present is bounded away from zero, this ratio (probability) satisfies

$$p(Q,R) \geq P_d > 0. \quad (4.1.27)$$

We want to modify the following expectation:

$$\Sigma := \mathbb{E} \sum_{Q\cap R=\emptyset,\, \ell(Q)\leq \ell(R),\, Q \text{ is good}} (Th_Q, h_R)(f, h_Q)(g, h_R).$$

This expectation is really a certain sum itself, namely, the sum over all lattices in Ω divided by $\sharp(\Omega) =: M$. Each time Q, R are not in a nice lattice we put zero in front of corresponding term. This changes very much the sum. However, we can make up for that, and we can *leave the sum unchanged* if for nice lattices we put the coefficient $1/p(Q,R)$ in front of corresponding terms (and keep 0 otherwise).

Then,

$$\frac{\text{number of lattices containing } Q, R}{M}$$

$$= \frac{1}{p(Q,R)} \frac{\text{number of nice lattices containing } Q, R}{M}.$$

Notice that in the original sum Σ terms Q, R are multiplied by the LHS. The modified sum will contain the same terms multiplied by the RHS. So it is not modified at all, it is the same sum exactly! We can write it again as

$$\mathbb{E} \sum_{Q\cap R=\emptyset,\, \ell(Q)\le \ell(R),\, Q \text{ is good}} m(Q,R,\omega)(Th_Q,h_R)(f,h_Q)(g,h_R),$$

where the random coefficients $m(Q,R,\omega)$ are either 0 (if the lattice $\mathcal{D}=\omega$ is not nice), or $1/p(Q,R)$ if the lattice is nice.

Now let us fix two positive integers $s \ge t$, fix a lattice, and consider this latter sum only for this lattice, and write it as

$$\sum_s \sum_t \sum_L \sum_{Q\subset L, R\subset L, \ell(Q)=2^{-t}\ell(L), \ell(R)=2^{-s}\ell(L)} m(Q,R)(Th_Q,h_R)(f,h_Q)(g,h_R) =: \sum_s \sum_t \sigma_{s,t}.$$

Each $\sigma_{s,t}$ is a dyadic shift of complexity (s,t). In fact, use (4.1.27) and (4.1.25) and easily see that the sum of squares of coefficients inside each L is bounded (we use again $\mu(B(x,r) \le r^m)$. Moreover, the terms $\left(\frac{\ell(Q)}{\ell(R)}\right)^{\varepsilon/2}$, $\left(\frac{\ell(R)}{D(Q,R)}\right)^{\varepsilon/2}$ from (4.1.25) gives us the desired exponentially small coefficient whose size is at most $2^{-\varepsilon(s-t)/2} \cdot 2^{-\varepsilon s/2} = 2^{-\varepsilon t/2} = 2^{-\varepsilon \max(s,t)/2}$.

Theorem 4.1.8 is completely proved.

4.2 Sharp Weighted Estimate of Dyadic Shifts

The goal of this section is to give the proof of the so-called A_2 conjecture using a) the decomposition of an arbitrary Calderón–Zygmund operator into dyadic shifts as in Theorem 4.1.8 of Section 4.1, b) the Bellman technique to estimate each individual shift. The A_2 conjecture asks for a sharp weighted estimate of the usual homogeneous Calderón–Zygmund operator in the usual Euclidean space with Lebesgue measure. But, as we already mentioned, that should be a weighted estimate. The main point of the conjecture is to give *sharp* in weight estimate for *all* Calderón–Zygmund operators.

4.2.1 Doubling Martingales

Consider a measure space X with σ-finite measure μ, let $\mathcal{L}_k = \{Q_j^k\}_j$, $k \in \mathbb{Z}$ (or $k \in \mathbb{Z}_+$) be partitions of X into disjoint sets Q_j^k, $0 < \mu(Q_j^k) < \infty$. We assume that the partition $\mathcal{L}_{k+1}$ is a refinement of $\mathcal{L}_k$. Let $A \subset B$ be two elements of $\mathcal{L}_{k+1}$ and $\mathcal{L}_k$ correspondingly. We assume that $\frac{\mu(A)}{\mu(B)} \geq c > 0$ for a constant c not depending on A, B. We call such situation doubling.

Let $\mathfrak{A}$ be the σ-algebra generated by all the partitions $\mathcal{L}_k$. All functions on X are assumed to be $\mathfrak{A}$-measurable. In this doubling martingale setting, one can define the shift of complexity (m, n) too. We restricted ourselves to the doubling martingale setting because we had in mind the application to Calderón–Zygmund operators on homogeneous metric spaces; see **[125]**.

We relate to $\mathcal{L}_k$ the projection operator $\mathcal{E}_k$ on $L^2(\mu)$:

$$\mathcal{E}_k f = \sum_j \langle f \rangle_{Q_k^j, \mu} \chi_{Q_k^j}.$$

We also consider the martingale difference operator $D_k f = E_k f - E_{k-1} f$. For each Q_{k-1}^j the martingale difference subspace $D_k(L^2(Q_{k-1}^j, d\mu))$ is finite dimensional with uniform estimate $\dim D_k(L^2(Q_{k-1}^j, d\mu)) = M_{k-1}(j) \leq M$. Fix an orthonormal basis in this space consisting of function supported on $Q := Q_{k-1}^j$, orthogonal to constants in $L^2(\mu)$, orthogonal to each other in $L^2(\mu)$, constant on each $Q_k^i \subset Q$, $Q_k^i \in \mathcal{E}_k$. Enumerate the function of these bases and call them $\{h^t_{Q_{k-1}^j}\}_{t=1}^{M_{k-1}^j}$. Notice that now we can write

$$D_k f = \sum_j \sum_{t=1}^{M_{k-1}(j)} (f, h^t_{Q_{k-1}^j})_\mu h^t_{Q_{k-1}^j},$$

This function will have the following bound, where C depends on the doubling constant of the martingale:

$$\|h_Q^j\|_{L^\infty(\mu)} \leq \frac{C}{\mu(Q)^{1/2}}. \tag{4.2.1}$$

DEFINITION We call $Q_k^i \subset Q_{k-1}^j$, $Q_k^i \in \mathcal{E}_k$ **sons** of Q_{k-1}^j. We of course call functions $h^j_{Q_k}$ Haar functions. We will use names Q, L, I, J for the elements of the partitions (sometimes with indices).

If $L \in \mathcal{L}_m$, $J \in \mathcal{L}_{m+n}$ we say $g(J) = g(L) + n$.
We call by $\mathbb{S}_{m,n}$ the operator given by the kernel

$$f \to \sum_{L \in \mathcal{D}} \int_L a_L(x, y) f(y) dy,$$

where

$$a_L(x, y) = \sum_{\substack{I \subset L, J \subset L \\ g(I) = g(L) + m,\, g(J) = g(L) + n}} c_{L,I,J} h_J^j(x) h_I^i(y),$$

where h^i_I, h^j_J are Haar functions normalized in $L^2(d\mu)$ and satisfying (4.2.1), and $|c_{L,I,J}| \le \frac{\sqrt{\mu(I)}\sqrt{\mu(J)}}{\mu(L)}$. Often we will skip superscripts i, j.

We are interested in sufficiently good estimate of

$$\|\mathbb{S}_{m,n}\|_w := \|\mathbb{S}_{m,n} : L^2(w\, d\mu) \to L^2(w\, d\mu)\|,$$

where $w \in A_2$. For such w, we put $\sigma = w^{-1}$ and introduce

DEFINITION 4.2.1 Put $[w]_{A_2} := \sup_I \langle w\rangle_{\mu,I}\langle\sigma\rangle_{\mu,I} < \infty$, and call it the A_2-norm of w (it is not a norm in the usual sense). Symbol $\|T\|_{w,\mu}$ will denote $\|T\|_{L^2(\mathbb{R}^d, w\, d\mu)\mapsto L^2(\mathbb{R}^d, w\, d\mu)}$. Sometimes we will drop μ abbreviating it to $\|T\|_w$ if it is clear what measure μ we are talking about.

In paper [**76**], the following theorem was proved:

THEOREM 4.2.2

$$\|\mathbb{S}_{m,n}\|_w \le C\,(m+n+1)[w]_{A_2}. \tag{4.2.2}$$

4.2.2 Applications

Here is the first application of this theorem. Let us notice that $\mathbb{R}^d$ is of course a geometrically doubling metric space with the usual Euclidean metric. The usual dyadic lattice in $\mathbb{R}^d$ and $\mu = m$, where m denote Lebesgue measure in $\mathbb{R}^d$, give us the example of doubling martingale setting introduced above. Now consider the definition of Calderón–Zygmund operators, namely, Definition 4.1.8. Specify the above definition to

$$\lambda(x,r) = r^d, \ |xy| = |x-y|,$$

then we have the classical Calderón–Zygmund kernel in $\mathbb{R}^d$ and Calderón–Zygmund operators in $\mathbb{R}^d$. Applying Section 4.1, we get the decomposition of any classical Calderón–Zygmund operator into shifts and generalized shifts (paraproducts) as in Theorem 4.1.8.

Now one applies Theorem 4.2.2 (and a corresponding result for generalized shifts) to get the following result, which was first established by T. Hytoñen [**72**], and which solves the so-called A_2 conjecture.

THEOREM 4.2.3 *Let T be a classical Calderón–Zygmund operator in $\mathbb{R}^d$ (in particular, we assume that T is bounded in $L^2(\mathbb{R}^d, m)$). Then*

$$\|T\|_w := \|T\|_{L^2(\mathbb{R}^d, w\, dm)\to L^2(\mathbb{R}^d, w\, dm)} \le C(d,T)[w]_{A_2}.$$

Another application of Theorem 4.2.2 is the generalization of this result to the geometrically doubling metric spaces $(X, |\cdot|, \mu)$ with their doubling measure μ mentioned in Section 4.1. The same result holds in this generality:

THEOREM 4.2.4 *Let $(X, |\cdot|, \mu)$ be a geometrically doubling metric space with doubling measure μ, $\operatorname{supp}\mu = X$. Let T be a Calderón–Zygmund operator on $(X, |\cdot|, \mu)$ in the sense of Definition 4.1.8, in particular the kernel of T and measure μ are related as in Definition 4.1.8 (and by this definition it is assumed that T is bounded in $L^2(X, \mu)$). Then,*

$$\|T\|_{w,\mu} := \|T\|_{L^2(\mathbb{R}^d, w\, d\mu) \to L^2(\mathbb{R}^d, w\, d\mu)} \leq C(X, T)[w]_{A_2}.$$

To prove Theorem 4.2.4 one needs to repeat the decomposition of Calderón–Zygmund operator into shifts, in other words one needs to repeat Theorem 4.1.8, but in metric setting. This is not a completely obvious exercise because one first needs to build the right probability space of "dyadic" lattices on the metric space consisting of Christ's cubes. So we refer the reader to paper [**125**], where this has been done. However, after this probability space of right kind of dyadic lattices of Christ's cubes is already built, one immediately has the same decomposition as in Theorem 4.1.8 into shifts, and one observes easily that by the doubling property of μ and the main properties of Christ's lattices, the corresponding shifts are martingale shifts of our Theorem 4.2.2, and they satisfy the doubling martingale property. So we apply Theorem 4.2.2, and this finishes the proof of the above weighted estimate on metric spaces with doubling measure.

We explained the role of Theorem 4.2.2 (and its counterpart for generalized shifts). Now we start to prove it using Bellman technique.

4.2.3 The Heart of the Matter: A Bilinear Embedding Estimate

Let I be an element of a given martingale partition. Let h_I^j denote some orthonormal basis in the space $\Delta(I) := \Delta_I(L^2(\mu))$ of functions $f \in L^2(\mu)$ supported on I and such that $\int f\, d\mu = 0$ and f is constant on each son $s(I)$ of I in our martingale partition.

To prove Theorem 4.2.2, we need the following simple lemma:

LEMMA 4.2.5

$$h_I^j = \alpha_I^j h_I^{w,j} + \beta_I^j \chi_I,$$

where

1) $|\alpha_I^j| \leq \sqrt{\langle w \rangle_{\mu,I}}$,
2) $|\beta_I^j| \leq \frac{|(h_I^j, w)_\mu|}{w(I)}$, *where* $w(I) := \int_I w\, d\mu$,
3) $\{h_I^{w,j}\}_I$ *is supported on* I, *orthogonal to constants in* $L^2(w\, d\mu)$,
4) $h_I^{w,j}$ *assumes on each son* $s(I)$ *a constant value,*
5) $\|h_I^{w,j}\|_{L^2(\mu)} = 1$.

The details are left to the reader. In fact, we have two unknown constants α_I^j, β_I^j and two conditions to determine them: 3) and 5). The rest is easy.

DEFINITION As before, let

$$\Delta_I w := \sum_{\text{sons of } I} |\langle w\rangle_{\mu,s(I)} - \langle w\rangle_{\mu,I}|.$$

It is a easy to see that the doubling property of our martingale implies

$$|(h_I^j, w)_\mu| \le C\,(\Delta_I w)\,\mu(I)^{1/2}. \tag{4.2.3}$$

Therefore, the property 2) above can be rewritten as 2')

$$|\beta_I^j| \le C\,\frac{|\Delta_I w|}{\langle w\rangle_{\mu,I}}\frac{1}{\mu(I)^{1/2}}.$$

Recall that

$$\sigma := w^{-1}.$$

To prove Theorem 4.2.2, we notice a trivial thing, namely, that to prove it is the same as proving

$$\|\mathbb{S}_{m,n}\|_\sigma \le C\,(m+n+1)[w]_{A_2}. \tag{4.2.4}$$

In fact, σ is w^{-1}, and A_2 norm is symmetric under reciprocation. To prove the last inequality we fix $\phi \in L^2(w\,d\mu), \psi \in L^2(\sigma\,d\mu)$. We need now to prove

$$|(\mathbb{S}_{m,n}\phi w, \psi\sigma)| \le C\,(n+m+1)\|\phi\|_w\|\psi\|_\sigma. \tag{4.2.5}$$

This is the same as (4.2.4) because $\|\phi w\|_\sigma = \|\phi\|_w$.

We first prove (4.2.4) only for $m = n = 0$. The proof will use the Bellman technique. We will explain a bit later why it is enough to consider $m = n = 0$ case. The passage to the general case is another use of the Bellman technique; this passage is quite sophisticated, but elementary. As the Bellman technique is, in its heart, just the use of certain convexity, this passage will be based on the tricks with convex functions.

Now we estimate $(\mathbb{S}\phi w, \psi\sigma) := (\mathbb{S}_{0,0}\phi w, \psi\sigma)$ as

$$\left|\sum_I c_I(\phi w, h_I)_\mu(\psi\sigma, h_I)_\mu\right|$$

$$\le \sum_I |(\phi w, h_I^w)_\mu|\sqrt{\langle w\rangle_{\mu,I}}|(\psi\sigma, h_I^\sigma)_\mu|\sqrt{\langle\sigma\rangle_{\mu,I}}$$

$$+\sum_I |\langle\phi w\rangle_{\mu,I}|\frac{|\Delta_I w|}{\langle w\rangle_{\mu,I}}|(\psi\sigma, h_I^\sigma)_\mu|\sqrt{\langle\sigma\rangle_{\mu,I}}\sqrt{\mu(I)}$$

$$+\sum_I |\langle\psi\sigma\rangle_{\mu,I}|\frac{|\Delta_I\sigma|}{\langle\sigma\rangle_{\mu,I}}|(\phi w, h_I^w)_\mu|\sqrt{\langle w\rangle_{\mu,I}}\sqrt{\mu(I)}$$

$$+\sum_I |\langle\phi w\rangle_{\mu,I}|\langle\psi\sigma\rangle_{\mu,I}|\frac{|\Delta_I w|}{\langle w\rangle_{\mu,I}}\frac{|\Delta_I\sigma|}{\langle\sigma\rangle_{\mu,I}}\mu(I) =: A+B+C+D.$$

We want to estimate each of these sums by $C\,[w]_{A_2}\|\phi\|_w\|\psi\|_\sigma$. The estimate

$$D \le C\,[w]_{A_2}\|\phi\|_w\|\psi\|_\sigma$$

is (and will be called) a bilinear embedding estimate. Notice that this fourth sum is the most difficult of all sums above.

We can put

$$R_I(\phi w) := |\langle\phi w\rangle_{\mu,I}|\frac{|\Delta_I w|}{\langle w\rangle_{\mu,I}}\sqrt{\mu(I)}, \tag{4.2.6}$$

$$S_I(\phi w) := |(\phi w, h_I^w)_\mu|\sqrt{\langle w\rangle_{\mu,I}}, \tag{4.2.7}$$

and the corresponding terms for $\psi\sigma$. So we have the mixture of S-terms, which are "cancellation" terms and R-terms, which are "non-cancellation" terms. The word cancellation here means that the test function ϕ got estimated against the function h_I^w whose average with respect to measure $wd\mu$ is zero. In non-cancellation terms, the test function gets integrated against a positive quantity, no cancellation is involved. Non-cancellation terms are usually more difficult to treat. We see that

$$A \le \sum_I S_I(\phi w)S_I(\psi\sigma),\ B \le \sum_I S_I(\phi w)R_I(\psi\sigma),$$

$$C \le \sum_I R_I(\phi w)S_I(\psi\sigma),\ \ D \le \sum_I R_I(\phi w)R_I(\psi\sigma).$$

Now

$$S_I(\phi w) \le \sqrt{|(\phi w, h_I^w)_\mu|^2}\sqrt{\langle w\rangle_{\mu,I}},\ \ S_I(\psi\sigma) \le \sqrt{|(\psi\sigma, h_I^\sigma)|^2}\sqrt{\langle\sigma\rangle_{\mu,I}} \tag{4.2.8}$$

Therefore, just by Hölder inequality

$$A \le \sup_I \sqrt{\langle w\rangle_{\mu,I}\langle\sigma\rangle_{\mu,I}}\|\phi\|_w\|\psi\|_\sigma \le [w]_{A_2}^{1/2}\|\phi\|_w\|\psi\|_\sigma. \tag{4.2.9}$$

Terms B, C are symmetric, so consider C.

Recall the following definition

DEFINITION 4.2.6 Consider a sequence of nonnegative numbers $\{\alpha_I\}_{I\in\mathcal{D}}$. We call this sequence the Carleson sequence if

$$\exists K<\infty \quad \forall J\in\mathcal{D} \quad \sum_{I\in\mathcal{D},\, I\subset J} \alpha_I \le K\,\mu(J).$$

The best constant K is called the Carleson constant of the sequence.

4.2.4 The First Use of Bellman Technique

Using function $B(x,y):=(xy)^\alpha$ one can prove now the following lemma

LEMMA 4.2.7 *The sequence*

$$\tau_I := \langle w\rangle_{\mu,I}^\alpha \langle\sigma\rangle_{\mu,I}^\alpha \left(\frac{|\Delta_I w|^2}{\langle w\rangle_{\mu,I}^2}+\frac{|\Delta_I\sigma|^2}{\langle\sigma\rangle_{\mu,I}^2}\right)\mu(I)$$

form a Carleson sequence with the Carleson constant at most $c_\alpha Q^\alpha$, where $Q:=[w]_{A_2}$ for any $\alpha\in(0,1/2)$.

PROOF To prove this, we need a very simple observation.

LEMMA 4.2.8 *Let $Q>1, 0<\alpha<\frac{1}{2}$. In domain $\Omega_Q:=\{(x,y):x>0, y>0, 1<xy\le Q\}$ function $B(x,y):=x^\alpha y^\alpha$ satisfies the following estimate of its Hessian matrix (of its second quadratic differential form)*

$$-d^2B(x,y)\ge \alpha(1-2\alpha)x^\alpha y^\alpha\left(\frac{(dx)^2}{x^2}+\frac{(dy)^2}{y^2}\right). \tag{4.2.10}$$

The second quadratic differential form $-d^2B(x,y)\ge 0$ everywhere in $x>0, y>0$. Also obviously $0\le B(x,y)\le Q^\alpha$ in Ω_Q.

PROOF Direct calculation. □

Put $Q:=[w]_{A_2}$. Fix now an element I of our martingale partition, and let $s_i(I), i=1,\dots,M$, be all its sons. Consider several points in Ω_Q: $a=(\langle w\rangle_{\mu,I},\langle\sigma\rangle_{\mu,I})$, $b_i=(\langle w\rangle_{\mu,s_i(I)},\langle\sigma\rangle_{\mu,s_i(I)})$, $i=1,\dots,M$. These are points of Ω_Q, because $Q=[w]_{A_2}$. Consider functions with values in $\mathbb{R}^2$: $c_i(t)=a(1-t)+b_it, 0\le t\le 1$ and $q_i(t):=B(c_i(t))$. We want to use Taylor's formula

$$q_i(0)-q_i(1)=-q_i'(0)-\int_0^1 dx\int_0^x q_i''(t)\,dt. \tag{4.2.11}$$

Notice two things: Lemma 4.2.8 shows that $-q_i''(t)\ge 0$ always. Moreover, it shows that if $t\in[0,1/2]$, then we have that the following qualitative estimate holds:

$$-q_i''(t) \geq c(\langle w\rangle_{\mu,I}\langle\sigma\rangle_{\mu,I})^{\alpha}\left(\frac{(\langle w\rangle_{\mu,s_i(I)} - \langle w\rangle_{\mu,I})^2}{\langle w\rangle_{\mu,I}^2} + \frac{(\langle\sigma\rangle_{\mu,s_i(I)} - \langle\sigma\rangle_{\mu,I})^2}{\langle\sigma\rangle_{\mu,I}^2}\right) \tag{4.2.12}$$

This requires a small explanation. We use the chain rule and (4.2.10) to estimate the (minus) second derivative of $q_i(t) = B(c_i(t))$. The number x in the right-hand side of (4.2.10) will be certain convex combinations of $\langle w\rangle_{\mu,I}$ and $\langle w\rangle_{\mu,s_i(I)}$. Similarly for y with changing w to σ. We need to understand why we can replace any $x(t)$ by $x(0) = \langle w\rangle_{\mu,I}$ and all $y(t)$ by $y(0) = \langle\sigma\rangle_{\mu,I}$ by paying only the price of a constant, which does not depend on w. This is true because of the following small reasoning.

If (x, y) is on the segment $[a, b_i]$, then the first coordinate of such a point cannot be larger than $C\,\langle w\rangle_{\mu,I}$, where C depends only on doubling of μ (not w). This control from above is obvious. In fact, at the "right" endpoint of $[a, b_i]$ the first coordinate is $\langle w\rangle_{\mu,s_i(I)} \leq \int_I w\,d\mu/\mu(s_i(I)) \leq C\,\int_I w\,d\mu/\mu(I)) = C\,\langle w\rangle_{\mu,I}$, with C only depending on the doubling of μ. The same is true for the second coordinate with the obvious change of w to σ. But there is no such type of control from below on this segment: the first coordinate cannot be smaller than $k\,\langle w\rangle_{\mu,I}$, but k may (and will) depend on the doubling of w (so ultimately on its $[w]_{A_2}$ norm). So the estimate from below will involve the doubling of w, which we must avoid.

This difficulty, however, does not exist if $(x, y) = (1-t)a + tb_i$ and $t \in [0, 1/2]$, so we are on the "left half" of interval $[a, b_i]$. In this case obviously, the first coordinate is $\geq \frac{1}{2}\langle w\rangle_{\mu,I}$ and the second coordinate is $\geq \frac{1}{2}\langle\sigma\rangle_{\mu,I}$.

We do not need to integrate $-q_i''(t)$ for all $t \in [0, 1]$ in (4.2.11). We can only use integration over $[0, 1/2]$ noticing that $-q_i''(t) \geq 0$ for $t \in [1/2, 1]$. Then the chain rule gives

$$q_i''(t) = (B(c_i(t)))'' = (d^2B(c_i(t))(b_i - a), b_i - a).$$

This immediately shows (4.2.12) for $t \in [0, 1/2]$ with constant c depending on the doubling of μ but *independent* of the doubling of w.

REMARK 4.2.9 We do not know whether c in (4.2.12) is an absolute constant. In the above reasoning, it does depend on the doubling constant of μ. However, in **[178]**, it is proved that $\mathbb{S}_{0,0}$ has the same linear estimate $c[w]_{A_2}$ when measure μ is not doubling anymore. In particular, our martingales need not be doubling. The proof again follows from the Bellman function technique, but more sophisticated than above. Notice also that immediately after **[178]**, the linear estimate of $\mathbb{S}_{0,0}$ by $c[w]_{A_2}$ was proved in [96]. The article [96] uses the stopping time technique instead of the Bellman technique.

The next step is to add all (4.2.11), with convex coefficients $\frac{\mu(s_i(I))}{\mu(I)}$, and to notice that $\sum_{i=1}^M \frac{\mu(s_i(I))}{\mu(I)} q_i'(0) = (\nabla B_Q(a), \sum_{i=1}^M (a-b_i)\frac{\mu(s_i(I))}{\mu(I)})_{\mathbb{R}^2} = 0$, because by definition

$$a = \sum_{i=1}^M \frac{\mu(s_i(I))}{\mu(I)} b_i .$$

Notice that the summation of all (4.2.11), with convex coefficients $\frac{\mu(s_i(I))}{\mu(I)}$ gives us now (we take into account (4.2.12) and positivity of $-q_i''(t)$)

$$B(a) - \sum_{i=1}^M \frac{\mu(s_i(I))}{\mu(I)} B(b_i)$$

$$\geq c(\langle w\rangle_{\mu,I}\langle \sigma\rangle_{\mu,I})^\alpha \sum_{i=1}^M \left(\frac{(\langle w\rangle_{\mu,s_i(I)} - \langle w\rangle_{\mu,I})^2}{\langle w\rangle_{\mu,I}^2} + \frac{(\langle \sigma\rangle_{\mu,s_i(I)} - \langle \sigma\rangle_{\mu,I})^2}{\langle \sigma\rangle_{\mu,I}^2} \right).$$

We used here the doubling of μ again, by noticing that $\frac{\mu(s_i(I))}{\mu(I)} \geq c$. We rewrite the previous inequality using our definition of $\Delta_I w, \Delta_I \sigma$ listed above as follows:

$$\mu(I)\, B(a) - \sum_{i=1}^M \mu(s_i(I))\, B(b_i)$$

$$\geq c\,(\langle w\rangle_{\mu,I}\langle \sigma\rangle_{\mu,I})^\alpha \left(\frac{(\Delta_I w)^2}{\langle w\rangle_{\mu,I}^2} + \frac{(\Delta_I \sigma)^2}{\langle \sigma\rangle_{\mu,I}^2} \right) \mu(I).$$

Notice that $B(a) = \left(\langle w\rangle_{\mu,I}\langle \sigma\rangle_{\mu,I}\right)^\alpha \leq Q^\alpha$. Now we iterate the above inequality and use the telescopic nature of it to get the following inequality for any Is in the martingale partition of ours:

$$\sum_{J\subset I,\, J\in\mathcal{D}} (\langle w\rangle_{\mu,J}\langle \sigma\rangle_{\mu,J})^\alpha \left(\frac{(\Delta_J w)^2}{\langle w\rangle_{\mu,J}^2} + \frac{(\Delta_J \sigma)^2}{\langle \sigma\rangle_{\mu,J}^2} \right) \mu(J) \leq C\, Q^\alpha \mu(I).$$

This is exactly the Carleson property of the sequence $\{\tau_I\}$ indicated in our Lemma 4.2.7, with the Carleson constant $C\,Q^\alpha$. The proof showed that C depended only on $\alpha \in (0, 1/2)$ and on the doubling constant of measure μ. $\square$

Now, using this lemma, we start to estimate the products of S_I's and R_I's defined above in (4.2.7), (4.2.6). We start with an obvious estimate for R_I. Let us fix $\alpha \in (0, 1/2)$, then (we skip subscript μ for a while)

$$R_I(\phi w) \leq \langle w\rangle_I^{-\alpha/2} \langle \sigma\rangle_I^{-\alpha/2} \sqrt{\tau_I}\, \langle |\phi| w\rangle_I .$$

Choose any $p \in (1,2)$ and write then

$$R_I(\phi w) \le \langle w\rangle_I^{-\alpha/2} \langle\sigma\rangle_I^{-\alpha/2} \sqrt{\tau_I}\, \langle|\phi|^p w\rangle_I^{1/p} \langle w\rangle_I^{1-1/p}.$$

Now we need a classical object, the martingale maximal function.

DEFINITION 4.2.10 Given the martingale partition on X, and measure ν, we define

$$M_\nu f(x) := \sup_{I\in\mathfrak{A}:\ x\in I} \frac{1}{\nu(I)} \int_I |f|\, d\nu.$$

We abbreviate: $M_w := M_{w\,d\mu}$.

Now we continue the estimate of R_I as follows:

$$R_I(\phi w) \le \langle w\rangle_I^{1-\alpha/2} \langle\sigma\rangle_I^{-\alpha/2} \sqrt{\tau_I}\, \inf_I M_w(|\phi|^p)(x)^{1/p}. \tag{4.2.13}$$

Similarly,

$$R_I(\psi\sigma) \le \langle\sigma\rangle_I^{1-\alpha/2} \langle w\rangle_I^{-\alpha/2} \sqrt{\tau_I}\, \inf_I M_\sigma(|\psi^P)(x)^{1/p}. \tag{4.2.14}$$

Combining these two estimates we get

$$R_I(\phi w) R_I(\psi\sigma) \le \langle w\rangle_I^{1-\alpha} \langle\sigma\rangle_I^{1-\alpha} \inf_I M_w(|\phi|^p)(x)^{1/p} \inf_I M_\sigma(|\psi^p)(x)^{1/p}\, \tau_I.$$

From (4.2.7) and (4.2.13), we get

$$R_I(\phi w) S_I(\psi\sigma) \le \langle w\rangle_I^{1-\alpha/2} \langle\sigma\rangle_I^{1-\alpha/2} \frac{\inf_I M_w(|\phi|^p)(x)^{1/p}}{\langle\sigma\rangle_I^{1/2}} \sqrt{\tau_I}\, |(\psi\sigma, h_I^\sigma)|.$$

Therefore,

$$R_I(\phi w) R_I(\psi\sigma) \le [w]_{A_2}^{1-\alpha} \inf_I M_w(|\phi|^p)(x)^{1/p} \inf_I M_\sigma(|\psi|^p)(x)^{1/p}\, \tau_I, \tag{4.2.15}$$

and

$$R_I(\phi w) S_I(\psi\sigma) \le [w]_{A_2}^{1-\alpha/2} \frac{\inf_I M_w(|\phi|^p)(x)^{1/p}}{\langle\sigma\rangle_I^{1/2}} \sqrt{\tau_I}\, |(\psi\sigma, h_I^\sigma)|. \tag{4.2.16}$$

The symmetric estimate is also valid, where we exchange w and σ as well as ϕ and ψ. Let us first sum up (4.2.16) over all I in the lattice of partition. Then we get by Hölder inequality

$$\sum_{I\in\mathfrak{A}} R_I(\phi w) S_I(\psi\sigma) \le [w]_{A_2}^{1-\alpha/2} \|\psi\|_\sigma \left(\sum_{I\in\mathfrak{A}} \frac{\inf_I M_w(|\phi|^p)(x)^{2/p}}{\langle\sigma\rangle_I} \tau_I \right)^{1/2}. \tag{4.2.17}$$

Again the symmetric estimate is also valid.

Similarly adding up terms in (4.2.15), we get

$$\sum_{I\in\mathfrak{A}} R_I(\phi w)R_I(\psi\sigma) \le [w]_{A_2}^{1-\alpha}\sum_{I\in\mathfrak{A}} \inf_I M_w(|\phi|^p)(x)^{1/p}\inf_I M_\sigma(|\psi|^p)(x)^{1/p}\,\tau_I, \tag{4.2.18}$$

To continue to estimate (4.2.17) and (4.2.18), we need now the following lemma.

LEMMA 4.2.11 *Let $\{\alpha_L\}_{L\in\mathcal{D}}$ be a Carleson sequence with Carleson constant B related to partition lattice $\mathcal{D}$. Let σ be any positive μ-integrable function. Let F be a positive function on X measurable with respect to a partition. Then*

$$\sum_L \frac{\inf_L F}{\langle\sigma\rangle_L}\alpha_L \le C\,B\int_X \frac{F}{\sigma}\,d\mu. \tag{4.2.19}$$

$$\sum_L (\inf_L F)\,\alpha_L \le 2B\int_X F\,d\mu. \tag{4.2.20}$$

REMARK 4.2.12 Notice that one cannot replace in this lemma $\inf_L F$ by bigger quantities $\langle F\rangle_L$.

PROOF Let us prove, for example, (4.2.19). We can think that Carleson constant of $\{\alpha_\ell\}$ is equal to $1/4$. Obviously for every L, $\mu\{x\in L : \sigma(x)\le 2\langle\sigma\rangle_{\mu,L}\}\ge \frac12\mu(L)$. We denote $\mathcal{B}_L := \{x\in L : \sigma(x)\le 2\langle\sigma\rangle_{\mu,L}\}$. We want to choose for every L a subset $E_L\subset\mathcal{B}_L$ in such a way, that all these sets are disjoint and $\mu(E_L)=\alpha_L$. We can think that the number of elements of partition, where $\alpha_L>0$, is finite.

We start with the smallest L (smallest by inclusion). They are disjoint, and the choice of measurable sets E_L in these smallest cubes is arbitrary with $\mu(E_L)=\alpha_L$. Suppose we want to make a choice in a certain element of partition Q, and the choice in all its descendants ℓ is already made. Notice that $\sum_{\ell \text{ descendants of } Q}\mu(E_\ell)=\sum\alpha_\ell\le\frac14\mu(Q)$. We said that $\mu(\mathcal{B}_Q)\ge\frac12\mu(Q)$. So $\mu(\mathcal{B}_Q\setminus\cup_{\ell \text{ descendants of } Q}E_\ell)\ge\frac14\mu(Q)$. Hence, it has enough space for the set of measure α_Q less or equal than $\frac14\mu(Q)$. So choose an arbitrary measurable $E_Q\subset\mathcal{B}_L\setminus\cup_{\ell \text{ descendants of } Q}E_\ell)$ of measure α_Q. Notice that on this set (which **is disjoint** with E_ℓ for all descendants), we also have

$$\frac{1}{\langle\sigma\rangle_Q}\le\frac{2}{\sigma(x)},\ \forall x\in E_Q.$$

Now it is obvious that

$$\sum_L \frac{\inf_L F}{\langle\sigma\rangle_L}\alpha_L \le 2\sum_L\int_{E_L}\frac{F(x)}{\sigma(x)}\,dx.$$

We are done with (4.2.19), where $C=8$. The lemma is proved. □

Now we apply the first part of the lemma to (4.2.17) with $F = M_w(|\phi|^p)(x)^{2/p}$. Then we can continue the estimate as follows:

$$\sum_{I\in\mathfrak{A}} R_I(\phi w)S_I(\psi\sigma) \le [w]_{A_2} \|\psi\|_\sigma \left(\int_X M_w(|\phi|^p)(x)^{2/p}\, w\, d\mu\right)^{1/2}$$
$$\le C(p)[w]_{A_2} \|\psi\|_\sigma \|\phi\|_w \tag{4.2.21}$$

because martingale maximal operator is bounded in $L^r(X, w\, d\mu)$, $r = 2/p$, with a constant that depends only on p, if $p < 2$. Again the symmetric estimate is also valid.

Similarly from (4.2.18) and the second part of Lemma 4.2.11 with $F = \inf_I \left[M_w(|\phi|^p)(x)^{1/p}\, M_\sigma(|\psi|^p)(x)^{1/p}\right]$, we get

$$\sum_{I\in\mathfrak{A}} R_I(\phi w)R_I(\psi\sigma) \le [w]_{A_2} \int_X M_w(|\phi|^p)(x)^{1/p}\, M_\sigma(|\psi|^p)(x)^{1/p}\, d\mu$$
$$= [w]_{A_2} \int_X M_w(|\phi|^p)(x)^{1/p}\, M_\sigma(|\psi|^p)(x)^{1/p}\, w^{1/2}\sigma^{1/2} d\mu$$
$$\le [w]_{A_2} \left(\int_X M_w(|\phi|^p)(x)^{2/p}\, w\, d\mu\right)^{1/2} \left(\int_X M_\sigma(|\psi|^p)(x)^{2/p}\, \sigma\, d\mu\right)^{1/2}$$
$$\le C(p)[w]_{A_2}, \|\phi\|_w \|\psi\|_\sigma. \tag{4.2.22}$$

We again used the boundedness of martingale maximal function independent of the measure involved in its definition. In the second inequality, we used a trivial (but indispensable) observation that

$$1 = w \cdot \sigma.$$

Finally, Theorem 4.2.2 is proved for shifts of complexity 0. To prove it for all shifts with the correct estimate in complexity, we will use the Bellman function to transfer the result.

4.2.5 Bellman Transfer Argument

In fact, we have built a funny concave function. Let us show what kind of concave function we have built. Then we use this function to transfer the estimate from $\mathbb{S}_{0,0}$ to $\mathbb{S}_{m,n}$ with arbitrary m, n.

Let us consider the usual dyadic filtration in $\mathbb{R}^1$ (just any dyadic lattice $\mathcal{D}$ on the real line). Let measure μ be just the Lebesgue measure on $\mathbb{R}^1$. Of course, the 0 shift with respect to this lattice and with respect to the Lebesgue measure is a particular case of what we just proved. We denote below as $|I|$ the Lebesgue measure of interval I, and Δ_I is the difference of averages over the left and the right halves of I.

Above, we proved the following inequality for all ϕ, ψ, w, and $\sigma = w^{-1}$ defined on the real line:

$$\frac{1}{|J|}\sum_{I\in\mathcal{D}:\, I\subset J} |\Delta_I(\phi w)||\Delta_I(\psi\sigma)||I| \le C\,[w]_{A_2}\,\langle \phi^2 w\rangle_I^{1/2}\,\langle \psi^2\sigma\rangle_I^{1/2}. \tag{4.2.23}$$

Recall that the symbol A_2^d denotes the dyadic A_2 class. Here is the second meeting with the Bellman function. We put

Definition 4.2.13

$$B_Q(x,y,X,Y,F,G) := \sup\left\{\frac{1}{|J|}\sum_{I\in\mathcal{D}:\, I\subset J} |\Delta_I(\phi w)||\Delta_I(\psi\sigma)||I| :\right.$$

$$w\in A_2^d,\, [w]_{A_2^d} = Q,\, \langle w\rangle_J = x,\, \langle \sigma\rangle_J = y,$$

$$\left.\langle \phi w\rangle_J = X,\, \langle \psi\sigma\rangle_J = y,\, \langle \phi^2 w\rangle_J = F,\, \langle \psi^2\sigma\rangle_J = G\right\}.$$

It is a well-defined function about which we know several things collected in the following theorem.

Theorem 4.2.14 *Function B is defined in the domain*

$$\Omega_Q := \{(x,y,X,Y,F,G)\ge 0\colon 1\le x\cdot y\le Q, X^2\le Fx, Y^2\le Gy\}.$$

Function $B_Q = B$ does not depend on J. Also

$$0\le B_Q\le C\,Q\,\sqrt{FG}.$$

And finally, B_Q is concave in the domain Ω, and, moreover, the following main inequality holds. If ζ, ζ_-, ζ_+ are three points in Ω, $\zeta = (x,y,X,Y,F,G), \zeta_\pm = (x_\pm, y_\pm, X_\pm, Y_\pm, F_\pm, G_\pm)$ such that $\zeta = (\zeta_+ + \zeta_-)/2$, then

$$B_Q(\zeta) - \left(\frac{B_Q(\zeta_+) + B_Q(\zeta_-)}{2}\right) \ge |X_+ - X_-||Y_+ - Y_-|. \tag{4.2.24}$$

Proof Only the property (4.2.24) requires the proof. Let $\varepsilon > 0$. Let $w_+ \in A_2^d, \phi_+, \psi_+$ almost realize the supremum in Definition 4.2.13 for data ζ_+. Here "almost" means that

$$\frac{1}{|J_+|}\sum_{I\in\mathcal{D}:\, I\subset J_+} |\Delta_I(\phi w)||\Delta_I(\psi\sigma)||I| \ge B_Q(\zeta_+) - \varepsilon.$$

Similarly, let $w_- \in A_2^d, \phi_-, \psi_-$ almost realize the supremum in Definition 4.2.13 for data ζ_-. Here "almost" means that

$$\frac{1}{|J_-|} \sum_{I \in \mathcal{D}\colon I \subset J_-} |\Delta_I(\phi w)||\Delta_I(\psi\sigma)||I| \geq B_Q(\zeta_-) - \varepsilon.$$

Notice that we already used the simple but important fact that $B_Q(\cdot)$ does not depend on dyadic interval J. Now we consider the following candidate for giving the supremum in Definition 4.2.13:

$$w = \begin{cases} w_+, \text{ on } J_+, \\ w_-, \text{ on } J_-, \end{cases} \quad \phi = \begin{cases} \phi_+, \text{ on } J_+, \\ \phi_-, \text{ on } J_-, \end{cases} \quad \psi = \begin{cases} \psi_+, \text{ on } J_+, \\ \psi_-, \text{ on } J_-. \end{cases}$$

One point should be clarified, namely, why the new w is in $A_2^d(J)$ and why $[w]_{A_2^d} \leq Q$? This statement involves checking A_2 condition on all dyadic intervals inside J. But for dyadic interval $I \in \mathcal{D}, I \subset J$, only three things may happen a) $I \subset J_+$, b) $I \subset J_-$, c) $I = J$. In cases a) and b), we use the facts that $w_\pm \in A_2^d(J_\pm)$ and $[w_\pm]_{A_2^d(J_\pm)} \leq Q$. So we are left with checking only c): $\langle w \rangle_J \langle \sigma \rangle_J \leq Q$. But $\langle w \rangle_J \langle \sigma \rangle_J = x \cdot y$ and this quantity is bounded by Q because $\zeta \in \Omega_Q$ just by the definition of Ω_Q.

Therefore, (w, ϕ, ψ) is a candidate for realizing $B_Q(\zeta)$. In particular,

$$\begin{aligned} B_Q(\zeta) &\geq \frac{1}{|J|} \sum_{I \in \mathcal{D}\colon I \subset J} |\Delta_I(\phi w)||\Delta_I(\psi\sigma)||I| \\ &|\Delta_J(\phi w)||\Delta_J(\psi\sigma)| \\ &\quad + \frac{1}{2} \cdot \frac{1}{|J_+|} \sum_{I \in \mathcal{D}\colon I \subset J_+} |\Delta_I(\phi w)||\Delta_I(\psi\sigma)||I| \\ &\quad + \frac{1}{2} \cdot \frac{1}{|J_-|} \sum_{I \in \mathcal{D}\colon I \subset J_-} |\Delta_I(\phi w)||\Delta_I(\psi\sigma)||I| \\ &\geq |\Delta_J(\phi w)||\Delta_J(\psi\sigma)| + \frac{1}{2} B(\zeta_+) + \frac{1}{2} B(\zeta_-) - 2\varepsilon \\ &= |X_+ - X_-||Y_+ - Y_-| + \frac{1}{2} B(\zeta_+) + \frac{1}{2} B(\zeta_-) - 2\varepsilon, \end{aligned}$$

which exactly means (4.2.24) if one tends ε to zero. □

Now we will prove that automatically such a function has even more interesting properties. And this will allow us to pass to the right estimate of (m, n) shifts with arbitrary m, n.

4.2.5.1 General Shifts. The theorem below was proved in [**161**] for $n = 2$ and then in [**181**] in general. We mention this just to emphasize that it is not at all that evident even for the case $n = 2$.

Let B_Q be as above.

THEOREM 4.2.15 *Let $\zeta, \zeta_k \in \Omega_Q$ satisfy*

$$\zeta = \frac{1}{2^n}\sum_{k=1}^{2^n} \zeta_k, .$$

Assume that $\{\zeta_k\}_{k=1}^{2^n}$ can be split into disjoint sets $\{\zeta_k^-\}_{k=1}^{2^{n-1}}, \{\zeta_k^+\}_{k=1}^{2^{n-1}}$, such that barycenters $\zeta^- := \frac{1}{2^{n-1}}\sum_{k=1}^{2^{n-1}} \zeta_k^-$, $\zeta^+ := \frac{1}{2^{n-1}}\sum_{k=1}^{2^{n-1}} \zeta_k^+$ also belong to Ω_Q. Moreover, assume that we can split $\{\zeta_k^\pm\}_{k=1}^{2^{n-1}}$ again into two sets of cardinality 2^{n-2} with barycenters in Ω_Q, etc.

Then with an absolute positive constant c, which does not depend on n, we have

$$B_{9Q}(\zeta) - \frac{1}{2^n}\sum_{k=1}^{2^n} B_{9Q}(\zeta_k) \ge c \cdot \left(\frac{1}{2^n}\sum_{k=1}^{2^n}|X_k - X|\right)\left(\frac{1}{2^n}\sum_{k=1}^{2^n}|Y_k - Y|\right). \tag{4.2.25}$$

First we need the following simple lemma about large collections of vectors in $\mathbb{R}^2$.

LEMMA 4.2.16 *Let $\{z_k\}_{k=1}^N$ be a collection of vectors in $\mathbb{R}^2$, $z_k = (x_k, y_k)$, and let $(0,0)$ be in their convex hull Z, namely $\sum_{k=1}^N \alpha_k x_k = 0$, $\sum_{k=1}^N \alpha_k y_k = 0$, where $\{\alpha_k\}$ are some convex coefficients. Then one can find a point $z = (x, y)$ such that $z, -z \in Z$ and such that, $|x| \ge c\sum_{k=1}^N \alpha_k|x_k|$ and $|y| \ge c\sum_{k=1}^N \alpha_k|y_k|$, where c is an absolute positive constant.*

PROOF It will be convenient to identify the vectors in $\mathbb{R}^2$ with complex numbers. Consider number $v_+ := \sum_{k:\, y_k \ge 0} \alpha_k z_k$. The fact that $(0,0) \in Z$ implies easily that $v_+ \in Z$. Notice that $\Im v_+ = \frac{1}{2}\sum_{k=1}^N \alpha_k|y_k|$. This is easy, because

$$\sum_{k=1}^N \alpha_k|y_k| = \sum_{k:\, y_k \ge 0} \alpha_k y_k - \sum_{k:\, y_k \le 0} \alpha_k y_k =: A - B,$$

but $A + B = 0$. Consider similarly, $v_- := \sum_{k:\, y_k \le 0} \alpha_k z_k = -v_+$. We also introduce $h_+ := \sum_{k:\, x_k \ge 0} \alpha_k z_k, h_- := \sum_{k:\, y_k \le 0} \alpha_k z_k$. Again $h_- = -h_+$. As before $h_-, h_+ \in Z$. This time, we can say that $\Re h_+ = \frac{1}{2}\sum_{k=1}^N \alpha_k|x_k|$. If we suddenly have that $|\Im h_+| \ge \Im v_+$, then we can put $z = h_+$ and finish the proof. If $|\Re v_+| \geq \Re h_+$, we can put $z = v_+$. Otherwise, we have

$$\Im v_+ > |\Im h_+|, \Re h_+ > |\Re v_+|.$$

In this case, consider two figures: one is parallelogram $\mathcal{P}$ with vertices in v_+, v_-, h_+, h_-, another is the rectangle $\mathcal{R} := [-\Re h_+, \Re h_+] \times [-\Im v_+, \Im v_+]$. Notice that $\mathcal{P} \subset \mathcal{R}$, and that each side of the rectangle contains one vertex of parallelogram $\mathcal{P}$. Consider the cross C defined as follows. At each vertex of $\mathcal{R}$ draw a dilated rectangle, whose sides have lengths equal to $\frac{5}{12}$ of the lengths of the corresponding sides of $\mathcal{R}$. We have four rectangles inside $\mathcal{R}$, which are similar to $\mathcal{R}$ with coefficient $\frac{5}{12}$. Delete them from $\mathcal{R}$. Then we get what we call cross C.

Here is an observation: any parallelogram such as $\mathcal{P}$, that has one vertex on each side of $\mathcal{R}$ is necessarily not covered by C. This is an easy geometric observation, immediately clear from the figure, and the fact that without loss of generality $\mathcal{R}$ can be a square.

This means that one of the deleted rectangles intersects $\mathcal{P}$. Consider a point z in the intersection. As $\mathcal{P} \subset Z$ (because all the vertices of $\mathcal{P}$ belong to Z), so $z \in Z$. On the other hand, the fact that z belongs to one of the deleted rectangles means, in particular, that

$$|\Re z| \geq \frac{1}{12}\Re h_+ \geq \frac{1}{24}\sum_{k=1}^{N} \alpha_k |x_k|, \; |\Im z| \geq \frac{1}{12}\Re v_+ \geq \frac{1}{24}\sum_{k=1}^{N} \alpha_k |y_k|.$$

This is what the lemma claims. □

We want to notice that $z \in Z$ can be written as follows:

$$z = \sum_{k=1}^{N} \alpha_k (1 + \beta_k) z_k, \; |\beta_k| \leq 1, \; \sum_{k=1}^{N} \alpha_k \beta_k = 0, \tag{4.2.26}$$

To explain this, let us remind that z is a convex combination of v_+, v_-, h_+, h_-. But, for example, v_+ is just the sum $\sum_{k=1}^{N} \alpha'_k z_k$, where $\alpha'_k = \alpha_k$ or 0. Therefore, the coefficients of v_+ in its decomposition over z_k are at most α_k, and, thus can be written as the sum of $\alpha_k p_k$, $0 \leq p_k \leq 1$. The same is true for v_-, h_+, h_-. So their convex combination z can be written as

$$z = \sum_{k=1}^{N} \alpha_k \gamma_k z_k, 0 \leq \gamma_k \leq 1.$$

Next we give another representation of z. We denote $c = \sum_{k=1}^N \alpha_k \gamma_k$. Then $0 < c \leq 1$. We use now that $0 = \sum_{k=1}^N \alpha_k z_k$. Then

$$z = (1-c)\sum_{k=1}^N \alpha_k z_k + \sum_{k=1}^N \alpha_k \gamma_k z_k = \sum_{k=1}^N [(1-c)+\gamma_k]\alpha_k z_k.$$

We notice now two facts: 1) $[(1-c)+\gamma_k] \in [0,2]$ and so can be written as $(1+\beta_k)$, where $|\beta_k| \leq 1$; 2) the latest representation of z has $\sum_{k=1}^N [(1-c)+\gamma_k]\alpha_k = 1$. In fact,

$$\sum_{k=1}^N [(1-c)+\gamma_k]\alpha_k = (1-c)\sum_{k=1}^N \alpha_k + \sum_{k=1}^N \alpha_k \gamma_k = 1-c+c = 1.$$

Gathering things together, we get $\sum_{k=1}^N (1+\beta_k)\alpha_k = 1 = \sum_{k=1}^N \alpha_k$ that implies

$$\sum_{k=1}^N \beta_k \alpha_k = 0,$$

and this is what we wanted to show.

Then using that $0 = \sum_{k=1}^N \alpha_k z_k$, we can also write

$$-z = \sum_{k=1}^N \alpha_k (1-\beta_k) z_k,\ |\beta_k| \leq 1,\ \sum_{k=1}^N \alpha_k \beta_k = 0.$$

It is convenient to use the notations

$$\Delta_x := \sum_{k=1}^N \alpha_k |x_k|,\ \Delta_y := \sum_{k=1}^N \alpha_k |y_k|.$$

We know already that

$$\left|\sum_{k=1}^N \alpha_k \beta_k x_k\right| = |\Re z| \geq \frac{1}{24}\Delta_x\,,\ \left|\sum_{k=1}^N \alpha_k \beta_k y_k\right| = |\Im z| \geq \frac{1}{24}\Delta_y. \quad (4.2.27)$$

Let us introduce

$$P := \sum_{k=1}^N \alpha_k (1+\beta_k/5) z_k.$$

Then, of course $-P = \sum_{k=1}^N \alpha_k (1-\beta_k/5) z_k$. And still $P, -P \in Z$. This is because $\{\alpha_k (1 \pm \beta_k/5)\}_{k=1}^N$ still form collections of convex coefficients.

It also follows immediately from (4.2.27) that

$$|\Re P| = \frac{1}{5}|\sum_{k=1}^{N} \alpha_k \beta_k x_k|| \geq \frac{1}{120}\Delta_x, \ |\Im P| = \frac{1}{5}|\sum_{k=1}^{N} \alpha_k \beta_k y_k| \geq \frac{1}{120}\Delta_y. \tag{4.2.28}$$

COROLLARY 4.2.17 *Let* $\{v_k = (x_k, y_k, X_k, Y_k, \dots, \dots)\}_{k=1}^N$ *be points in the non-convex domain* $\Omega_Q \subset \mathbb{R}^6$ *introduced before Theorem 4.2.15. And let* $v = (x, y, X, Y, \dots, \dots)$ *be the barycenter:* $v = \sum_{k=1}^N \alpha_k v_k$. *Then one can find points* $v^\pm$ *in the convex hull of* $\{v_k\}_{k=1}^N$ *such that* $v^\pm \in \Omega_{4Q}$ *and* $|X - X^\pm| \geq \frac{1}{120}\sum_{k=1}^N \alpha_k |X - X_k|$, $|Y - Y^\pm| \geq \frac{1}{120}\sum_{k=1}^N \alpha_k |Y - Y_k|$.

PROOF We just apply Lemma 4.2.16 to $z_k := (X_k, Y_k) - (x, y)$, $k = 1, \dots, N$. This lemma defines special convex combination coefficients and two points $(X^\pm, Y^\pm)$ that are very special convex combinations of points (X_k, Y_k), such that $(X, Y) = \frac{1}{2}((X^+, Y^+) + (X^-, Y^-))$

$$|X - X^\pm| \geq \frac{1}{120}\sum_{k=1}^{N} \alpha_k |X - X_k|, \ |Y - Y^\pm| \geq \frac{1}{120}\sum_{k=1}^{N} \alpha_k |Y - Y_k|.$$

Define $(x^\pm, y^\pm) = \sum_{k=1}^N \alpha_k (1 \pm \beta_k)(x_k, y_k)$, where $|\beta_k| < 1, k = 1, \dots, N$. Then $x^\pm \leq 2x, y^\pm \leq 2y$ (we used here that all x_k, y_k are nonnegative). Corollary is proved if we put

$$v^\pm := \sum_{k=1}^{N} \alpha_k (1 \pm \beta_k) v_k,$$

as it is easy to check that $v^\pm \in \Omega_{4Q}$ just because $x^\pm y^\pm \leq 4xy \leq 4Q$. □

Now we are ready to prove Theorem 4.2.15. The idea is taken from S. Treil's [**181**], but the reasoning is somewhat modified.

PROOF For any vector $\zeta \in \mathbb{R}^6$, we will denote $(\zeta)_i, i = 1, \dots, 6$, its i-th coordinate. We specify the previous construction to $N = 2^n, \alpha_k = \frac{1}{2^n}$, and we found special $\beta_k, |\beta_k| \leq 1$, such that points $P_+ := \sum_{k=1}^N \frac{1}{2^n}(1 + \beta_k/5)\zeta_k$, $P_- := \sum_{k=1}^N \frac{1}{2^n}(1 - \beta_k/5)\zeta_k$ lie in the convex hull Z of $\{\zeta_k\}$, $\zeta = (P_+ + P_-)/2$, $\sum_{k=1}^{2^n} \beta_k = 0$, and

$$|(P_+ - P_-)_3| \geq \frac{1}{120}\Delta_x, \ |(P_+ - P_-)_4| \geq \frac{1}{120}\Delta_y, \tag{4.2.29}$$

where

$$\Delta_x = \frac{1}{2^n}\sum_{k=1}^{n} |X_k - X|, \ \Delta_y = \frac{1}{2^n}\sum_{k=1}^{n} |Y_k - Y|, \tag{4.2.30}$$

where X_k, Y_k are $(\zeta_k)_3, (\zeta_k)_4$, correspondingly, $k = 1, \ldots, 2^n$, X, Y are $(\zeta)_3, (\zeta)_4$ correspondingly.

We will write

$$B_{9Q}(\zeta) - \frac{1}{2^n}\sum_{k=1}^{2^n} B_{9Q}(\zeta_k) = B_{9Q}(\zeta) - \frac{1}{2}(B_{9Q}(P_+) + B_{9Q}(P_-))$$
$$+ \frac{1}{2}(B_{9Q}(P_+) - \frac{1}{2^n}\sum_{k=1}^{2^n}(1+\beta_k/5)B_{9Q}(\zeta_k)) + \frac{1}{2}(B_{9Q}(P_-)$$
$$- \frac{1}{2^n}\sum_{k=1}^{2^n}(1-\beta_k/5)B_{9Q}(\zeta_k)) = A_0 + A_1 + A_2. \tag{4.2.31}$$

Easily combining (4.2.24), (4.2.29), (4.2.30), we get the estimate of term A_0:

$$A_0 \ge \frac{1}{6^4}\left(\frac{1}{2^n}\sum_{k=1}^{n}|X_k - X|\right)\left(\frac{1}{2^n}\sum_{k=1}^{n}|Y_k - Y|\right). \tag{4.2.32}$$

Let us see now that

$$A_1 \ge 0, \tag{4.2.33}$$

(and symmetrically $A_2 \ge 0$). That would be easy if Ω_Q would be convex. This convexity and the fact that $P_+ = \frac{1}{2^n}\sum_{k=1}^{2^n}(1+\beta_k/5)\zeta_k$ would immediately prove (4.2.33), even with B_Q instead of B_{9Q} in the definition of A_1. However, Ω_Q is not convex at all, and so the convex hull Z of points $\{\zeta_k\}$ can go very much outside Ω_Q breaking the possibility to use the concavity of B_Q (or even B_{9Q}) on Z.

However, let us start to split vectors v_k into groups as in the statement of the theorem: all of them is group G_n^0, groups G_{n-1}^1, G_{n-1}^2 are two groups of 2^{n-1} cardinality each, then G_{n-1}^1 is split to G_{n-2}^1, G_{n-2}^2, G_{n-1}^2 is split to G_{n-2}^3, G_{n-2}^4, etc. Groups G_0^s will consist of one element each. Let us denote a generic ζ_k by symbol u. Then the barycenter in G_m^j is $\zeta_m^j = \frac{1}{2^m}\sum_{u\in G_m^j} u$. By one of the assumption of our theorem, barycenter, ζ_m^j ($j = 1, \ldots, 2^{n-m}$) belongs to Ω_Q for all m, j.

The coefficient β_k corresponding to $u = \zeta_k$ will be called $\beta(u)$. Let us notice that for any group $G = G_m^j$

$$s_G := \sum_{u\in G}\frac{1}{2^n}(1+\beta(u)/5) \in \frac{\operatorname{card}(G)}{2^n}[4/5, 6/5]. \tag{4.2.34}$$

For any group G we already considered its barycenter $\zeta_G = \frac{2^n}{\text{card}(G)} \sum_{u \in G} \frac{1}{2^n} u$. Now we set

$$b_G := \frac{1}{s_G} \sum_{u \in G} \frac{1}{2^n} (1 + \beta(u)/5) u.$$

It is an assumption of the theorem that $\zeta_G \in \Omega_Q$. Let us see that

$$b_G \in \Omega_{\frac{9}{4}Q}, \ \forall G. \tag{4.2.35}$$

For that, we need to check how big is the product of the first two coordinates $(b_G)_1 \cdot (b_G)_2$. But by (4.2.34) and by the fact that $|\beta(u)| \leq 1$ for all our points u, we establish that

$$(b_G)_1 \leq \frac{5}{4} \cdot \frac{6}{5} (\zeta_G)_1, \ (b_G)_2 \leq \frac{5}{4} \cdot \frac{6}{5} (\zeta_G)_2.$$

Therefore, $(b_G)_1 \cdot (b_G)_2 \leq \frac{9}{4} Q$, and so (4.2.35) is valid.

Now notice that if a group G is split to two groups E, F, then $s_G b_G = s_E b_E + s_F b_F$, therefore,

$$B_{9Q}(b_G) - \left[\frac{s_E}{s_G} B_{9Q}(b_E) + \frac{s_F}{s_G} B_{9Q}(b_F) \right] \geq 0.$$

In fact, points b_G, b_E, b_F all lie in $\Omega_{\frac{9}{4}Q}$, as has been just shown. On the other hand, coefficients $\frac{s_E}{s_G}, \frac{s_E}{s_G}$ are convex coefficients and we know from (4.2.34) that each of them lies in $[1/4, 3/4]$. Consider the segment $\mathcal{L}$ on which all three points b_G, b_E, b_F lie. It is not obliged to lie in $\Omega_{\frac{9}{4}Q}$. But the simple geometry shows that $\mathcal{L} \subset \Omega_{9Q}$. Then function B_{9Q} is concave on the whole $\mathcal{L}$ and the previous estimate is proved.

Let us rewrite what we just proved as follows:

$$s_G B_{9Q}(b_G) - [b_E B_{9Q}(b_E) + s_F B_{9Q}(s b_F)] \geq 0. \tag{4.2.36}$$

Apply this inequality first with $G = G_n^0$, $E = G_{n-1}^1, F = G_{n-1}^2$, then $G = G_{n-1}^1$, $E = G_{n-2}^1$, $F = G_{n-2}^2$, G_{n-1}^2 and $G = G_{n-1}^2$, $E = G_{n-2}^3$, $F = G_{n-2}^2$, G_{n-1}^4, etc. Finally adding the telescopic sums, we will get

$$B_{9Q}(\zeta) - \sum_{k=1}^{2^n} \frac{1}{2^n} (1 + \beta(\zeta_k)/5) B_{9Q}(\zeta_k) \geq 0. \tag{4.2.37}$$

This is exactly (4.2.33). Absolutely symmetrically we would prove the same for A_2 (see the definition of A_2 in (4.2.31)):

$$B_{9Q}(\zeta) - \sum_{k=1}^{2^n} \frac{1}{2^n} (1 - \beta(\zeta_k)/5) B_{9Q}(\zeta_k) \geq 0. \tag{4.2.38}$$

Notice that we used a simple remark that $s_{G_n^0} = \sum_{u\in G_n^0} \frac{1}{2^n}(1 + \beta(u)/5) = \sum_{k=1}^{2^n} \frac{1}{2^n}(1 + \beta(\zeta_k)/5) = 1$. This is where we used the condition $\sum_{k=1}^{2^n} \beta_k = 0$. This latter condition is just (4.2.26) for our case $\alpha_k = \frac{1}{2^n}, k = 1, \dots, 2^n$.

Adding two last inequalities and taking into account (4.2.31) and (4.2.32) we finish the proof of Theorem 4.2.15. □

This theorem easily proves a linear in complexity and linear in $[w]_{A_2}$ estimate of weighted dyadic shifts, thus proving the A_2 conjecture. Let us do that.

4.2.5.2 *Sharp Estimates for a Wide Class of Dyadic Singular Operators.* Notice first of all that convex coefficients $\frac{1}{2^n}$ played no role in the proof of Theorem 4.2.15. We can prove a more general result. Let B_Q be as above.

THEOREM 4.2.18 *Let $\zeta, \zeta_k \in \Omega_Q$, and ζ is a convex combination of $\{\zeta_k\}_{k=1}^{2^n}$:*

$$\zeta = \sum_{k=1}^{2^n} \alpha_k \zeta_k.$$

Assume that $\{\zeta_k\}_{k=1}^{2^n}$ can be split into disjoint sets $\{\zeta_k^-\}_{k=1}^{2^{n-1}}, \{\zeta_k^+\}_{k=1}^{2^{n-1}}$ such that "barycenters"

$$\zeta^- := \frac{1}{\sum_{k=1}^{2^{n-1}} \alpha_k^-} \sum_{k=1}^{2^{n-1}} \alpha_k^- \zeta_k^-, \; \zeta^+ := \frac{1}{\sum_{k=1}^{2^{n-1}} \alpha_k^+} \sum_{k=1}^{2^{n-1}} \alpha_k^+ \zeta_k^+$$

also belong to Ω_Q. Here $\{\alpha_k^-\}, \{\alpha_k^+\}$ is the splitting of convex coefficients $\{\alpha_k\}_{k=1}^{2^n}$. Moreover, assume that we can split $\{\zeta_k^\pm\}_{k=1}^{2^{n-1}}$ again into two sets of cardinality 2^{n-2} with corresponding "barycenters" in Ω_Q, etc. ….

Then with an absolute positive constant c, which does not depend on n and does not depend on convex coefficients $\{\alpha_k\}_{k=1}^{2^n}$, we have

$$B_{9Q}(\zeta) - \sum_{k=1}^{2^n} \alpha_k B_{9Q}(\zeta_k) \geq c \cdot \left(\sum_{k=1}^{2^n} \alpha_k |X_k - X| \right) \left(\sum_{k=1}^{2^n} \alpha_k |Y_k - Y| \right). \tag{4.2.39}$$

PROOF The main part of the proof of Theorem 4.2.15, where two special points $P_\pm = \sum_k \alpha_k (1 \pm \beta_k)\zeta_k$ are constructed has been written with general convex coefficients. So we need to check only that the last part of the proof of Theorem 4.2.15 can be carried out. This is the part where the estimate (4.2.36) is proved. We again have the groups $G = G_n^0, G_{n-1}^1, G_{n-1}^2, \dots, G_0^s, s = 1, \dots, 2^n$, but now for a generic group G, we define ζ_G, s_G, b_G as follows (we call by letter u a generic element of the set $\{\zeta_k\}_{k=1}^{2^n}$, and $\alpha(u)$, $\beta(u)$ stand for corresponding α_k, β_k):

$$\zeta_G = \frac{1}{\sum_{u \in G} \alpha(u)} \sum_{u \in G} \alpha(u) u,$$

$$s_G = \sum_{u \in G} \alpha(u)(1 + \beta(u)/5),$$

$$b_G = \frac{1}{\sum_{u \in G} \alpha(u)(1 + \beta(u)/5)} \sum_{u \in G} \alpha(u)(1 + \beta(u)/5) u.$$

If for every group G, the geometric property (4.2.35) for these new defined b_G holds, then the proof finishes verbatim as before. To establish (4.2.35), let us compare the first coordinate $(b_G)_1$ of $\mathbb{R}^6$ vector b_G with the first coordinate $(\zeta_G)_1$ of the $\mathbb{R}^6$ vector ζ_G. For that, let us notice that

$$\frac{\sum_{u \in G} \alpha(u)}{\sum_{u \in G} \alpha(u)(1 + \beta(u)/5)} \in \left[\frac{5}{6}, \frac{5}{4}\right]$$

just because $|\beta(u)| \leq 1$ for any u from our set $\{\zeta_k\}_{k=1}^{2^n}$. Also

$$\frac{\sum_{u \in G} \alpha(u)(1 + \beta(u)/5)(u)_1}{\sum_{u \in G} \alpha(u)(u)_1} \in \left[\frac{4}{5}, \frac{6}{5}\right]$$

by the same reason and by the fact that all first coordinates $(u)_1$ are positive. We conclude then that

$$(b_G)_1 \leq \frac{3}{2}(\zeta_G)_1.$$

Symmetrically,

$$(b_G)_2 \leq \frac{3}{2}(\zeta_G)_2.$$

Automatically, we conclude that for any group G, the vector b_G lies in $\Omega_{\frac{9}{4}} Q$. As we noticed, this is enough to finish the proof of Theorem 4.2.18. □

Theorem 4.2.18 will give us now the linear in weight and linear in complexity estimate of the so called L^1-normalized shifts. Let us consider $(X, \mathfrak{A}, \mu)$, the space with σ-algebra and measure. The σ-algebra will be generated by the filtrations $\mathcal{L}_k$ as at the beginning of this section.

We will not even assume the doubling property of the filtrations with respect to μ. For the sake of simplicity, we will assume that each set $A \in \mathcal{L}_k$ has exactly two sons A_+, A_- in $\mathcal{L}_{k+1}$. This is not essential, the result of this section is true with any martingale filtration, but the proof will be given for "non-doubling" dyadic filtrations.

Recall that at the beginning of Section 4.2.1, we defined certain martingale shift operators. We call by $\mathbb{S}_{m,n}$ the operator given by the kernel

$$f \to \sum_{L\in\mathfrak{A}} \int_L a_L(x,y) f(y) d\mu(y),$$

where

$$a_L(x,y) = \sum_{\substack{I\subset L, J\subset L\\ g(I)=g(L)+m,\, g(J)=g(L)+n}} c_{L,I,J} h_J(x) h_I(y),$$

where h_I, h_J are Haar functions normalized in $L^2(d\mu)$ with supports on atoms $I \in \mathcal{L}_{g(L)+m}, I \in \mathcal{L}_{g(L)+n}$.

Definition 4.2.19 Below we denote by $\mathfrak{A}_k(L)$ the collection of all atoms I such that $g(I) = g(L) + k$.

We assumed (for simplicity only) the martingales to be dyadic, so there is only one Haar function per atom, but we did not assume any doubling from μ on our dyadic martingale, so there is no property (4.2.1).

One can consider the different classes of martingale shifts $\mathbb{S}_{m,n}$ by imposing different requirements on the kernels a_L, $L \in \mathfrak{A}$, or on the coefficients $c_{L,I,J}$.

One possible class is when we require that operators with kernels a_L be bounded operators from $L^2(L,\mu)$ to itself with norm at most 1. We call such (m,n) shifts L^2-normalized martingale shifts of complexity (m,n).

An obvious subclass is given by normalization

$$\forall L \in \mathfrak{A} \qquad \sum_{I\in\mathfrak{A}_m(L), J\in\mathfrak{A}_n(L)} |c_{L,I,J}|^2 \le 1. \tag{4.2.40}$$

In fact, this gives us that the Hilbert–Schmidt norm of the operators with kernels a_L is at most 1, or

$$\int_{L\times L} |a_L(x,y)|^2 d\mu(x) d\mu(y) \le 1. \tag{4.2.41}$$

In particular, there is an even more narrow class of shifts, which we will call L^1-normalized martingale shifts of complexity (m,n). This class is given by the requirement

$$|a_L(x,y)| \le \frac{1}{\mu(L)}. \tag{4.2.42}$$

REMARK 4.2.20 It turns out that L^1-normalized martingale shifts are all we need for decomposition of CZ operators on spaces of homogeneous type into shifts. In fact, Theorem 4.2.4 proved in [**125**] (above we proved a particular Euclidean case of it, Theorem 4.1.8) decomposes CZ operator into shifts of a special form; they look like above, but have the following estimate on coefficients:

$$|c_{L,I,J}| \leq \frac{\sqrt{\mu(I)}\sqrt{\mu(J)}}{\mu(L)}. \tag{4.2.43}$$

But it is immediately clear that estimate (4.2.1) holds for $\|h_I\|_\infty, \|h_J\|_\infty$ (with the constant depending on the doubling of μ). Hence, automatically, the kernel $a_L(x,y) = \sum_{I\in\mathfrak{A}_m(L),J\in\mathfrak{A}_n(L)} c_{L,I,J}h_J(x)h_I(y)$ has the estimate (4.2.42) up to a constant that depends only on doubling of μ.

The conclusion is that as soon as we give the estimate linear in weight and complexity for any L^1-normalized martingale shift of complexity (m,n), we would get the right (linear in weight) estimate of any CZ operator on spaces of homogeneous type.

Later, there will be given a simple example of L^2-normalized martingale shift that in nonhomogeneous situation will not have the right estimate.

REMARK 4.2.21 It is interesting to note that "paradoxically" one can prove a linear in weight estimate of CZ operators on weighted spaces. This is proved in [**195**] by using ideas of [**96**]. This technique goes beyond the scope of the present text.

We will give now the right estimate of any L^1-normalized martingale shift of complexity (m,n)–even with non-doubling situation!

In the proof below, the σ-algebra $\mathfrak{A}$ is assumed to be dyadic (but not necessarily doubling), but this is only for simplicity, the result holds for any martingale filtration.

THEOREM 4.2.22 *Let us have* $(X,\mathfrak{A},\mu)$ *built by filtrations. Let* $Tf = \sum_{L\in\mathfrak{A}} T_L := \sum_{L\in\mathfrak{A}} \int_L a_L(x,y)f(y)d\mu(y)$, *where*

$$a_L(x,y) = \sum_{I\in\mathfrak{A}_m(L),J\in\mathfrak{A}_n(L)} c_{L,I,J}h_J(x)h_I(y),$$

and let it be L^1*-normalized, namely, let* (4.2.42) *holds. Then*

$$\|T\|_w \leq C(m+n+1)[w]_{A_2}.$$

PROOF We can think that the number of terms in the sum that defines T is finite. We need to prove the desired estimate with C independent of this number. Recall that we have a convention to denote $\sigma := w^{-1}$. We need to estimate the bilinear form with two test functions ϕ, ψ $|(T(\phi w), \psi\sigma)|$ by $C(m+n+1)[w]_{A_2}\|\phi\|_w\|\psi\|_\sigma$. Let us think that test functions have compact support in $L_0 \in \mathfrak{A}$.

Fix $L \in \mathfrak{A}$, let $\ell = g(L)$, and consider the following projection operator $\Delta_L = 1_L \cdot (\mathcal{E}_{\ell+m+n+1} - \mathcal{E}_\ell)$. Notice two things: 1) $(T_L(\phi w), \psi\sigma) = (T_L\Delta_L(\phi w), \Delta_L(\psi\sigma))$, 2) the image of operator Δ_L consists of functions supported on L that are constant on each atom from $\mathfrak{A}_{m+n+1}(L)$, and having zero average with respect to integration by μ. The first property just follows from $L^2(\mu)$ orthogonality of all $h_K, K \in \mathfrak{A}$, the second one is obvious.

Therefore, by (4.2.42), we obviously have

$$|(T_L(\phi w), \psi\sigma)| \leq \frac{1}{\mu(L)}\|\Delta_L(\phi w)\|_{L^1(\mu)}\|\Delta_L(\psi\sigma)\|_{L^1(\mu)}.$$

If we denote a generic atom lying in L whose generation is $\ell + m + n + 1$ by letter I (there are 2^{m+n+1} of them), then we have

$$\Delta_L(\phi w) = \sum_{I \in \mathfrak{A}_{m+n+1}(L)} (\langle \phi w\rangle_I - \langle \phi w\rangle_L) \cdot 1_I.$$

Denote $\alpha(I) = \frac{\mu(I)}{\mu(L)}$ for each $I \in \mathfrak{A}_{m+n+1}(L)$. Then

$$|(T_L(\phi w), \psi\sigma)| \leq \mu(L)\left(\sum_{I \in \mathfrak{A}_{m+n+1}(L)} |\langle \phi w\rangle_I - \langle \phi w\rangle_L|\alpha(I)\right) \cdot \left(\sum_{I \in \mathfrak{A}_{m+n+1}(L)} |\langle \psi\sigma\rangle_I - \langle \psi\sigma\rangle_L|\alpha(I)\right). \tag{4.2.44}$$

We are practically under assumptions of Theorem 4.2.18.

For every L, we have a vector ζ_L in $\mathbb{R}^6$ defined coordinate wise like that:

$$(\zeta_L)_1 = \langle w\rangle_L, (\zeta_L)_2 = \langle \sigma\rangle_L, (\zeta_L)_3 = \langle \phi w\rangle_L,$$
$$(\zeta_L)_4 = \langle \psi\sigma\rangle_L, (\zeta_L)_5 = \langle |\phi|^2 w\rangle_L, (\zeta_L)_6 = \langle |\psi|^2\sigma\rangle_L.$$

It lies in Ω_Q from Theorem 4.2.18 because the weight is in A_2 and $[w]_{A_2} = Q$. We have also the collection of vectors $\zeta_I \in \mathbb{R}^6$, $I \in \mathfrak{A}_{m+n+1}(L)$ defined exactly as ζ_L above but with I replacing L. There are 2^{m+n+1} such vectors. Vector ζ_L is the convex combination of ζ_I's with coefficients $\alpha(I)$ just introduced.

Theorem 4.2.18 has also the requirement of splitting to groups and having the "barycenter" of each group inside Ω_Q. This requirement is fulfilled easily if the groups are organized as follows. Let $k = m + n + 1$. Then G_k^0 is collection of all $I \in \mathfrak{A}_k(L)$. To understand G_{k-1}^1, G_{k-1}^2, we consider L_+, L_- two atoms that are sons of L. Then all $I \in \mathfrak{A}_{k-1}(L_+)$ form G_{k-1}^1, and all $I \in \mathfrak{A}_{k-1}(L_-)$ form G_{k-1}^2. The subsequent splitting is organized by the same rule. It is obvious now that "barycenter" for G_{k-1}^1 will be vector ζ_{L_+} obtained by the rule above with L_+ replacing L. The same for G_{k-1}^2, where we will get ζ_{L_-}. The pattern says that the groups will correspond to Is from $\mathfrak{A}_k(L)$ that lie in a certain descendant A of L. Then, the corresponding "barycenter" is just ζ_A and it lies in Ω_Q just by definition.

We denoted $m + n + 1$ by k. If we now apply Theorem 4.2.18, we get

$$\mu(L)\left(\sum_{I\in\mathfrak{A}_k(L)} |\langle\phi w\rangle_I - \langle\phi w\rangle_L|\alpha(I)\right)\left(\sum_{I\in\mathfrak{A}_k(L)} |\langle\psi\sigma\rangle_I - \langle\psi\sigma\rangle_L|\alpha(I)\right)$$
$$\leq \mu(L)B_{9Q}(\zeta_L) - \sum_{I\in\mathfrak{A}_k(L)} \mu(I)B(\zeta_I).$$

But by (4.2.44), this represents the estimate of $|(T_L(\phi w), \psi\sigma)|$.

Now we start to add these inequalities together for all $L \in \mathfrak{A}(L_0)$. They are not exactly telescopic, but obviously they split to k telescopic sums.

Finally, we will get the following estimate:

$$\sum_{L\in\mathfrak{A}(L_0)} |(T_L(\phi w), \psi\sigma)| \leq k B_{9Q}(\zeta_{L_0}).$$

By Theorem 4.2.18, the right-hand side is at most $CQ\sqrt{\langle|\phi|^2 w\rangle_{L_0}}\sqrt{\langle|\psi|^2\sigma\rangle_{L_0}}$. We finished the proof of Theorem 4.2.22. We did not use the doubling property to prove the weighted estimate for L^1 normalized martingale shifts. □

4.2.5.3 An Example of Martingale Shift That Does Not Obey the Linear Weighted Estimate. In Section 4.3.13.1 we give an example of a martingale shift, which is L^2 (but not L^1) normalized, and, as a result, does not obey the right estimate.

4.2.5.4 Repeating the Scheme of the Proof: Transference Using Bellman Function. It is instructive to repeat the scheme of the proof of the linear estimate of all L^1-normalized martingale shifts (Theorem 4.2.22). We first gave a proof of the boundedness of the simplest shift of complexity $(0, 0)$. We did this in a general doubling martingale situation. The proof involved a certain convexity argument that was organized as the use of a certain concrete concave function of two real variables: $(x, y) \to (xy)^\alpha$.

We start with the linear estimates of such $(0, 0)$ shift for a simplest possible martingale: the dyadic homogeneous martingales with respect to the Lebesgue measure on the real line, namely the martingale on $(\mathbb{R}, \mathcal{D}, dx)$, where $\mathcal{D}$ is any fixed dyadic lattice on the real line. Then we use the fact that such dyadic martingale transforms have linear in $[w]_{A_2}$ characteristic estimate in weighted space $L^2(\mathbb{R}, w\,dx)$, $w \in A_2$, to prove that there exists *another* concave function.

This time, it was a concave function of six real variables. We proved its existence in Theorem 4.2.14, but notice that Theorem 4.2.14 is the existence theorem, we do not give an explicit formula for this special function. It depends on $Q = [w]_{A_2}$ and was called B_Q, it is defined for $\Omega_Q \subset \mathbb{R}^6$. It may be called *the Bellman function of the weighted homogeneous dyadic martingale problem*.

It is important to notice that the existence of B_Q of Theorem 4.2.14 is *necessary and sufficient* for dyadic martingale transforms to have linear in $[w]_{A_2}$ characteristic estimate in weighted space $L^2(\mathbb{R}, w\,dx)$. We do not lose any information by replacing the fact of the correct estimate by the fact of the existence of a certain concave function.

It is very essential – but unfortunate – that Ω_Q is not a convex domain. Exactly this fact presents us with difficulties in proving Theorem 4.2.15. This non-convexity is a signature property of *weighted* problems.

Theorem 4.2.15 can be called *the transference using Bellman function*. This theorem transfers the proof for homogeneous $(0, 0)$ shifts to any homogeneous (doubling) martingale shift of any complexity (m, n). This transference is done with the right estimate in complexity and in weight.

Moreover, the same Bellman function transfers the right estimate to any shift of any complexity in *nonhomogeneous* martingale situation as well. But shift should be L^1-normalized (otherwise the desired estimate just cannot hold). This transference is done in Theorem 4.2.18.

4.3 Universal Sufficient Condition: Boundedness of All Calderón–Zygmund Operators in Two Different Weighted Spaces

4.3.1 Bump Conjecture and Entropy Conjecture

Two-weight estimates for singular integrals of the CZ type and their close relatives, dyadic singular operators are important in various applications. They also represent a challenging problem, which has been completely solved only in a handful of situations.

When the necessary and sufficient conditions were obtained they were in terms of the so-called two-weight $T1$ theorem. This means that the two-weight boundedness of the corresponding operator was proved to be equivalent to the collection of testing conditions on the operator and its adjoint. These testing conditions are 1) not so easy to check and 2) they are not directly in terms of weights.

This is why the need for a simple universal and at the same time sharp sufficient condition on the pair of weights was always existent. This need got formulated as the so-called *bump conjecture.*

We formulate it here and we solve it here using the Bellman technique, and using the same general Bellman function/transference approach as in the previous section; see Section 4.2.5.4 for listing the steps of this approach.

In fact, here we propose another way to look at the bump conjecture. This way uses local entropies of the weights. The bump conjecture follows from the entropy approach. The entropy approach was suggested in [**182**] and our exposition below follows this work.

For the interesting operators and measures, the following joint two-weight analog of the A_2 condition is necessary for estimating (4.3.4):

$$\sup_I \left(|I|^{-1}\int_I v\,dx\right)\left(|I|^{-1}\int_I u\,dx\right) < \infty. \tag{4.3.1}$$

Simple counterexamples built in Section 4.3.14 show that this condition is not sufficient for the boundedness. So a natural way to get a sufficient condition is to "bump" the norms, i.e., to replace the L^1 norms of u and v in (4.3.1) by some stronger norms.

The first idea that comes to mind is to replace L^1 norms by $L^{1+\varepsilon}$ norms. Namely, it was proved by C. J. Neugebauer [**138**] that (4.3.1) with $v^{1+\varepsilon}$, $u^{1+\varepsilon}$ instead of v and u is sufficient for the boundedness of Calderón–Zygmund operators.

More generally, given a Young function Φ and a cube I, one can consider the normalized on I Orlicz space $L^\Phi(I)$ with the norm given by

$$\|f\|_{L^\Phi(I)} := \inf\left\{\lambda > 0 : \int_I \Phi\left(\frac{f(x)}{\lambda}\right)\frac{dx}{|I|} \le 1\right\}.$$

And it was conjectured that if the reciprocals to the Young functions Φ_1 and Φ_2 are integrable near infinity, namely,

$$\int^\infty \frac{dx}{\Phi_i(x)} < \infty,\ i = 1, 2, \tag{4.3.2}$$

then the condition

$$\sup_I \|v\|_{L^{\Phi_1}(I)} \|u\|_{L^{\Phi_2}(I)} < \infty \tag{4.3.3}$$

implies that for any bounded Calderón–Zygmund operator T, the operator $M_{v^{1/2}} T M_{u^{1/2}}$ is bounded in L^2.

This hypothesis attracted much attention of many mathematicians; Carlos Pérez, who invented it, gave it a name of "the bump conjecture."

We prove it in this section, and actually we prove a genuinely more general "entropy bump conjecture," which we are going to formulate now. Notice that after the first Bellman proof in [**122**] the bump conjecture was proved by an entirely different method by Andrei Lerner [**108**], who also proved a corresponding statement in the weighted metric L^p, $p \neq 2, 1 < p < \infty$.

4.3.2 The Entropy Bump Conjecture

Let us introduce for a weight u

$$\mathbf{u}_I := \langle u \rangle_I = \|u\|_{L^1(I)}, \qquad \mathbf{u}_I^* := \|u\|_{L\log L(I)} \approx \|M\mathbf{1}_I u\|_{L^1(I)},$$

and similarly for a weight v.

The condition on two weights

$$\sup_I \mathbf{v}_I^* \mathbf{u}_I^* < \infty$$

is not a sufficient condition for the estimate (4.3.4) for classical Calderón–Zygmund operators T:

$$\int |T(uf)|^2 v d\mu \leq C \int |f|^2 u d\mu. \tag{4.3.4}$$

The counterexample is built in Section 4.3.14.

To make up for the "smallness" of $L \log L$-bumps $\sup_I \mathbf{v}_I^* \mathbf{u}_I^*$, we introduce the "penalty term," or the entropy bump term.

Namely, let $\alpha : [1, \infty) \to \mathbb{R}_+$ be a function such that $t \mapsto t\alpha(t)$ is increasing and

$$C_\alpha := \frac{1}{\alpha(1)} + \int_1^\infty \frac{1}{t\alpha(t)} dt < \infty.$$

The entropy bump of the weight u is the following quantity:

$$\mathcal{E}_I^\alpha(u) := \mathbf{u}_I^* \alpha\left(\mathbf{u}_I^* / \mathbf{u}_I\right). \tag{4.3.5}$$

We will prove that the entropy bump condition

$$\sup_I \mathcal{E}_I^\alpha(u)\mathcal{E}_I^\alpha(v) < \infty \tag{4.3.6}$$

is already sufficient for the boundedness of $M_{v^{1/2}}TM_{u^{1/2}}$ for any Calderón–Zygmund operator, as soon as $C_\alpha < \infty$.

Moreover, given two Orlicz functions Φ_1, Φ_2 (with a mild extra regularity) satisfying (4.3.2) and two weights u, v satisfying (4.3.3), one can show the existence of α, with $C_\alpha < \infty$ such that the entropy bump condition (4.3.6) is satisfied, see Lemma 4.3.20 below.

4.3.3 Universal Two-Weight Estimate for Martingale Shifts and Calderón–Zygmund Operators

As at the beginning of Section 4.2.1, we consider the σ-finite measure space $(X, \mathfrak{A}, \mu)$ with filtration $\mathcal{L}_k$, which is just the sequence of increasing σ-algebras, generating $\mathfrak{A}$. We assume that each σ-algebra $\mathcal{L}_k$ is atomic, meaning that there is an at most countable disjoint collection $\mathcal{A}_n$ of sets (called atoms of $\mathcal{L}_n$) such that $\mathcal{L}_n$ consists of unions of atoms. Then each atom from $\mathcal{A}_n$ is the union of at most countable disjoint collection $\mathcal{A}_{n+1}$ of atoms of σ-algebra $\mathcal{L}_{n+1}$.

The typical example is any martingale lattice $\mathcal{D}$ in $\mathbb{R}^n$. However, notice that, unlike in Section 4.2.1, we do not assume any doubling (homogeneity) of μ with respect to filtrations. So the reader can think that we work with $(\mathbb{R}^n, \mathcal{D}, \mu)$ with the Lebesgue measure μ, but with non-doubling filtration.

We shall not preclude that an atom I belongs to several different $\mathcal{A}_n$'s. However, for the sake of simplicity, we will think that such n is unique. In what follows, we usually do not assign a special symbol for measure μ, we use $|A|$ for $\mu(A)$ and dx for $d\mu$.

Definition 4.3.1 Let I be an atom, let n be such that $I \in \mathcal{D}_n$, then elements of $\mathcal{A}_{n+1}$ such that $I' \subset I$ are called children of I and their collection is denoted by $ch(I)$. Recall that the collection of atoms that are the descendants of I after k generations is denoted by $\mathfrak{A}_k(I)$. In particular, $ch(I) = \mathfrak{A}_1(I)$.

4.3.4 Martingale Differences, Martingale Shifts, and Paraproducts

For an atom I, the martingale difference operator is

$$\Delta_I f = -\mathcal{E}_I f + \sum_{J \in ch(I)} \mathcal{E}_J f,$$

where $\mathcal{E}_I f = \langle f \rangle_I \cdot 1_I$.

The n-th order martingale difference is

$$\Delta_I^n f = -\mathcal{E}_I f + \sum_{J \in \mathfrak{A}_n(I)} \mathcal{E}_J f,$$

These are orthogonal projections in $L^2(\mu)$. We denote the images by $D_I := \Delta_I(L^2(\mu))$, $D_I^n = \Delta_I^n(L^2(\mu))$.

DEFINITION 4.3.2 In these notations, the familiar L^1-normalized martingale shift of order n is the operator T such that

$$Tf = \sum_{L \in \mathfrak{A}} T_L(\Delta_L^n f),$$

where T_L are operators from D_L^n to itself such that

$$|(T_L f, g)| \leq \frac{1}{\mu(L)} \|f\|_{L^1(\mu)} \|g\|_{L^1(\mu)} \ \forall f, g \in D_L^n. \tag{4.3.7}$$

DEFINITION 4.3.3 The L^2-normalized martingale shift of order n is the operator T such that

$$Tf = \sum_{L \in \mathfrak{A}} T_L(\Delta_L^n f),$$

where T_L are operators from D_L^n to itself such that

$$|(T_L f, g)| \leq \|f\|_{L^2(\mu)} \|g\|_{L^2(\mu)} \ \forall f, g \in D_L^n. \tag{4.3.8}$$

Both types of shifts are automatically bounded in unweighted $L^2(\mu)$ of course.

We leave this statement as an exercise for the reader.

Notice that L^1-normalized martingale shifts on complexity (m, n), $m \leq n$, are particular cases of such shifts of order $n + 1$.

The importance of martingale shifts comes from the fact that in the classical case of dyadic lattices in $\mathbb{R}^n$, any Calderón–Zygmund operator can be represented as a weighted average (over all translations of the standard dyadic lattice in $\mathbb{R}^n$) of the paraproducts and the martingale shifts, with the coefficients decreasing exponentially in the order of the shifts. We proved this in Theorem 4.1.8 of Section 4.1; see [**72, 73, 76**]. This means that the uniform boundedness of the paraproducts (see definition below) together with bounds on the martingale shifts that grow sub-exponentially in the order of the shifts imply the boundedness of the Calderón–Zygmund operators.

In this section, we only consider the martingale shifts of order 1 because by modifying the filtration, every martingale shift of order n can be represented as a sum of n martingale shifts of complexity 1 (each with respect to its own filtration).

Let us explain this. Consider filtrations $\mathcal{F}^r$, $r = 0, 1, \dots, n-1$ defined by generations $\mathcal{D}_{r+k}$, $k \in \mathbb{Z}$, then a Haar shift T can be represented as the sum $T = \sum_{r=1}^{n} T_r$, where T_r is a martingale shift of order 1 (martingale transform) with respect to the filtration $\mathcal{F}^r$. This splitting is trivial if each $I \in \mathcal{D}$ has a nontrivial collection of children: In this case, each interval I belongs to a unique generation $\mathcal{D}_j$ (this case, we assumed for simplicity), and thus the block $T_{_I}$ can be canonically assigned to a unique filtration.

So, an estimate for the martingale shifts of complexity 1 gives the estimate for general martingale shifts that grow linearly in order, which is more than enough for the estimates of Calderón–Zygmund operators.

DEFINITION 4.3.4 Let $b = (b_{_I})_{I\in\mathcal{D}}$, $b_{_I} \in D_{_I}$. A *paraproduct* $\Pi = \Pi_b$ is an operator on L^2 given by

$$Tf := \sum_{I\in\mathcal{D}} \langle f\rangle_{_I} b_{_I}.$$

Often $b_I = \Delta_{_I} b$ for some function b, but we will not distinguish between the cases when b is a sequence and when it is generated by a function.

It follows immediately from the martingale Carleson embedding theorem that the paraproduct Π_b is bounded in L^2 if and only if the sequence $\{\|b_{_I}\|_2^2\}_{I\in\mathcal{D}}$ satisfies the *Carleson measure condition*

$$\|b\|_{Carl} := \sup_{I_0\in\mathcal{D}} |I_0|^{-1} \sum_{I\in\mathcal{D},\, I\subset I_0} \|b_{_I}\|_2^2 < \infty \tag{4.3.9}$$

In what follows, we always normalize the paraproducts by assuming $\|b\|_{Carl} \le 1$.

The main result of this section is the following theorem.

THEOREM 4.3.5 *Let T be either any martingale shift or a paraproduct, normalized by the condition*

$$\sup_{I\in\mathcal{D}} |I|^{-1} \sum_{I'\in\mathcal{D},I'\subset I} \|b_{_{I'}}\|_\infty^2 |I'| \le 1, \tag{4.3.10}$$

and let weights u, v satisfy the entropy bump condition

$$\sup_{I\in\mathcal{D}} \alpha\left(\mathbf{u}_{_I}^*/\mathbf{u}_{_I}\right)\mathbf{u}_{_I}^*\alpha\left(\mathbf{v}_{_I}^*/\mathbf{v}_{_I}\right)\mathbf{v}_{_I}^* := A < \infty, \tag{4.3.11}$$

where $\alpha : [1, \infty) \to \mathbb{R}_+$ is such that the function $t \mapsto t\alpha(t)$ is increasing and

$$C_\alpha := \frac{1}{\alpha(1)} + \int_1^\infty \frac{1}{t\alpha(t)} dt < \infty. \tag{4.3.12}$$

Then the operator $M_v^{1/2} T M_u^{1/2}$ is bounded in L^2, or equivalently

$$\int_{\mathcal{X}} |T(fu)|^2 v dx \le AC_\alpha \int_{\mathcal{X}} |f|^2 u dx \qquad \forall f \in L^2(u), \tag{4.3.13}$$

where A is an absolute constant.

The proof of this result will more or less follow the scheme made explicit in Section 4.2.5.4. We will construct the proof for the simplest dyadic homogeneous martingale on filtration generated by a dyadic lattice in $\mathbb{R}$. And then we transfer the result to any martingale shift. There will also be differences with the proof of Section 4.2.5.4. One thing will work for us: We will not be using bilinear estimates; as a result, our Bellman function will be defined in convex domains. This is a big relief because most of the difficulties we encountered in Section 4.2.5.4 were created by non-convexity of the domain of definition of the Bellman function of weighted dyadic homogeneous martingale transform. However, there will be a new feature in the proof. Namely, the Bellman functions below will be defined on infinitely dimensional spaces (spaces of distribution functions). So, in a sense, these will be Bellman functionals. As before, their main feature will be the combination of specific concavity properties with size properties.

4.3.5 An Embedding Theorem and Entropy Bumping

In this section, it is more convenient to define $\mathbf{u}_I^*$ as

$$\begin{aligned}\mathbf{u}_I^* := \|M(u\mathbf{1}_I)\|_{L^1(I)} &= \langle M(u\mathbf{1}_I)\rangle_I \le \|u\|_{\Lambda_{\psi_0}(I)} \\ &= \|u\mathbf{1}_I\|_{\Lambda_{\psi_0}(I)} \approx \|u\|_{L\log L(I)},\end{aligned}$$

where $M = M_{\mathcal{D}}$ is the martingale maximal function,

$$Mf(x) = \sup_{I\in\mathcal{D}:\, x\in I} |I|^{-1}\int_I |f| dx,$$

and $\psi_0(s) := s\ln(e/s)$, and $\Lambda_\psi(I) = \Lambda_\psi(I, \frac{dx}{|I|})$ is the Lorentz space.

The reader who wants to refresh the memory about basic properties of Lorentz spaces can address Section 4.3.11 below.

Recall that a sequence $a = \{a_I\}_{I\in\mathcal{D}}$ is called *Carleson* if

$$\sup_{I_0\in\mathcal{D}} |I_0|^{-1}\sum_{I\in\mathcal{D}} |a_I|\cdot|I| =: \|a\|_{\text{Carl}} < \infty.$$

THEOREM 4.3.6 *Let $a = \{a_I\}_{I\in\mathcal{D}}$: be a Carleson sequence and let T be the operator*

$$Tf = \sum_{I\in\mathcal{D}} \langle f\rangle_I a_I \mathbf{1}_I.$$

Let $\alpha : [1,\infty) \to \mathbb{R}_+$ be an increasing function satisfying $\int_1^\infty \frac{dt}{t\alpha(t)} < \infty$, and let

$$\alpha\left(\mathbf{u}_I^*/\mathbf{u}_I\right) \mathbf{u}_I^* \mathbf{v}_I^* \alpha\left(\mathbf{v}_I^*/\mathbf{v}_I\right) \le A < \infty \tag{4.3.14}$$

Then the operator $f \mapsto T(fu)$ is a bounded operator $L^2(u) \to L^2(v)$,

$$\|T(fu)\|_{L^2(v)} \le C\|a\|_{\text{Carl}} A^{1/2} \|f\|_{L^2(u)}.$$

where $C = 4C_\alpha$,

$$C_\alpha = \frac{1}{\alpha(1)} + \int_1^\infty \frac{dt}{t\alpha(t)}. \tag{4.3.15}$$

Theorem 4.3.6 follows trivially and immediately from the following “embedding theorem” via Cauchy–Schwarz inequality.

THEOREM 4.3.7 *Let $a = \{a_I\}_{I\in\mathcal{D}}$, $a_I \ge 0$ be a Carleson sequence. Then for any $f \in L^2(u)$*

$$\sum_{I\in\mathcal{D}} \frac{|\langle fu\rangle_I|^2}{\alpha(\mathbf{u}_I^*/\mathbf{u}_I)\mathbf{u}_I^*} a_I |I| \le C\|a\|_{\text{Carl}} \|f\|_{L^2(u)}^2,$$

where $C = 4C_\alpha$,

$$C_\alpha = \frac{1}{\alpha(1)} + \int_1^\infty \frac{dt}{t\alpha(t)}.$$

REMARK 4.3.8 Here is the difference with the A_2 conjecture proved in the previous section. There we could not reduce the proof to embedding theorems with one weight. We had to treat a bilinear embedding where both weights $u = w$ and $v = w^{-1} = \sigma$ were present.

PROOF To prove this theorem, we can assume without loss of generality that $\|a\|_{\text{Carl}} = 1$. Let us use the notation $d\nu := u\, d\mu$. The angular brackets always mean averaging with respect to the underlying measure μ. If we want to have an averaging with respect to new measure ν we denote this as follows: $\langle\cdot\rangle_{I,\nu}$. Let us introduce the following weighted *Carleson potential*

$$A_J := \nu(J)^{-1} \sum_{I\in\mathcal{D}: I\subset J} a_I \nu(I), \tag{4.3.16}$$

$\nu(I) = \langle u\rangle_I |I|$. Obviously $\frac{\langle fu\rangle_I}{\langle u\rangle_I} = \langle f\rangle_{I,\nu}$.

Suppose we constructed a function $\mathcal{B}(x, y)$, $x \in \mathbb{R}$, $y \geq 0$, such that

(1) $\mathcal{B}$ is convex;
(2) $-\dfrac{\partial \mathcal{B}}{\partial y} \geq x^2 \varphi(y)$, where

$$\varphi(y) := \begin{cases} 1/(y\alpha(y)), & y \geq 1, \\ 1/\alpha(1), & 0 < y < 1; \end{cases}$$

(3) $0 \leq \mathcal{B}(x, y) \leq Cx^2$, where $C = 4C_\alpha$

Then the embedding theorem will follow by our usual telescopic sum argument and convexity. Let us see how it works.

Define

$$x_I := \langle f \rangle_{I,\nu}.$$

Notice that

$$x_I = \frac{1}{\nu(I)} \sum_{J \in \mathrm{ch}(I)} \nu(J) x_J$$

and that

$$A_I = a_I + \frac{1}{\nu(I)} \sum_{J \in \mathrm{ch}(I)} \nu(J) y_J =: a_I + \bar{A}_I.$$

Now let us estimate $-\mathcal{B}(x_I, A_I) + \frac{1}{\nu(I)} \sum_{J \in \mathrm{ch}(I)} \nu(J) \mathcal{B}(x_J, A_J)$. Clearly

$$-\mathcal{B}(x_I, \bar{A}_I) + \frac{1}{\nu(I)} \sum_{J \in \mathrm{ch}(I)} \nu(J) \mathcal{B}(x_j, A_J) \geq 0,$$

this follows from convexity of function $\mathcal{B}$. Hence, by mean value theorem for some $c_I \in [\bar{A}_I, A_I]$, we have

$$\begin{aligned} -\mathcal{B}(x_I, A_I) + \frac{1}{\nu(I)} \sum_{J \in \mathrm{ch}(I)} \nu(J) \mathcal{B}(x_J, A_J) &\geq \mathcal{B}(x_I, \bar{A}_I) - \mathcal{B}(x_I, A_I) \\ &= \langle f \rangle_{I,\nu}^2 \left(-\frac{\partial \mathcal{B}}{\partial y} \right) (x_I, c_I)(A_I - \bar{A}_I) \geq \langle f \rangle_{I,\nu}^2 \varphi(A_I) a_I \end{aligned}$$

The last inequality follows because function φ is decreasing (as $t\alpha(t)$ is increasing), so $\varphi(c_I) \geq \varphi(A_I)$.

Notice one more thing, namely, that

$$A_I \leq \mathbf{u}_I^* / \mathbf{u}_I. \tag{4.3.17}$$

In fact,

$$A_I = \nu(I)^{-1} \sum_{I' \in \mathcal{D}, I' \subset I} a_{I'} \langle u \rangle_{I'} |I'|$$
$$\leq \nu(I)^{-1} \|a\|_{\text{Carl}} \langle M \mathbf{1}_I u \rangle_I |I| \leq \|a\|_{\text{Carl}} \mathbf{u}_I^* / \mathbf{u}_I ;$$

Multiplying inequality

$$-\mathcal{B}(x_I, A_I) + \frac{1}{\nu(I)} \sum_{J \in \text{ch}(I)} \nu(J) \mathcal{B}(x_J, A_J) \geq \langle f \rangle_{I,\nu}^2 \varphi(\mathbf{u}_I^* / \mathbf{u}_I) a_I$$

by $\nu(I) = \langle u \rangle_I \mu(I) = \mathbf{u}_I \mu(I)$, we can rewrite this new estimate as

$$\sum_{J \in \text{ch}(I)} \nu(I') B(x_J, A_J) \; - \nu(I) B(x_I, A_I) \geq \frac{\langle fu \rangle_I^2}{\mathbf{u}_I \alpha(\mathbf{u}_I^* / \mathbf{u}_I) \mathbf{u}_I^* / \mathbf{u}_I} a_I |I|$$

Writing this estimate for $I = I_0$ and going $n-1$ generations down, we get using $\mathcal{B} \geq 0$

$$\sum_{\substack{I \in \text{ch}_k(I_0), \\ 0 \leq k < n}} \frac{\langle fu \rangle_I^2}{\alpha(\mathbf{u}_I^* / \mathbf{u}_I) \mathbf{u}_I^*} a_I |I| \leq \sum_{I \in \text{ch}_n(I_0)} \nu(I) B(x_I, A_I) - \nu(I_0) B(x_{I_0}, A_{I_0})$$
$$\leq \sum_{I \in \text{ch}_n(I_0)} \nu(I) B(x_I, A_I).$$

The estimate $\mathcal{B}(x, y) \leq Cx^2$ implies

$$\nu(I) B(x_I, A_I) \leq C \langle f \rangle_{I,\nu}^2 \nu(I) \leq C \int_I f^2 u \, dx;$$

the last estimate is just the Cauchy–Schwarz inequality. Therefore,

$$\sum_{k=0}^{n-1} \sum_{I \in \text{ch}_n(I_0)} \frac{\langle f \rangle_{I,\nu}^2}{\alpha(\mathbf{u}_I^* / \mathbf{u}_I) \mathbf{u}_I^*} a_I |I| \leq C \int_{I_0} f^2 u \, dx,$$

and letting $n \to \infty$, we get that

$$\sum_{I \in \mathcal{D}, I \subset I_0} \frac{\langle f \rangle_{I,\nu}^2}{\alpha(\mathbf{u}_I^* / \mathbf{u}_I) \mathbf{u}_I^*} a_I |I| \leq C \int_{I_0} f^2 u \, dx.$$

Summing the above estimate for all $I_0 \in \mathcal{D}_m$ and letting $m \to -\infty$, we get the conclusion of Theorem 4.3.7. □

4.3.6 Constructing the Bellman Function

We will look for a function $\mathcal{B} : \mathbb{R} \times \mathbb{R}_+ \to [0, \infty)$ of form $\mathcal{B}(x, y) = x^2 m(y)$, where m is a C^1 convex function. The Hessian of $\mathcal{B}$ is easy to compute:

$$\begin{pmatrix} \mathcal{B}_{xx} & \mathcal{B}_{xy} \\ \mathcal{B}_{xy} & \mathcal{B}_{yy} \end{pmatrix} = \begin{pmatrix} 2m(y) & 2xm'(y) \\ 2xm'(y) & x^2 m''(y) \end{pmatrix}.$$

It is clear that the Hessian is positive semidefinite (i.e., the function $\mathcal{B}$ is convex) if and only if

$$\begin{pmatrix} 2m(y) & 2m'(y) \\ 2m'(y) & m''(y) \end{pmatrix} \geq 0, \tag{4.3.18}$$

$$2(m')^2 \leq mm''.$$

Note, that property 2 of $\mathcal{B}$ is equivalent to the estimate

$$-m'(t) \geq \frac{1}{t\alpha(t)} =: \varphi(t). \tag{4.3.19}$$

Lemma 4.3.9 *Let $\varphi : \mathbb{R}_+ \to [0, \infty)$ be a bounded decreasing function such that*

$$\|\varphi\|_1 = \int_0^\infty \varphi(t)dt < \infty.$$

Then there exists a bounded decreasing convex C^1-function $m : [0, \infty) \to [0, \infty)$, $m(0) \leq 4\|\varphi\|_1$, satisfying (4.3.18) *and such that*

$$-m'(t) \geq \varphi(t) \qquad \forall t > 0.$$

Moreover, $\|m\|_\infty \leq (2 + \log 2) \int_0^\infty \varphi(t)dt$.

Proof Decreasing function φ on $[0, \infty]$ can be approximated from above by the linear combination of characteristic functions $\mathbf{1}_{[0,c_k]}$, moreover, the integrability of φ implies that we can choose the coefficients $\{a_k\}$ of this linear combinations to satisfy $\sum_k a_k c_k \leq (1 + \varepsilon)\|\varphi\|_1$. We will come to this statement a bit later, let us think now that φ *is* such a linear combination, so $\sum_k a_k c_k = \|\varphi\|_1$.

Consider the simplest case of just one characteristic function $\varphi = \mathbf{1}_{[0,1]}$. Choose

$$m_0(t) = \begin{cases} A_0 - 2\log(1 + t),\ 0 \leq t \leq 1, \\ A_0 - \log 2,\ t \geq 1. \end{cases}$$

Obviously, $-m'(t) \geq \mathbf{1}_{[0,1]}$ for all t. It is easy to see that $2(m')^2 \leq mm''$ holds if $A_0 = 2 + \log 2$.

Now for $\varphi_k = \mathbf{1}_{[0,c_k]}$ the corresponding m_k is just $m_k(t) = c_k m_0(t/c_k)$. Consider now $\varphi = \sum_k \mathbf{1}_{[0,c_k]}$ and put $m = \sum_k a_k m_0(x/c_k)$. The condition $-m' \geq \varphi$ is then obvious. The condition $2(m')^2 \leq mm''$ is less obvious, but it follows from the fact that it is equivalent to convexity of the function of two variables $x^2 m(y)$. Convexity is preserved under the summation with positive coefficients and this remark finishes the construction of m in the case when φ is a linear combination of characteristic functions.

In the general case, the fact that φ is bounded and decreasing allows us to conclude that it has only countably many jumps, all of the first order. The sum of the jumps converges. Let us replace φ by a bigger function $t \to \varphi(t-)$. It is easy to see that this new φ (we will call it by the same name) is upper semicontinuous.

It is easy to see that for any decreasing upper semicontinuous function φ, there exists a nonnegative measure γ on $\mathbb{R}_+$ of total mass $\|\varphi\|_1$ such that for all $t > 0$

$$\varphi(t) = \int_{\mathbb{R}_+} r^{-1}\mathbf{1}_{[0,r]}(t)d\gamma(r) \tag{4.3.20}$$

Consider

$$m(t) := \int_{\mathbb{R}_+} m_r(t)d\gamma(r)$$

We will prove now that it satisfies the conclusion of the lemma. Function $x^2 m(y)$ again is convex as an integral of convex functions with respect to finite positive measure. So we need to check only that

$$-m'(t) \geq \int_{\mathbb{R}_+} r^{-1}\mathbf{1}_{[0,r]}(t)d\gamma(r). \tag{4.3.21}$$

To justify this, we first define

$$\tilde{m}(y) := \int_{\mathbb{R}_+} m_r'(y)d\gamma(r) \tag{4.3.22}$$

and then conclude that by Tonelli's theorem

$$\begin{aligned}\int_y^\infty -\tilde{m}(x)dx &= \int_y^\infty \int_{\mathbb{R}_+} -m_r'(x)d\gamma(r)dx \\ &= \int_{\mathbb{R}_+}\int_y^\infty -m_r'(x)dxd\gamma(r) = \int_{\mathbb{R}_+} m_r(y)d\gamma(r) = m(y).\end{aligned}$$

This means that $m'y) = \tilde{m}(y)$. On the other hand, by the fact that $m'_r(y) \geq r^{-1}\mathbf{1}_{[0,r]}(y)$, we conclude that

$$\tilde{m}(y) := \int_{\mathbb{R}_+} m'_r(y)d\gamma(r) \geq \int_{\mathbb{R}_+} r^{-1}\mathbf{1}_{[0,r]}(y)d\gamma(r) = \varphi(y).$$

Therefore, (4.3.21) holds, and the lemma is proved. □

4.3.7 Bellman Functionals on Distribution Functions and Differential Embedding Theorem

Recall that for $u \geq 0$, any Lorentz norm $\|u\|_{\Lambda_\psi(I)}$ can be computed using the normalized distribution function N_I^u,

$$N_I^u(t) := \mu(I)^{-1}\mu\{x \in I : u(x) > t\}, \qquad t \geq 0, \tag{4.3.23}$$

namely,

$$\|u\|_{\Lambda_\psi(I)} := \int_0^1 \psi(N_I(t))dt.$$

The distribution functions N_I^u possess very simple martingale behavior,

$$N_{I_0}^u = \frac{1}{\mu(I_0)} \sum_{I \in \mathrm{ch}(I_0)} \mu(I)N_I^u, \tag{4.3.24}$$

so it is convenient to treat $\mathbf{u}^*$ as a functional on normalized distribution functions N, i.e., on decreasing functions $N : \mathbb{R}_+ \to [0,1]$,

$$\mathbf{u}^*(N) := \int_0^1 \psi_0(N(t))dt, \qquad \psi_0(s) = s\ln(e/s). \tag{4.3.25}$$

We will also need the functional $N \mapsto \mathbf{u}(N)$,

$$\mathbf{u}(N) := \int_0^\infty N(t)dt. \tag{4.3.26}$$

Clearly if $N = N_I^u$, then $\mathbf{u}(N_I^u) = \langle u \rangle_I =: \mathbf{u}_I$.

We want to understand first how concave is the function $N \to \mathbf{u}^*(N)$. For that we fix normalized distribution functions N, N_1, denote $\Delta N := N_1 - N$, and consider $N_\theta := N + \theta\Delta N$, which, for every $\theta \in [0,1]$ is obviously a normalized distribution function.

LEMMA 4.3.10 *Let N, N_1 be compactly supported distribution functions taking finitely many values. Then*

$$-\frac{d^2\mathbf{u}^*(N_\theta)}{d\theta^2} \geq \frac{(\int |\Delta N(t)|dt)^2}{\int N_\theta(t)dt} =: \frac{|\Delta \mathbf{u}|^2}{\mathbf{u}_\theta}$$

PROOF Recall that

$$\mathbf{u}^*(N_\theta) = \int_0^\infty \psi_0(N(t) + \theta \Delta N(t))\, dt,$$

where $\psi_0(s) = s \log(e/s)$. Since $\psi''(s) = -1/s$, we get, differentiating under the integral that

$$-\frac{d^2\mathbf{u}^*(N_\theta)}{d\theta^2} = \int_0^\infty \frac{\Delta N(t)^2}{N_\theta(t)} dt. \tag{4.3.27}$$

Note that under the assumptions of the lemma, there is no problem in justifying differentiating the integral.

Now lemma follows from the Cauchy–Schwartz inequality:

$$\begin{aligned}(\mathbf{u}_\Delta)^2 = \left(\int_0^\infty |\Delta N(t)|\, dt\right)^2 &\le \left(\int_0^\infty \frac{\Delta N(t)^2}{N_\theta(t)} dt\right)\left(\int_0^\infty N_\theta(t) dt\right) \\ &= \left(\int_0^\infty \frac{\Delta N(t)^2}{N_\theta(t)} dt\right) \mathbf{u}_\theta.\end{aligned}$$

□

COROLLARY 4.3.11 *Let N, N_1, N_2 be the distribution functions such that $N = (N_1 + N_2)/2$ and $\mathbf{u}(N_{1,2}) < \infty$. Denote $\Delta N := N_1 - N$ and let $\mathbf{u}_\Delta = \int_0^\infty |\Delta N(t)| dt$. Let $\mathbf{u} = \int_0^\infty N(t) dt$, $\mathbf{u}_i = \int_0^\infty N_i(t) dt$, $i = 1, 2$. Then*

$$\mathbf{u}^*(N) - \frac{\mathbf{u}^*(N_1) + \mathbf{u}^*(N_2)}{2} \ge \frac{1}{2} \cdot \frac{(\mathbf{u}_\Delta)^2}{\mathbf{u}} \ge \frac{1}{2} \cdot \frac{(\mathbf{u}_1 - \mathbf{u})^2}{\mathbf{u}} = \frac{1}{8} \cdot \frac{(\mathbf{u}_1 - \mathbf{u}_2)^2}{\mathbf{u}}.$$

PROOF It is sufficient to prove this corollary only for compactly supported distribution functions taking finitely many values: The general case is then obtained by approximation.

So, let N_1, N_2 be compactly supported distribution functions, taking finitely many values. Introducing function

$$F(\tau) = \mathbf{u}^*(N) - \frac{\mathbf{u}^*(N + \tau \Delta N) + \mathbf{u}^*(N - \tau \Delta N)}{2},$$

we see that $F'(0) = 0$, and we get using Taylor's formula that

$$\mathbf{u}^*(N) - \frac{\mathbf{u}^*(N_1) + \mathbf{u}^*(N_2)}{2} = F(1) - F(0) = \frac{F''(\theta)}{2}$$

for some $\theta \in (0, 1)$. Recalling that by Lemma 4.3.10 and by symmetry

$$F''(\theta) \ge \frac{(\mathbf{u}_\Delta)^2}{2} \left(\frac{1}{\mathbf{u}_\theta} + \frac{1}{\mathbf{u}_{-\theta}}\right)$$

and noticing that

$$\frac{1}{2}\left(\frac{1}{\mathbf{u}_\theta}+\frac{1}{\mathbf{u}_{-\theta}}\right) \geq \frac{1}{(\mathbf{u}_\theta+\mathbf{u}_{-\theta})/2}=\frac{1}{\mathbf{u}},$$

we get the conclusion. □

In what follows, we normalize the bump function α to have

$$C_\alpha = \frac{1}{\alpha(1)} + \int_1^\infty \frac{1}{t\alpha(t)}dt = 1.$$

Let us denote by $\mathcal{N}$ the set of all compactly supported distribution functions taking finitely many values.

Recall that we found a convex function $\mathcal{B}(x,y) = x^2 m(y)$ with properties (1)–(3) listed in the previous section.

Define a new (Bellman) function $\widetilde{\mathcal{B}} : \mathbb{R} \times \mathcal{N} \to [0,\infty)$ by formula

$$\widetilde{\mathcal{B}}(\mathbf{f}, N) = 2\mathbf{u}(N)\mathcal{B}\left(\frac{\mathbf{f}}{\mathbf{u}(N)}, \frac{\mathbf{u}^*(N)}{\mathbf{u}(N)}\right) + \frac{\mathbf{f}^2}{\mathbf{u}(N)} =: 2\widetilde{\mathcal{B}}_1(\mathbf{f}, N) + \widetilde{\mathcal{B}}_2(\mathbf{f}, N).$$

It follows from property 3 of $\mathcal{B}$ that

$$0 \leq \widetilde{\mathcal{B}}(\mathbf{f}, N) \leq 9\frac{\mathbf{f}^2}{\mathbf{u}(N)}. \tag{4.3.28}$$

Clearly, $\widetilde{\mathcal{B}}_1(\mathbf{f}, N) = \mathcal{B}_1(\mathbf{f}, \mathbf{u}(N), \mathbf{u}^*(N))$, where $\mathcal{B}_1$ is a function of three scalar arguments,

$$\mathcal{B}_1(\mathbf{f}, \mathbf{u}, \mathbf{u}^*) = \frac{\mathbf{f}^2}{\mathbf{u}} m(\mathbf{u}^*/\mathbf{u}). \tag{4.3.29}$$

The following theorem implies *the differential embedding* for the homogeneous dyadic martingales.

THEOREM 4.3.12 *Let*

$$\mathbf{f} = \frac{\mathbf{f}_+ + \mathbf{f}_-}{2}, \qquad N(t) = \frac{N_+(t) + N_-(t)}{2}.$$

Then the function $\widetilde{\mathcal{B}}$ *introduced above satisfies*

$$\begin{aligned} \frac{1}{2}\left(\widetilde{\mathcal{B}}(\mathbf{f}_+, N_+) + \widetilde{\mathcal{B}}(\mathbf{f}_-, N_-)\right) - \widetilde{\mathcal{B}}(\mathbf{f}, N) &\geq \frac{1}{72}\cdot\frac{(\mathbf{f}_+ - \mathbf{f})^2}{\alpha(\mathbf{u}^*/\mathbf{u})\mathbf{u}^*} \\ &= \frac{1}{4\cdot 72}\cdot\frac{(\mathbf{f}_+ - \mathbf{f}_-)^2}{\alpha(\mathbf{u}^*/\mathbf{u})\mathbf{u}^*}, \end{aligned} \tag{4.3.30}$$

where $\mathbf{u} = \mathbf{u}(N)$, $\mathbf{u}^* = \mathbf{u}^*(N)$. *(Note that* $\mathbf{f}_+ - \mathbf{f} = \mathbf{f} - \mathbf{f}_-$, *so we can replace* $(\mathbf{f}_+ - \mathbf{f})^2$ *in the right side by* $(\mathbf{f}_- - \mathbf{f})^2$*)*

We recall that for a weight u we denoted $\mathbf{u}_I := \langle u \rangle_I = \|u\|_{L^1(I)}$, $\mathbf{u}_I^* := \|u\|_{\Lambda_{\psi_0}(I)}$. The differential embedding theorem for homogeneous dyadic martingales is the following statement, where this time dx does mean the Lebesgue measure on the real line, and Is are the usual dyadic intervals.

THEOREM 4.3.13 *Let $\alpha : [1, \infty) \to \mathbb{R}_+$ be an increasing function such that*

$$C_\alpha := \frac{1}{\alpha(1)} + \int_1^\infty \frac{1}{t\alpha(t)} dt < \infty.$$

Then for any $f \in L^2(\mathbb{R}, udx)$,

$$\sum_{I \in \mathcal{D}} \frac{\|\Delta_I(fu)\|^2_{L^1(I)}}{\alpha(\mathbf{u}_I^*/\mathbf{u}_I)\mathbf{u}_I^*} |I| \le AC_\alpha \|f\|^2_{L^2(u)}, \tag{4.3.31}$$

where A is an absolute constant.

PROOF Recall that we can normalize α to have $C_\alpha = 1$. Without loss of generality, one may think that function f is supported on $I_0 \in \mathcal{D}$. Let $(\mathbb{R}, \mathcal{D}, dx)$ be a usual dyadic σ-algebra with the Lebesgue measure on the real line. For any $I \in \mathcal{D}$, we put

$$\mathbf{f}_I := \langle fu \rangle_I, N_I(t) : \frac{1}{|I|}\{x \in I : u(x) > t\}.$$

$$\mathbf{f}_\pm I := \langle fu \rangle_{I_\pm}, N_I(t) : \frac{1}{|I_\pm|}\{x \in I_\pm : u(x) > t\}.$$

Following our pervasive Bellman scheme, we now use inequality (4.3.30) with $\mathbf{f} = \mathbf{f}_I, \mathbf{f}_\pm = \mathbf{f}_{I_\pm}, N = N_I, N_\pm = N_{I_\pm}$. We do this for every $I \in \mathcal{D}(I_0)$.

Notice that by (4.3.28), we have

$$|J|\widetilde{\mathcal{B}}(\langle fu \rangle_J, N_J) \le 9C_\alpha \frac{\langle fu \rangle_J^2}{\langle u \rangle_J} |J| \le 9C_\alpha \langle f^2 u \rangle_J |J| = 9 \int_J f^2 u \, dx.$$

The telescopic nature of the sum in inequality (4.3.30) and that inequality (and of course the positivity of $\widetilde{\mathcal{B}}$ now) immediately imply

$$\frac{1}{72} \sum_{J \in \mathcal{D}(I_0)} \frac{\|\Delta_J(fu)\|^2_{L^1(I)}}{\alpha(\mathbf{u}_J^*/\mathbf{u}_J)\mathbf{u}_I^*} |J| \le 9 \int_{I_0} f^2 u \, dx.$$

Theorem 4.3.13 is proved. □

We already noticed that such embedding theorem proves the entropy bump conjecture. So the entropy bump conjecture is already proved for homogeneous dyadic shift of order 0 on the real line. Of course if Theorem 4.3.12 gets proved.

To prove convexity Theorem 4.3.12, we need the following lemma.

LEMMA 4.3.14 *Let again* $\mathbf{u} = \mathbf{u}(N)$, $\mathbf{u}^* = \mathbf{u}^*(N)$, *and let*

$$\Delta\mathbf{u} := \int_0^\infty \Delta N(t)dt$$

If $N_\pm = N \pm \Delta N$, $\mathbf{f} = (\mathbf{f}_+ + \mathbf{f}_-)/2$ *then*

$$\frac{\widetilde{\mathcal{B}}_1(\mathbf{f}_+, N_+) + \widetilde{\mathcal{B}}_1(\mathbf{f}_-, N_-)}{2} - \widetilde{\mathcal{B}}_1(\mathbf{f}, N) \geq \frac{1}{72} \cdot \frac{(\Delta\mathbf{u})^2\mathbf{f}^2}{\mathbf{u}^2\alpha(\mathbf{u}^*/\mathbf{u})\mathbf{u}^*}.$$

Let us postpone the proof of this lemma, and let us show how it implies Theorem 4.3.12.

Put

$$\mathbf{f}_\theta = \mathbf{f} + \theta(\mathbf{f}_+ - \mathbf{f}_-)/2,\ N_\theta = N + \theta\Delta N,\ \theta \in [-1, 1].$$

Denote $\mathbf{u} = \int_0^\infty N(t)dt$, $\mathbf{u}_\pm = \int_0^\infty (N(t) \pm 1\Delta N(t))dt$, $\Delta\mathbf{f} = \mathbf{f}_+ - \mathbf{f}_-$, $\Delta\mathbf{u} = \mathbf{u}_+ - \mathbf{u}_-$. Consider $b(\theta) = 2b_1(\theta) + b(\theta)$, where

$$b_1(\theta) := \mathcal{B}_1(\mathbf{f}_\theta, N_\theta),\ b_2(\theta) := \mathcal{B}_2(\mathbf{f}_\theta, N_\theta).$$

Theorem 4.3.12 asks to estimate $\frac{b(1)+b(-1)}{2} - b(0) = 2\int_{-1}^1 (1 - |t|)b''(t)dt$. We will see below that functions b_1, b_2 are convex, so it is enough to estimate from below the right-hand side in

$$\begin{aligned} \frac{b(1) + b(-1)}{2} - b(0) \geq 2\left(\frac{b_1(1) + b_1(-1)}{2} - b_1(0)\right) \\ + 2\int_{-1/2}^{1/2} (1 - |t|)b''(t)dt. \end{aligned} \tag{4.3.32}$$

We use Lemma 4.3.14 that claims

$$\left(\frac{b_1(1) + b_1(-1)}{2} - b_1(0)\right) \geq \frac{1}{2} \cdot \frac{(\Delta\mathbf{u})^2\mathbf{f}^2}{\mathbf{u}^2\alpha(\mathbf{u}^*/\mathbf{u})\mathbf{u}^*}. \tag{4.3.33}$$

In particular, b_1 is convex. Function b_2 is also convex as a superposition of convex function of two variable $(X, Y) \to \frac{X^2}{Y}$ and two linear functions in θ : $X(\theta) = \mathbf{f}_\theta$, $Y(\theta) = \mathbf{u}(N_\theta) = \int_0^\infty N(t)dt + \theta \int_0^\infty \Delta N(t)dt = \mathbf{u} + \theta(\mathbf{u}_+ - \mathbf{u}_-)/2$. Moreover, as function $b_2(\theta)$ is given by a simple formula

$$b_2(\theta) = \frac{\mathbf{f}_\theta^2}{\mathbf{u} + \theta(\mathbf{u}_+ - \mathbf{u}_-)/2} =: \frac{\mathbf{f}_\theta^2}{\mathbf{u}_\theta}$$

and it is easy to estimate $b_2''(\theta)$ from below:

$$b_2''(\theta) = \frac{1}{2}\frac{\mathbf{f}_\theta^2}{\mathbf{u}_\theta}\left(\left(\frac{\Delta\mathbf{f}}{\mathbf{f}_\theta}\right) - \left(\frac{\Delta\mathbf{u}}{\mathbf{u}_\theta}\right)\right)^2 \geq \frac{1}{2}\frac{\mathbf{f}_\theta^2}{\mathbf{u}_\theta}\left(\left|\frac{\Delta\mathbf{f}}{\mathbf{f}_\theta}\right| - \left|\frac{\Delta\mathbf{u}}{\mathbf{u}_\theta}\right|\right)^2 \tag{4.3.34}$$

Now notice that $\mathbf{u}, \mathbf{u}_\pm, \mathbf{f}, \mathbf{f}_\pm$ are all nonnegative and $\mathbf{u} = (\mathbf{u}_+ + \mathbf{u}_-)/2, \mathbf{f} = (\mathbf{f}_+ + \mathbf{f}_-)/2$. This elementary gives us inequalities:

$$\theta \in [-1/2, 1/2] \Rightarrow \frac{1}{2}\mathbf{f} \leq \mathbf{f}_\theta \leq \frac{3}{2}\mathbf{f},\ \frac{1}{2}\mathbf{u} \leq \mathbf{u}_\theta \leq \frac{3}{2}\mathbf{u}. \tag{4.3.35}$$

If we assume temporarily that

$$\left|\frac{\Delta\mathbf{f}}{\mathbf{f}}\right| \geq 6\left|\frac{\Delta\mathbf{u}}{\mathbf{u}}\right|, \tag{4.3.36}$$

then elementary combination of (4.3.34), (4.3.35), and (4.3.36) implies

$$\theta \in [-1/2, 1/2] \Rightarrow b_2''(\theta) \geq \frac{1}{108}\frac{\Delta\mathbf{f}^2}{\mathbf{u}}. \tag{4.3.37}$$

Automatically (4.3.32) and (4.3.36) then give

$$\frac{b(1) + b(-1)}{2} - b(0) \geq 2\int_{-1/2}^{1/2} (1 - |t|) b''(t)dt \geq \frac{1}{72}\frac{\Delta\mathbf{f}^2}{\mathbf{u}}$$

$$\geq \frac{1}{72}\frac{\Delta\mathbf{f}^2}{\alpha\,(\mathbf{u}^*/\mathbf{u})\,\mathbf{u}^*},$$

because our normalization $C_\alpha = 1$ in particular gives us $t\alpha(t) \geq 1$ if $t \geq 1$.

So the inequality of Theorem 4.3.12 is proved if (4.3.36) holds. What if it does not hold? Namely, suppose that

$$\left|\frac{\Delta\mathbf{f}}{\mathbf{f}}\right| \leq 6\left|\frac{\Delta\mathbf{u}}{\mathbf{u}}\right|. \tag{4.3.38}$$

In this case, we use (4.3.33), which implies under the assumption (4.3.38) that

$$\frac{b(1) + b(-1)}{2} - b(0) \geq \frac{1}{72}\frac{\Delta\mathbf{f}^2}{\alpha\,(\mathbf{u}^*/\mathbf{u})\,\mathbf{u}^*}.$$

Theorem 4.3.12 is proved modulo the proof of Lemma 4.3.14.

4.3.7.1 The Proof of Lemma 4.3.14. Denote

$$\mathbf{u}_\pm = \mathbf{u}(N \pm \Delta N), \qquad \mathbf{u}^*_\pm = \mathbf{u}^*(N \pm \Delta N),$$

and let

$$\mathbf{u}^*_0 := (\mathbf{u}^*_+ + \mathbf{u}^*_-)/2, \qquad \mathbf{u}_0 = (\mathbf{u}_+ + \mathbf{u}_-)/2;$$

note that $\mathbf{u}_0 = \mathbf{u} = \mathbf{u}(N) = \int_0^\infty N(t)dt$.

However, we can only say that $\mathbf{u}^*_0 \le \mathbf{u}^*$. See Corollary 4.3.11.

Note also that $\mathbf{u}_\pm = \mathbf{u} \pm \Delta\mathbf{u}$, where $\Delta\mathbf{u} = \int_0^\infty \Delta N(t)dt$. Another symbol to recall is $\mathbf{u}_\Delta := \int_0^\infty |\Delta N(t)|dt$.

Recall that $\widetilde{\mathcal{B}}_1(\mathbf{f}, N) = \mathcal{B}_1(\mathbf{f}, \mathbf{u}, \mathbf{u}^*)$ (see (4.3.29)), so

$$\widetilde{\mathcal{B}}_1(\mathbf{f}_+, N_+) + \widetilde{\mathcal{B}}_1(\mathbf{f}_-, N_-) - 2\widetilde{\mathcal{B}}_1(\mathbf{f}, N) \tag{4.3.39}$$

$$= \widetilde{\mathcal{B}}_1(\mathbf{f}_+, N_+) + \widetilde{\mathcal{B}}_1(\mathbf{f}_-, N_-) - 2\mathcal{B}_1(\mathbf{f}, \mathbf{u}, \mathbf{u}^*_0) \tag{4.3.40}$$

$$+ 2\left(\mathcal{B}_1(\mathbf{f}, \mathbf{u}, \mathbf{u}^*_0) - \mathcal{B}_1(\mathbf{f}, \mathbf{u}, \mathbf{u}^*)\right) \tag{4.3.41}$$

Denoting $d^2\mathbf{u}^* := \mathbf{u}^*_0 - \mathbf{u}^* \le 0$, we can estimate the term (4.3.41) by applying mean value theorem to (4.3.29):

$$\mathcal{B}_1(\mathbf{f}, \mathbf{u}, \mathbf{u}^*_0) - \mathcal{B}_1(\mathbf{f}, \mathbf{u}, \mathbf{u}^*) = \frac{\mathbf{f}^2}{\mathbf{u}} m'\left(\frac{\mathbf{u}^* + \theta d^2\mathbf{u}^*}{\mathbf{u}}\right)\frac{d^2\mathbf{u}^*}{\mathbf{u}},$$

with a certain θ, $0 < \theta < 1$. By Corollary 4.3.11

$$-d^2\mathbf{u}^* \ge \frac{1}{2}\cdot\frac{(\mathbf{u}_\Delta)^2}{\mathbf{u}} \ge \frac{1}{2}\cdot\frac{(\Delta\mathbf{u})^2}{\mathbf{u}},$$

so, recalling that m' is negative, we get

$$\begin{aligned}\mathcal{B}_1(\mathbf{f}, \mathbf{u}, \mathbf{u}^*_0) - \mathcal{B}_1(\mathbf{f}, \mathbf{u}, \mathbf{u}^*) &\ge -\frac{1}{2}\cdot\frac{(\Delta\mathbf{u})^2\mathbf{f}^2}{\mathbf{u}^3} m'\left(\frac{\mathbf{u}^* + \theta d^2\mathbf{u}^*}{\mathbf{u}}\right)\\ &\ge -\frac{1}{2}\cdot\frac{(\Delta\mathbf{u})^2\mathbf{f}^2}{\mathbf{u}^3} m'\left(\frac{\mathbf{u}^*}{\mathbf{u}}\right);\end{aligned}$$

the last inequality holds because $-m'$ is decreasing (m is convex) and $d^2\mathbf{u}^* \le 0$ (recall also that $m' < 0$). Recalling that $-m'(t) \ge 1/(t\alpha(t))$, we get from there

$$\mathcal{B}_1(\mathbf{f}, \mathbf{u}, \mathbf{u}^*_0) - \mathcal{B}_1(\mathbf{f}, \mathbf{u}, \mathbf{u}^*) \ge \frac{1}{2}\cdot\frac{(\Delta\mathbf{u})^2\mathbf{f}^2}{\mathbf{u}^2\alpha(\mathbf{u}^*/\mathbf{u})\mathbf{u}^*} \tag{4.3.42}$$

Next, we want to show that the term (4.3.40) is nonnegative. To do that, we use the convexity of the function $\mathcal{B}(x, y) = x^2 m(y)$, defined in Section 4.3.6.

Denoting

$$x = \mathbf{f}/\mathbf{u}, \qquad y = \mathbf{u}_0^*/\mathbf{u}$$

$$x_\pm = \mathbf{f}_\pm/\mathbf{u}_\pm \qquad y_\pm = \mathbf{u}_\pm^*/\mathbf{u}_\pm$$

we can write (4.3.40) as

$$\mathbf{u}_+\mathcal{B}(x_+, y_+) + \mathbf{u}_-\mathcal{B}(x_-, y_-) - 2\mathbf{u}\mathcal{B}(x, y) \tag{4.3.43}$$

Since $\mathbf{u} = (\mathbf{u}_+ + \mathbf{u}_-)/2$, we have that the right-hand sides of the obvious formulae below are actually convex combinations:

$$x = \frac{\mathbf{u}_+}{2\mathbf{u}}x_+ + \frac{\mathbf{u}_-}{2\mathbf{u}}x_-, \qquad y = \frac{\mathbf{u}_+}{2\mathbf{u}}y_+ + \frac{\mathbf{u}_-}{2\mathbf{u}}y_-,$$

so the convexity of $\mathcal{B}$ implies that

$$\frac{\mathbf{u}_+}{2\mathbf{u}}\mathcal{B}(x_+, y_+) + \frac{\mathbf{u}_-}{2\mathbf{u}}\mathcal{B}(x_-, y_-) - \mathcal{B}(x, y) \geq 0.$$

But this means that (4.3.43), and so (4.3.40) are nonnegative.

Combining the estimate (4.3.42) for (4.3.41) with the nonnegativity of (4.3.40) we get the conclusions of Lemma 4.3.14.

4.3.8 The Bellman Functional for Differential Embedding for Homogeneous Dyadic Martingales

This section is dedicated to an introduction of *the true Bellman function* $\mathbf{B}$ for the differential embedding Theorem 4.3.13. There are two subtle points, which we wish to explain: 1) it is not a function, it is necessarily a functional in the sense that it has to be defined on an infinite dimensional space of distribution functions, 2) we really do not need this functional neither for Theorem 4.3.13 nor for what follows. In fact, Theorem 4.3.13 has been already proved, and the reader can ask the question why we need its exact Bellman function(al)? Let us start with this explanation. We wish to emphasize that Theorem 4.3.13 is equivalent to the existence of a certain convex functional with specific properties of convexity and size. This equivalence (which we will see shortly) gives rise to the following reasoning.

Suppose we proved Theorem 4.3.13 not by use of function $\widetilde{\mathcal{B}}$ (it is one of multitude possible Bellman functions of the problem, and not the exact one) but by some other method, for example, by some more or less sophisticated stopping time argument (which is possible, see [**108**]).

If we have the above-mentioned equivalence of the main estimate (4.3.31) and the existence of a certain convex functional $\mathbf{B}$, then we can use from now on this functional in transference argument, which allows us to prove new and sometimes quite different results only vaguely resembling differential embedding inequality (4.3.31).

So let us start with (4.3.31) and let us pretend that we do not care by which method it has been proved. Fix an interval $I_0 = [0,1]$ and consider all strictly positive integrable function u on I_0 (weights) such that their distribution function assumes only finitely many values. Such distribution functions form the class $\mathcal{N}$, for $N \in \mathcal{N}, N(0-) = 1$. We define the normalized distribution function of u on any dyadic interval $I \in \mathcal{D}(I_0)$ as usual:

$$N_I(t) := \frac{1}{|I|}|\{x \in I : u(x) > t\}|.$$

On convex domain $\Omega = \{(\mathbf{F}, \mathbf{f}, N) \in \mathbb{R}_+ \times \mathbb{R} \times \mathcal{N} : \mathbf{f}^2 \leq \mathbf{F} \cdot \int_0^\infty N(t)dt\}$, we define the following functional:

$$\mathbf{B}(\mathbf{F}, \mathbf{f}, N) := \sup \left\{ \frac{1}{|I_0|} \sum_{I \in \mathcal{D}(I_0)} \frac{\|\Delta_I (fu)\|^2_{L^1(I)}}{\alpha(\mathbf{u}^*_I/\mathbf{u}_I)\mathbf{u}^*_I} |I| \right\},$$

where the supremum is taken over all test functions f and all weights u such that

- the distribution function of u equals N,
- $\langle fu \rangle_{I_0} = \mathbf{f}$,
- $\langle f^2 u \rangle_{I_0} = \mathbf{F}$,

and where $\mathbf{u}_I = \int_0^\infty N_I(t)dt$, $\mathbf{u}^*_I = \int_0^\infty \psi_0(N_I(t))dt$, $\psi_0(s) := s\log(e/s)$.

Theorem 4.3.13, or rather its main inequality (4.3.31), proves the following theorem.

THEOREM 4.3.15 *Let* $\mathbf{u}$ *denote* $\mathbf{u}_{I_0}$, $\mathbf{u}^*$ *denote* $\mathbf{u}^*_{I_0}$. *Let the increasing positive function* α *on* $[1, \infty)$ *be such that* $\int_1^\infty \frac{1}{t\alpha(t)}dt < \infty$. *Denote*

$$C_\alpha = \frac{1}{\alpha(1)} + \int_1^\infty \frac{1}{t\alpha(t)}dt$$

.

Functional $\mathbf{B}$ *is defined on convex set*

$$\Omega = \{(\mathbf{F}, \mathbf{f}, N) \in \mathbb{R}_+ \times \mathbb{R} \times \mathcal{N} : \mathbf{f}^2 \leq \mathbf{F} \cdot \int_0^\infty N(t)dt\}$$

and satisfies the following properties of size and concavity:

$$0 \le \mathbf{B}(\mathbf{F}, \mathbf{f}, N) \le AC_\alpha \mathbf{F},$$

$$\mathbf{B}(\mathbf{F}, \mathbf{f}, N) - \frac{\mathbf{B}(\mathbf{F}_+, \mathbf{f}_+, N_+) + \mathbf{B}(\mathbf{F}_-, \mathbf{f}_-, N_-)}{2} \ge \frac{(\mathbf{f}_+ - \mathbf{f})^2}{\alpha(\mathbf{u}^*/\mathbf{u})\mathbf{u}^*}.$$

PROOF The proof of Theorem 4.3.15 follows verbatim many proofs above, in particular, the proof of Theorem 4.2.14. So we leave the proof to the reader as an exercise. The main step in the proof is to observe an (obvious) independence of $\mathbf{B}$ from the interval (or atom) I_0. □

So we can see that the main inequality (4.3.31) of Theorem 4.3.13 implies the existence of such a functional $\mathbf{B}$. It does not matter how (4.3.31) happened to be proved. In our case, we proved (4.3.31) by producing *some* functional that has exactly two properties of size and concavity listed above. In fact, let us recall that we gave a certain formula for *convex* functional $\widetilde{\mathcal{B}}(\mathbf{f}, N)$. It is immediate that one can choose two positive absolute constants c, C such that the functional $C\mathbf{F} - c\widetilde{\mathcal{B}}(\mathbf{f}, N)$ would satisfy the properties of size and concavity listed for $\mathbf{B}$.

This does not mean that $\mathbf{B}(\mathbf{F}, \mathbf{f}, N) = C\mathbf{F} - c\widetilde{\mathcal{B}}(\mathbf{f}, N)$ of course. One can prove that (with appropriate absolute strictly positive constants c, C)

$$\mathbf{B}(\mathbf{F}, \mathbf{f}, N) \le C\mathbf{F} - c\widetilde{\mathcal{B}}(\mathbf{f}, N).$$

We leave the proof of this inequality to the reader.

REMARK 4.3.16 One more remark is in order. For transferring the result of differential embedding Theorem 4.3.13 to other situations, we do not need to know the exact formula of $\mathbf{B}$. Moreover, we do not need $\mathbf{B}$, we rather need the existence of *any* functional with size and convexity properties of Theorem 4.3.15. In particular, functional $C\mathbf{F} - c\widetilde{\mathcal{B}}(\mathbf{f}, N)$ works perfectly for our future goals. But its concrete formula does not matter. Notice that *any* proof of Theorem 4.3.13 automatically gives us a functional with the correct size and concavity properties that will be used for transference in the next section.

4.3.9 Transference of (4.3.31) to Arbitrary Martingale Shifts

Now we will use $\mathbf{B}$ (or any other functional with properties listed in Theorem 4.3.15) to transfer the result of Theorem 4.3.13 to prove the correct estimate of any martingale transform on any σ-algebra generated by atomic filtrations and provided with any (may be non-doubling) measure μ. We refer the reader to Section 4.3.3 for the definition of such martingale transform. We prove the result for martingale transforms of order 1 only because in Section 4.3.3 we showed that then the result for martingale transforms of order n will be basically the same, only n appears in the right-hand side of the estimate.

THEOREM 4.3.17 *Let $\alpha : [1, \infty) \to \mathbb{R}_+$ be an increasing function such that*

$$C_\alpha := \frac{1}{\alpha(1)} + \int_1^\infty \frac{1}{t\alpha(t)} dt < \infty.$$

Let $(X, \mathfrak{A}, \mu)$ be any σ-algebra generated by atomic filtrations $\mathfrak{A}_k$ with atoms $\mathcal{A} = \cup_k \mathcal{A}_k$. Let u, v be weights on X, that is nonnegative functions from $L^1(\mu)$.

Then for any $f \in L^2(\mathbb{R}, ud\mu)$,

$$\sum_{I \in \mathcal{A}} \frac{\|\Delta_I (fu)\|^2_{L^1(I,\mu)}}{\alpha(\mathbf{u}^*_I/\mathbf{u}_I)\mathbf{u}^*_I} \mu(I) \leq AC_\alpha \|f\|^2_{L^2(u)}, \tag{4.3.44}$$

where A is an absolute constant.

In particular, for any martingale transform T of order 1 on arbitrary $(X, \mathfrak{A}, \mu)$ has the two-weight estimate

$$\int_X |T(uf)|^2 \, v d\mu \leq C \int_X |f|^2 \, v d\mu, \tag{4.3.45}$$

as soon as two-sided entropy bump condition 4.3.14 holds.

PROOF Let $\mathcal{B}$ be any functional satisfying the concavity condition of Theorem 4.3.15. Then it will automatically satisfy a more general concavity condition. Namely, let $(\mathbf{F}, \mathbf{f}, N)$ be a convex combination of $\{(\mathbf{F}_k, \mathbf{f}_k, N_k)\}_{k=1}^m$, where m can be even infinity. Let the coefficients of this convex combination be $\{\alpha_k\}_{k=1}^m$. Then the following inequality holds:

$$\mathbf{B}(\mathbf{F}, \mathbf{f}, N) - \sum_{k=1}^m \alpha_k \mathbf{B}(\mathbf{F}_k, \mathbf{f}_k, N_k) \geq \frac{1}{25} \frac{(\sum_{k=1}^m \alpha_k |\mathbf{f}_k - \mathbf{f}|)^2}{\alpha(\mathbf{u}^*/\mathbf{u})\mathbf{u}^*}, \tag{4.3.46}$$

where as before $\mathbf{u} = \int_0^\infty N(t)dt$, $\mathbf{u}^* = \int_0^\infty \psi_0(N(t))dt$, and ψ_0 is the entropy function $\psi_0(s) = s\log(e/s)$.

To prove (4.3.46), use the standard duality to find $\beta' = \{\beta'_k\}_{k=1}^m \in \ell^\infty_m$, $\|\beta'\|_\infty \leq 1$ such that $\sum_{k=1}^m \beta'_k(\mathbf{f}_k - \mathbf{f})\alpha_k = \sum_{k=1}^m |\mathbf{f}_k - \mathbf{f}|\alpha_k$. Then notice that the left-hand side here will not change if we replace β'_k by $\beta_k = \beta'_k - \sum_{k=1}^m \alpha_k \beta'_k$. This is of course because we assumed that $0 = \sum_{k=1}^m \alpha_k(\mathbf{f}_k - \mathbf{f})$.

We get the sequence $\beta = \{\beta_k\}_{k=1}^m \in \ell_m^\infty$, $\|\beta\|_\infty \le 2$ such that

$$\sum_{k=1}^m \beta_k(\mathbf{f}_k - \mathbf{f})\alpha_k = \sum_{k=1}^m |\mathbf{f}_k - \mathbf{f}|\alpha_k, \qquad \sum_{k=1}^m \beta_k\alpha_k = 0. \tag{4.3.47}$$

Let us consider two new collections of coefficients: $\{\alpha_k(1+\beta_k/5)\}$ and $\{\alpha_k(1-\beta_k/5)\}$. By the aforementioned properties of β, these are again the collections of convex coefficients.

Now instead of the whole cloud of points $\{(\mathbf{F}_k, \mathbf{f}_k, N_k)\}_{k=1}^m$, consider just two points:

$$(\mathbf{F}_+, \mathbf{f}_+, N_+) = \sum_{k=1}^m \alpha_k(1+\beta_k/5)(\mathbf{F}_k, \mathbf{f}_k, N_k),$$

$$(\mathbf{F}_-, \mathbf{f}_-, N_-) = \sum_{k=1}^m \alpha_k(1-\beta_k/5)(\mathbf{F}_k, \mathbf{f}_k, N_k).$$

Then the left-hand side of (4.3.46) can be rewritten as follows:

$$\begin{aligned}
&\mathbf{B}(\mathbf{F}, \mathbf{f}, N) - \sum_{k=1}^m \alpha_k \mathbf{B}(\mathbf{F}_k, \mathbf{f}_k, N_k) \\
&= \mathbf{B}(\mathbf{F}, \mathbf{f}, N) - \frac{\mathbf{B}(\mathbf{F}_+, \mathbf{f}_+, N_+) + \mathbf{B}(\mathbf{F}_-, \mathbf{f}_-, N_-)}{2} \\
&\quad + \frac{1}{2}\left(\mathbf{B}(\mathbf{F}_+, \mathbf{f}_+, N_+) - \sum_{k=1}^m \alpha_k(1+\beta_k/5)\mathbf{B}(\mathbf{F}_k, \mathbf{f}_k, N_k)\right) \\
&\quad \frac{1}{2}\left(\mathbf{B}(\mathbf{F}_-, \mathbf{f}_-, N_+) - \sum_{k=1}^m \alpha_k(1-\beta_k/5)\mathbf{B}(\mathbf{F}_k, \mathbf{f}_k, N_k)\right) \\
&=: B + A_+ + A_-.
\end{aligned}$$

Both terms $A_\pm$ are nonnegative because function $\mathbf{B}$ is concave by Theorem 4.3.15. To estimate the term B from below, we use again Theorem 4.3.15 to obtain $B \ge \frac{(\mathbf{f}_+ - \mathbf{f})^2}{\alpha(\mathbf{u}^*/\mathbf{u})\mathbf{u}^*}$. We are left to see that $\mathbf{f}_+ - \mathbf{f} = \frac{1}{5}\sum_{k=1}^m \beta_k(\mathbf{f}_k - \mathbf{f})\alpha_k = \frac{1}{5}\sum_{k=1}^m \alpha_k|\mathbf{f}_k - \mathbf{f}|$, we used here (4.3.47).

Now we can finish the proof of Theorem 4.3.15 by a standard Bellman technique scheme. Namely, given any test function f and a weight u on $(X, \mathfrak{A}, \mu)$, we define for every atom $I \in \mathcal{A}$, the argument of $\mathbf{B}$ as follows:

$$\mathbf{F}_I := \langle f^2 u\rangle_I,\ \mathbf{f}_I := \langle fu\rangle_I,\ N_I(t) = \frac{1}{\mu(I)}\mu\{x \in I : u(x) > t\}.$$

Bellman scheme prescribes us to consider for each atom $I \in \mathcal{A}_k$, its children, atoms $J \in \mathcal{A}_{k+1}$, and to consider the collection of convex coefficients $\alpha(J) = \frac{\mu(J)}{\mu(I)}$, $J \in \mathrm{ch}(I)$. Then it prescribes to estimate from below the term

$$\mathbf{B}(\mathbf{F}_I, \mathbf{f}_I, N_I) - \sum_{J \in \mathrm{ch}(I)} \alpha(J)\mathbf{B}(\mathbf{F}_J, \mathbf{f}_J, N_J)$$

This is exactly what inequality (4.3.46) does, and we obtain

$$\begin{aligned} &\frac{1}{25} \frac{\left(\sum_{J\in \mathrm{ch}(I)} \alpha(J)|\mathbf{f}_J - \mathbf{f}_I|\right)^2 \mu(I)}{\alpha(\mathbf{u}_I^*/\mathbf{u}_I)\mathbf{u}_I^*} \\ &\quad \leq \mu(I)\mathbf{B}(\mathbf{F}_I, \mathbf{f}_I, N_I) - \sum_{J\in \mathrm{ch}(I)} \mu(J)\mathbf{B}(\mathbf{F}_J, \mathbf{f}_J, N_J). \end{aligned} \tag{4.3.48}$$

We can assume that the support of f lies in an atom I_0, and then sum up inequalities (4.3.48) for all $I \in \mathfrak{A}(I_0)$. The telescopic nature of the sum in the right-hand side shows that we need to estimate only one term: $\mathbf{B}(\mathbf{F}_{I_0}, \mathbf{f}_{I_0}, N_{I_0})$. But the size condition of Theorem 4.3.15 (we use it for the first time now) claims that

$$\mathbf{B}(\mathbf{F}_{I_0}, \mathbf{f}_{I_0}, N_{I_0}) \leq AC_\alpha \mathbf{F}_{I_0} = AC_\alpha \langle f^2 u\rangle_{I_0}.$$

Gathering these observations together, we conclude that

$$\sum_{I\in \mathcal{A}(I_0)} \frac{\|\Delta_I(fu)\|^2_{L^1(I,\mu)}\, \mu(I)}{\alpha(\mathbf{u}_I^*/\mathbf{u}_I)\mathbf{u}_I^*} \leq AC_\alpha \|f\|^2_{L^2(u)},$$

because $\sum_{J\in \mathrm{ch}(I)} \alpha(J)|\mathbf{f}_J - \mathbf{f}_I| = \Delta_I(fu)\|_{L^1(I,\mu)}$. Theorem 4.3.17 is fully proved. □

4.3.10 Two-Sided Bumps for Paraproducts

Theorem 4.3.5 for paraproducts is immediately obtained by combining embedding Theorems 4.3.7 and 4.3.15 via Cauchy–Schwarz.

Indeed, denote

$$\mathbf{u}_I^* := \|u\|_{\Lambda_{\psi_0}(I)} \asymp \|M\mathbf{1}_I\|_1, \qquad \psi_0(s) = s\ln(e/s),$$

and similarly for v. For $f \in L^2(u)$, $g \in L^2(v)$ we get, assuming (4.3.11) with $A = 1$,

$$\begin{aligned}
\left|(\Pi_b fu, g)_{L^2(v)}\right| &\le \sum_{I\in\mathcal{D}} |\langle fu\rangle_I| \cdot \left|(b_I u, \Delta_I(gv))_{L^2(\mu)}\right| \\
&\le \sum_{I\in\mathcal{D}} \frac{|\langle fu\rangle_I| \cdot \|b_I\|_\infty \mu(I)^{1/2}}{\left(\alpha(\mathbf{u}_I^*/\mathbf{u}_I)\mathbf{u}_I^*\right)^{1/2}} \cdot \frac{\|\Delta_I(gv)\|_{L^1(I,\mu)}\mu(I)^{1/2}}{\left(\alpha(\mathbf{v}_I^*/\mathbf{v}_I)\mathbf{v}_I^*\right)^{1/2}} \\
&\le \left(\sum_{I\in\mathcal{D}} \frac{|\langle fu\rangle_I|^2 \|b_I\|_\infty^2 \mu(I)}{\alpha(\mathbf{u}_I^*/\mathbf{u}_I)\mathbf{u}_I^*}\right)^{1/2} \\
&\quad \times \left(\sum_{I\in\mathcal{D}} \frac{\|\Delta_I(gv)\|^2_{L^1(I,\mu)}\mu(I)}{\alpha(\mathbf{v}_I^*/\mathbf{v}_I)\mathbf{v}_I^*}\right)^{1/2};
\end{aligned}$$

the second inequality holds because of (4.3.11) with $A = 1$ (note also that $\|gv\|_1 = \|gv\|_{L^1(I,\mu)}\mu(I)$), and the last one is just the Cauchy–Schwarz.

Applying to the sums in parentheses Theorems 4.3.7 and 4.3.15, respectively, we get Theorem 4.3.5 for paraproducts.

Gathering together estimates, we can get the constant C in (4.3.13) to be equal to $(24C_\alpha A^{1/2})^2$, where C_α is defined by (4.3.12) and A is the supremum in (4.3.11). □

4.3.11 The Discussion of Lorentz Spaces, Orlicz Spaces, and the Assumptions of the Theorems

Let us recall that the Lorentz space $\Lambda_\psi = \Lambda_\psi(\mathcal{X}, \mu)$ is defined as the set of measurable functions f on $\mathcal{X}$ such that

$$\|f\|_{\Lambda_\psi} := \int_0^{\mu(\mathcal{X})} f^*(s)d\psi(s) < \infty; \tag{4.3.49}$$

here f^* is the nondecreasing rearrangement of the function $|f|$, and ψ is a quasiconcave function.

Recall, see **[12, chapter 2, definition 5.6]** that a function $\psi : [0, \infty) \to [0, \infty)$ *quasiconcave* if

(1) ψ is increasing;
(2) $\psi(s) = 0$ iff $s = 0$;
(3) $s \mapsto \psi(s)/s$ is decreasing.

Note, that conditions (1) and (3) imply that the function ψ is continuous for all $s > 0$; it can have a jump discontinuity at $s = 0$, but in this paper we will only consider continuous functions ψ. Note also, that an increasing concave ψ satisfying $\psi(0) = 0$ is also quasiconcave.

The following simple proposition, see [**12, chapter 2, proposition 5.10**], shows that without loss of generality, we can assume that the function ψ is concave.

PROPOSITION 4.3.18 *If ψ is quasiconcave, then the least concave majorant $\widetilde{\psi}$ of ψ satisfies*

$$\frac{1}{2}\widetilde{\psi}(s) \leq \psi(s) \leq \widetilde{\psi}(s) \qquad \forall s \geq 0. \tag{4.3.50}$$

If the function ψ is concave, the expression (4.3.49) indeed defines a norm on the space of functions f satisfying (4.3.49).

In this paper, μ will always be a probability measure without atoms, particularly the normalized Lebesgue measure on an interval.

If $N = N_f$ is the distribution function of f,

$$N_f(t) := \mu\{x \in \mathcal{X} : |f(x)| > t\},$$

then making the change of variables $s = N(t)$ and integrating by parts, we can rewrite (4.3.49) as

$$\|f\|_{\Lambda_\psi} = \int_0^\infty \psi(N(t))dt = \int_0^\infty N(t)\Psi(N(t))dt; \tag{4.3.51}$$

here we define $\Psi(s) := \psi(s)/s$. This calculation is definitely justified for continuous f^*, so N is the inverse of f^*; approximating a general f^* by an increasing sequence of continuous decreasing functions gives us (4.3.51) for all f.

4.3.12 Comparison of Lorentz Spaces with Other Rearrangement Invariant Spaces

We present here a well-known fact helping to put the results in perspective, so it could be still beneficial for the reader.

Recall that for a rearrangement invariant Banach function space X, its *fundamental function* $\psi = \psi_X$ is defined as

$$\psi_X(s) = \|\mathbf{1}_E\|_X, \qquad \mu(E) = s. \tag{4.3.52}$$

Note that the fundamental function for the Lorentz space Λ_ψ is exactly ψ.

It is an easy calculation (see also [**12, chapter 4, Lemma 8.17**]) that for the Orlicz space L^ϕ equipped with the Luxemburg norm, its fundamental function ψ is given by

$$\psi(s) = 1/\Phi^{-1}(1/s); \tag{4.3.53}$$

in particular, the fundamental function for L^p spaces is given by $t^{1/p}$.

For "bumping" the Muckenhoupt condition, the spaces Λ_ψ are easier to work with than the Orlicz spaces L^Φ traditionally used for this purpose.

To be able to replace the Orlicz norm, one has to estimate the norm in L^Φ below by the norm in an appropriate Lorentz space Λ_ψ.

In **[122]**, the following comparison of the Lorentz and Orlicz norms was obtained.

LEMMA 4.3.19 *Let the underlying measure space $(\mathcal{X}, \mu)$ be a probability space, i.e. $\mu(\mathcal{X}) = 1$. Let Φ be a Young function such that*

$$\int^\infty \frac{dt}{\Phi(t)} < \infty,$$

and let the function Ψ on $(0, 1]$ be defined parametrically

$$\Psi(s) := \Phi'(t) \qquad \textit{for} \quad s = \frac{1}{\Phi(t)\Phi'(t)}.$$

Then the function ψ, $\psi(s) = s\Psi(s)$ is quasiconcave and satisfies

$$\int_0^1 \frac{ds}{\psi(s)} < \infty,$$

and there exists $C < \infty$ such that $\|f\|_{\Lambda_\psi} \le C\|f\|_{L^\Phi}$ for all measurable f.

In **[122]**, it is shown that for Young's functions with extra regularity the norms are equivalent.

Space $L \log L$ as a Lorentz space.

Recall that the space $L \log L$ is usually defined as the Orlicz space with the Young function $\Phi_0(t) = t \log^+(t)$ (the function $\Phi(t) = t \log(1 + t)$ is also used and gives us an equivalent norm). If the underlying measure space $(\mathcal{X}, \mu)$ satisfies $\mu(\mathcal{X}) = 1$, the space $L \log L$ can also be defined as the Lorentz space Λ_{ψ_0} with $\psi_0(s) = s \ln(e/s)$, $s \in [0, 1]$, see **[12, chapter 4, sections 6, 8]**.

If the underlying space is a unit interval I_0 (with Lebesgue measure), then it is well known that $\|Mf\|_1$, where M is the Hardy–Littlewood maximal function, defines an equivalent norm on $L \log L$. Recall that the Hardy–Littlewood maximal function $M = M_{I_0}$ on and interval I_0 is defined by

$$Mf(s) = \sup_{I: s \in I} |I|^{-1} \int_I |f(x)| dx,$$

where the supremum is taken over all intervals $I \subset I_0$, $s \in I$.

Moreover, it is well known, see **[12, chapter 4, eqn. (6.3)]**, that for $\psi_0(s) = s\ln(e/s)$ and $\mu(\mathcal{X}) = 1$

$$\|f\|_{\Lambda_{\psi_0}} = \int_0^1 Mf^*(s)ds. \tag{4.3.54}$$

This implies that for any collection $\mathcal{F}$ of measurable subsets of $\mathcal{X}$, we have the estimate for the corresponding maximal function $M_{\mathcal{F}}$,

$$\int_{\mathcal{X}} M_{\mathcal{F}} f \le \|f\|_{\Lambda_{\psi_0}};$$

here

$$M_{\mathcal{F}} f(x) := \sup_{I \in \mathcal{F}:\, x \in I} \mu(I)^{-1} \int_I |f| d\mu.$$

4.3.13 Comparison of Orlicz and Entropy Bumps

Recall that a Young function Φ is called *doubling* (or satisfying Δ_2 condition) if there is a constant $C < \infty$ such that $\Phi(2t) \le C\Phi(t)$ for all sufficiently large t. We will use this definition for arbitrary increasing functions, without requiring the convexity.

In this section, we assume that the underlying measure space is a probability space, i.e., that $\mu(\mathcal{X}) = 1$.

Fix a norm $\|\cdot\|_*$ in $L\log L$, namely let $\|f\|_* := \|f\|_{\Lambda_{\psi_0}}$, $\psi_0(s) = s\ln(e/s)$.

Lemma 4.3.20 *Let Φ be a doubling Young function satisfying*

$$\int^{\infty} \frac{dt}{\Phi(t)} < \infty \tag{4.3.55}$$

and such that the function $t \mapsto \Phi(t)/(t\ln t)$ is increasing for sufficiently large t. Then there exists an increasing function $t \mapsto t\alpha(t) > 0$ on $[1,\infty)$ such that

$$\int_1^{\infty} \frac{dt}{t\alpha(t)} < \infty$$

and

$$\alpha\left(\frac{\|f\|_*}{\|f\|_1}\right)\|f\|_* \le \|f\|_{L^\Phi}.$$

The assumption that the function $t \mapsto \Phi(t)/(t\ln t)$ is increasing seems not too restrictive, especially in the light of Lemma 4.3.21 below. It is also satisfied by the standard logarithmic bumps of t, namely, for the Young functions

$$\Phi(t) = t \ln t \ln_2 t \dots \ln_{n-1} t (\ln_n t)^{1+\varepsilon}, \qquad \varepsilon > 0;$$

here $\ln_{k+1}(t) = \ln(\ln_k(t))$ and $\ln_1(t) := \ln t$.

LEMMA 4.3.21 *Let a Young function Φ satisfy* (4.3.55). *Then there exists $c > 0$ such that*

$$\Phi(t) \geq c \cdot t \ln t$$

for all sufficiently large t.

PROOF By convexity, we have $\Phi(t) \leq t\Phi'(t)$. Therefore, (4.3.55) implies

$$\int_1^t \frac{dt}{t\Phi'(t)} \leq C_1 < \infty$$

independently of t. As $1/\Phi'(t)$ decreases, we write that the left-hand side is at least $\frac{\ln t}{\Phi'(t)}$. Then $\Phi'(t) \geq c_1 \ln t$, where $c_1 = C_1^{-1}$. Lemma 4.3.21 follows. □

If Φ is a Young function, it is an easy corollary of convexity that

$$\Phi(t) \leq t\Phi'(t) \tag{4.3.56}$$

If Φ is doubling, it is an easy corollary of the representation

$$\Phi(t) = \int_0^t \Phi'(x)dx \tag{4.3.57}$$

that there exists $c > 0$ such that

$$\Phi(t) \geq ct\Phi'(t) \tag{4.3.58}$$

for all sufficiently large t. If a Young function Φ is doubling, we can conclude from (4.3.56) and (4.3.58) that Φ' is also doubling. On the other hand, if Φ' is doubling, then representation (4.3.57) implies that Φ is doubling as well.

LEMMA 4.3.22 *Let $\Phi_{1,2}$ be doubling Young functions such that*

$$\int^\infty \frac{dt}{\Phi_{1,2}(t)} < \infty.$$

Then there exists a doubling Young function Φ, $\Phi(t) \leq \min\{\Phi_1(t), \Phi_2(t)\}$ such that

$$\int^\infty \frac{dt}{\Phi(t)} < \infty. \tag{4.3.59}$$

Moreover, there exists $c > 0$ such that

$$\Phi(t) \geq c \min\{\Phi_1(t), \Phi_2(t)\} \tag{4.3.60}$$

for all sufficiently large t.

PROOF Let $\phi_{1,2} := \Phi'_{1,2}$, so

$$\Phi_{1,2}(t) = \int_0^t \phi_{1,2}(x)dx.$$

Define

$$\phi(t) := \min\{\phi_1(t), \phi_2(t)\}, \qquad \Phi(t) := \int_0^t \phi(x)dx.$$

Clearly ϕ is increasing and doubling, so Φ is a doubling Young function. Since $\phi(t) \leq \phi_{1,2}(t)$, we can conclude that $\Phi \leq \min\{\Phi_1, \Phi_2\}$.

Finally, since all Young functions are doubling, we get using (4.3.56) and (4.3.58) that

$$\Phi(t) \geq ct\phi(t) = ct \min\{\phi_1(t), \phi_2(t)\} \geq c \min\{\Phi_1(t), \Phi_2(t)\}.$$

Therefore,

$$\frac{1}{\Phi(t)} \leq \frac{1}{c} \max\left\{\frac{1}{\Phi_1(t)}, \frac{1}{\Phi_2(t)}\right\} \leq \frac{1}{c}\left(\frac{1}{\Phi_1(t)} + \frac{1}{\Phi_2(t)}\right);$$

integrating this inequality gives us (4.3.59). □

LEMMA 4.3.23 *Let $\Psi : (0, 1] \to \mathbb{R}_+$, $\Psi(1) > 0$ be a decreasing function such that $\psi(s) := s\Psi(s)$ is increasing, $\psi(0_+) = 0$, and such that*

$$\int_0 \frac{ds}{s\Psi(s)} < \infty.$$

Assume that $t \mapsto \Psi(e^{-t})$ is convex near ∞. Then there exists an increasing function $t \mapsto t\alpha(t) > 0$ on $[1, \infty)$ satisfying

$$\int_1^\infty \frac{dt}{t\alpha(t)} < \infty$$

and such that

$$\alpha\left(\frac{\|f\|_*}{\|f\|_1}\right) \|f\|_* \leq \|f\|_{\Lambda_\psi}. \tag{4.3.61}$$

PROOF First, we can assume without loss of generality that $t \mapsto \Psi(e^{-t})$ is convex for all $t \in [1, \infty)$. Indeed, let that function be convex for $t > a$. Defining

$$\Psi_1(s) := \begin{cases} \Psi(s), & s < e^{-a} := b; \\ \Psi(b_-) & s \geq e^{-a} \end{cases},$$

we can immediately see that the function $t \mapsto \Psi_1(e^{-t})$ is convex for all $t \geq 1$. And replacing Ψ by Ψ_1, we get an equivalent norm on Λ_ψ.

So, let us assume that $t \mapsto \Psi(e^{-t})$ is convex for all $t \geq 1$. Define

$$\alpha(t) := \frac{\Psi(ee^{-t})}{t}, \qquad \gamma(t) := t\alpha(t) = \Psi(ee^{-t}),$$

so for $\Psi_0(s) := \ln(e/s)$, we have

$$\Psi(s) = \gamma(\Psi_0(s)). \tag{4.3.62}$$

Change of variables $s = e^{1-t}$ gives us that

$$\int_1^\infty \frac{dt}{t\alpha(t)} = \int_1^\infty \frac{dt}{\Psi(e^{1-t})} = \int_0^1 \frac{ds}{s\Psi(s)} < \infty.$$

To prove (4.3.61), let us first notice that because of homogeneity, we can assume without loss of generality that

$$\|f\|_1 = \int_0^\infty N(t)dt = 1.$$

Then defining probability measure μ by $d\mu = N(t)dt$ and using (4.3.62) can rewrite (4.3.61) as

$$\gamma\left(\int_0^\infty \Psi_0(N(t))d\mu(t)\right) \leq \int_0^\infty \gamma\left(\Psi_0(N(t))\right)d\mu(t) \tag{4.3.63}$$

(recall that $\Psi_0(s) = \ln(e/s)$).

But the function γ is convex, so (4.3.63) follows immediately from Jensen inequality. □

Proof of Lemma 4.3.20 Let $\Phi(t)/(t\ln t)$ be increasing for all $t \geq t_0 \geq e^e$.

Applying Lemma 4.3.22 to functions $\Phi_1 = \Phi$ and Φ_2, where $\Phi_2(t) = t\ln^2 t$ for $t \geq t_0$, we get a function Φ_0 such that $\int^\infty 1/\Phi_0 < \infty$, and such that $\Phi_0(t) \leq t\ln^2 t$ (for $t \geq t_0$).

The function $\Phi_0(t)/(t\ln t)$ is not necessarily increasing, but Φ_0 is equivalent to the function $\Phi_{\min} := \min\{\Phi_1, \Phi_2\}$, and $\Phi_{\min}(t)/(t\ln t)$ is increasing (for $t \geq t_0$) as minimum of increasing functions. Since $\Phi_{\min}(t)/(t\ln t)$ is increasing and Φ_0 is equivalent to $\Phi_{\min}$ near ∞, we conclude that for all $t \geq t_0$

$$ct\ln t \leq \Phi_0(t) \leq Ct\ln^2 t. \tag{4.3.64}$$

Since Φ_0 is doubling, (4.3.56) and (4.3.58) imply that

$$c\ln t \leq \Phi_0'(t) \leq C\ln^2 t \qquad \forall t \geq t_0. \tag{4.3.65}$$

Therefore, replacing Φ by Φ_0 we can assume without loss of generality that

$$ct\ln t \leq \Phi(t) \leq Ct\ln^2 t, \qquad c\ln t \leq \Phi'(t) \leq C\ln^2 t \qquad \forall t \geq t_0. \tag{4.3.66}$$

Now let us recall Lemma 4.3.19. We construct a function Ψ_0 dominated by the function Ψ from the lemma, but still such that $\int_0 \frac{ds}{s\Psi_0(s)} < \infty$. Recall that Ψ in Lemma 4.3.19 was given by

$$\Psi(\tilde{s}) := \Phi'(t) \qquad \text{for} \quad \tilde{s} = \frac{1}{\Phi(t)\Phi'(t)}.$$

We can see from (4.3.66) that

$$\frac{c}{t\ln^4 t} \le \frac{1}{\Phi(t)\Phi'(t)} \le \frac{C}{t\ln^2 t}.$$

So defining

$$\Psi_0(s) := \frac{\Phi_{\min}(t)}{t} \qquad \text{for} \quad s = \frac{c}{t\ln^4 t}.$$

we get that $\Psi_0 \le C\Psi$. Indeed, the functions Ψ and Ψ_0 are decreasing, $\Phi_{\min}(t)/t \le C\Phi(t)/t \le C\Phi'(t)$, and we have two relationships: $s \le \tilde{s}$ and $\Psi_0(s) \le C\Psi(\tilde{s})$.

We can pick c sufficiently small, so $s \in (0,1]$ correspond to $t \in [t_1, \infty)$, $t_1 \ge t_0 \ge e^e$.

It is also clear that $\psi_0(s) := s\Psi_0(s)$ is increasing, $\psi_0(0_+) = 0$. Let us check that $1/(s\Psi_0(s))$ is integrable near 0: since $-ds = (t^{-2}\ln^{-2} t - t^{-2}\ln^{-5} t)dt \le Ct^{-2}\ln^{-2} t dt$

$$\int_0 \frac{ds}{s\Psi_0(s)} \le C\int^{\infty} \frac{t}{\Phi(t)} t\ln^4 t \frac{dt}{t^2\ln^4 t} = C\int^{\infty} \frac{dt}{\Phi(t)} < \infty.$$

Finally, we claim that $-\Psi_0(s)/\ln s$ is decreasing. For this, we need to show that

$$-\frac{\ln s}{\Psi_0(s)} = \frac{t\cdot(\ln t + 4\ln\ln t - \ln c)}{\Phi_{\min}(t)}$$

is a decreasing function of t (and so the increasing function of s). But the term $t\ln t/\Phi_{\min}(t)$ is decreasing by the assumption, the second term is decreasing because $(\ln\ln t)/\ln t$ is decreasing for $t \ge e^e$, and the last term is decreasing if $c \le 1$, which we always can assume without loss of generality.

Define $\varphi(t) := \Psi_0(e^{-t})$, $t > 0$, $\varphi(0) := 0$. Clearly, φ is an increasing function on $[0, \infty)$. Change of variable in the integral shows that

$$\int_0 \frac{ds}{s\Psi_0(s)} < \infty \qquad \Longleftrightarrow \qquad \int^{\infty} \frac{dt}{\varphi(t)} < \infty. \tag{4.3.67}$$

The fact that $-\Psi_0(s)/\ln s$ is decreasing translates to the statement that $\varphi(t)/t$ is increasing. Then for the inverse φ^{-1}, we get that $\varphi^{-1}(\tau)/\tau$ is decreasing. Also, $\varphi^{-1}(\tau)$ is clearly increasing, so φ^{-1} is a pseudoconcave function.

Therefore, by Proposition 4.3.18 the least concave majorant $\widetilde{\varphi}^{-1}$ of φ^{-1} satisfies

$$\frac{1}{2}\widetilde{\varphi}^{-1}(\tau) \le \varphi^{-1}(\tau) \le \widetilde{\varphi}^{-1}(\tau).$$

Therefore,

$$\widetilde{\varphi}(t) \le \varphi(t) \le \widetilde{\varphi}(2t). \tag{4.3.68}$$

The function $\widetilde{\varphi}$ is convex. Note that

$$\int^{\infty} \frac{dt}{\widetilde{\varphi}(t)} = 2\int^{\infty} \frac{dt}{\widetilde{\varphi}(2t)} \le 2\int^{\infty} \frac{dt}{\varphi(t)} < \infty.$$

Therefore, the function $\widetilde{\Psi}$, $\widetilde{\Psi}(s) := \widetilde{\varphi}(-\ln s)$ satisfies the assumptions of Lemma 4.3.23. Since $\widetilde{\Psi} \le \Psi_0$, applying Lemma 4.3.23 to $\widetilde{\Psi}$ give us the conclusion of Lemma 4.3.20. □

4.3.13.1 What if Martingale Shift on Nonhomogeneous Filtration is Only L^2-Normalized? Here we present a simple example of a (bounded in a non-weighted $L^2(\mu)$) martingale transform T and a dyadic A_2-weight w, such that T is not bounded in $L^2(w)$. The example below presents a shift T on $(X, \mathfrak{A}, \mu)$, where σ-algebra $\mathfrak{A}$ is generated by atomic filtration, whose blocks are uniformly bounded in $L^2(d\mu)$ but the transform fails to be bounded in $L^2(wd\mu)$.

Take a small $\varepsilon > 0$. Consider an interval I, $|I| = 2$, and split it into four subintervals (children) I_k, $|I_1| = |I_3| = 1 - \varepsilon$, $|I_2| = |I_4| = \varepsilon$. Denote $J_1 = I_1 \cup I_2$, $J_2 = I_3 \cup I_4$. Defining atoms: $\mathcal{A}_0$ consists of one atom I, $\mathcal{A}_1$ consists of two atoms J_1, J_2, and $\mathcal{A}_2$ consist of four atoms $I_i, i = 1, \ldots, 4$. Measure $\mu = dx$. (Here we make a small remark that we could have chosen all $I_i, i = 1, \ldots, 4$ to be the usual dyadic grandchildren of I and define a new measure μ such that $\mu(I_1) = \mu(I_3) = 1 - \varepsilon, \mu(I_2) = \mu(I_4) = \varepsilon$. This would be a totally equivalent construction.)

Define

$$h_1 := 2^{-1/2}(\mathbf{1}_{J_1} - \mathbf{1}_{J_2}), \qquad h_2 := \varepsilon^{-1/2}\mathbf{1}_{I_2} - \varepsilon^{1/2}(1-\varepsilon)^{-1}\mathbf{1}_{I_1}.$$

The functions $h_{1,2}$ are Haar functions, i.e., they are constant on children of I and orthogonal to constants. Note also that

$$\|h_1\|_{L^2} = 1, \qquad \|h_2\|_{L^2} \le 2^{1/2}$$

(if $\varepsilon < 1/2$). Then the operator T

$$Tf = (f, h_1)_{L^2} h_2$$

is a bounded operator in L^2, $\|T\| \le 2^{1/2}$. It is a shift of order 1 in the definition of Section 4.3.4 (or a shift of complexity $(0,1)$ in the terminology of the previous section).

Define a weight w,

$$w(x) := \begin{cases} 1, & x \in I_1 \cup I_3, \\ \varepsilon^{-1} & x \in I_2 \cup I_4. \end{cases}$$

Then w satisfies the A_2 condition and $[w]_{A_2} \le 2$. Here in the definition of A_2 condition we checked the averages over I and over its children I_k. Note, that if we also check the A_2 condition on intervals $J_{1,2}$, we still have the same estimate $[w]_{A_2} \le 2$. But even if we consider averages over all possible unions of intervals I_k, we still have the estimate $[w]_{A_2} \le 3$.

Since $Th_1 = h_2$ and

$$\|h_1\|_{L^2(w)} \le 2, \qquad \|h_2\|_{L^2(w)} \ge \varepsilon^{-1/2}$$

we get that

$$\|T\|_{L^2(w)\to L^2(w)} \ge \varepsilon^{-1/2}/2.$$

Considering a sequence of $\varepsilon_n \searrow 0$ and taking a direct sum of the above examples, we get a bounded martingale transform T, which is a shift of order 1 in the definition of Section 4.3.4, and an A_2 weight w such that T is not bounded in $L^2(w)$.

4.3.13.2 Why We Impose Condition (4.3.10) for Paraproducts?

REMARK 4.3.24 Note, that in the homogeneous case, when

$$|I'|/|I| \ge \delta > 0 \qquad \forall I \in \mathcal{D},\ \forall I' \in \mathrm{ch}(I),$$

$\|\Delta_I f\|_\infty^2 |I| \asymp \|\Delta_I f\|_2^2$ uniformly for all $f \in L^2$ and for all $I \in \mathcal{D}$ (this in fact can be used as a definition of homogeneous lattices), so in the homogeneous cases, the paraproduct normalization condition (4.3.10) is equivalent up to a constant to the classical normalization condition (4.3.9).

Condition (4.3.9), as we mentioned before, is equivalent to the bound on the norm of the paraproduct in the non-weighted L^2; our condition (4.3.10) is a bit stronger in the general nonhomogeneous case.

A natural question arises: why not impose in our weighted results a weaker (and more natural) condition (4.3.9), equivalent to the bound on the norm of the paraproduct in the non-weighted $L^2(\mu)$? We just said that for homogeneous filtrations there is no difference; two normalization conditions – our one (4.3.10) and a natural one (4.3.9) – are equivalent. However, the following counterexample shows that natural condition turns out to be too weak in a nonhomogeneous situation. We will construct now a measure space $(X, \mathfrak{A}, \mu)$ generated by a simple (but nonhomogeneous) filtration such that a paraproduct with $\{b_I\}$s satisfying (4.3.9) will not be bounded in $L^2(wd\mu)$ even though w will be from martingale A_2 class. Then we will give a corresponding counterexample showing that even entropy bump condition (4.3.11) cannot make up for the curse of nonhomogeneity.

In fact, this is almost the same example as in Section 4.3.13.1. Let us use the same $I, I_1, I_2, I_3, I_4, J_1, J_2$ as before, $\mu = dx$, and $b_I = h_2$, The block of our paraproduct operator is just

$$T_I f := \langle f \rangle_I h_2.$$

The weight w is also exactly as before. As $T\mathbf{1}_I = h_2$, we can see again that $\|T\|_w \geq \varepsilon^{-1/2}$.

Now we take $\varepsilon_n \searrow 0, n = 1, \dots,$ and we repeat the same construction on intervals $L_n = I + 2n$. On each, there will be its own weight w_n and function h_2^n. Let $\mathcal{A}_0 = \{I\} \cup_{n=1}^{\infty} \{L_n\}$, let $\mathcal{A}_1$ consist of J_1, J_2 and their corresponding counterparts on L_ns, we assume the same for $\mathcal{A}_2$.

The paraproduct operator is just

$$Tf := \langle f \rangle_I h_2 + \sum_{n=1}^{\infty} \langle f \rangle_{L_n} h_2^n.$$

The assumption (4.3.9) is trivially satisfied (it sufficient to check it only once, namely on I). Weight $w = w|I + \sum_n w_n|L_n$ is in martingale A_2, because $w|I$ was in A_2 independently of ε. However, the norm of this paraproduct operator is at least $\varepsilon_n^{-1/2}$ for all n. So it is unbounded in $L^2(wd\mu)$.

To see that the entropy bump conjecture also fails under L^2-normalization of the paraproduct, notice that if consider again T_I and put $u := w, v = c(\varepsilon)w^{-1}$, then

$$\sup_{\ell \in \mathcal{A}(I)} \alpha(\mathbf{u}_\ell^*/\mathbf{u}_\ell)\mathbf{u}_\ell^* \cdot \alpha(\mathbf{u}_\ell^*/\mathbf{u}_\ell)\mathbf{u}_\ell^* \asymp \alpha(\log 1/\varepsilon) \log 1/\varepsilon \cdot c(\varepsilon).$$

On the other hand, as $v = c(\varepsilon)w^{-1}, u = w$, we have $\|(T_I)_v : L^2(v) \to L^2(u)\|^2 = c(\varepsilon)\|(T_I)_{w^{-1}} : L^2(w^{-1}) \to L^2(w)\|^2$. The latter expression is equal to $c(\varepsilon)\|T_I : L^2(w) \to L^2(w)\|^2$, which, as we see, is at least $c(\varepsilon)/4\varepsilon$. Hence, for most interesting α, satisfying $\int_1^\infty \frac{dt}{t\alpha(t)} < \infty$, we can choose $c(\varepsilon)$ such that

$$\alpha(\log 1/\varepsilon) \log 1/\varepsilon \cdot c(\varepsilon) \leq 1,$$

and therefore, the entropy bump assumption

$$\sup_{\ell\in\mathcal{A}(I)} \alpha(\mathbf{u}_\ell^*/\mathbf{u}_\ell)\mathbf{u}_\ell^* \cdot \alpha(\mathbf{u}_\ell^*/\mathbf{u}_\ell)\mathbf{u}_\ell^* \asymp 1$$

is satisfied, but the norm of the corresponding very simple paraproduct T_I from $L^2(I, vdx)$ to $L^2(I, udx)$ is as large as we wish. For any sub-exponential α this choice of $c(\varepsilon)$ is obviously possible.

Then again we can choose the sequence $\varepsilon_n \searrow 0$ and repeat the construction above to see that the entropy bump assumption with any sub-exponential α can have the entropy bump condition (4.3.11) satisfied, but a paraproduct operator is unbounded. This explains why in a homogeneous situation, any paraproduct normalization is fine, but in a nonhomogeneous situation, one should be careful, the condition (4.3.10) works, and an L^2-normalization condition (4.3.9) does not.

REMARK 4.3.25 One can state condition (4.3.14) with two different functions $\alpha_1(\mathbf{u}_I^*/\mathbf{u}_I)$ and $\alpha_2(\mathbf{v}_I^*/\mathbf{v}_I)$, such that $\int_1^\infty \frac{dt}{t\alpha_k(t)} < \infty$ for $k = 1, 2$. However, defining $\alpha(t) = \min\{\alpha_1(t), \alpha_2(t)\}$, one can easily see that α is an increasing function satisfying $\int_1^\infty \frac{dt}{t\alpha(t)} < \infty$. So, replacing $\alpha_{1,2}$ with α, we get a weaker condition, which is exactly (4.3.14). Thus, there is nothing to gain considering different function α.

4.3.13.3 What if in Theorem 4.3.7 the Carleson Potential Were Bounded?

REMARK 4.3.26 As before $\nu = u\,d\mu$. If the sequence A_J defined in (4.3.16) of Theorem 4.3.7 (Carleson potential) were bounded: $A_J \le K < \infty$, then the martingale Carleson embedding theorem would imply that

$$\sum_{I\in\mathcal{D}} \frac{|\langle fu\rangle_I|^2}{\langle u\rangle_I} a_I \mu(I) = \sum_{I\in\mathcal{D}} \left|\nu(I)^{-1}\int_I f d\nu\right|^2 a_I \mu(I) \le 4K\|f\|^2_{L^2(\nu)},$$

and Theorem 4.3.7 would be proved in this simple way.

Unfortunately, the potential A is unbounded, so the above estimate is not true under the assumptions of the theorem: We need to put something bigger in the denominator to get a true estimate. And the proof becomes more involved.

Namely, we can only guarantee that

$$\begin{aligned} A_I &= \nu(I)^{-1} \sum_{I'\in\mathcal{D}, I'\subset I} a_{I'} \langle u\rangle_{I'} \mu(I') \\ &\le \nu(I)^{-1}\|a\|_{\text{Carl}} \langle M\mathbf{1}_I u\rangle_I \mu(I) \le \|a\|_{\text{Carl}} \mathbf{u}_I^*/\mathbf{u}_I; \end{aligned}$$

the first inequality here is the standard estimate of the Carleson embedding via maximal function.

4.3.14 A Counterexample with $L \log L$ Bumping

As we mentioned before, the $L \log L$ bump (even two-sided) is not sufficient for the boundedness of Calderón–Zygmund operators. Moreover, we will show that for any rearrangement invariant Banach spaces X and Y on a unit interval, where the fundamental functions ψ of X is such that

$$\int_0 \frac{ds}{\psi(s)} = +\infty$$

there exists a pair of weights u, v on $\mathbb{R}$ such that for all intervals I

$$\|u\|_{X(I)} \|v\|_{Y(I)} \leq B < \infty \tag{4.3.69}$$

but

$$\|T(1_{[-1,1]}u)\|_{L^2(v)} = \infty; \tag{4.3.70}$$

here T is the Hilbert Transform, $X(I) = X(I, \frac{dx}{|I|})$ and similarly for $Y(I)$.

Thus, the operator $f \mapsto T(uf)$ does not act $L^2(u) \to L^2(v)$. Moreover, even a weak type estimate (for the adjoint operator) fails. Recall that the adjoint operator (acting $(L^2(v) \to L^2(u))$ is given by $g \mapsto -T(gv)$: and (4.3.70) implies that the operator $g \mapsto T(gv)$ does not act $L^2(v) \to L^{2,\infty}(u)$ (i.e., it is not even of weak type 2–2).

This follows from a well-known reasoning: if the operator $T(v\cdot)$ is bounded $L^2(v) \to L^{2,\infty}(u)$ by constant C, then by definition

$$t^2 u\{|T(vg)| > t\} \leq C\|g\|^2_{L^2(v)} \qquad \forall g \in L^2(v).$$

Now consider any measurable set E such that $u(E) = u\{T(vg) > t\}$. We can call the collection of such sets $\mathcal{E}_t$. We wish to obtain from this the following inequality:

$$\begin{aligned} \int \mathbf{1}_E |T(vg)| u dx &\leq C_1 \|g\|_{L^2(v)} (u\{|T(vg)| > t\})^{1/2} \\ &= C_1 \|g\|_{L^2(v)} u(E)^{1/2}, \quad \forall E \in \mathcal{E}_t. \end{aligned} \tag{4.3.71}$$

From the previous inequality, it follows immediately that

$$\||T^*(\mathbf{1}_E u)\|^2 v \leq C\|\mathbf{1}_E\|^2_{L^2(u)} = C \int_E u$$

for all measurable E. But in our example below, this fails even for E being the interval $[-1, 1]$, so we cannot have the weak type estimate of the type $T(v\cdot) : L^2(v) \to L^{2,\infty}(u)$.

It is left to explain why (4.3.71) holds. To do that we just write the integral in the right-hand side using the distribution function:

$$\int_t^\infty u\{b|T(vg)| > s\}ds \le \int_t^\infty \min\left(\frac{\|g\|^2_{L^2(v)}}{s^2}, u\{|T(vg)| > t\}\right) ds.$$

The latter integral is bounded by

$$\int_{\|g\|_{L^2(v)}/u\{|T(vg)|>t\}^{1/2}}^\infty \frac{u\{|T(vg)| > t\}}{s^2} ds$$
$$+ u\{|T(vg)| > t\} \int_t^{\|g\|_{L^2(v)}/u\{|T(vg)|>t\}^{1/2}} ds,$$

which is bounded by $2\|g\|_{L^2(v)} u\{|T(vg)| > t\}^{1/2}$, and this is exactly (4.3.71).

To construct the example, define $u := \mathbf{1}_{[-1,1]}$. It follows from the definition of the fundamental function $\psi = \psi_X$ that for $I = [0, a]$ or $I = [-a, a]$, $a \ge 1$

$$\|u\|_{X(I)} = \psi(1/a) \tag{4.3.72}$$

Defining

$$v(x) := \begin{cases} 1/\psi(1/|x|), & |x| \ge 1, \\ 1/\psi(1), & |x| < 1, \end{cases}$$

we can see that (4.3.69) is satisfied.

Indeed, if $I = [0, a]$ or $I = [-a, a]$, then the estimate (4.3.69) follows immediately from (4.3.72), definition of v and the property of Banach function spaces $\|v\|_{Y(I)} \le C\|v\|_{L^\infty(I)}$.

If $I \cap [-1, 1] = \varnothing$, then (4.3.69) is trivial (left-hand side is zero), so one needs to show that (4.3.69) holds uniformly for all I, $I \cap [-1, 1] \ne \varnothing$.

If $|I| > 2$ (and $I \cap [-1, 1] \ne \varnothing$) then, denoting $I_0 = [-1, 1]$

$$\begin{aligned} \|u\|_{X(I)} = \psi(|I \cap I_0|/|I|) &\le \psi(|I_0|/|I|) && \text{because } \psi(s) \uparrow \\ &\le 4\psi(|I_0|/(4|I|)) && \text{because } s\psi(s) \downarrow \\ &= 4\psi(1/(2|I|)), \end{aligned}$$

and

$$\|v\|_{Y(I)} \le C\|v\|_{L^\infty(I)} \le C\psi(1/(|I| + 1))^{-1} \le C\psi(1/(2|I|))^{-1}.$$

Combining these two inequalities, we get (4.3.69) with $B = 4C$.

If $|I| \le 2$ (and still $I \cap [-1, 1] \ne \varnothing$), then

$$v(x) \le 1/\psi(1/3), \qquad u(x) \le 1 \quad \forall x \in I,$$

so (4.3.69) trivially holds for such intervals.

Thus, (4.3.69) holds for the weights u, v.

Let now $I = [-1, 1]$. Then $\mathbf{1}_I u = u$ and $\|\mathbf{1}_I u\|^2_{L^2(u)} = 2$. On the other hand, for $|x| > 1$

$$|T\mathbf{1}_I u(x)| \geq 1/|x|,$$

so

$$\|T\mathbf{1}_I u\|^2_{L^2(v)} \geq 2\int_1^\infty \frac{1}{x^2} \cdot \frac{1}{\psi(1/x)} dx = 2\int_0^1 \frac{ds}{\psi(s)} = \infty,$$

and (4.3.70) is proved. □

REMARK 4.3.27 A similar construction shows the necessity of the integrability condition

$$\int^\infty \frac{dt}{t\alpha(t)} < \infty \tag{4.3.73}$$

in Theorem 4.3.5.

Namely, suppose that this condition fails for a penalty function α, and β is an arbitrary penalty function (that could satisfy (4.3.73)). As usual, we assume that $t \mapsto t\alpha(t)$ and $t \mapsto t\beta(t)$ are increasing; let us also assume that $t \mapsto e^t/(t\alpha(t))$ increases for $t \geq 1$.

Then it is possible to construct a pair of weights u, v satisfying the bump condition

$$\sup_I \alpha(\mathbf{u}^*_I/\mathbf{u}_I)\mathbf{u}^*_I \mathbf{v}^*_I \beta(\mathbf{v}^*_I/\mathbf{v}_I) < \infty \tag{4.3.74}$$

and such that (4.3.70) holds.

To do that, we again put $u = \mathbf{1}_{[-1,1]}$. Recalling that $\mathbf{u}^*_I = \|u\|_{\Lambda_{\psi_0}(I)}$, $\psi_0(s) = s\ln(e/s)$, we compute using the definition of the fundamental function that for $I = [0, a]$, $a \geq 1$

$$\mathbf{u}^*_I = \|u\|_{\Lambda_{\psi_0}(I)} = a^{-1}\ln(ea), \qquad \mathbf{u}_I = \|u\|_{L^1(I)} = a^{-1}.$$

Defining for $|x| \geq 1$,

$$v(x) = \frac{|x|}{\ln(e|x|)\alpha(\ln(e|x|))}$$

and putting $v(x) = v(1)$ for $|x| < 1$, we get the weights satisfying (4.3.74) and such that (4.3.70) holds.

The fact that (4.3.70) holds follows elementarily from the failure of the integrability condition (4.3.73). The fact that the bump condition (4.3.74) is satisfied can be proved similar to how it was done in the previous example.

The crucial fact there is that the weight v is clearly doubling, $v(x) \le 2v(2x)$, and it is also increasing by the assumption that $t \mapsto e^t/(t\alpha(t))$ increases for $t \ge 1$. Then for any interval $I = [0, a]$

$$\|v\|_{L^1(I)} \asymp \|v\|_{\Lambda_{\psi_0}(I)} \asymp \|v\|_{L^\infty(I)} = \psi(a),$$

and so it is easy to check that (4.3.74) holds for intervals I of form $[0, a]$ and $[-a, a]$.

The general case can be reduced to this case. We need only to check that the intervals I such that $I \cap [-1, 1] \neq \varnothing$. Let us fix $|I|$ and consider the worst possible cases, i.e., the maximal possible values of $\mathbf{u}^*_I$ and $\mathbf{v}^*_I$, and the minimal possible values of $\mathbf{u}$ and $\mathbf{v}$ (assuming that $|I|$ is fixed and $I \cap [-1, 1] \neq \varnothing$).

Then using doubling property of v, we get the conclusion.

We leave the detail as an exercise for the reader.

REMARK 4.3.28 For general p, $1 < p < \infty$, the bumping condition reads

$$\|u\|^{1/p'}_{X(I)} \|v\|^{1/p}_{Y(I)} \le B < \infty \tag{4.3.75}$$

uniformly on all intervals I; here $1/p + 1/p' = 1$.

One can see that exponents are correct by investigating the homogeneity. Note also that the case $X = Y = L^1$ gives us the joint A_p condition for two weights.

If the fundamental function $\psi = \psi_X$ is such that

$$\int_0^1 \psi(s)^{-p/p'} s^{p-2} ds = \infty \tag{4.3.76}$$

(nothing is assumed about X), one can construct a pair of weights u, v satisfying (4.3.75) and such that for the Hilbert Transform T

$$\|T(\mathbf{1}_{[-1,1]} u)\|_{L^p(v)} = \infty. \tag{4.3.77}$$

To do that we define $u := \mathbf{1}_{[-1,1]}$ and $v(x) = \max\{1/\psi(1), 1/\psi(|x|)\}^{-p/p'}$. All the calculations are similar to the presented above for the L^2 case; we leave them as an exercise for the reader.

REMARK 4.3.29 In the above examples, (4.3.70) and (4.3.77) do hold if we replace the Hilbert Transform T by the maximal function.

An example that for $\psi = \psi_X$ satisfying (4.3.76) and for $Y = L^1$, the bumping condition (4.3.75) is not sufficient for the weighted estimate for the maximal function was presented in [**146**]; see Proposition 4.3.2 there.

This proposition is stated in a slightly different language, but after translation one can see that integrability of the left-hand side in (4.3.76) is equivalent to the integrability condition on $\varphi_{X'}$ in there; note that our ψ and $\varphi_{X'}$ in [**146**] are related as $\varphi_{X'}(s) = s\psi^{-1/p'}(s)$.

As for the Hilbert Transform, the only counterexample (previous to ours) we are aware of is the example in [**46**] showing that the condition (4.3.69) with $X = L\log L$, $Y = L^1$ is not sufficient for the operator $g \mapsto T(fv)$ to be acting $L^2(v) \to L^{2,\infty}(u)$.

5

Application of Bellman Technique to the Endpoint Estimates of Singular Integrals

5.1 Endpoint Estimate

The endpoint estimates play an important part in the theory of singular integrals (weighted or unweighted). They are usually the most difficult estimates in the theory, and the most interesting of course. It is a general principle that one can extrapolate the estimate from the endpoint situation to all other situations. We refer the reader to the book [**45**] that treats this subject of extrapolation in depth.

On the other hand, it happens quite often that the singular integral estimates exhibit a certain "blow-up" near the endpoint. To catch that blow-up can be a difficult task. We demonstrate this hunt for blow-ups by the examples of weighted dyadic singular integrals and their behavior in $L^p(w)$. The endpoint p will be naturally 1, and slightly unnaturally 2, depending on the martingale singular operator. The singular integrals in this chapter are the easiest possible. They are dyadic martingale operators on σ-algebra generated by usual homogeneous dyadic lattice on the real line. We do not consider any nonhomogeneous situations, and this standard σ-algebra generated by a dyadic lattice $\mathcal{D}$ will be provided with Lebesgue measure.

Our goal will be to show how the technique of Bellman function gives us the proof of the blow-up of the weighted estimates of the corresponding weighted dyadic singular operators. This blow-up will be demonstrated by certain estimates from below of the Bellman function of a dyadic problem. Interestingly, then one could bootstrap the correct estimates from below of a dyadic operators to the estimate from below of such classical operators as, e.g., the Hilbert transform. The same rate of blow-up persists for the classical operators.

As to the Bellman function part to be described below, that part will be reduced to the task of finding the lower estimates for the solutions of specific Monge–Ampère differential equations with concrete first-order drifts.

5.1.1 Endpoint Estimate for Weighted Martingale Transform

We will work with a standard dyadic filtration $\mathcal{D} = \cup_k \mathcal{D}_k$ on $\mathbb{R}$, where

$$\mathcal{D}_k \stackrel{\text{def}}{=} \left\{ \left[\frac{m}{2^k}, \frac{m+1}{2^k} \right) : m \in \mathbb{Z} \right\}.$$

We consider the martingale transform related to this homogeneous dyadic filtration.

As before, for an average, we will use the notation

$$\langle \varphi \rangle_{_I} \stackrel{\text{def}}{=} \frac{1}{|I|} \int_I \varphi \, dt.$$

In Section 1.5 of the Introduction, we defined martingale differences

$$\Delta_{_J} \varphi \stackrel{\text{def}}{=} \sum_{I \in \text{ch}(J)} \chi_{_I} \big(\langle \varphi \rangle_{_I} - \langle \varphi \rangle_{_J} \big).$$

Recall that the symbol $\text{ch}(J)$ denotes the dyadic children of J.

For our case of dyadic lattice on the line $|\Delta_{_J} \varphi|$ is constant on J, the set $\text{ch}(J)$ consists of two halves of J (J^+ and J^-), and

$$\Delta_{_J} \varphi = \tfrac{1}{2} \big(\langle \varphi \rangle_{_{J+}} - \langle \varphi \rangle_{_{J-}} \big) (\chi_{_{J+}} - \chi_{_{J-}}).$$

We consider the dyadic A_1 class of weights. We omit the word "dyadic" in what follows, since we consider only dyadic operators. A positive function w is called an A_1-weight if

$$[w]_{A_1} \stackrel{\text{def}}{=} \sup_{J \in \mathcal{D}} \frac{\langle w \rangle_{_J}}{\inf_J w} < \infty.$$

By Mw, we will denote the dyadic maximal function of w, that is,

$$Mw(x) = \sup\{ \langle w \rangle_{_J} : J \in \mathcal{D},\ J \ni x \}. \tag{5.1.1}$$

Then the fact that $w \in A_1$ has "norm" Q, will mean that

$$Mw \le Q \cdot w \quad a.\,e.,$$

and $Q = [w]_{A_1}$ is the best constant in this inequality.

Recall that a martingale transform is an operator given by the formula $T_\varepsilon \varphi = \sum_{J \in \mathcal{D}} \varepsilon_{_J} \Delta_{_J} \varphi$. It is convenient to use Haar function $h_{_J}$ associated with the dyadic interval J.

As in Section 1.7, we use the following form for martingale differences:

$$\Delta_{_J} \varphi = (\varphi, h_{_J}) h_{_J}.$$

Sometimes it is more convenient to use the Haar functions H_J normalized in L^∞: $H_J = |J|^{1/2}h_J$. In this notation, the martingale transform ψ of a function φ is

$$\psi = T_\varepsilon\varphi = \sum_{J\in\mathcal{D}} \frac{\varepsilon_J}{|J|}(\varphi, H_J)H_J = \sum_{J\in\mathcal{D}} \varepsilon_J(\varphi, h_J)h_J,$$

where $(\cdot\,,\,\cdot)$ stands for the scalar product in L^2. In all our calculations, we always think that the sum has unspecified, but finite number of terms, so we do not have to worry about the convergence of this series. Nevertheless, approximation arguments give us the final estimates for an arbitrary L^1 function φ. As to the values of the coefficients of multiplicator, we consider the class $|\varepsilon_J| \le 1$ or its important subclass $\varepsilon_J = \pm 1$.

We are interested in the weak estimate for the martingale transform T in the weighted space $L^1(\mathbb{R}, w\,dx)$, where $w \in A_1$. The endpoint exponent is naturally $p = 1$, and we wish to understand the order of magnitude of the constant $C([w]_{A_1})$ in the weak type inequality for the dyadic martingale transform:

$$\frac{1}{|I|}\sup_{\varepsilon} w\{t \in I\colon \sum_{J\in\mathcal{D}(I)} \varepsilon_J(\varphi, h_J)h_J(t) \ge \lambda\} \le C([w]_{A_1})\frac{\langle|\varphi|w\rangle_I}{\lambda}. \tag{5.1.2}$$

Here φ runs over all functions, such that $\operatorname{supp}\varphi \subset I$ and $\varphi \in L^1(I, w\,dt)$, $w \in A_1$; $\varepsilon = \{\varepsilon_J\}$ and $|\varepsilon_J| \le 1$. For a set S, we write $w(S)$ for $\int_S w(t)\,dt$. This section is devoted to the study of the asymptotically "sharp" order of magnitude of constants $C([w]_{A_1})$ in terms of $[w]_{A_1}$ if $[w]_{A_1}$ is large. We are primarily interested in the estimate of $C([w]_{A_1})$ from below, that is, in finding the worst possible A_1 weight in terms of weak type estimate (of course, this involves finding the worst test function φ as well).

We will prove the following result.

THEOREM 5.1.1 *For any Q, $Q \ge 4$, there is a weight $w \in A_1$ with $[w]_{A_1} = Q$, such that constant $C(Q)$ from* (5.1.2) *satisfies*

$$C(Q) \ge \frac{1}{515}Q(\log Q)^{1/3}.$$

In [**111**], the following estimate from above has been proved.

THEOREM 5.1.2 *There is a positive absolute constant c, such that for any weight $w \in A_1$, estimate* (5.1.2) *holds with*

$$C(\,[w]_{A_1}) = c\,[w]_{A_1}\log[w]_{A_1}.$$

5.1.2 Two Problems of Muckenhoupt

Theorem 5.1.1 is a subtle result, and it will take some space below to prove. Recall that Muckenhoupt conjectured that for the Hilbert transform $\mathcal{H}$ and any weight $w \in A_1$, the following two estimates hold on a unit interval I:

$$w\{x \in I : |\mathcal{H}f(x)| > \lambda\} \le \frac{C}{\lambda}\int_I |f| Mw dx, \tag{5.1.3}$$

$$w\{x \in I : |\mathcal{H}f(x)| > \lambda\} \le \frac{C\,[w]_{A_1}}{\lambda}\int_I |f| w dx. \tag{5.1.4}$$

Obviously, if (5.1.3) holds, then (5.1.4) is valid as well. It took many years to *disprove* (5.1.3). This was done by M. Reguera [**157**] for the martingale transform, and together with C. Thiele in [**158**] for the Hilbert transform. The constructions involve a very irregular weight w (almost a sum of delta measures), so there was hope that such an effect cannot appear when the weight is regular in the sense that $w \in A_1$.

Theorem 5.1.1 gives us a counterexample to this hope for the case when the Hilbert transform is replaced by the martingale transform on a usual homogeneous dyadic filtration. The reader can consult [**123**] to see that for the Hilbert transform a counterexample also exists, and so (5.1.4) fails as well. The counterexample for the Hilbert transform is the transference of a counterexample we build here for the martingale transform. The blow-up estimate $Q(\log Q)^{1/3}$ holds for the Hilbert transform as well.

Another construction, which is due to A. Lerner, F. Nazarov, S. Ombrosi [**110**], gives us the sharp blow-up estimate $cQ\log Q$. It is explained in the next section. It has a Bellman function origin, even though it does not use Bellman function directly.

5.2 The Bellman Function of Weak Weighted Estimate of the Martingale Transform and Its Logarithmic Blow-Up

As in Section 1.7, we shall look at pairs of functions φ, ψ, where φ is a dyadic martingale starting at x_1, and ψ is a dyadic martingale transform of φ starting at x_2, i.e.,

$$\varphi = x_1 + \sum_{J\in\mathcal{D}(I)} (\varphi, h_J)h_J\,, \qquad \psi = x_2 + \sum_{J\in\mathcal{D}(I)} \varepsilon_J(\varphi, h_J)h_J\,.$$

We consider here the class of martingale transforms, where ψ is differentially subordinated to φ, i.e., we assume that $|\varepsilon_J| \le 1$.

Passing to the weighted case, we need to investigate a Bellman function of a few additional variables. Now two additional variables x_4 and x_5 appear describing a test weight w. Namely, we put

$$x_4 = \langle w \rangle_I \qquad \text{and} \qquad x_5 = \inf_I w.$$

The test weight w will run over the set of all A_1-weights with $[w]_{A_1} \le Q$ and with the prescribed parameters x_4 and x_5. This, by the way, means that these parameters must satisfy the following condition: $x_4 \le Qx_5$.

The coordinates x_1 and x_2 were introduced above; they are the same as in the unweighted case of Section 1.7, but we need to change slightly the third coordinate:

$$x_3 = \langle |\varphi| w \rangle_I$$

because now we fix a weighted norm of the test function $\varphi \in L^1(I, w\,dx)$. A Bellman point $x = (x_1, x_2, x_3, x_4, x_5) = \mathfrak{b}_{\varphi,\psi,w}$ is defined by a dyadic martingale φ started at x_1, by a subordinated to φ martingale ψ started at x_2, and by A_1-weight w. The Bellman function at this point is defined as follows:

$$\mathbf{B}(x) \stackrel{\text{def}}{=} \mathbf{B}(x; Q) \stackrel{\text{def}}{=} \sup \frac{1}{|I|} w(\{t \in I : \psi(t) \ge 0\}), \tag{5.2.1}$$

where the supremum is taken over all admissible triples φ, ψ, w.

Remark 5.2.1 We found the formula for $\mathbf{B}(x_1, x_2, x_3, 1, 1)$ in Section 1.7.

Sometimes we mark the Bellman function by the index Q to emphasize that it depends on a fixed parameter Q. And, in fact, we are interested just in the dependence of $\mathbf{B}$ on this parameter. However, during our calculations, we will omit this index.

This Bellman function is defined in the following subdomain of $\mathbb{R}^5$:

$$\Omega \stackrel{\text{def}}{=} \{x \in \mathbb{R}^5 : x_3 \ge |x_1| x_5,\ 0 < x_5 \le x_4 \le Qx_5\}. \tag{5.2.2}$$

Note that formally, the Bellman function is defined on the whole $\mathbb{R}^5$, but in the domain Ω, we include only the points for which the set of test functions is not empty and therefore $\mathbf{B}(x) \ne -\infty$ (we would like to assume that $\mathbf{B} \ge 0$).

5.2.1 The Properties of $\mathbf{B}_Q$

5.2.1.1 The First Property: Boundary Conditions. On the boundary $x_4 = x_5$ the weight is a constant function $w = x_4 = x_5$, and therefore, by Theorem 1.7.4, we have the following:

$$\mathbf{B}_Q(x) \stackrel{\text{def}}{=} \mathbf{B}(x_1, x_2, x_3, x_5, x_5)$$

$$= \begin{cases} x_5, & \text{if } x_3 + x_2 x_5 \geq 0, \\ x_5 \left(1 - \dfrac{\left(\frac{x_3}{x_5} + x_2\right)^2}{x_2^2 - x_1^2}\right), & \text{if } x_3 + x_2 x_5 < 0. \end{cases}$$

As we already mentioned, we will usually skip subscript Q and write simply $\mathbf{B}$ instead of $\mathbf{B}_Q$.

5.2.1.2 The Second Property: the Homogeneity. It is clear that if $\{\varphi, \psi, w\}$ is the set of admissible triples for a point $x \in \Omega$, then the set of triples $\{s_1\varphi, s_1\psi, s_2 w\}$ is admissible for the point

$$\tilde{x} = (s_1 x_1, s_1 x_2, s_1 s_2 x_3, s_2 x_4, s_2 x_5).$$

This is true for an arbitrary pair of positive numbers s_1, s_2. Then by definition of the Bellman function, we have

$$\mathbf{B}(\tilde{x}) = s_2 \mathbf{B}(x). \tag{5.2.3}$$

In what follows, we deal mainly with the restriction $\mathcal{B}$ of $\mathbf{B}$ to the intersection of Ω with the three-dimensional affine plane

$$\{(x_1, -1, x_3, x_4, 1) \colon |x_1| \leq x_3,\ 1 \leq x_4 \leq Q\}, \tag{5.2.4}$$

i. e., function

$$\mathcal{B}(x_1, x_3, x_4) = \mathbf{B}(x_1, -1, x_3, x_4, 1). \tag{5.2.5}$$

We will identify the above mentioned part of the three-dimensional affine plane with a subdomain of $\mathbb{R}^3$:

$$G \stackrel{\text{def}}{=} \{(x_1, x_3, x_4) \colon |x_1| \leq x_3,\ 1 \leq x_4 \leq Q\}. \tag{5.2.6}$$

For $x_2 \geq 0$, we always have $\mathbf{B}(x) = x_4$, because for any such point the constant test function $\psi = x_2$ is admissible, and for $x_2 < 0$ we can reconstruct $\mathbf{B}$ from $\mathcal{B}$, due to homogeneity (5.2.3): choosing $s_1 = -x_2^{-1}$ and $s_2 = x_5^{-1}$, we get

$$\mathbf{B}(x) = x_5 \mathbf{B}\left(-\frac{x_1}{x_2}, -1, -\frac{x_3}{x_2 x_5}, \frac{x_4}{x_5}, 1\right) = x_5 \mathcal{B}\left(-\frac{x_1}{x_2}, -\frac{x_3}{x_2 x_5}, \frac{x_4}{x_5}\right). \tag{5.2.7}$$

5.2.1.3 The Third Property: Special Form of Concavity. Here we state our main inequality, the weighted analog of Lemma 1.7.1.

LEMMA 5.2.2 *Let $x^\pm$ be two points in Ω, such that $|x_2^+ - x_2^-| \le |x_1^+ - x_1^-|$ and let point x with $x_i = \frac{1}{2}(x_i^+ + x_i^-)$ for $1 \le i \le 4$ and $x_5 = \min\{x_5^+, x_5^-\}$ be in Ω as well. Then*

$$\mathbf{B}(x) - \frac{\mathbf{B}(x^+) + \mathbf{B}(x^-)}{2} \ge 0. \tag{5.2.8}$$

PROOF We repeat almost verbatim the proof of Lemma 1.7.1. Fix $x^\pm \in \Omega$, and take two triples of test functions $\varphi^\pm$, $\psi^\pm$, $w^\pm$, giving the supremum in $\mathbf{B}(x^+)$, $\mathbf{B}(x^-)$, respectively, up to a small number $\eta > 0$. Using the fact that function $\mathbf{B}$ does not depend on the interval where test functions are defined, we assume that φ^+, ψ^+, w^+ live on I^+ and φ^-, ψ^-, w^- live on I^-, i.e.,

$$\varphi^\pm = x_1^\pm + \sum_{J \in \mathcal{D}(I)} a_J h_J, \qquad \psi^\pm = x_2^\pm + \sum_{J \in \mathcal{D}(I)} \varepsilon_J a_J h_J, \quad |\varepsilon_J| \le 1.$$

Consider

$$\varphi(t) \stackrel{\text{def}}{=} \begin{cases} \varphi^+(t), & \text{if } t \in I^+ \\ \varphi^-(t), & \text{if } t \in I^- \end{cases} = \frac{x_1^+ + x_1^-}{2} + \frac{x_1^+ - x_1^-}{2} h_I + \sum_{J \in \mathcal{D}(I)} a_J h_J,$$

$$\psi(t) \stackrel{\text{def}}{=} \begin{cases} \psi^+(t), & \text{if } t \in I^+ \\ \psi^-(t), & \text{if } t \in I^- \end{cases} = \frac{x_2^+ + x_2^-}{2} + \frac{x_2^+ - x_2^-}{2} h_I + \sum_{J \in \mathcal{D}(I)} \varepsilon_J a_J h_J.$$

and

$$w(t) \stackrel{\text{def}}{=} \begin{cases} w^+(t), & \text{if } t \in I^+, \\ w^-(t), & \text{if } t \in I^-. \end{cases}$$

Since $|x_2^+ - x_2^-| \le |x_1^+ - x_1^-|$ and all $|\varepsilon_J| \le 1$, ψ is subordinated to φ. Moreover, according to hypothesis of the Lemma, point x is in Ω, whence $x_4 \le Q x_5$, i.e., $[w]_{A_1} \le Q$. Therefore, the triple φ, ψ, w is an admissible triple of test functions corresponding to point x, and

$$\begin{aligned}
\mathbf{B}(x) &\ge \frac{1}{|I|} w(\{t \in I_0 \colon \psi(t) \ge 0\}) \\
&= \frac{1}{2|I^+|} w^+(\{t \in I^+ \colon \psi(t) \ge 0\}) + \frac{1}{2|I^-|} w^-(\{t \in I^- \colon \psi(t) \ge 0\}) \\
&\ge \tfrac{1}{2}\mathbf{B}(x^+) + \tfrac{1}{2}\mathbf{B}(x^-) - 2\eta.
\end{aligned}$$

Since this inequality holds for an arbitrarily small η, we can pass to the limit as $\eta \to 0$, which gives us the required assertion. □

5.2.1.4 The Fourth Property: B Decreases in x_5. This is a corollary of the preceding property, i.e., it follows from the main inequality. Indeed, if we put in the hypotheses of Lemma 5.2.2 $x_i^+ = x_i^-$ for $1 \le i \le 4$ and $x_5^+ > x_5^-$, then $x_5 = x_5^-$, and inequality (5.2.8) turns into

$$\mathbf{B}(x_1, x_2, x_3, x_4, x_5^-) - \mathbf{B}(x_1, x_2, x_3, x_4, x_5^+) \ge 0, \tag{5.2.9}$$

which means that B is monotone in x_5.

5.2.1.5 The Fifth Property: Function $t \mapsto \frac{1}{t}\mathcal{B}(x_1, tx_3, tx_4)$ is Increasing. Function $\mathcal{B}$ was defined in (5.2.5). This property of $\mathcal{B}$ is in fact the preceding property rewritten in terms of $\mathcal{B}$. Indeed, if we put $x_2 = -1$ and use (5.2.7) and (5.2.9) we get the required monotonicity (we just rewrite (5.2.7) and (5.2.9) and use the notation $t^+ = \frac{1}{x_5^-}$ and $t^- = \frac{1}{x_5^+}$).

5.2.1.6 The Sixth Property: Function $\mathcal{B}$ is Concave. Lemma 5.2.2 applied to the case $x_2^+ = x_2^- = -1$ and $x_5^+ = x_5^- = 1$ guarantees the stated concavity.

5.2.1.7 The Seventh Property: The Symmetry and Monotonicity in x_1. It is easy to see from the definition that $\mathbf{B}$, and hence $\mathcal{B}$ as well, is even in its variable x_1.

Concavity of $\mathcal{B}$ (in x_1) and this symmetry together imply that $x_1 \mapsto \mathcal{B}(x_1, x_3, x_4)$ is increasing on $[-x_3, 0]$ and decreasing on $[0, x_3]$.

5.2.2 The Goal and the Idea of the Proof

It would be natural now to solve the corresponding boundary value problem for the Monge–Ampère equation, to find the function $\mathbf{B}$, as it was done in the unweighted case, and then to find the constant we are interested in:

$$C(Q) = \sup\Big\{\frac{|x_2|\mathbf{B}(x)}{x_3} : x_2 < 0,\ x_3 \ge |x_1|x_5,\ x_5 \le x_4 \le Qx_5\Big\}.$$

However, for now this task is too difficult for us. So, we use the listed properties of $\mathbf{B}$ to prove the following estimate from below on function $\mathcal{B}$.

THEOREM 5.2.3 *If $Q \ge 4$ then*

$$\mathcal{B}(x_1, x_3, x_4) \ge \frac{1}{515} Q(\log Q)^{1/3} x_3. \tag{5.2.10}$$

at some point $(x_1, x_3, x_4) \in G$.

Now a couple of words about the idea of the proof of Theorem 5.2.3. Ideally, we would like to find the formula for $\mathcal{B}$ (and therefore for $\mathbf{B}$ because of (5.2.7)). To proceed, we rewrite the third property of $\mathbf{B}$ (see Section 5.2.1.3) as a PDE on $\mathcal{B}$. Then, using the boundary conditions on $\mathcal{B}$ on ∂G (the domain G is defined in (5.2.6)), we may hope to solve this PDE. Unfortunately there are many roadblocks on this path, starting with the fact that the third property of **B** is not a PDE, it is rather a partial differential inequality in finite difference form. In the unweighted case, we pay no attention to this important fact. We simply assume the required smoothness of our function to find a smooth candidate. After such a candidate was found, we have proved that it coincides with the required Bellman function. Now we cannot find a candidate and we will work with the abstractly defined Bellman function whose smoothness is unknown. We will write the inequality in finite difference form as a pointwise partial differential inequality, but for that we will need a subtle result of Aleksandrov.

5.2.3 From Finite Difference Inequality to Differential Inequality via Aleksandrov's Theorem

As it was mentioned in Section 5.2.1.6, the function $\mathcal{B}$ is concave on its domain of definition G. By the result of Aleksandrov, see Theorem 6.9 of [**60**], $\mathcal{B}$ has all second derivatives almost everywhere in G. Second property (homogeneity) of function **B** (see (5.2.7)) implies that the function **B** has all second derivatives almost everywhere in Ω.

First, using this fact, we rewrite the homogeneity condition (see Section 5.2.1.2) in the following differential form:

$$x_1\mathbf{B}_{x_1} + x_2\mathbf{B}_{x_2} + x_3\mathbf{B}_{x_3} = 0; \tag{5.2.11}$$

$$x_3\mathbf{B}_{x_3} + x_4\mathbf{B}_{x_4} + x_5\mathbf{B}_{x_5} = \mathbf{B}. \tag{5.2.12}$$

We have got these equalities by differentiating (5.2.3) with respect to s_1 and with respect to s_2 and taking the result for $s_1 = s_2 = 1$.

Our second step is to replace the main inequality in finite difference form by the inequality in the form of a pointwise partial differential inequality. Lemma 5.2.2 implies that the quadratic form

$$\sum_{i,j=1}^{4} \mathbf{B}_{x_i x_j}\xi_i\xi_j \tag{5.2.13}$$

is nonpositive at almost any interior point of Ω and for all vectors $\xi \in \mathbb{R}^4$ such that $|\xi_2| \le |\xi_1|$.

We consider three partial cases of (5.2.13) with $\xi_1 = \xi_2$, with $\xi_1 = -\xi_2$, and with $\xi_2 = 0$. Moreover, to reduce our investigation to consideration of 2×2

matrices, we choose some special relation between ξ_3 and ξ_4. In the first case, we consider the quadratic form on the vector ξ with

$$\xi_1 = \xi_2 = -\delta_1, \quad \xi_3 = x_3(\delta_1 + \delta_2), \quad \xi_4 = x_4\delta_2.$$

In the second case, we put

$$\xi_1 = -\xi_2 = \delta_1, \quad \xi_3 = x_3(\delta_1 + \delta_2), \quad \xi_4 = x_4\delta_2.$$

Then we get two quadratic forms:

$$\sum_{i,j=1}^{4} \mathbf{B}_{x_i x_j} \xi_i \xi_j = \sum_{i,j=1}^{2} K_{ij}^{\pm} \delta_i \Delta_J,$$

where we denote by $K^{\pm}$ two 2×2 unpleasant (on the first glance) matrices:

$$K^{\pm} = \begin{pmatrix} \begin{matrix} \mathbf{B}_{x_1x_1} \pm \mathbf{B}_{x_1x_2} \mp x_3\mathbf{B}_{x_1x_3} \pm \mathbf{B}_{x_1x_2} \\ + \mathbf{B}_{x_2x_2} - 2x_3\mathbf{B}_{x_2x_3} \mp x_3\mathbf{B}_{x_1x_3} + x_3^2\mathbf{B}_{x_3x_3} \end{matrix} & \begin{matrix} \mp x_3\mathbf{B}_{x_1x_3} - x_3\mathbf{B}_{x_2x_3} + x_3^2\mathbf{B}_{x_3x_3} \\ \mp x_4\mathbf{B}_{x_1x_4} - x_4\mathbf{B}_{x_2x_4} + x_3x_4\mathbf{B}_{x_3x_4} \end{matrix} \\ \begin{matrix} \mp x_3\mathbf{B}_{x_1x_3} - x_3\mathbf{B}_{x_2x_3} + x_3^2\mathbf{B}_{x_3x_3} \\ \mp x_4\mathbf{B}_{x_1x_4} - x_4\mathbf{B}_{x_2x_4} + x_3x_4\mathbf{B}_{x_3x_4} \end{matrix} & x_3^2\mathbf{B}_{x_3x_3} + 2x_3x_4\mathbf{B}_{x_3x_4} + x_4^2\mathbf{B}_{x_4x_4} \end{pmatrix}.$$

These matrices are nonpositively defined and their half sum is the following nonpositively defined matrix:

$$\begin{pmatrix} \mathbf{B}_{x_1x_1} + \mathbf{B}_{x_2x_2} - 2x_3\mathbf{B}_{x_2x_3} + x_3^2\mathbf{B}_{x_3x_3} & -x_3\mathbf{B}_{x_2x_3} + x_3^2\mathbf{B}_{x_3x_3} - x_4\mathbf{B}_{x_2x_4} + x_3x_4\mathbf{B}_{x_3x_4} \\ -x_3\mathbf{B}_{x_2x_3} + x_3^2\mathbf{B}_{x_3x_3} - x_4\mathbf{B}_{x_2x_4} + x_3x_4\mathbf{B}_{x_3x_4} & x_3^2\mathbf{B}_{x_3x_3} + 2x_3x_4\mathbf{B}_{x_3x_4} + x_4^2\mathbf{B}_{x_4x_4} \end{pmatrix}. \tag{5.2.14}$$

Before proceeding further, we rewrite this matrix in terms of the function $\mathcal{B}$. For this aim, we have to get rid of the derivatives with respect to x_2 in this matrix. We are able to do this by using (5.2.11):

$$\begin{aligned} -x_2\mathbf{B}_{x_2x_3} &= \mathbf{B}_{x_3} + x_1\mathbf{B}_{x_1x_3} + x_3\mathbf{B}_{x_3x_3}; \\ -x_2\mathbf{B}_{x_2x_4} &= x_1\mathbf{B}_{x_1x_4} + x_3\mathbf{B}_{x_3x_4}; \\ x_2^2\mathbf{B}_{x_2x_2} &= 2x_1\mathbf{B}_{x_1} + 2x_3\mathbf{B}_{x_3} + x_1^2\mathbf{B}_{x_1x_1} + 2x_1x_3\mathbf{B}_{x_1x_3} + x_3^2\mathbf{B}_{x_3x_3}. \end{aligned}$$

Using these expressions at the point $x = (x_1, -1, x_3, x_4, 1)$, we can rewrite the matrix (5.2.14) as follows:

$$\begin{pmatrix} (1 + x_1^2)\mathcal{B}_{x_1x_1} + 2x_1\mathcal{B}_{x_1} & -x_1\mathcal{B}_{x_3} - x_1x_3\mathcal{B}_{x_1x_3} - x_1x_4\mathcal{B}_{x_1x_4} \\ -x_1\mathcal{B}_{x_3} - x_1x_3\mathcal{B}_{x_1x_3} - x_1x_4\mathcal{B}_{x_1x_4} & x_3^2\mathcal{B}_{x_3x_3} + 2x_3x_4\mathcal{B}_{x_3x_4} + x_4^2\mathcal{B}_{x_4x_4} \end{pmatrix}. \tag{5.2.15}$$

Now we consider the matrix K^0 that appears if we take $\xi_1 = x_1\delta_1$, $\xi_2 = 0$, $\xi_3 = x_3\delta_2$, and $\xi_4 = x_4\delta_2$ in our quadratic form

$$\sum_{i,j=1}^{4} \mathbf{B}_{x_ix_j}\xi_i\xi_j = \sum_{i,j=1}^{2} K^0_{ij}\delta_i\Delta_J.$$

In result we get

$$K^0 = \begin{pmatrix} x_1^2\mathbf{B}_{x_1x_1} & x_1x_3\mathbf{B}_{x_1x_3} + x_1x_4\mathbf{B}_{x_1x_4} \\ x_1x_3\mathbf{B}_{x_1x_3} + x_1x_4\mathbf{B}_{x_1x_4} & x_3^2\mathbf{B}_{x_3x_3} + 2x_3x_4\mathbf{B}_{x_3x_4} + x_4^2\mathbf{B}_{x_4x_4} \end{pmatrix}.$$

The same matrix at the point $x = (x_1, -1, x_3, x_4, 1)$ is

$$\begin{pmatrix} x_1^2\mathcal{B}_{x_1x_1} & x_1x_3\mathcal{B}_{x_1x_3} + x_1x_4\mathcal{B}_{x_1x_4} \\ x_1x_3\mathcal{B}_{x_1x_3} + x_1x_4\mathcal{B}_{x_1x_4} & x_3^2\mathcal{B}_{x_3x_3} + 2x_3x_4\mathcal{B}_{x_3x_4} + x_4^2\mathcal{B}_{x_4x_4} \end{pmatrix}. \quad (5.2.16)$$

Taking the sum of (5.2.15) and (5.2.16), we get the following nonpositive matrix:

$$\begin{pmatrix} (1 + 2x_1^2)\mathcal{B}_{x_1x_1} + 2x_1\mathcal{B}_{x_1} & -x_3\mathcal{B}_{x_3} \\ -x_3\mathcal{B}_{x_3} & 2(x_3^2\mathcal{B}_{x_3x_3} + 2x_3x_4\mathcal{B}_{x_3x_4} + x_4^2\mathcal{B}_{x_4x_4}) \end{pmatrix} \le 0. \quad (5.2.17)$$

DEFINITION 5.2.4 Consider a subdomain of G,

$$G_1 \stackrel{\text{def}}{=} \{(x_1, x_3, x_4) \in G \colon x_3 > 2|x_1|,\ 2 < x_4 < Q\}.$$

Fix now $x = (x_1, x_3, x_4) \in G_1$ (now x is a three-vector, not a five-vector as above, see (5.2.6)) and a parameter $t \in [1/2, 1]$. Consider inequality (5.2.17) at the point $x^t = (x_1, tx_3, tx_4)$.

Let us introduce a new function β, which is a certain averaging of $\mathcal{B}$, namely, for any $x \in G_1$, we put

$$\beta(x) \stackrel{\text{def}}{=} 2\int_{1/2}^{1} \mathcal{B}(x^t)\,dt.$$

Notice several simple facts. First of all, as $\mathcal{B}$ is concave, the differentiation under the integral sign is easily justified, and we get

$$x_i\beta_{x_i}(x) = 2\int_{1/2}^{1} x_i^t\mathcal{B}_{x_i}(x^t)\,dt, \qquad x_i^2\beta_{x_ix_i} = 2\int_{1/2}^{1} (x_i^t)^2\mathcal{B}_{x_ix_i}(x^t)\,dt.$$

For every function F on domain G, we introduce the notation,

$$\gamma_F(x) \stackrel{\text{def}}{=} x_3^2F_{x_3x_3} + 2x_3x_4F_{x_3x_4} + x_4^2F_{x_4x_4},$$

then

$$\gamma_{\beta}(x) = 2\int_{1/2}^{1} \gamma_{\mathcal{B}}(x^t)dt. \tag{5.2.18}$$

Now integrate (5.2.17) on the interval $t \in [1/2, 1]$. The previous simple observations allow us now to rewrite our reduced concavity condition in the form

$$\begin{pmatrix} (1+2x_1^2)\beta_{x_1x_1} + 2x_1\beta_{x_1} & -x_3\beta_{x_3} \\ -x_3\beta_{x_3} & 2\gamma_{\beta} \end{pmatrix} \leq 0. \tag{5.2.19}$$

The reader may wonder, why we are so keen to replace (5.2.17) by a virtually same matrix (5.2.19)? The answer is because we can give a very good pointwise estimate on $\gamma_{\beta}(x)$, $x \in G_1$. Unfortunately, we cannot give any pointwise estimate on $\gamma_{\mathcal{B}}(x)$, $x \in G$.

Our reduced concavity condition (5.2.19) is equivalent to the assertion that $\gamma_{\beta} \leq 0$ and the determinant of the matrix in (5.2.19) is nonnegative, i.e.,

$$-2\gamma_{\beta} \cdot [-(1+2x_1^2)\beta_{x_1x_1} - 2x_1\beta_{x_1}] \geq x_3^2\beta_{x_3}^2. \tag{5.2.20}$$

Let us denote

$$R \stackrel{\text{def}}{=} \sup \frac{\mathcal{B}(x)}{x_3}, \qquad x = (x_1, x_3, x_4) \in G.$$

Our goal formulated in (5.2.10) is to prove $R \geq cQ(\log Q)^{\varepsilon}$ (with $c = \frac{1}{515}$ and $\varepsilon = \frac{1}{3}$). We are still not too close, but notice that automatically $\mathcal{B}(x) \leq Rx_3$, $x \in G$.

5.2.4 Logarithmic Blow-Up

First we find a pointwise estimate on γ_{β}.

LEMMA 5.2.5 *If $x = (x_1, x_3, x_4)$ is such that $|x_1| \leq \frac{1}{4}x_3$ and $x_4 \geq 4$, then*

$$-\gamma_{\beta}(x) \leq 8R\left(|x_1| + \frac{x_3}{x_4}\right).$$

PROOF Consider the following functions:

$$\rho(t) \stackrel{\text{def}}{=} \mathcal{B}(x^t), \quad x \in G_1, \qquad \text{and} \qquad r(t) \stackrel{\text{def}}{=} \rho(1)t - \rho(t)$$

on the interval $[t_0, 1]$, where $t_0 = \max\left(\frac{|x_1|}{x_3}, \frac{1}{x_4}\right)$.

Recall that the function $\rho(t)/t$ is increasing (see property five of **B** in Section 5.2.1). Therefore, $\rho(t)/t \leq \rho(1)$, i.e. $r(t) \geq 0$. Since r is convex (because

ρ is concave) and $r(1) = 0$, r is a decreasing function on $[t_0, 1]$, in particular $r'(1) \le 0$. Let us estimate the maximal value of r in the following way:

$$r(t_0) < \rho(1)t_0 \le Rx_3t_0 < R\left(|x_1| + \frac{x_3}{x_4}\right). \tag{5.2.21}$$

Under the hypotheses of the Lemma, we have $t_0 \le \frac{1}{4}$, and therefore

$$-\int_{1/2}^1 \rho''(t)\,dt \le \int_{1/2}^1 r''(t)\,dt \le 4\int_{1/2}^1 (t-t_0)r''(t)\,dt$$
$$\le 4\int_{t_0}^1 (t-t_0)r''(t)\,dt = 4r'(1)(1-t_0) - 4r(1) + 4r(t_0).$$

Using estimate (5.2.21) and the properties of r ($r'(1) \le 0$ and $r(1) = 0$), we get

$$-\int_{1/2}^1 \rho''(t)\,dt \le 4R\left(|x_1| + \frac{x_3}{x_4}\right).$$

The equality $\gamma_{\mathcal{B}}(x^t) = t^2\rho''(t)$ implies

$$-\int_{1/2}^1 \gamma_{\mathcal{B}}(x^t)\,dt \le 4R\left(|x_1| + \frac{x_3}{x_4}\right).$$

So, by (5.2.18), this is the stated in the Lemma estimate. □

Now we would like to get an estimate for β_{x_3} from below. For this aim, we construct a pair of test functions φ, ψ and a test weight w, which supply us with the following estimate for the function $\mathcal{B}$.

LEMMA 5.2.6 *If $x = (x_1, x_3, x_4)$ is such that $2x_3 + x_1 \ge 1$, then*

$$\mathcal{B}(x) \ge \frac{2x_4 - 1}{4}.$$

PROOF Below h_I stands for the L^∞-normalized Haar function of interval I. Let us take the following test functions on the interval $[0, 1]$:

$$\varphi = x_1 + x_3h_{(0,1)} + (x_3 - x_1)h_{\left(0,\frac{1}{2}\right)} + (x_3 + x_1)h_{\left(\frac{1}{2},1\right)};$$

$$\psi = -1 + x_3h_{(0,1)} + (x_3 - x_1)h_{\left(0,\frac{1}{2}\right)} - (x_3 + x_1)h_{\left(\frac{1}{2},1\right)};$$

$$w = 1 + 2(x_4 - 1)\chi_{\left(\frac{1}{4},\frac{3}{4}\right)}.$$

The Bellman point corresponding to this triple is $(x_1, -1, x_3, x_4, 1)$. The function ψ on the interval $\left(\frac{1}{2}, \frac{3}{4}\right)$ has the value $2x_3 + x_1 - 1$, where the weight w is $2x_4 - 1$. Therefore, if $2x_3 + x_1 \geq 1$, then by the definition $\mathcal{B}(x) \geq (2x_4 - 1)/4$. □

COROLLARY 5.2.7 *If $x_3 + x_1 \geq 1$, then*

$$\beta(x) \geq \frac{3x_4 - 2}{8}.$$

PROOF If $x_3 + x_1 \geq 1$, then $2tx_3 + x_1 \geq 1$ for all $t \in \left[\frac{1}{2}, 1\right]$. And therefore,

$$\beta(x) = 2\int_{1/2}^{1} \mathcal{B}(x^t)\, dt \geq \tfrac{1}{2}\int_{1/2}^{1} (2tx_4 - 1)dt = \frac{3x_4 - 2}{8}.$$

□

COROLLARY 5.2.8 *If $x_3 + x_1 \geq 1$ and $x_4 \geq 2$, then*

$$\beta(x) \geq \frac{x_4}{4}.$$

COROLLARY 5.2.9 *If $x_4 \geq 2$, then*

$$\beta(x_1, 1, x_4) \geq \frac{x_4}{4}.$$

PROOF Since the function $\mathbf{B}$ is even in x_1, the functions $\mathcal{B}$ and β are even as well. Therefore, without loss of generality, we can assume that $x_1 \geq 0$. Hence for $x_3 = 1$, the condition $x_3 + x_1 \geq 1$ holds, and we have the required estimate. □

COROLLARY 5.2.10 *If $x_4 \geq 2$, then there exists an $a = a(x_4) \in (0, 1]$ such that $\beta(0, a, x_4) = \frac{x_4}{8}$.*

PROOF Since the function β is continuous, the conditions

$$\beta(0, 0, x_4) = 0 \quad \text{and} \quad \beta(0, 1, x_4) \geq \frac{x_4}{4}$$

guarantee the existence of the required a. □

REMARK 5.2.11 The function β is increasing in x_3 because it is positive, concave, and defined on an infinite interval $(0, \infty)$. Therefore, the root a is unique.

LEMMA 5.2.12 *For any $x \in G$, we have*

$$\beta(x) \geq \left(1 - \frac{2|x_1|}{x_3}\right) \beta(0, x_3, x_4). \tag{5.2.22}$$

PROOF Since β is even in x_1, we can assume $x_1 > 0$. The stated estimate is the immediate consequence of the following two facts, β is nonnegative and concave in x_1:

$$\beta(x) \geq \left(1 - \frac{2x_1}{x_3}\right)\beta(0, x_3, x_4) + \frac{2x_1}{x_3}\beta\left(\frac{x_3}{2}, x_3, x_4\right).$$

□

LEMMA 5.2.13 *Let $a = a(x_4)$ be the function described in Corollary 5.2.10. If $x = (x_1, x_3, x_4)$ is such that $4x_1 \leq x_3 \leq a$, $2 \leq x_4 \leq Q$, then*

$$\beta_{x_3}(x) \geq \max\left\{\frac{x_4 - 16Rx_3}{16a}, \frac{x_4}{8}\right\}. \tag{5.2.23}$$

PROOF Since β is concave with respect to x_3, and $\beta_{x_3} \geq 0$ for $x_3 \in (0, a)$ we can write

$$a\beta_{x_3}(x) \geq (a - x_3)\beta_{x_3}(x) \geq \beta(x_1, a, x_4) - \beta(x_1, x_3, x_4).$$

Assuming that $x_1 \geq 0$, we can use Lemma 5.2.12:

$$\beta(x_1, a, x_4) \geq \left(1 - \frac{2x_1}{a}\right)\beta(0, a, x_4) = \left(1 - \frac{2x_1}{a}\right)\frac{x_4}{8} \geq \frac{1}{16}x_4.$$

Together, with the general estimate $\beta(x) \leq Rx_3$, we obtain

$$\beta_{x_3}(x) \geq \frac{x_4 - 16Rx_3}{16a}.$$

To get the second inequality, we estimate $\beta_{x_3}(a)$:

$$\beta_{x_3}(a) \geq \frac{\beta(x_1, 1, x_4) - \beta(x_1, a, x_4)}{1 - a} \geq \beta(x_1, 1, x_4) - \beta(x_1, a, x_4).$$

Now we use Corollary 5.2.9 together with the property of β to decrease with respect to x_1 for $x_1 > 0$:

$$\beta(x_1, 1, x_4) \geq \frac{x_4}{4} \qquad \text{and} \qquad \beta(x_1, a, x_4) \leq \beta(0, a, x_4) = \frac{x_4}{8}.$$

In result, we get the required estimate:

$$\beta_{x_3}(a) \geq \frac{x_4}{4} - \frac{x_4}{8} = \frac{x_4}{8}.$$

□

Let us denote the function on the right-hand side of (5.2.23) by m. We can rewrite it in the following form:

$$m(x_3, x_4) = \begin{cases} \frac{x_4 - 16Rx_3}{16a}, & \text{if } x_3 \leq \frac{(1-2a)x_4}{16R}; \\ \frac{x_4}{8}, & \text{if } x_3 \geq \frac{(1-2a)x_4}{16R}. \end{cases} \tag{5.2.24}$$

All preparations are made and we are ready to prove Theorem 5.2.3.

Proof of **Theorem 5.2.3.** Now we combine Lemmas 5.2.5 and 5.2.13 to deduce from (5.2.20) the following inequality:

$$-(1+2x_1^2)\beta_{x_1x_1} - 2x_1\beta_{x_1} \geq \frac{x_3^2(\beta_{x_3})^2}{-\gamma_\beta} \geq \frac{x_3^2m^2}{8R(|x_1| + \frac{x_3}{x_4})},$$

that holds under assumptions $4|x_1| \leq x_3 \leq a \leq 1$ and $4 \leq x_4 \leq Q$. Dividing both part of this inequality over $\sqrt{1+2x_1^2}$, we can rewrite it in the form

$$-\frac{\partial}{\partial x_1}\left(\sqrt{1+2x_1^2}\,\beta_{x_1}\right) \geq \frac{x_3^2m^2}{8R(|x_1| + \frac{x_3}{x_4})\sqrt{1+2x_1^2}}.$$

Integrating this inequality and taking into account that β is even in x_1 (i.e., $\beta_{x_1}(0, x_3, x_4) = 0$), we get

$$-\sqrt{1+2x_1^2}\,\beta_{x_1} \geq \frac{x_3^2m^2}{8R}\int_0^{x_1} \frac{dt}{\left(t + \frac{x_3}{x_4}\right)\sqrt{1+2t^2}}$$

$$\geq \frac{x_3^2m^2}{8R\sqrt{1+2x_1^2}}\int_0^{x_1}\frac{dt}{t+\frac{x_3}{x_4}} = \frac{x_3^2m^2}{8R\sqrt{1+2x_1^2}}\log\left(1+\frac{x_4}{x_3}x_1\right).$$

Once more we divide over the square root and integrate in x_1:

$$\beta(0, x_3, x_4) - \beta(x_1, x_3, x_4) \geq \frac{x_3^2m^2}{8R}\int_0^{x_1}\log\left(1+\frac{x_4}{x_3}t\right)\frac{dt}{1+2t^2}$$

$$\geq \frac{x_3^2m^2}{8R(1+2x_1^2)}\int_0^{x_1}\log\left(1+\frac{x_4}{x_3}t\right)dt$$

$$= \frac{x_3^3m^2}{8Rx_4(1+2x_1^2)}\left[\left(1+\frac{x_4}{x_3}x_1\right)\log\left(1+\frac{x_4}{x_3}x_1\right) - \frac{x_4}{x_3}x_1\right]$$

$$\geq \frac{x_3^3m^2}{9Rx_4}\left[\left(1+\frac{x_4}{x_3}x_1\right)\log\left(1+\frac{x_4}{x_3}x_1\right) - \frac{x_4}{x_3}x_1\right].$$

In the last estimate, we use the restriction $4|x_1| \leq x_3 \leq 1$, whence $1+2x_1^2 \leq \frac{9}{8}$.

Now we use inequality (5.2.22) from Lemma 5.2.12 and the general inequality $\beta(x) \leq Rx_3$:

$$\beta(0, x_3, x_4) - \beta(x_1, x_3, x_4) \leq \frac{2x_1}{x_3}\beta(0, x_3, x_4) \leq 2x_1R.$$

Combining with the preceding inequality, we come to the following estimate:

$$\frac{x_3^3m^2}{18R^2x_1x_4}\left[\left(1+\frac{x_4}{x_3}x_1\right)\log\left(1+\frac{x_4}{x_3}x_1\right) - \frac{x_4}{x_3}x_1\right] \leq 1.$$

Recall that this estimate we obtained in the following domain of variables:

$$0 \leq 4x_1 \leq x_3 \leq a \quad \text{and} \quad 4 \leq x_4 \leq Q.$$

Let us now choose the values of these variables. Since the function $t \mapsto \frac{1+t}{t}\log(1+t)$ monotonously increases, we get the best possible estimate when we take the maximal possible value of x_1, i.e., $x_1 = \frac{1}{4}x_3$:

$$\frac{x_3^2 m^2}{18R^2 x_4}\left[\left(1+\frac{x_4}{4}\right)\log\left(1+\frac{x_4}{4}\right)-\frac{x_4}{4}\right] \le 1.$$

Since the behavior of the function $a(x_4)$ is unknown, we cannot choose the best possible value of x_4, we take the largest value $x_4 = Q$:

$$\frac{x_3^2 m^2}{18R^2 Q}\left[\left(1+\frac{Q}{4}\right)\log\left(1+\frac{Q}{4}\right)-\frac{Q}{4}\right] \le 1,$$

where, of course, $a = a(Q)$ and $m = m(x_3, Q)$. To simplify this expression, we use the following elementary estimate:

$$\left(1+\frac{t}{4}\right)\log\left(1+\frac{t}{4}\right)-\frac{t}{4} \ge \frac{t}{16}\log t \quad \text{for} \quad t \ge 4.$$

To check this inequality, we consider the function

$$f(t) \overset{\text{def}}{=} 16\left(1+\frac{t}{4}\right)\log\left(1+\frac{t}{4}\right) - 4t - t\log t$$

and check that $f(t) \ge 0$ for $t \ge 4$. Also we have the following:

$$f(4) = 32\log 2 - 16 - 4\log 4 = 8\log\frac{8}{e^2} > 0;$$

$$f'(t) = 4\log\left(1+\frac{t}{4}\right) - \log t - 1;$$

$$f'(4) = 4\log 2 - \log 4 - 1 = \log\frac{4}{e} > 0;$$

$$f''(t) = \frac{4}{t+4} - \frac{1}{t} = \frac{3t-4}{t(t+4)} > 0 \quad \text{for} \quad t \ge 4.$$

In result, we get

$$\frac{x_3^2 m^2}{288R^2}\log Q \le 1 \qquad \text{for any} \quad x_3 \in [0, a]. \tag{5.2.25}$$

Now we need to investigate the function $x_3 \mapsto x_3 m(x_3, Q)$ on the interval $[0, a]$. If $a \ge \frac{1}{4}$, then this function is increasing and takes its maximal value at the point $x_3 = a$, and (5.2.25) yields

$$\frac{a^2 Q^2}{288R^2 \cdot 8^2}\log Q \le 1,$$

or

$$R \geq \frac{a}{96\sqrt{2}} Q\big(\log Q\big)^{1/2} \geq \frac{(\log 4)^{1/6}}{4 \cdot 96\sqrt{2}} Q\big(\log Q\big)^{1/3} \geq \frac{1}{515} Q\big(\log Q\big)^{1/3}. \tag{5.2.26}$$

Notice that we replaced $(\log Q)^{1/2}$ by $(\log Q)^{1/3}$ in the estimate above. This is, of course, super rough. But next we will see that exponent $1/3$ appears in the estimate below anyway: We can get only such exponent for other values of the unknown parameter a.

From now on, we assume that $a < \frac{1}{4}$. In this case, the function has a local maximum at the point $x_3 = \frac{Q}{32R}$. Indeed, since $aR \geq \beta(0, a, Q) = \frac{Q}{8}$, we have $a \geq \frac{Q}{8R} > \frac{Q}{32R}$, therefore the point $x_3 = \frac{Q}{32R}$ is in the domain. The value of the function $x_3 m(x_3, Q)$ at this the point is $\frac{Q}{32R} \cdot \frac{Q}{32a}$. On the other hand, at the end of the interval for $x_3 = a$ we have the value $am(a, Q) \geq \frac{aQ}{8}$. If $a^2 < \frac{Q}{128R}$ then we use the first estimate:

$$1 \geq \left(\frac{Q}{32R} \cdot \frac{Q}{32a}\right)^2 \frac{1}{288R^2} \log Q \geq \frac{Q^4}{9 \cdot 2^{25} R^4} \cdot \frac{2^7 R}{Q} \log Q \geq \left(\frac{Q}{134R}\right)^3 \log Q,$$

or

$$R \geq \frac{1}{134} Q\big(\log Q\big)^{1/3}.$$

In the case if $a^2 \geq \frac{Q}{128R}$, we use the second estimate:

$$1 \geq \left(\frac{aQ}{8}\right)^2 \frac{1}{288R^2} \log Q \geq \frac{Q^3}{9 \cdot 2^{18} R^3} \log Q \geq \left(\frac{Q}{134R}\right)^3 \log Q,$$

and again

$$R \geq \frac{1}{134} Q\big(\log Q\big)^{1/3}. \tag{5.2.27}$$

Therefore, if $a < \frac{1}{4}$ estimate 5.2.27 holds.

Comparing the estimates we got for different possible values of the unknown parameter a, namely, (5.2.26) and (5.2.27), we see that the estimate

$$R \geq \frac{1}{515} Q\big(\log Q\big)^{1/3}$$

is true in all cases. This completes the proof of Theorem 5.2.3, and therefore, the proof of Theorem 5.1.1.

5.2.5 Our Bellman Function B as a Viscosity Super Solution of a Degenerate Elliptic Equation

Let us recall to the reader that we defined in (5.2.1) function $\mathbf{B}$ on domain Ω introduced in (5.2.2). We want to demonstrate in this short section that $\mathbf{B}$ is a supersolution in viscosity sense of a certain degenerate elliptic equation.

We never used this before, but this knowledge might happen to be important. In particular, it may happen to be true that the reader more familiar with viscosity (super)solutions can simplify a bit our proof of Theorem 5.2.3, which we just finished proving. In this section, D^2u denotes the Hessian matrix of u.

DEFINITION 5.2.14 An equation $H(x, u, Du, D^2u) = 0,\ x \in \Omega \subset \mathbb{R}^d$, on a function u defined in an open domain Ω is called degenerate elliptic, if function H satisfies the following condition: for any point $(x, u, p) \in \Omega \times \mathbb{R} \times \mathbb{R}^d$ and any two $d \times d$ real symmetric matrices X and Y, we have that from $Y \geq X$ it follows that $H(x, u, p, X) \geq H(x, u, p, Y)$.

For example, $H(x, u, p, X) = -\operatorname{trace} X$ gives us a degenerate elliptic equation $-\Delta u = 0$.

DEFINITION 5.2.15 A lower semicontinuous function u is called a viscosity supersolution of $H(x, u, Du, D^2u) = 0$ (a degenerate elliptic equation) if for every point $x_0 \in \Omega$ and for every C^2 function v such that 1) $v(x_0) = u(x_0)$, 2) $v(x) \leq u(x)$ for x in a small neighborhood of x_0 inside Ω, one has inequality $H(x_0, v(x_0), Dv(x_0), D^2v(x_0)) \geq 0$.

To define viscosity subsolution, one changes lower to upper semicontinuous, require $v(x) \geq u(x)$ for x in a small neighborhood of x_0 inside Ω, and get the conclusion that

$$H(x_0, v(x_0), Dv(x_0), D^2v(x_0)) \leq 0.$$

To define the degenerate elliptic equation, whose viscosity super solution is $\mathbf{B}$ in Ω from (5.2.2), we consult Theorem 5.2.2.

Our function $H(x, u, p, X)$ will depend only on matrices X that run over 5×5 real symmetric matrices. Let $\mathcal{A}$ be the set of vectors e such that $e = (0, e_2, e_3, e_4, 0)$, $\|e\| = 1$. We consider the following H_{wmt}, where subscript stands for "weak martingale transform."

$$H_{wmt}(X) \stackrel{\text{def}}{=} -\sup_{e \in \mathcal{A}}[(Xe, e) + X_{11}e_2^2].$$

It is very easy to check that if $Y \geq X$ are two real symmetric matrices, then $H(X) \geq H(Y)$.

Let us see that $\mathbf{B}$ from (5.2.1) satisfies all conditions of viscosity supersolution of $H_{wmt}(D^2u) = 0$ in Ω from (5.2.2). The lower semicontinuity of $\mathbf{B}$ follows easily from its definition. Now let us fix $x = (x_1, x_2, x_3, x_4, x_5)$. If a smooth v satisfies $v(x) \le \mathbf{B}(x)$ in a neighborhood of this x, then

$$\begin{aligned} &v(x_1 \pm dx \pm dx_2, x_3 \pm dx_3, x_4 \pm dx_4, x_5) \\ \le &\mathbf{B}(x_1 \pm dx \pm dx_2, x_3 \pm dx_3, x_4 \pm dx_4, x_5) \end{aligned}$$

for all sufficiently small real numbers dx_i. We also have $v(x) = \mathbf{B}(x)$. Automatically, Lemma 5.2.2 gives us now that for all sufficiently small real numbers dx_i the following holds

$$\begin{aligned} v(x) \ge \tfrac{1}{4}\big(&v(x_1 - dx_2, x_2 - dx_2, x_3 - dx_3, x_4 - dx_4, x_5) \\ &+ v(x_1 + dx_2, x_2 - dx_2, x_3 - dx_3, x_4 - dx_4, x_5) \\ &+ v(x_1 - dx_2, x_2 + dx_2, x_3 + dx_3, x_4 + dx_4, x_5) \\ &+ v(x_1 + dx_2, x_2 + dx_2, x_3 + dx_3, x_4 + dx_4, x_5)\big). \end{aligned} \tag{5.2.28}$$

Function v is smooth. Let us use Taylor's formula for all terms in the right-hand side of (5.2.28). We can easily see that $v(x)$ will disappear together with all terms having the first derivatives of v. After simple algebra, which we leave to the reader, we can see that (5.2.28) implies a "infinitesimal" version of itself, which holds for any triple (dx_2, dx_3, dx_4)

$$\begin{aligned} -\big(&v_{x_3x_3}(dx_3)^2 + 2v_{x_3x_4}dx_3dx_4 + v_{x_4x_4}(dx_4)^2 \\ &+ v_{x_2x_2}(dx_2)^2 + 2v_{x_3x_2}dx_3dx_2 + 2v_{x_4x_2}dx_4dx_2 \\ &+ v_{x_1x_1}(dx_2)^2\big) \ge 0. \end{aligned} \tag{5.2.29}$$

From the definition of H_{wmt}, the reader can immediately see that we just proved that

$$H_{wmt}(D^2v(x_0)) \ge 0.$$

This means exactly that $\mathbf{B}$ is a viscosity super solution of a degenerate elliptic equation $H_{wmt}(D^2u) = 0$.

5.3 Sharp Weak Weighted Estimate for the Martingale Transform

In the previous section, we presented the approach via the Bellman function technique for the estimate from below of the weak norm $L^1(w) \to L^{1,\infty}(w)$ of the martingale transform. This approach, however, does not give the sharp estimate $[w]_{A_1} \log(1 + [w]_{A_1})$ from below, instead, the estimate from below had the form $[w]_{A_1} (\log(1 + [w]_{A_1}))^{1/3}$.

In the present section, we transfer the construction of A. Lerner, F. Nazarov, and S. Ombrosi [**110**] from the Hilbert transform to martingale transform. Rubio de Francia technique is used in proving the existence of weights that give the sharp asymptotic estimate $[w]_{A_1}\log(1+[w]_{A_1})$ from below. As a result of the application of Rubio de Francia extrapolation technique, the weights are given by rather implicit construction. The reader is warned not to think that the weights w_n below provide the desired sharp estimate of the type $[w]_{A_1}\log(1+[w]_{A_1})$. This is not the case. The constructive weights w_n below provide us with the starting point for a certain iterative procedure, which, after being finished, gives us the desired "worst possible" weights.

Everywhere in what follows in this section, A_1, A_2 denote dyadic classes of weights, all operators below including the maximal operator are dyadic too. So we omit the superscript d having this in mind.

5.3.1 Construction of Special Weights $w_n \in A_2$

In this section, we adapt the proof of [**110**] for the case of the martingale transform, the main issue is to choose the signs of the martingale transform. This should be done consistently simultaneously for all points, where we estimate the transform from below.

For dyadic interval I, we denote I^-, I^+ its left and right children.

Definition 5.3.1 We also denote

$$I_0 = I, I_1 = I^{++}, I_2 = I_1^{++}, \dots . I_{m-1} = I_{m-2}^{++},\ m = 2, \dots, k,$$

and $C(I) \stackrel{\text{def}}{=} \{I_m\}_{m=0}^{k-2}$, and we put

We put

$$\varepsilon = 4^{-k}.$$

We fix a large number p (will be $\asymp 1/\varepsilon$), and we build the sequence of weights by the rule: let ω, σ be two numbers such that $\omega\sigma = p$, we put

$$w_0(\omega,\sigma,I) = \frac{\omega}{\sqrt{p}}\Big((\sqrt{p}-\sqrt{p-1})\chi_{I_-} + (\sqrt{p}+\sqrt{p-1})\chi_{I_+}\Big),$$

$$\begin{aligned} w_n(\omega,\sigma,I) &= \sum_{m=0}^{k-2} w_{n-1}(3\omega,\sigma/3,I_m^{+-}) \\ &\quad + \frac{\omega}{p}\left(\sum_{m=0}^{k-2}\chi_{I_m^-} + \chi_{I_{k-1}^-} + \tau(\varepsilon)\chi_{I_{k-1}^+}\right), \end{aligned} \tag{5.3.1}$$

where $\tau(\varepsilon) = \frac{9\varepsilon}{1+5\varepsilon}$ is chosen to satisfy Lemma 5.3.2 below.

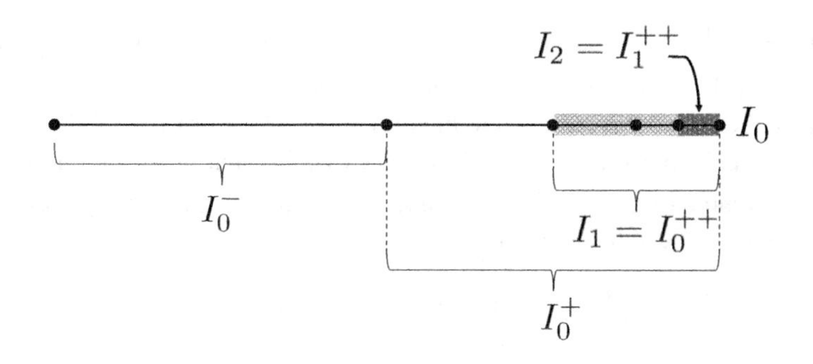

Figure 5.1 Intervals I_m.

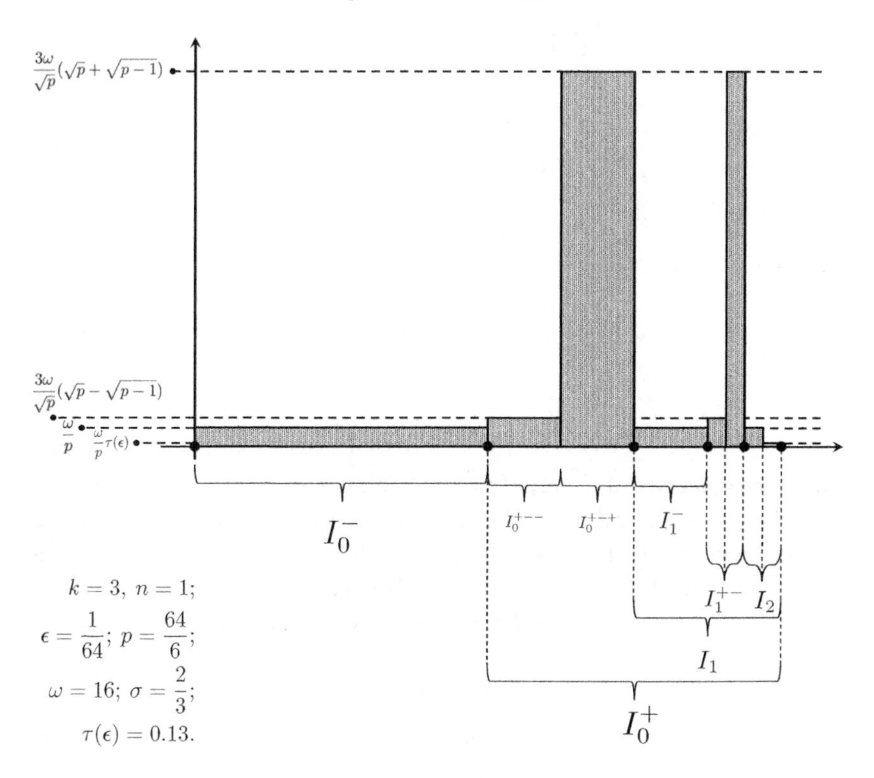

Figure 5.2 Weight w_1.

Lemma 5.3.2 $\langle w_n(\omega, \sigma, I)\rangle_I = \omega,\ \langle w_n^{-1}(\omega, \sigma, I)\rangle_I = \sigma.$

Proof The proof is by induction. For $n = 0$ the statement is clearly valid. Let the statement be proved for $n - 1$. To prove it for n, we notice that the value $\frac{\omega}{p}$ of w_n happens in different places: 1) it happens on measure

$\frac{1}{2}(|I_0| + \dots |I_{k-2}|) = \frac{1}{2}(1 + 1/4 + \cdots + 1/4^{k-2})|I|$, that is on $1/2\frac{1-4\varepsilon}{3/4}|I|$, which is $\frac{2}{3}(1-4\varepsilon)|I|$ and 2) $\frac{\omega}{p}$ happens also on measure $\frac{1}{2}|I_{k-1}| = 2\varepsilon|I|$. So, in total, the value $\frac{\omega}{p}$ is assigned to measure $\frac{2}{3}(1-4\varepsilon+3\varepsilon)|I| = \frac{2}{3}(1-\varepsilon)|I|$. Value $\frac{\omega\tau}{p}$ is assigned to measure $2\varepsilon|I|$. The total amount of measure left after these assignments is $\left(\frac{1}{3} - \frac{4}{3}\varepsilon\right)|I|$. Using induction hypothesis, we get that the average of w_n over I is equal to ω if

$$3\left(\frac{1}{3} - \frac{4}{3}\varepsilon\right) + \frac{2}{3}(1-\varepsilon)\frac{1}{p} + 2\varepsilon\frac{\tau}{p} = 1,$$

which gives us the first equation on τ, p:

$$\left(\frac{2}{3}(1-\varepsilon) + 2\varepsilon\tau\right)\frac{1}{p} - 4\varepsilon = 0. \tag{5.3.2}$$

Now we do the same with weight $\sigma_n \stackrel{\text{def}}{=} w_n^{-1}$:

$$\frac{\sigma}{3}\left(\frac{1}{3} - \frac{4}{3}\varepsilon\right) + \sigma\frac{2}{3}(1-\varepsilon) + \frac{\sigma}{\tau}2\varepsilon = \sigma. \tag{5.3.3}$$

Equations (5.3.2), (5.3.3) give

$$\tau = \frac{9\varepsilon}{1+5\varepsilon},\ p \approx \frac{1}{6\varepsilon}. \tag{5.3.4}$$

□

Definition 5.3.3 Consider $w_n(\omega, \sigma, I)$ introduced above. Interval I is called the forming interval of w_n, $\Phi_0(I) \stackrel{\text{def}}{=} \{I\}$. We denote by $\Phi_1(I)$ the collection $\{I_m^{+-}\}_{I_m \in C(I)}$. We call this collection the intervals forming w_{n-1}. Now we define the $\Phi_k(I)$ as the collection $\Phi_1(J)$, where J runs over the collection $\Phi_{k-1}(I)$. We call $\Phi_k(I)$ the collection forming w_{n-k}. Also we denote by $\operatorname{supp}(w_{n-k})$ the set $\cup_{J\in\Phi_k(I)} J$, that is the union of intervals forming w_{n-k}.

In what follows, n will be chosen

$$n = 4^k. \tag{5.3.5}$$

Notice that

$$p \asymp 4^k. \tag{5.3.6}$$

Lemma 5.3.4 *We have* $[w_n(\omega, \sigma, I)]_{A_2^d(I)} \approx p^2$.

Proof Without loss of generality, we can assume that $I = [0, 1)$. Take any dyadic subinterval J of I. First, we consider the case when the right endpoint of J coincides with the right endpoint of $[0,1)$. Suppose $|J| = 2^{-\ell}$ where $\ell \geq 1$ is even, say $\ell = 2r$. If $r = k-1$, then

$$\langle w_n \rangle_J \langle w_n^{-1} \rangle_J = \frac{1}{4}(1+\tau)\left(1+\frac{1}{\tau}\right) \approx \frac{1}{\varepsilon} \approx p$$

Further, assume $r < k-1$. Then the only m's that participate in (5.3.1) are $m \geq r$. Let $c \stackrel{\text{def}}{=} \varepsilon 4^r$. Then $\varepsilon \leq c \leq 1/16$. Clearly,

$$\begin{aligned}\langle w_n \rangle_J &= \frac{\omega}{p} \cdot \tfrac{1}{2} \cdot \frac{1-\varepsilon 4^{r+1}}{3/4} \\ &\quad + \frac{\omega}{p} \cdot 2\varepsilon 4^r + \frac{\omega\tau}{p} \cdot 2\varepsilon 4^r + 3\omega \cdot \frac{1}{4} \cdot \frac{1-\varepsilon 4^{r+1}}{3/4} \leq 100\frac{\omega}{p}(1+c); \\ \langle w_n^{-1} \rangle_J &= \frac{p}{\omega} \cdot \tfrac{1}{2} \cdot \frac{1-\varepsilon 4^{r+1}}{3/4} \\ &\quad + \frac{p}{\omega} \cdot 2\varepsilon 4^r + \frac{p}{\omega\tau} \cdot 2\varepsilon 4^r + \frac{\sigma}{3} \cdot \frac{1}{4} \cdot \frac{1-\varepsilon 4^{r+1}}{3/4} \leq 100\frac{p}{\omega}(1+\frac{c}{\tau})\end{aligned}$$

and we obtain that $\langle w_n \rangle_J \langle w_n^{-1} \rangle_J$ is at most of the order of $\varepsilon^{-1} \asymp p$.

Now suppose $J = 2^{-\ell}$ where ℓ is odd. Choose the smallest integer r so that $2r > \ell$, and acting as before we can estimate

$$\langle w_n \rangle_J \langle w_n^{-1} \rangle_J \leq \tfrac{1}{2}\left(3\omega + 100\frac{\omega}{p}(1+c)\right)\tfrac{1}{2}\left(\frac{\sigma}{3} + 100\frac{p}{\omega}(1+\frac{c}{\tau})\right)$$

and we see that in the cross-product there is a term $p(1+\frac{c}{\tau})$. It is at most Cp^2. □

Intervals of type I_{k-1}^+ play a special role. We call them special. Assume that I is an interval involved in forming $\omega_{n-\ell}$. Then there is only one special interval in $I \setminus \operatorname{supp} w_{n-\ell-1}$. Its length is $2\frac{1}{4^k}|I|$. But there are $k-2$ such special intervals in $I \cap (\operatorname{supp} w_{n-\ell-1} \setminus \operatorname{supp} w_{n-\ell-2})$. Their total length is

$$2\frac{1}{4^k}\sum_{m=1}^{k-2}\frac{1}{4^m}|I| = \frac{1}{3}(1-\frac{1}{4^{k-2}})\frac{2}{4^k}|I|.$$

Similarly, the length of the union of special intervals in

$$I \cap (\operatorname{supp} w_{n-\ell-2} \setminus \operatorname{supp} w_{n-\ell-3})$$

is

$$\left(\frac{1}{3}\left(1-\frac{1}{4^{k-2}}\right)\right)^2\frac{2}{4^k}|I|$$

etc.

If we denote the family of such special intervals in

$$[0,1) \cap (\operatorname{supp} w_{n-\ell} \setminus \operatorname{supp} w_{n-\ell-1}),$$

by $\mathcal{A}_\ell$, and their union by by A_ℓ, we then have

$$|A_\ell| = \left(\frac{1}{3}\left(1 - \frac{1}{4^{k-2}}\right)\right)^\ell \frac{2}{4^k}. \tag{5.3.7}$$

5.3.2 Martingale Transform Estimate

We are going to find A_1 weights such that

$$\|T\colon L^1(w) \to L^{1,\infty}(w)\| \geq c[w]_{A_1} \log[w]_{A_1}.$$

It turns out that it is enough to construct weights $w \in A_2$ such that dyadic maximal function (see the definition in (5.1.1)) has the norm in $L^2(w^{-1})$ bounded as follows:

$$\|M\|_{w^{-1}} \leq Ap, \tag{5.3.8}$$

and such that at the same time, the following holds:

$$\|T(\mathbf{1}_{I_0} w)\|^2_{w^{-1}} \geq cp^2(\log p)^2 \|\mathbf{1}_{I_0}\|_w. \tag{5.3.9}$$

The above is explained in [**110**] by using the extrapolation of Rubio de Francia. The reader may find this explanation repeated below during the proof of Theorem 5.3.6.

Take the sequence w_n of weights from Section 5.3.1 above. Now, consider a collection of special intervals J as was introduced above. This family splits into $\mathcal{A}_\ell$ collections, $A_\ell = \cup_{J \in \mathcal{A}_\ell} J$. Let $J \in \mathcal{A}_\ell$, $x \in J$. First, we want to estimate from below the following:

$$T_J w_n(x) = \sum_{R \in \mathcal{D}\colon R \in \operatorname{row}(J)} \varepsilon_R(w_n, h_R) h_R(x),\ x \in J.$$

Let us explain what is $\operatorname{row}(J)$. Let the interval forming $w_{n-\ell}$ (see Definition 5.3.3) and containing J be called K. Then $J = K^+_{k-1}$. Consider also $K = K_0, K_1 = K^{++}, \ldots, K_m, \ldots, K_{k-2}$. J is the right child of K_{k-1}. In the sum forming $Tw_n(x), x \in J$ above, we choose first $R = K^+, K^+_1, \ldots, K^+_m, \ldots, K^+_{k-2}$.

Call them $\operatorname{row}(J) = \{K^+(J), K^+_1(J), \ldots, K^+_m(J), \ldots, K^+_{k-2}(J)\}$. In terms of the collection $C(K)$ (see Definition 5.3.1), the row of J is defined as the right children of intervals from $C(K)$.

We did not choose yet the signs ε_R. Here is the choice:

$$\begin{aligned} &\forall J \in \mathcal{A}_\ell, \text{ and for even } \ell, \\ &\varepsilon_R = -1, \text{ if } R \in \operatorname{row}(J), \text{ otherwise } \varepsilon_R = 0. \end{aligned} \tag{5.3.10}$$

In other words, intervals I in the union of all rows of all special intervals in $\mathcal{A}_\ell$ with even ℓ, got $\varepsilon_I = -1$, all other intervals $I \in \mathcal{D}$ got $\varepsilon_I = 0$.

Recall that we fixed an interval $J \in \mathcal{A}_\ell$, ℓ being even, and $x \in J$. We have then

$$(w_n, h_{K_m^+(J)})h_{K_m^+(J)}(x) = (\langle w_n \rangle_{K_m^{+-}(J)} - \langle w_n \rangle_{K_m^{++}(J)}), \; x \in J.$$

By construction, the interval K_m^{+-} is a forming interval of $w_{n-\ell-1}$, thus by (5.3.1) the average over it is $3^{\ell+1}\omega$. On the other hand, the average $\langle w \rangle_{K_m^{++}}$ will be some average of $3^{\ell+1}\omega$ on $k-2-m$ intervals forming $w_{n-\ell-1}$ and lying in K to the right of K_m^{+-} and of $\frac{\omega}{p}$ on the rest of K_m^{++}. The total mass of $k-2-m$ intervals forming $w_{n-\ell-1}$ and lying in K to the right of K_m^{+-} is at most $\frac{1}{3}|K_m^{+-}|$. Thus, this second average is by a fixed small constant smaller than $3^{\ell+1}\omega$. So, with positive absolute constant c_1

$$(w, h_{K_m^+(J)})h_{K_m^+(J)}(x) \geq c_1 3^{\ell+1}\omega. \tag{5.3.11}$$

Hence, if ℓ is even, we will have positive contributions of order $3^{\ell+1}$ from its row $\operatorname{row}(J)$. Therefore,

$$x \in J, \; J \in \mathcal{A}_\ell \Rightarrow T_J w_n(x) \geq c_1 k 3^{\ell+1}\omega, \tag{5.3.12}$$

where one can see that $c_1 = \frac{2}{3} - O(\frac{1}{k})$.

Now we need to bookkeep the contribution of $T_{\tilde{J}} w_n(x)$ at the same point $x \in J$, where we need to take into account all special $\tilde{J} \neq J$. That contribution is formed by intervals from $\operatorname{row}(\tilde{J})$, $\tilde{J} \in \mathcal{A}_{\ell'}$, ℓ' is even, $\ell' = \ell-2, \ell-4, \ldots, 0$. All other contributions are zero (recall that x in the special interval $J \in \mathcal{A}_\ell$).

As an example, consider the tower of intervals $J \subset K_{m_\ell}^{+-}(\tilde{J}_\ell) \subset \cdots \subset K_{m_0}^{+-}(\tilde{J}_0)$. Interval $K_{m_{\ell'}}^{+-}(\tilde{J}_{\ell'})$ is a forming interval of $w_{n-\ell'-1}$, $\ell' < \ell$, ℓ' is even. But the contribution will be not only from this tower, but also from all the intervals lying in the same rows as the intervals in the tower above.

The contribution to $Tw_n(x), x \in J$, of the rows assigned to other special intervals is zero.

We need to consider only the contribution of the rows of intervals to which the intervals in the tower above belong. That contribution will be (by absolute value) at most

$$k(3^{\ell+1-2} + 3^{\ell+1-4} + \cdots) \leq \tfrac{1}{2} k 3^\ell,$$

and, therefore, the total contribution of those $\tilde{J}$s can be estimated by $|\sum_{i=0, i\, even}^{\ell-2} T_{\tilde{J}_i} w_n(x)| \leq \frac{1}{2} k 3^\ell$, $x \in J$, and cannot spoil the number $c_1 k 3^{\ell+1} \omega$ from (5.3.12) too much.

Also σ_n-measure of such a $J \in \mathcal{A}_\ell$ as above is $\frac{1+\varepsilon}{9\varepsilon} \frac{\sigma}{3^\ell} |J|$. Combining this, (5.3.12), and estimate of $|\mathcal{A}_\ell|$ from (5.3.7), we get (ℓ is even)

$$\int_{\mathcal{A}_\ell} (T w_n)^2 \sigma_n dx \geq c \omega^2 k^2 3^{2\ell} \frac{1+\varepsilon}{9\varepsilon} \frac{\sigma}{3^\ell} \left(\frac{1}{3} \left(1 - \frac{1}{4^{k-2}} \right) \right)^\ell \frac{1}{4^k}.$$

Or, using that $\omega \sigma = p$, $4^k = 1/\varepsilon$, we get

$$\begin{aligned} \int_{\mathcal{A}_\ell} (T w_n)^2 \sigma_n dx &\geq c \omega \, k^2 \, p \frac{1+\varepsilon}{9\varepsilon} \left(\left(1 - \frac{1}{4^{k-2}} \right) \right)^\ell \frac{1}{4^k} \\ &\geq c \omega \, k^2 \, p \left(\left(1 - \frac{1}{4^{k-2}} \right) \right)^\ell . \end{aligned}$$

Now,

$$\int_0^1 (T w_n)^2 \sigma_n dx \geq c \omega \, k^2 \, p \sum_{\ell=0,\, \ell\, even}^{4^k} \left(\left(1 - \frac{1}{4^{k-2}} \right) \right)^\ell \geq c \omega \, k^2 \, p 4^k .$$

But $4^k \approx p$, $k \approx \log p$. So we get

$$\int_0^1 (T w_n)^2 \sigma_n dx \geq c \omega \, p^2 (\log p)^2 = c \, p^2 (\log p)^2 \int_0^1 w_n dx. \tag{5.3.13}$$

THEOREM 5.3.5 *We have the following estimate of dyadic maximal operator M: $\|M\|_{w_n^{-1}} \leq A p$.*

We prove this theorem in Section 5.3.3. It is a dyadic situation, so the proof is somewhat easier than in [**110**].

Given Theorem 5.3.5 and what was done above, we can now prove the following theorem, the analog of the main result of [**110**], but for the martingale transform instead of the Hilbert transform.

THEOREM 5.3.6 *There is a sequence of weights $W \in A_1$ such that their dyadic A_1 norms $[W]_{A_1} \to \infty$, and martingale transforms T such that $\|T \colon L^1(W) \to L^{1,\infty}(W)\| \geq c [W]_{A_1} \log [W]_{A_1}$ with an absolute positive c.*

The proof of Theorem 5.3.6 is verbatim the same as in [**110**], the corresponding weights $W_n \in A_1$ are obtained from $w_n \in A_2$ constructed above by the method of Rubio de Francia. For these weights one uses the same dyadic martingale transforms as above. We repeat the proof for the convenience of the reader. The proof also shows how complicated are weights W_n built with the help of w_n.

PROOF We think of T as of linear operator with nice kernel, in practice T will be, say, a martingale transform with only finitely many ε_I non-zero, but the estimates will not depend on how many non-zeros T has.

We always think that T has symmetric or antisymmetric kernel.

By T_w, we understand the operator that acts on test functions as follows:

$$T_w f = T(wf).$$

Let w be a weight (for our goals it will be one of w_n built above), and $\alpha > 0$. Let g be a function from $L^2(w^{-1})$ to be chosen soon, $\|g\|_{w^{-1}} = 1$, and we use Rubio de Francia function

$$\mathcal{R}g = \sum_{k=0}^{\infty} \frac{M^k g}{2^k \|M^k\|_{w^{-1}}}.$$

Then,

$$\begin{aligned} \|\mathcal{R}g\|_{w^{-1}} &\le 2, \; g \le Rg, \\ [\mathcal{R}g]_{A_1} &\le \|M\|_{w_n^{-1}}. \end{aligned} \tag{5.3.14}$$

Then, choosing appropriate g, $\|g\|_{w^{-1}} = 1$ and denoting $F \stackrel{\text{def}}{=} wf$, we can write

$$\begin{aligned} \alpha(w\{T_{w^{-1}}F > \alpha\})^{1/2} &= \alpha(w\{Tf > \alpha\})^{1/2} \\ &\le 2\alpha \int_{Tf>\alpha} g dx \\ &\le 2\alpha \int_{Tf>\alpha} \mathcal{R}g\, dx \le 2N([\mathcal{R}g]_{A_1}) \int |f| \mathcal{R}g\, dx \\ &\le 2N([\mathcal{R}g]_{A_1}) \|f\|_w \|\mathcal{R}g\|_{w^{-1}} \\ &\le 4N([\mathcal{R}g]_{A_1}) \|f\|_w = 4N([\mathcal{R}g]_{A_1}) \|F\|_{w^{-1}}. \end{aligned} \tag{5.3.15}$$

Here $N([\mathcal{R}g]_{A_1})$ denotes the estimate from above of the weak norm $T\colon L^1(Rg) \to L^{1,\infty}(Rg)$.

This weight Rg is the future weight W_n mentioned before the start of the proof.

Henceforth, one obtains the estimate

$$N([\mathcal{R}g]_{A_1}) \ge \tfrac{1}{4} \|T_{w^{-1}} \colon L^2(w^{-1}) \to L^{2,\infty}(w)\|. \tag{5.3.16}$$

By duality and (anti) symmetry of T, the latter norm is

$$\|T_{w^{-1}}\colon L^2(w^{-1}) \to L^{2,\infty}(w)\| = \|T_w\colon L^{2,1}(w) \to L^2(w^{-1})\|$$
$$\geq \frac{\|T(\mathbf{1}_{I_0} w)\|_{w^{-1}}}{\|\mathbf{1}_{I_0}\|_w}. \tag{5.3.17}$$

In the last inequality, we used the fact that in the norms of characteristic function in $L^{2,1}(w)$ and in $L^2(w)$ are the same.

Use (5.3.14) and (5.3.16), (5.3.17):

$$\frac{\|T(\mathbf{1}_{I_0} w)\|_{w^{-1}}}{\|\mathbf{1}_{I_0}\|_w} \leq 4N(\|M\|_{w^{-1}}). \tag{5.3.18}$$

Now we plug into this inequality (with $w = w_n$) inequality (5.3.9) and the result of Theorem 5.3.5. Then we obtain for all large p that

$$p \log p \leq CN(p).$$

This is what we wanted. □

5.3.3 The Proof of Theorem 5.3.5

We put initial numbers ω, σ to be

$$\omega = 1,\ \sigma = p. \tag{5.3.19}$$

It is well known that for maximal function, one has the $T1$ theorem; see [**165**]. Hence, it is sufficient to check that with an absolute constant C

$$\forall J \in \mathcal{D}(I) \quad \int_J \left[M(w\chi_J)\right]^2 w^{-1} dx \leq Cp^2 w(J). \tag{5.3.20}$$

Using Definition 5.3.3 and following [**110**], define a function (of course we put $\omega = 1$, but it is convenient to keep writing it):

$$\tilde{w}(x) \stackrel{\text{def}}{=} \omega \sum_{\ell=1}^{n} 3^{\ell-1} \chi_{\text{supp}\, w_{n-\ell+1} \setminus \text{supp}\, w_{n-\ell}} + \omega 3^n \chi_{\text{supp}\ w_0}. \tag{5.3.21}$$

Lemma 5.3.7 *With an absolute constant C, $Mw \leq C\tilde{w}$.*

Proof If $x \in$ supp w_0, we have (taking into account the normalization in (5.3.19))

$$w(x) = w_0\left(3^n\omega, \frac{\sigma}{3^n}; x\right)$$
$$\leq \frac{3^n}{\sqrt{p}}\left((\sqrt{p} - \sqrt{p-1})\chi_{I_-} + (\sqrt{p} + \sqrt{p-1})\chi_{I_+}\right) \leq 2 \cdot 3^n.$$

On the complement of $\operatorname{supp} w_0$, $w \le \frac{3^n}{p}$. So the claim of lemma is obvious for $x \in \operatorname{supp} w_0$.

If $x \in \operatorname{supp} w_{n-\ell+1} \setminus \operatorname{supp} w_{n-\ell}$, $\ell = 1, 2, \dots, n$, then

$$w(x) \le \frac{3^{\ell-1}\omega}{p}$$

and outside of $\operatorname{supp} w_{n-\ell+1}$, $w(x) \le \frac{3^{\ell-2}\omega}{p}$. If we average w over a dyadic interval J containing $x \in \operatorname{supp} w_{n-\ell+1} \setminus \operatorname{supp} w_{n-\ell}$, and if this J intersects dyadic intervals forming $\operatorname{supp} w_{n-\ell}$, then each of this dyadic interval I is inside J. Then, by Lemma 5.3.2, for each such I, we have $w(I) = 3^\ell |I|$. Combining all this together, we get

$$w(J) \le 2 \cdot 3^\ell |J|,$$

which proves the lemma. □

LEMMA 5.3.8 *Let $I_0 = [0,1]$ and let $w = w_n$, as in Section 5.3.1. Then with a finite absolute constant C we have*

$$\int_{I_0} \tilde{w}^2 w^{-1}\, dx \le C p^2 w(I_0).$$

PROOF Let us consider first $m = 0$. Denote

$$F_j = [0,1] \cap (\operatorname{supp} w_{n-j} \setminus \operatorname{supp} w_{n-j-1}).$$

Let A_j be the union of all special intervals (see Section 5.3.1) contained in F_j.

On the complement of $\operatorname{supp} w_{n-j-1}$, $w(x) \le \frac{3^j}{p}$. If in addition $x \notin A_j$, then we have $w \le \frac{3^j}{p}$. Hence, here $\tilde{w}(x) = pw(x)$. This is convenient, because together with $w(x)w^{-1}(x) = 1$, this implies

$$\int_{F_j \setminus A_j} \tilde{w}^2 w^{-1} dx = p^2 \int_{F_j \setminus A_j} w(x) dx = p^2 w(F_j \setminus A_j).$$

On A_j, $w^{-1}(x) = \frac{p}{3^j} \frac{1+5\varepsilon}{9\varepsilon}$. We saw in (5.3.7) that

$$|A_j| = \left(\frac{1}{3}\left(1 - \frac{1}{4^{k-2}}\right)\right)^j \frac{2}{4^k} \le \frac{1}{3^j}\frac{2}{4^k}.$$

By definition of $\tilde{w}$ we conclude now

$$\int_{\cup_{j=0}^{n-1} A_j} \tilde{w}^2 w^{-1} dx \le p \sum_{j=0}^{n-1} 3^{2j} \frac{1}{3^{2j}} \frac{2}{4^k} \frac{1+5\varepsilon}{9\varepsilon} \le \frac{np}{4^k \varepsilon}.$$

As $n = 4^k$ and $p \asymp \varepsilon^{-1}$, we get by using Lemma 5.3.2 in the last inequality

$$\int_{\cup_{j=0}^{n-1} A_j} \tilde{w}^2 w^{-1} dx \leq \frac{np}{4^k \varepsilon} \leq Cp^2 w(I).$$

Therefore,

$$\int_I \tilde{w}^2 w^{-1} dx \leq \int_{\cup_{j=0}^{n-1} F_j \setminus A_j} \tilde{w}^2 w^{-1} dx$$

$$+ \int_{\cup_{j=0}^{n-1} A_j} \tilde{w}^2 w^{-1} dx + \int_{\operatorname{supp} w_0} \tilde{w}^2 w^{-1} dx$$

$$\leq p^2 \sum_{j=0}^{n-1} w(F_j \setminus A_j) + Cp^2 w(I) + \int_{\operatorname{supp} w_0} \tilde{w}^2 w^{-1} dx$$

$$\leq Cp^2 w(I) + \int_{\operatorname{supp} w_0} \tilde{w}^2 w^{-1} dx$$

on $\operatorname{supp} w_0$, $\tilde{w}(x) = 3^n = w(x)$, hence the last integral is just at most $w(I)$. Finally, we get ($I_0 = I = [0,1]$)

$$\int_{I_0} \tilde{w}^2 w^{-1} dx \leq Cp^2 w(I_0) = Cp^2 w(I). \tag{5.3.22}$$

□

LEMMA 5.3.9 *Let $I = [0,1]$ and $I_0 = I, I_1 = I^{++}, I_2 = I_1^{++}, \ldots, I_{k-2} = I_{k-3}^{++}$ as before. Let $w = w_n$. Then with a finite absolute constant C, we have for $m = 1, \ldots, k-2$*

$$\int_{I_m} \tilde{w}^2 w^{-1}\, dx \leq Cp^2 w(I_m).$$

PROOF Notice that $I_m \backslash I_{m+1}$ consists of an interval $F \stackrel{\text{def}}{=} I_m^{+-}$, which is one of the intervals forming w_{n-1}, and of interval G, such that G belongs to $\operatorname{supp} w_n \backslash \operatorname{supp} w_{n-1}$. On such intervals, by the definition (5.3.21) of $\tilde{w}$, $\tilde{w} = \omega$, and $w = \frac{\omega}{p}$ (of course we can remember that ω is normalized in (5.3.19), but it does not matter in the calculations below). Thus, on G, $\tilde{w} = pw$.

Hence,

$$\int_{I_m \backslash I_{m+1}} \tilde{w}^2 w^{-1} dx \leq p^2 \int_G w^2 w^{-1} dx$$

$$+ \int_F \tilde{w}^2 w^{-1} dx \leq p^2 w(I) + \int_F \tilde{w}^2 w^{-1} dx.$$

Notice that we can estimate the last integral by Lemma 5.3.8. In fact, interval F plays the role of I, and weight w on F (and so $\tilde{w}$) is constructed exactly as

$w = w_n$ on I, only it starts not with ω but with 3ω and takes $n-1$ steps to be constructed. By scale invariance, we get by using Lemma 5.3.8

$$\int_F \tilde{w}^2 w^{-1} dx \le Cp^2 w(F) \le Cp^2 w(I).$$

Together, two last displayed inequalities give the following:

$$\int_{I_m \setminus I_{m+1}} \tilde{w}^2 w^{-1} dx \le (C+1)p^2 w(I_m \setminus I_{m+1}). \tag{5.3.23}$$

We can write now ($m = 1, \dots, k-2$) using (5.3.23):

$$\int_{I_m} \tilde{w}^2 w^{-1} dx = \sum_{j=m}^{k-2} \int_{I_j \setminus I_{j+1}} \tilde{w}^2 w^{-1} dx$$
$$+ \int_{I_{k-1}} \tilde{w}^2 w^{-1} dx \le (C+1)p^2 w(I_m) + \int_{I_{k-1}} \tilde{w}^2 w^{-1} dx.$$

But on I_{k-1}, we have $\tilde{w} = \omega$ (see (5.3.21)) and so

$$\int_{I_{k-1}} \tilde{w}^2 w^{-1} dx \le C\omega^2 \frac{(1+5\varepsilon)p}{9\varepsilon\omega} |I_{k-1}| \le C\omega |I_{k-1}| p^2. \tag{5.3.24}$$

On the other hand, as $m \le k-2$, we have by virtue of Lemma 5.3.2

$$w(I_m) \ge \omega |I_m| \ge \omega |I_{k-1}|. \tag{5.3.25}$$

Combining (5.3.24) and (5.3.25), we get ($m \le k-2$)

$$\int_{I_{k-1}} \tilde{w}^2 w^{-1} dx \le Cp^2 w(I_m).$$

We finally get that for $m = 1, \dots, k-2$, the following holds:

$$\int_{I_m} \tilde{w}^2 w^{-1} dx \le Cp^2 w(I_m). \tag{5.3.26}$$

□

In the next lemma, we stop working with $\tilde{w}$.

LEMMA 5.3.10 *Let $I = [0,1]$ and $I_0 = I,\ I_1 = I_0^{++}, \dots, I_{k-1} = I_{k-2}^{++}$ as before. Let $w = w_n$. With an absolute constant C, we have the following:*

$$\int_{I_{k-1}} (M(\chi_{I_{k-1}} w))^2 w^{-1}\, dx \le Cpw(I_{k-1}).$$

PROOF Clearly, by construction of w, we have

$$M(\chi_{I_{k-1}} w) \leq \frac{\omega}{p}.$$

Hence,

$$\int_{I_{k-1}} (M(\chi_{I_{k-1}} w))^2 w^{-1}\, dx \leq \frac{\omega^2}{p^2} \frac{(1+5\varepsilon)p}{9\varepsilon\omega} |I_{k-1}| \leq C\omega |I_{k-1}|.$$

On the other hand,

$$w(I_{k-1}) \geq \tfrac{1}{2} \frac{\omega}{p} |I_{k-1}|.$$

Therefore,

$$\int_{I_{k-1}} (M(\chi_{I_{k-1}} w))^2 w^{-1}\, dx \leq Cpw(I_{k-1}).$$

□

LEMMA 5.3.11 *Let* $I = [0,1]$ *and* $I_0 = I, I_1 = I_0^{++}, \ldots, I_{k-1} = I_{k-2}^{++}$ *as before. Let* J *be the dyadic father of* I_{k-1}*, and* $w = w_n$*. Then with a finite absolute constant* C*, we have*

$$\int_J (M(\chi_J w))^2 w^{-1}\, dx \leq Cp^2 w(J).$$

PROOF Interval J consists of J_- and $J_+ = I_{k-1}$. We have by Lemma 5.3.2 and by $w \leq \frac{\omega}{p}$ on I_{k-1} (see (5.3.1)):

$$x \in J_+ = I_{k-1} \Rightarrow M(\chi_J w) \leq \omega.$$

And we also know by Lemma 5.3.2 that $w(J) \geq w(J_-) = \frac{1}{2}\omega |J|$. So

$$\int_{J_+} (M(\chi_J w))^2 w^{-1} dx \leq \omega^2 \frac{(1+5\varepsilon)p}{9\varepsilon\omega} |J| \leq Cp^2 w(J).$$

Now notice that we can estimate the integral $\int_{J_-} (M(w\chi_J))^2 w^{-1} dx$ by Lemma 5.3.8. In fact, interval J_- plays the role of I, and weight w on J_- (and so $\tilde{w}$) is constructed exactly as $w = w_n$ on I, only it starts not with ω but with 3ω and takes $n-1$ steps to be constructed. By scale invariance, we get by using Lemma 5.3.8

$$\int_{J_-} \tilde{w}^2 w^{-1} dx \leq Cp^2 w(J_-) \leq Cp^2 w(J).$$

Combining two last display inequalities, we get the lemma's claim. □

We are ready to combine the lemmas of this section with the fact that for $x \in \operatorname{supp} w_{n-\ell+1} \setminus \operatorname{supp} w_{n-\ell}$, $w(x) \le \omega \frac{3^{\ell-1}}{p}$ in order to see that we obtained the following theorem.

THEOREM 5.3.12 *Let $I = [0,1]$, $w = w_n$. Let J be any dyadic subinterval of I, such that it is not contained in any dyadic interval forming* $\operatorname{supp} w_{n-1}$. *There exists an absolute constant C such that*

$$\int_J (M(\chi_J w))^2 w^{-1} dx \le Cp^2 w(J). \tag{5.3.27}$$

Now we shall observe that if interval J is inside or equal to one of the intervals forming $\operatorname{supp} w_{n-1}$, say K, then K plays the role of $I = [0,1]$ and w_{n-1} is the same type of weight as w_n, only starting with 3ω instead of ω, but the value of ω was immaterial in the above considerations.. Therefore, we can extend the claim (5.3.27) of Theorem 5.3.12 to dyadic intervals that are not contained in any dyadic interval forming $\operatorname{supp} w_{n-2}$. The constant C is exactly the same. We can continue to reason this way, and we obtain (5.3.20).

5.4 Obstacle Problems for Unweighted Square Function Operator: Burkholder–Gundy–Davis Function

Recall that h_J denotes the normalized in L^2 Haar function supported on interval J. Now let g be a test function on an interval I, then

$$g = \langle g \rangle_I \mathbf{1}_I + \sum_{J \in \mathcal{D}(I)} \Delta_J g$$

with $\Delta_J g = (g, h_J) h_J$. The square function of g is the following aggregate:

$$Sg(x) \stackrel{\text{def}}{=} \left(\sum_{\substack{J \in \mathcal{D}(I) \\ x \in J}} (\Delta_J g)^2(x) \right)^{1/2}.$$

Marcinkiewicz–Paley inequalities [**114**] relate the norms of $g - \langle g \rangle_I$ and Sg, claiming that for certain situations, these norms are equivalent.

Let $W(t)$ be the standard Brownian motion starting at zero, and T be any stopping time. Below $\|f\|_\alpha$ stands for L^α norm.

D. Burkholder [**21**], P. Millar [**118**], A. A. Novikov [**139**], D. Burkholder and R. Gundy [**32**], and B. Davis [**50**], found the following norm estimates:

$$c_\alpha \|T^{1/2}\|_\alpha \le \|W(T)\|_\alpha, \quad 1 < \alpha < \infty, \quad \|T^{1/2}\|_\alpha < \infty; \tag{5.4.1}$$

$$\|W(T)\|_\alpha \le C_\alpha \|T^{1/2}\|_\alpha, \quad 0 < \alpha < \infty. \tag{5.4.2}$$

Davis [**50**] found the best possible values of constants above.

It was explained in [**50**] that the same sharp estimates (5.4.3) and (5.4.4) below hold with $W(T)$ replaced by an integrable function g on $[0, 1]$, and $T^{1/2}$ replaced by the dyadic square function of g.

More precisely, Davis proved that

$$c_\alpha \|Sg\|_\alpha \le \|g\|_\alpha, \quad 2 \le \alpha < \infty; \tag{5.4.3}$$

$$\|g\|_\alpha \le C_\alpha \|Sg\|_\alpha, \quad 0 < \alpha \le 2 \tag{5.4.4}$$

with the same constants as above, and these constants are sharp in those ranges of α. Inequality (5.4.4) with the same sharp constant as in (5.4.2) but for the range $\alpha \ge 3$ was proved by G. Wang [**196**]. In the range $\alpha \in (2, 3)$, the sharp constant in (5.4.4) is not known to the best of our knowledge. The same can be said about (5.4.3) in the range $\alpha \in (1, 2)$. Notice also that Wang's results are proved for square functions of conditionally symmetric martingales. So Wang's setting is more general than the dyadic setting presented here.

We will present here the original proof by B. Davis of estimates (5.4.1), (5.4.3) based on the construction of a corresponding Bellman function. But we wish to cast the proofs in the language of obstacle problems. To prepare the ground, we start with an explanation of the obstacle problems related to square function estimates.

5.4.1 Obstacle Problems Related to Square Function Estimates

We will always work with functions on some interval I, and $\mathcal{T} \stackrel{\text{def}}{=} \mathcal{T}(I)$ is the class of test functions. We say that $f \in \mathcal{T}$ if f is constant on each dyadic interval from $\mathcal{D}_N(I)$ for some finite N.

The main players will be an "arbitrary" function $O\colon \mathbb{R} \times \mathbb{R}_+ \to \mathbb{R}$ (an obstacle) and a function $U\colon \mathbb{R} \times \mathbb{R}_+ \to \mathbb{R}$, $U \ge O$, satisfying the following inequality:

$$2U(p, q) \ge U(p + a, \sqrt{a^2 + q^2}) + U(p - a, \sqrt{a^2 + q^2}). \tag{5.4.5}$$

We will call this *the main inequality*, functions U satisfying the main inequality will be precisely Bellman functions of various estimates concerning square function operator.

Of course, the existence of U majorizing O and satisfying (5.4.5) is not at all ensured.

Notice that (5.4.5) is preserved by taking infimum over the family of functions U.

DEFINITION 5.4.1 We call the smallest U satisfying the main inequality and majorizing O *the heat envelope* of O.

We would like to find the heat envelope of some specific O.

THEOREM 5.4.2 *Let U satisfy main inequality* (5.4.5). *Then for any $f \in \mathcal{T}(I)$ and any value of a parameter q*

$$\langle U(f, \sqrt{q^2 + (Sf)^2}) \rangle_I \le U(\langle f \rangle_I, q). \tag{5.4.6}$$

Here is a corollary relating the main inequality with square function estimates.

COROLLARY 5.4.3 *Let U satisfy main inequality* (5.4.5). *Then for any $f \in \mathcal{T}(I)$*

$$\langle U(f, Sf) \rangle_I \le U(\langle f \rangle_I, 0). \tag{5.4.7}$$

Before proving Theorem 5.4.2, we wish to answer the question, when, given O, one can find a finite-valued function majorizing O and satisfying the main inequality.

THEOREM 5.4.4 *Let*

$$\mathbf{U}(p, q) \stackrel{\text{def}}{=} \sup_{\substack{f \in \mathcal{T}(I) \\ \langle f \rangle_I = p}} \langle O(f, \sqrt{q^2 + S^2 f}) \rangle_I. \tag{5.4.8}$$

If this function is finite valued, then it satisfies the main inequality.

Now we wish to formulate results that can be considered as converse to Theorem 5.4.2. They concern the obstacle problem for (5.4.5).

As was already mentioned, by this we understand finding U satisfying (5.4.5) and majorizing a given function (obstacle) $O\colon \mathbb{R} \times \mathbb{R}_+ \to \mathbb{R}$. It turns out that one can give "simple" conditions necessary and sufficient for the solvability of the obstacle problem.

THEOREM 5.4.5 *Let an obstacle function O, and a function $F\colon \mathbb{R} \to \mathbb{R}$ satisfying $F(p) \ge O(p, 0)$, be given. A finite-valued function U satisfying*

- *main inequality* (5.4.5)
- $U \ge O$
- $U(p, 0) \le F(p)$

exists if and only if

$$\langle O(f, Sf) \rangle_I \le F(\langle f \rangle_I), \quad \forall f \in \mathcal{T}. \tag{5.4.9}$$

It will be especially important to use this result with one special F: $F = 0$.

THEOREM 5.4.6 *Given an obstacle function O, to find U satisfying main inequality* (5.4.5) *and such that $U \geq O$ and $U(p,0) \leq 0$, it is necessary and sufficient to have*

$$\langle O(f,Sf)\rangle_{_I} \leq 0, \qquad \forall f \in \mathcal{T}. \tag{5.4.10}$$

Proof of theorem 5.4.2 In the following, we denote by $\mathbb{E}_k$ the expectation with respect to σ-algebra generated by dyadic intervals of family $\mathcal{D}_k$. Fix a function $f \in \mathcal{T}$, and put $f_k \stackrel{\text{def}}{=} \mathbb{E}_k f$, $k \geq 0$. Since for any $f \in \mathcal{T}$, there exists a number N such that f is constant on each $J \in \mathcal{D}_N(I)$, we have $f_N = f$. Then inequality (5.4.6) will be proved if we check that for any k, we have

$$\langle U(f_{k+1}, \sqrt{q^2 + (Sf_{k+1})^2})\rangle_{_I} \leq \langle U(f_k, \sqrt{q^2 + (Sf_k)^2})\rangle_{_I}, \tag{5.4.11}$$

because for $k = 0$, we have $f_0(x) = \mathbb{E}_0 f = \langle f\rangle_{_I} \mathbf{1}_{_I}$ and $Sf_0 = 0$, and therefore $\langle U(f_0, \sqrt{q^2 + (Sf_0)^2})\rangle_{_I} = U(\langle f\rangle_{_I}, q)$.

Let us now fix $J \in \mathcal{D}_k$. Denote $p \stackrel{\text{def}}{=} \langle f\rangle_{_J}$ and let $\langle f\rangle_{_{J_+}} = p + a$, then $\langle f\rangle_{_{J_-}} = p - a$, and $f_{k+1}(x) = p \pm a$ for all $x \in J_\pm$ correspondingly, $f_k = p$ is a constant function on the whole J. Furthermore, Sf_{k+1} and Sf_k are constant functions on J as well and $S^2 f_{k+1} = S^2 f_k + a^2$. Therefore, by the main inequality, we have

$$\begin{aligned}
\langle U(f_{k+1}, \sqrt{q^2 + (Sf_{k+1})^2})\rangle_{_J} &= \tfrac{1}{2} U(p + a, \sqrt{q^2 + (Sf_k)^2 + a^2}) \\
&\quad + \tfrac{1}{2} U(p - a, \sqrt{q^2 + (Sf_k)^2 + a^2}) \\
&\leq U(p, \sqrt{q^2 + (Sf_k)^2}) \\
&= \langle U(f_k, \sqrt{q^2 + (Sf_k)^2})\rangle_{_J}.
\end{aligned}$$

In remains to multiply this inequality by $|J|$ and take a sum over all $J \in \mathcal{D}_k(I)$, and after dividing by $|I|$, we come to the desired inequality. □

Proof of Theorem 5.4.4 It is clear by its definition and by rescaling that $\mathbf{U}$ does not depend on the interval I, where test functions are defined. Therefore, given the data $(p + a, \sqrt{a^2 + q^2})$, we can find a function f_+ optimizing $\mathbf{U}(p + a, \sqrt{a^2 + q^2})$ up to ε, and we can think as well that it lives on I_+. Similarly, given the data $(p - a, \sqrt{a^2 + q^2})$, we can find a function f_- optimizing $\mathbf{U}(p - a, \sqrt{a^2 + q^2})$ up to ε, and we can think as well that it lives on I_-.

Concatenate functions $f_\pm$ on $I_\pm$ to the following function:

$$f(x) = \begin{cases} f_+(x), & x \in I_+ \\ f_-(x), & x \in I_- \end{cases}$$

Since $\langle f\rangle_I = p$, we have

$$
\begin{aligned}
\mathbf{U}(p,q) &\geq \langle O(f, \sqrt{q^2 + S^2 f})\rangle_I \\
&= \tfrac{1}{2}\langle O(f, \sqrt{q^2 + S^2 f})\rangle_{I_+} + \tfrac{1}{2}\langle O(f, \sqrt{q^2 + S^2 f})\rangle_{I_-} \\
&= \tfrac{1}{2}\langle O(f_+, \sqrt{q^2 + a^2 + S^2 f_+})\rangle_{I_+} + \tfrac{1}{2}\langle O(f_-, \sqrt{q^2 + a^2 + S^2 f_-})\rangle_{I_-} \\
&\geq \tfrac{1}{2}\mathbf{U}(p+a, \sqrt{a^2+q^2}) + \tfrac{1}{2}\mathbf{U}(p-a, \sqrt{a^2+q^2}) - \varepsilon.
\end{aligned}
$$

As ε is an arbitrary positive number, we are done. □

Now we prove Theorem 5.4.5.

PROOF First we prove the "if" part. We are given an obstacle O and a function F such that $F(p) \geq O(p,0)$. We defined

$$
\mathbf{U}(p,q) = \sup_{\substack{f\in\mathcal{T}(I)\\ \langle f\rangle_I = p}} \langle O(f, \sqrt{q^2 + S^2 f})\rangle_I .
$$

It is obvious that $\mathbf{U}(p,q) \geq O(p,q)$, one just plugs the constant function $f = p\mathbf{1}_I$.

It is also clear that $\mathbf{U}(p,0) \leq F(p)$. Indeed,

$$
\mathbf{U}(p,0) = \sup_{\substack{f\in\mathcal{T}(I)\\ \langle f\rangle_I = p}} \langle O(f, Sf)\rangle_I \leq F(\langle f\rangle_I) = F(p)
$$

by assumption (5.4.9). Hence, $\mathbf{U}(p,0)$ is finite valued.

The fact that function $\mathbf{U}$ defined as above satisfies the main inequality (5.4.5) follows from the reasoning of Theorem 5.4.4. Then by (5.4.5) with $q = 0$ it is finite valued.

Now we prove the "only if" part. We need to prove that

$$
\langle O(f, Sf)\rangle_I \leq F(\langle f\rangle_I)
$$

if there exists a majorant U of O satisfying the main inequality and satisfying $U(p,0) \leq F(p)$. This is easy:

$$
\langle O(f, Sf)\rangle_I \leq \langle U(f, Sf)\rangle_I \leq U(\langle f\rangle_I, 0) \leq F(\langle f\rangle_I),
$$

where the second inequality follows from Corollary 5.4.3. □

The following theorem sums up the results of this section.

THEOREM 5.4.7 *There exists a finite-valued function U majorizing O and satisfying the main inequality if and only if* $\mathbf{U}$ *from* (5.4.8) *is finite valued.*

Moreover, if $\mathbf{U}$ *is finite valued, then the infimum of functions* U *majorizing* O *and satisfying the main inequality is equal to* $\mathbf{U}$.

PROOF We already saw in Theorem 5.4.4 that $\mathbf{U}$ from (5.4.8) (if finite valued) is one of those functions U that majorizes O and satisfies the main inequality.

On the other hand, for any function U that majorizes O and satisfies the main inequality, we know from Theorem 5.4.2 that for any test function f and any nonnegative q the following holds:

$$U(\langle f\rangle_I, q) \geq \langle U(f, \sqrt{q^2 + S^2 f})\rangle_I \geq \langle O(f, \sqrt{q^2 + S^2 f})\rangle_I.$$

Take now the supremum over test functions in the right-hand side. By definition, we obtain $\mathbf{U}(\langle f\rangle_I, q)$. Theorem is proved. □

We will consider the following examples.

EXAMPLE 0 Davis function that gives us the proof of (5.4.3) for $\alpha \geq 2$. Here the obstacle function will be

$$O_0(p,q) = c_\alpha^\alpha |q|^\alpha - |p|^\alpha, \tag{5.4.12}$$

where the best-value of c_α was found by Davis [**50**].

EXAMPLE 1 Bollobás function. Here the obstacle function will be

$$O_1(p,q) = \mathbf{1}_{q\geq 1} - C|p|, \tag{5.4.13}$$

where the best value of C was suggested by B. Bollobás [**15**]. This was verified by A. Osękowski [**140**], see also [**70**].

EXAMPLE 2 Bollobás function. Here the obstacle function will be

$$O_2(p,q) = \mathbf{1}_{p^2+q^2\geq 1} - C|p|, \tag{5.4.14}$$

where the best value of C was suggested by B. Bollobas [**15**] and also verified by A. Osękowski [**140**], see also [**70**].

EXAMPLE 3 Bellman function associated with the Chang–Wilson–Wolff theorem.

$$O_3(p,q;\lambda) = \mathbf{1}_{[\lambda,\infty)}(p)\mathbf{1}_{[0,1]}(q). \tag{5.4.15}$$

Function U is not fully known in the case, it has been partially found above in Section 1.9. We will explain its significance in Section 5.8 of this chapter.

5.4.2 Davis Obstacle Problem

In this section, we want to find the minimal value c_α for which there exists a function $\mathbf{U}\colon \mathbb{R}^2 \to \mathbb{R}$ that solves the problem with the obstacle function of Example 0, i.e.,

$$\mathbf{U}(p,q) \stackrel{\text{def}}{=} \sup\{\langle c_\alpha^\alpha (q^2 + (Sf)^2)^{\alpha/2} - |f|^\alpha\rangle_I : \langle f\rangle_I = p\}. \tag{5.4.16}$$

In other words, we want to find the heat envelope of O_0. Let $\alpha \geq 2$ and let $\beta = \frac{\alpha}{\alpha-1} \leq 2$ be the conjugate exponent of α. Let

$$\begin{aligned} N_\alpha(x) &\stackrel{\text{def}}{=} {}_1F_1\left(-\frac{\alpha}{2}, \frac{1}{2}, \frac{x^2}{2}\right) \\ &= \sum_{m=0}^{\infty} \frac{(-2x^2)^m}{(2m)!} \frac{\alpha}{2}\left(\frac{\alpha}{2}-1\right)\cdots\left(\frac{\alpha}{2}-m+1\right) \\ &= 1 - \frac{\alpha}{2}x^2 + \frac{\alpha}{12}\left(\frac{\alpha}{2}-1\right)x^4 \ldots \end{aligned} \tag{5.4.17}$$

be the confluent hypergeometric function. $N_\alpha(x)$ satisfies the Hermite differential equation

$$N_\alpha''(x) - xN_\alpha'(x) + \alpha N_\alpha(x) = 0 \quad \text{for} \quad x \in \mathbb{R} \tag{5.4.18}$$

with initial conditions $N_\alpha(0) = 1$ and $N_\alpha'(0) = 0$. Let c_α be the smallest positive zero of N_α.

Set

$$u_\alpha(x) \stackrel{\text{def}}{=} \begin{cases} -\dfrac{\alpha c_\alpha^{\alpha-1}}{N_\alpha'(c_\alpha)} N_\alpha(x), & 0 \leq |x| \leq c_\alpha; \\ c_\alpha^\alpha - |x|^\alpha, & c_\alpha \leq |x|. \end{cases} \tag{5.4.19}$$

Clearly, $u_\alpha(x)$ is $C^1(\mathbb{R}) \cap C^2(\mathbb{R} \setminus \{c_\alpha\})$, a smooth even concave function. The concavity follows from Lemma 5.4.9 on page 375 and the fact that $N_\alpha'(c_\alpha) < 0$. Finally, we define

$$U(p,q) \stackrel{\text{def}}{=} \begin{cases} |q|^\alpha u_\alpha\left(\frac{|p|}{|q|}\right), & q \neq 0, \\ -|p|^\alpha, & q = 0. \end{cases} \tag{5.4.20}$$

In this section, we are going to prove the following result.

THEOREM 5.4.8 *Function* $\mathbf{U}$ *from* (5.4.16) *is equal to* U *written above in* (5.4.20).

For the first time, the function $U(p,q)$ appeared in [**50**]. Later it was also used in [**196, 197**] in the form $\widetilde{u}(p,t) = U(p,\sqrt{t})$, $t \geq 0$. Since we want to prove that

$$U = \mathbf{U},$$

at first we will verify the following properties:

$$U(p,q) \geq |q|^\alpha c_\alpha^\alpha - |p|^\alpha, \quad (p,q) \in \mathbb{R}^2, \tag{5.4.21}$$

$$2U(p,q) \geq U(p+a, \sqrt{a^2+q^2}) + U(p-a, \sqrt{a^2+q^2}), \ (p,q,a) \in \mathbb{R}^3. \tag{5.4.22}$$

When these two properties get proved, Theorem 5.4.7 ensures that

$$\mathbf{U} \leq U. \tag{5.4.23}$$

We called (5.4.21) the *obstacle condition*, and (5.4.22) the *main inequality*. The infinitesimal form of (5.4.22) is

$$\frac{1}{q} U_q + U_{pp} \leq 0, \tag{5.4.24}$$

which follows from the main inequality by expanding it into Taylor's series with respect to a and comparing the second-order terms.

First, we check (5.4.24). On domain $p/q \in (-c_\alpha, c_\alpha)$, $q > 0$, this follows from (5.4.20) and the first line of (5.4.19). Moreover, on this domain, we have equality $U_q/q + U_{pp} = 0$, which easily follows from (5.4.18). On the complementary domain, where $|p| \geq c_\alpha q$, we have

$$\begin{aligned}\frac{1}{q} U_q + U_{pp} &= \alpha(c_\alpha^\alpha q^{\alpha-2} - (\alpha-1)|p|^{\alpha-2}) \\ &= \alpha q^{\alpha-2} c_\alpha^{\alpha-2}\left(c_\alpha^2 - (\alpha-1)\left(\frac{|p|}{c_\alpha q}\right)^{\alpha-2}\right) < 0\end{aligned}$$

because $\alpha \geq 2$ and, as we will see below in Lemma 5.4.9, $c_\alpha \leq 1$. In fact, we need more, we need also to check that in the sense of distributions (5.4.24) is also satisfied, but this calculation we leave for the reader.

Inequality (5.4.24) guarantees that

$$X_t = U(W(t), \sqrt{t}) \quad \text{is a supermartingale for} \quad t \geq 0.$$

In fact, using Itô's formula, see (2.4.2) from Chapter 2, we get

$$dX(t) = \frac{1}{2\sqrt{t}} \frac{\partial U}{\partial q} dt + \frac{1}{2} \frac{\partial^2 U}{\partial p^2} dt + \frac{\partial U}{\partial p} dW(t),$$

and therefore, (5.4.24) implies that $dX(t) - \frac{\partial U}{\partial p} dW(t) \leq 0$, so $X(t)$ is a supermartingale.

Finally, the supermartingale property gives us the second inequality below

$$\mathbb{E}(T^{\frac{\alpha}{2}}c_\alpha^\alpha - |B_T|^\alpha) \overset{(5.4.21)}{\leq} \mathbb{E}U(B_T, \sqrt{T}) \leq U(0,0) = 0,$$

which yields (5.4.3).

Now we are going to prove that $U(p,q)$ is the minimal function with properties (5.4.21) and (5.4.22).

The next step is to go from infinitesimal version (5.4.24) to finite difference inequality (5.4.22). For that, we need several lemmas.

LEMMA 5.4.9 *The minimal positive root c_α of N_α has the following properties.*

1) *The estimate $0 < c_\alpha \leq 1$ is valid for $\alpha \geq 2$.*
2) *c_α is decreasing in $\alpha > 0$.*
3) *$N'_\alpha(t) \leq 0,\ N''_\alpha(t) \leq 0$ on $[0, c_\alpha]$ for $\alpha > 0$.*

PROOF Consider $G_\alpha(t) \overset{\text{def}}{=} e^{-t^2/4}N_\alpha(t)$. Notice that the zeros of G_α and N_α are the same. It follows from (5.4.18) that

$$G''_\alpha + \left(\alpha + \frac{1}{2} - \frac{t^2}{4}\right) G_\alpha = 0, \quad G_\alpha(0) = 1 \quad \text{and} \quad G'_\alpha(0) = 0. \quad (5.4.25)$$

Besides, we know that the solution is even. Consider the critical case $\alpha = 2$. In this case, $G_2(t) = e^{-t^2/4}(1 - t^2)$ and the smallest positive zero is $s_2 = 1$. Therefore, it follows from the Sturm comparison principle that $0 < c_\alpha < 1$ for $\alpha > 2$ (see below). Moreover, the same principle applied to G_{α_1} and G_{α_2} with $\alpha_1 > \alpha_2$ implies that G_{α_1} has a zero inside the interval $(-s_{\alpha_2}, s_{\alpha_2})$. Thus, we conclude that c_α is decreasing in α.

To verify that $N'_\alpha, N''_\alpha \leq 0$ on $[0, c_\alpha]$, first we claim that

$$N_{\alpha_2} \geq N_{\alpha_1} \quad \text{on} \quad [0, s_{\alpha_1}]$$

for $\alpha_1 > \alpha_2 > 0$. Indeed, the proof works in the same way as the proof of Sturm's comparison principle. For the convenience of the reader, we decided to include the argument. As before, consider $G_{\alpha_j} = e^{-t^2/4}N_{\alpha_j}$. It is enough to show that $G_{\alpha_2} \geq G_{\alpha_1}$ on $[0, s_{\alpha_1}]$. It follows from (5.4.25) that $G''_{\alpha_2}(0) > G''_{\alpha_1}(0)$. Therefore, using the Taylor series expansion at the point 0, we see that the claim is true at some neighborhood of zero, say $[0, \varepsilon)$ with ε sufficiently small. Next we assume the contrary, i.e., that there is a point $a \in [\varepsilon, s_{\alpha_1}]$ such that $G_{\alpha_2} \geq G_{\alpha_1}$ on $[0, a]$, $G_{\alpha_2}(a) = G_{\alpha_1}(a)$ and $G'_{\alpha_2}(a) < G'_{\alpha_1}(a)$ (notice that the case $G'_{\alpha_2}(a) = G'_{\alpha_1}(a)$, by the uniqueness theorem for ordinary differential equations, would imply that $G_{\alpha_2} = G_{\alpha_1}$ everywhere, which is impossible). Consider the Wronskian

$$W = G'_{\alpha_1}G_{\alpha_2} - G_{\alpha_1}G'_{\alpha_2}.$$

We have $W(0) = 0$ and $W(a) = G_{\alpha_1}(a)(G'_{\alpha_1}(a) - G'_{\alpha_2}(a)) \geq 0$. On the other hand, we have

$$W' = (\alpha_2 - \alpha_1)G_{\alpha_1}G_{\alpha_2} < 0 \quad \text{on} \quad [0, a),$$

which is a clear contradiction, and this proves the claim.

It follows from (5.4.17) that

$$N''_\alpha = -\alpha N_{\alpha-2}, \tag{5.4.26}$$

and inequalities $N_{\alpha-2} \geq N_\alpha \geq 0$ on $[0, c_\alpha]$ imply that

$$N''_\alpha \leq 0 \text{ on } [0, c_\alpha].$$

Since $N'_\alpha(0) = 0$, and $N''_\alpha \leq 0$ on $[0, c_\alpha]$, we must have $N'_\alpha \leq 0$ on $[0, c_\alpha]$. □

LEMMA 5.4.10 *For any $p \in \mathbb{R}$, the function*

$$t \mapsto U(p, \sqrt{t}) \quad \textit{is convex for} \quad t \geq 0. \tag{5.4.27}$$

PROOF Without loss of generality, assume that $p \geq 0$. We recall that $U(p, \sqrt{t}) = t^{\alpha/2}u_\alpha(p/\sqrt{t})$. Since $\alpha \geq 2$, the only interesting case to consider is when $p/\sqrt{t} < c_\alpha$ (otherwise we just use that $t^{\alpha/2}$ is convex). In this case, we have $U(p, \sqrt{t}) = \kappa_\alpha t^{\alpha/2} N_\alpha(p/\sqrt{t})$, where κ_α is a positive constant. In particular, by (5.4.18), we have $U(p, \sqrt{t})_t + \frac{1}{2}U(p, \sqrt{t})_{pp} = 0$. Using (5.4.26), we obtain

$$U(p, \sqrt{t})_t = -\frac{U(p, \sqrt{t})_{pp}}{2} = -\frac{\kappa_\alpha}{2} t^{\frac{\alpha}{2}-1} N''_\alpha(p/\sqrt{t}) = \frac{\alpha\kappa_\alpha}{2} t^{\frac{\alpha-2}{2}} N_{\alpha-2}(p/\sqrt{t}).$$

Therefore, it would be enough to show that for any $\gamma \geq 0$, the function $x^{-\gamma}N_\gamma(x)$ is decreasing for $x \in (0, s_{\gamma+2})$. Differentiating, and using (5.4.18) again, we obtain

$$\frac{\mathrm{d}}{\mathrm{d}x}\left(\frac{N_\gamma(x)}{x^\gamma}\right) = \frac{N''_\gamma(x)}{x^{\gamma+1}},$$

which is nonpositive by Lemma 5.4.9. □

The next lemma, together with Lemma 5.4.10 and (5.4.24), implies that $U(p, q)$ satisfies (5.4.22).

LEMMA 5.4.11 (Barthe–Maurey [**11**]) *Let J be a convex subset of $\mathbb{R}$, and let $V(p, q)\colon J \times \mathbb{R}_+ \to \mathbb{R}$ be such that*

$$V_{pp} + \frac{V_q}{q} \leq 0 \quad \textit{for all} \quad (p, q) \in J \times \mathbb{R}_+; \tag{5.4.28}$$

$$t \mapsto V(p, \sqrt{t}) \quad \textit{is convex for each fixed} \quad p \in J. \tag{5.4.29}$$

Then for all (p, q, a) *with* $p \pm a \in J$ *and* $q \geq 0$, *we have*

$$2V(p, q) \geq V(p + a, \sqrt{a^2 + q^2}) + V(p - a, \sqrt{a^2 + q^2}). \tag{5.4.30}$$

The lemma says that the global finite difference inequality (5.4.30) is in fact implied by its infinitesimal form (5.4.28) under the extra condition (5.4.29).

PROOF The argument is borrowed from [**11**].

Without loss of generality assume $a \geq 0$. Consider the process

$$X_t = V(p + W(t), \sqrt{q^2 + t}), \quad t \geq 0.$$

Here $W(t)$ is the standard Brownian motion starting at zero. It follows from Itô's formula together with (5.4.28) that X_t is a supermartingale. Indeed, by Itô's formula, we have

$$X_t = X_0 + \int_0^t V_p \, dW(t) + \frac{1}{2} \int_0^t \left(V_{pp} + \frac{V_q}{\sqrt{q^2 + t}} \right) dt$$

and notice that the drift term is negative. Let τ be the stopping time such that $W(\tau)$ hits a or $-a$, i.e.,

$$\tau = \inf\{t \geq 0 \colon W(t) \notin (-a, a)\}.$$

The supermartingale property of X_t and concavity (5.4.29) yield the following chain of inequalities:

$$\begin{aligned}
V(p, q) &= X_0 \geq \mathbb{E}X_\tau = \mathbb{E}V(p + W(\tau), \sqrt{q^2 + \tau}) \\
&= P(W(\tau) = -a)\mathbb{E}(V(p - a, \sqrt{q^2 + \tau})|W(\tau) = -a) \\
&\quad + P(W(\tau) = a)\mathbb{E}(V(p + a, \sqrt{q^2 + \tau})|W(\tau) = a) \\
&= \tfrac{1}{2} \Big(\mathbb{E}(V(p - a, \sqrt{q^2 + \tau})|W(\tau) = -a) \\
&\quad + \mathbb{E}(V(p + a, \sqrt{q^2 + \tau})|W(\tau) = a) \Big) \\
&\geq \tfrac{1}{2} \Big(V\Big(p - a, \sqrt{q^2 + \mathbb{E}(\tau|W(\tau) = -a)}\Big) \\
&\quad + V\Big(p + a, \sqrt{q^2 + \mathbb{E}(\tau|W(\tau) = a)}\Big) \Big) \\
&= \tfrac{1}{2} \Big(V\Big(p - a, \sqrt{q^2 + a^2}\Big) + V\Big(p + a, \sqrt{q^2 + a^2}\Big) \Big).
\end{aligned}$$

Notice that we have used $P(W(\tau) = a) = P(W(\tau) = -a) = 1/2$, $\mathbb{E}(\tau|W(\tau) = a) = \mathbb{E}(\tau|W(\tau) = -a) = a^2$ and the fact that the map $t \mapsto V(p, \sqrt{t})$ is convex together with Jensen's inequality. □

5.4.3 Majorization of the Obstacle Function

We have finished the proof of inequality (5.4.22). Now we are going to check (5.4.21) from page 374. Let

$$\kappa_\alpha = -\frac{\alpha c_\alpha^{\alpha-1}}{N_\alpha'(c_\alpha)}.$$

Function $\kappa_\alpha N_\alpha$ in the first line of (5.4.19) is equal to function $g \stackrel{\text{def}}{=} c_\alpha^\alpha - x^\alpha$ at $x = c_\alpha$. To prove that $\kappa_\alpha N_\alpha \geq g$ on $[0, c_\alpha]$, it is enough to prove $\kappa_\alpha N_\alpha' \leq g'$ on this interval. At point c_α, these derivatives coincide by the choice of κ_α. Notice that $\kappa_\alpha > 0$ and that N_α' and g' are negative. Therefore, to check that $-\kappa_\alpha N_\alpha' \geq -g'$, it is enough to show that function $-N_\alpha'/x^{\alpha-1}$ is decreasing on $[0, c_\alpha]$, i.e.,

$$\left(\frac{-N_\alpha'}{x^{\alpha-1}}\right)' \leq 0. \tag{5.4.31}$$

But

$$\left(\frac{N_\alpha'}{x^{\alpha-1}}\right)' = \frac{xN_\alpha'' - (\alpha-1)N_\alpha'}{x^\alpha} = \frac{N_\alpha'''}{x^\alpha},$$

where the last equality follows from (5.4.18).

On the other hand, from (5.4.17), it follows that $N_\alpha''' = -\alpha N_{\alpha-2}'$. This expression is positive by Lemma 5.4.9. Hence, (5.4.31) is proved. This gives us that

$$u_\alpha \geq c_\alpha^\alpha - |x|^\alpha, \qquad x \in [-c_\alpha, c_\alpha].$$

We conclude that the function U from page 373 majorizes the obstacle:

$$U(p,q) \geq c_\alpha^\alpha |q|^\alpha - |p|^\alpha. \tag{5.4.32}$$

5.4.4 Why Constant c_α is Sharp?

The example, which shows that the value c_α given on page 373 cannot be replaced by larger value is based on results of A. Novikov [**139**] and L. Shepp [**166**]. Introduce the following stopping time:

$$T_a \stackrel{\text{def}}{=} \inf\{t > 0 \colon |W(t)| = a\sqrt{t+1}\}, \quad a > 0.$$

It was proved in [**166**] that $\mathbb{E}T_a^{\alpha/2} < \infty$ if $a < c_\alpha$ and that $\mathbb{E}t_{c_\alpha}^{\alpha/2} = \infty$, $\alpha > 0$. This gives us that $\mathbb{E}t_a^{\alpha/2} \to \infty$, when $a \to c_\alpha-$. From here, we get

$$\lim_{a \to c_\alpha -} \frac{\mathbb{E}(T_a+1)^{\alpha/2}}{\mathbb{E}T_a^{\alpha/2}} = 1.$$

By definition of T_a, we have $|W(T_a)| = a\sqrt{T_a + 1}$, and hence

$$\lim_{a \to c_\alpha -} \frac{\mathbb{E}|W(T_a)|^\alpha}{\mathbb{E}T_a^{\alpha/2}} \to c_\alpha^\alpha.$$

Now it follows immediately that the best constant in (5.4.1) cannot be larger than c_α defined on page 373. Davis in **[50]** extended this estimate for the case of dyadic square function estimate (5.4.3).

5.4.5 Why U from page 373 Is the Smallest Function Satisfying (5.4.21) and (5.4.22)?

We know that on $\{(p,q)\colon q \geq 0,\ |p|^\alpha \leq c_\alpha^\alpha q^\alpha\}$

$$|q|^\alpha c_\alpha^\alpha - |p|^\alpha \leq \mathbf{U}(p,q) \leq U(p,q). \tag{5.4.33}$$

Indeed, we proved that U satisfies the main inequality and that it majorizes the obstacle $|q|^\alpha c_\alpha^\alpha - |p|^\alpha$. We also proved that $\mathbf{U}$ is the smallest such function (this is true for any obstacle whatsoever, see Theorem 5.4.7). Hence, (5.4.33) is verified.

By definition on page 373 $\mathbf{U}$ is homogeneous of degree α. We introduce $\mathbf{b}(p) \stackrel{\text{def}}{=} \mathbf{U}(p,1)$, $b(p) \stackrel{\text{def}}{=} U(p,1)$. Thus we need to prove that

$$b(p) = \mathbf{b}(p), \qquad p \in [-c_\alpha, c_\alpha]. \tag{5.4.34}$$

One can easily rewrite (5.4.22) in terms of $\mathbf{b}$: for all $x \pm \tau \in [-c_\alpha, c_\alpha]$ the following holds:

$$2\mathbf{b}(x) \geq (1+\tau^2)^{\alpha/2}\left(\mathbf{b}\left(\frac{x+\tau}{\sqrt{1+\tau^2}}\right) + \mathbf{b}\left(\frac{x-\tau}{\sqrt{1+\tau^2}}\right)\right). \tag{5.4.35}$$

Since by construction $U(p,q) = 0$ if $|q|^\alpha c_\alpha^\alpha - |p|^\alpha = 0$, we conclude that $b(\pm c_\alpha) = \mathbf{b}(\pm c_\alpha) = 0$.

Inequality (5.4.35) implies that $\mathbf{b}'' - x\mathbf{b}' + \alpha\mathbf{b} \leq 0$ in the sense of distributions.

Combining (5.4.22) with a simple observation that $\mathbf{U}$ by definition increases in q, we can conclude that function $\mathbf{U}$ is concave in p for every fixed q, so $\mathbf{b}$ is concave.

Let us recall that for any concave function f, the following holds (see e.g., **[60]**):

$$f(x+h) = f(x) + f'(x)h + \frac{1}{2}f''(x)h^2 + o(h^2), \quad h \to 0, \quad \text{for a. e. } x. \tag{5.4.36}$$

Then (5.4.36) and inequality (5.4.35) implies that $\mathbf{b}'' - x\mathbf{b}' + \alpha\mathbf{b} \leq 0$ a. e. But function $\mathbf{b}$ is concave. In particular, it is continuous everywhere, and its derivative $\mathbf{b}'$ is also its distributional derivative, and it is everywhere defined as a decreasing function.

Let $(\mathbf{b})''$ denote the distributional derivative of decreasing function $\mathbf{b}'$. Thus, it is a nonpositive measure. We denote its singular part by symbol σ_s. Hence, in the sense of distributions

$$(\mathbf{b})'' - x\mathbf{b}'\,dx + \alpha\mathbf{b}\,dx = \big(\mathbf{b}'' - x\mathbf{b}' + \alpha\mathbf{b}\big)\,dx + d\sigma_s \le 0. \tag{5.4.37}$$

LEMMA 5.4.12 *Let $\alpha > 0$. Let even nonnegative concave function v defined on $[-c_\alpha, c_\alpha]$ satisfy $v(\pm c_\alpha) = 0$. Let v satisfy $v'' - xv' + \alpha v \le 0$ on $(-c_\alpha, c_\alpha)$ pointwise and in the sense of distributions. Assume also that v has finite derivative at c_α: $v'(c_\alpha) > -\infty$. Then $v'' - xv' + \alpha v = 0$ on $(-c_\alpha, c_\alpha)$ pointwise and in the sense of distributions. Also $v = cu_\alpha$ for some constant c, where u_α is from* (5.4.19).

PROOF Let $u \stackrel{\text{def}}{=} u_\alpha$ be from (5.4.19). It is a C^2 function and $u'' - xu' + \alpha u = 0$ on $[-c_\alpha, c_\alpha]$. Denote

$$g \stackrel{\text{def}}{=} v'' - xv' + \alpha v.$$

Function v is concave, so its second derivative is defined a. e., and we assumed that $g \le 0$.

Consider everywhere defined function

$$w \stackrel{\text{def}}{=} v'u - u'v.$$

Its derivative is defined almost everywhere, and let us first calculate it a. e.:

$$w' = v''u - u''v = (g + xv' - \alpha v)u - (xu' - \alpha u)v = gu + xw.$$

Also in distributional sense

$$(w)' = (v)''u - u''v\,dx = (gu + xw)\,dx + u\,d\sigma_s.$$

Hence,

$$\frac{d}{dx}e^{-x^2/2}w = gue^{-x^2/2}, \quad \text{for almost every } x, \tag{5.4.38}$$

and

$$\left(e^{-x^2/2}w\right)' = gue^{-x^2/2}\,dx + ue^{-x^2/2}\,d\sigma_s, \text{ in distribution sense.} \tag{5.4.39}$$

Measure σ_s is nonpositive, therefore, these two inequalities (5.4.38), (5.4.39) mean that for any two points $0 < a < b < c_\alpha$, we have

$$e^{-b^2/2}w(b) - e^{-a^2/2}w(a) \le \int_a^b gue^{-x^2/2}\,dx,$$

moreover, the inequality is strict, if $\sigma_s(a,b) \ne 0$.

Let us tend b to c_α. Looking at the definition $w = v'u - u'v$ and using the assumptions of lemma, we conclude that $e^{-b^2/2}w(b) \to 0$. Hence,

$$e^{-a^2/2}w(a) \geq \int_a^{c_\alpha} (-g)ue^{-x^2/2}\,dx. \tag{5.4.40}$$

Again the inequality is strict if $\sigma_s(a, c_\alpha) \neq 0$.

Now let us tend $a \to 0$. By smoothness and evenness $u'(a) \to 0$. But $u(a) > 0$ and $v'(a) \leq 0$ for a. e. $a > 0$. Therefore,

$$\limsup_{a\to 0+} e^{-a^2/2}w(a) \leq 0.$$

Combining this with (5.4.40), we conclude that

$$\int_0^{c_\alpha} (-g)ue^{-x^2/2}\,dx \leq 0$$

with the strict inequality if $\sigma_s(0, c_\alpha) \neq 0$. The strict inequality, of course, leads to contradiction (recall that $-g \geq 0, u > 0$), so we conclude that σ_s is a zero measure on $(0, c_\alpha)$. But also even a non-strict inequality implies that $g = 0$ a. e.

We conclude from (5.4.38), (5.4.39) that $e^{-x^2/2}w(x)$ is constant on $(0, c_\alpha)$. But we already saw that this function tends to 0 when x tends to c_α. Thus, identically on $(0, c_\alpha)$

$$u'v - v'u = w = 0.$$

This means that $v/u = const$. Lemma is proved. □

Now it is easy to prove (5.4.34): $\mathbf{b} = b$. Choose $v = \mathbf{b}$, the assumptions on ordinary differential inequality is easy to verify, see (5.4.37). Of course this function vanishes at $\pm c_\alpha$. Also by the definition of b, it is clear (see (5.4.19), (5.4.20)) that

$$b'(c_\alpha) = -\alpha c_\alpha^{\alpha-1} > -\infty.$$

We are left to see that the same is true for $\mathbf{b}'(c_\alpha)$.

Recall that $\mathbf{b}(\cdot) = \mathbf{U}(\cdot, 1)$, $b(\cdot) = U(\cdot, 1)$, then by (5.4.33), we definitely know that

$$c_\alpha^\alpha - |x|^\alpha \leq \mathbf{b}(x) \leq b(x), \quad x \in [-c_\alpha, c_\alpha].$$

The functions on the left and on the right vanish at c_α and have the same derivative $-\alpha c_\alpha^{\alpha-1}$ at C_α. Hence, $\mathbf{b}$ is in fact differentiable at c_α (the left derivative exists), and its (left) derivative satisfies

$$\mathbf{b}'(c_\alpha) = b'(c_\alpha) = -\alpha c_\alpha^{\alpha-1} > -\infty.$$

But now Lemma 5.4.12 says that $\mathbf{b} = const \cdot b$. Since we have the above relationship on derivatives, the constant has to be 1. We proved (5.4.34). This gives us the following equality:

$$U = \mathbf{U},$$

where U was defined in (5.4.19), (5.4.20). We found the Bellman function $\mathbf{U}$ for Burkholder–Gundy–Davis inequality, and we completely solved the obstacle problem with the obstacle $O(p,q) = c_\alpha^\alpha q^\alpha - |p|^\alpha$, $\alpha \ge 2$.

5.4.6 When Obstacle Coincides with Its Heat Envelope

The next corollary immediately follows, and it describes one possibility when the heat envelope coincides with its obstacle.

COROLLARY 5.4.13 *Let $O(p,q)$ be $C^2(\mathbb{R} \times [0,\infty))$ obstacle such that*

$$O_{pp} + \frac{O_q}{q} \le 0 \quad \textit{and}$$

$$t \mapsto O(p, \sqrt{t}) \quad \textit{is convex.}$$

Then the heat envelope U of O satisfies $U(p,q) = O(p,q)$.

The next proposition says that if O satisfies the "backward heat equation" then the convexity assumption $t \mapsto O(p,\sqrt{t})$ is necessary and sufficient for main inequality (5.4.5).

PROPOSITION 5.4.14 *Let $O(p,q) \in C^4(\mathbb{R} \times [0,\infty))$ be such that*

$$O_{pp} + \frac{O_q}{q} = 0$$

for all $(p,q) \in \mathbb{R} \times (0,\infty)$. Then the following conditions are equivalent

(i) The map $t \mapsto O(p,\sqrt{t})$ is convex for $t \ge 0$.
(ii) $2O(p,q) \ge O(p+a, \sqrt{q^2+a^2}) + O(p-a, \sqrt{q^2+a^2})$ for all $p, a \in \mathbb{R}$ and all $q \ge 0$.

PROOF The implication $(i) \Rightarrow (ii)$ follows from Lemma 5.4.11. It remains to show the implication $(ii) \Rightarrow (i)$. By Taylor's formula as $a \to 0$, we have

$$\begin{aligned} &O(p+a, \sqrt{q^2+a^2}) + O(p-a, \sqrt{q^2+a^2}) \\ &\quad = 2O(p,q) + \left(O_{pp} + \frac{O_q}{q}\right) a^2 \\ &\quad + \left(O_{pppp} + 6\frac{O_{ppq}}{q} + 3\frac{O_{qq}}{q^2} - 3\frac{O_q}{q^3}\right)\frac{a^4}{12} + o(a^4) \end{aligned}$$

Since $O_{pp} + \frac{O_q}{q} = 0$, we see that

$$O_{pppp} + 6\frac{O_{ppq}}{q} + 3\frac{O_{qq}}{q^2} - 3\frac{O_q}{q^3} = 2\left(\frac{O_q}{q^3} - \frac{O_{qq}}{q^2}\right).$$

Therefore,

$$0 \geq O(p+a, \sqrt{q^2+a^2}) + O(p-a, \sqrt{q^2+a^2}) - 2O(p,q) =$$
$$= \left(\frac{O_q}{q} - O_{qq}\right)\frac{a^4}{6q^2} + o(a^4).$$

Thus, we obtain that $\frac{O_q}{q} - O_{qq} \leq 0$. On the other hand, the latter inequality is equivalent to the fact that $t \to O(p, \sqrt{t})$ is convex. □

5.5 Bollobás Function

The classical Littlewood–Khintchine inequality states that

$$\left(\sum_{k=1}^n a_k^2\right)^{1/2} \leq L\int_0^1 \left|\sum_{k=1}^n a_k r_k(t) dt\right|, \tag{5.5.1}$$

where $\{r_k(t)\}$ are Rademacher functions. It was one of Littlewood's problems to find the best value for constant L. The problem was solved by S. Szarek [**175**]; see also [**65**]. The sharp constant is $L = \sqrt{2}$.

B. Bollobás [**15**] considered the following related problem, which we formulate in the form convenient for us. The problem considered by Bollobás is the following: What is the best value for the constant B for the following inequality:

$$\lambda\big|\{t \in (0,1)\colon Sf(t) \geq \lambda\}\big| \leq C\|f\|_1\,? \tag{5.5.2}$$

Consider $x_n \stackrel{\text{def}}{=} \sum_{k=1}^n a_k r_k(t)$. If we denote $\lambda \stackrel{\text{def}}{=} \left(\sum_{k=1}^n a_k^2\right)^{1/2} = Sx_n(x)$ (obviously Sx_n is a constant function), we get

$$\left(\sum_{k=1}^n a_k^2\right)^{1/2} = \lambda\big|\{t \in (0,1)\colon Sx_n(t) \geq \lambda\}\big|$$
$$\leq C\int_0^1 \big|\sum_{k=1}^n a_k r_k(t)\big| dt. \tag{5.5.3}$$

This says that $\sqrt{2} = L \leq B$. On the other hand, D. Burkholder in [**21**] proved that $B \leq 3$. B. Bollobás in [**15**] conjectured the best value of B, and in 2009 A. Osękowski [**140**] proved this conjecture. We will give a slightly different proof by solving the obstacle problem and finding the heat envelopes of two obstacles:

$$O_1(p,q) = \mathbf{1}_{q\ge 1} - C_1|p|, \tag{5.5.4}$$

$$O_2(p,q) = \mathbf{1}_{p^2+q^2\ge 1} - C_2|p|. \tag{5.5.5}$$

We are interested in the smallest possible values of C_1 and C_2 such that these functions have (finite) heat envelopes. The reader will see, in particular, that $C_1 = C_2 = C$ and that the heat envelopes of these two functions coincide.

Define the following Bellman function:

$$\mathbf{B}(x,\lambda) \overset{\text{def}}{=} \inf\{\langle|\varphi|\rangle_{_J} : \langle\varphi\rangle_{_J} = x;\ S_J^2\varphi \ge \lambda \text{ a.e. on } J\}. \tag{5.5.6}$$

Some of the obvious properties of $\mathbf{B}$ are the following:

- Domain: $\Omega_{\mathbf{B}} \overset{\text{def}}{=} \{(x,\lambda) : x\in\mathbb{R};\ \lambda > 0\}$;
- $\mathbf{B}$ is increasing in λ and even in x;
- Homogeneity: $\mathbf{B}(tx, t^2\lambda) = |t|\mathbf{B}(x,\lambda)$;
- Range/obstacle condition: $|x| \le \mathbf{B}(x,\lambda) \le \max\{|x|, \sqrt{\lambda}\}$;
- Main inequality:

$$2\mathbf{B}(x,\lambda) \le \mathbf{B}(x-a,\lambda-a^2) + \mathbf{B}(x+a,\lambda-a^2),\ \ \forall\, |a| < \sqrt{\lambda}. \tag{5.5.7}$$

- $\mathbf{B}$ is convex in x, and so it is easy to see that $\mathbf{B}$ is minimal at $x=0$:

$$\mathbf{B}(0,\lambda) \le \mathbf{B}(x,\lambda),\ \ \forall\, x; \tag{5.5.8}$$

  therefore, we can use that $\mathbf{B}$ is increasing in λ and also use the minimality at $x=0$ to obtain from (5.5.7) that $\mathbf{B}$ is nondecreasing in x for $x \ge 0$, and nonincreasing in x for $x \le 0$;
- Greatest subsolution: If $B(x,\lambda)$ is any continuous nonnegative function on $\Omega_{\mathbf{B}}$, which satisfies the main inequality (5.5.7), and the range condition $B(x,\sqrt{\lambda}) \le \max\{|x|,\sqrt{\lambda}\}$, then $B \le \mathbf{B}$.

5.5.1 Bellman Induction

THEOREM 5.5.1 *If B is any subsolution as defined above, then $B \le \mathbf{B}$.*

PROOF We must prove that $B(x,\lambda) \le \langle|\varphi|\rangle_{_J}$ for any function φ on J with $\langle\varphi\rangle_{_J} = x$, $|J| = |\{x \in J : S_J^2\varphi(x) \ge \lambda\}|$. As before, we may assume that there is some dyadic level $N \ge 0$ below which the Haar coefficients of φ are zero.

If $\lambda \le (\Delta_{_J}\varphi)^2$, then by the range/obstacle condition above

$$B(x,\lambda) \le \max\{|x|,\sqrt{\lambda}\} \le \max\{|x|, |\Delta_{_J}\varphi|\} \le \langle|\varphi|\rangle_{_J},$$

and we are done. Otherwise, put $\lambda_{J_\pm} = \lambda - (\Delta_{_J}\varphi)^2 > 0$, $x_{J_\pm} = \langle\varphi\rangle_{_{J_\pm}}$. Then by the main inequality:

$$|J|B(x,\lambda) \le |J_-|B(x_{J_-},\lambda_{J_-}) + |J_+|B(x_{J_+},\lambda_{J_+}).$$

If $\lambda_{J_-} \le (\Delta_{J_-}\varphi)^2$, it follows as before that $|J_-|B(x_{J_-},\lambda_{J_-}) \le \int_{J_-}|\varphi|$, and otherwise we iterate further on J_-.

Continuing this way down to the last level N and putting $\lambda_I \stackrel{\text{def}}{=} \lambda - (\Delta_{I^{(1)}}\varphi)^2 - \cdots - (\Delta_{_J}\varphi)^2$ for every $I \in \mathcal{D}_N(J)$, where $I^{(1)}$ denotes the dyadic father of I, the previous iterations have covered all cases where $\lambda_I \le 0$, and we have (with $x_I I$)

$$|J|B(x,\lambda) \le \sum_{\substack{I\in\mathcal{D}_N(J)\\ \lambda_I\le 0}} \int_I |\varphi| + \sum_{\substack{I\in\mathcal{D}_N(J)\\ \lambda_I> 0}} |I|B(x_I,\lambda_I). \tag{5.5.9}$$

Now note that for all $I \in \mathcal{D}_N(J)$, we must have $\lambda_I \le (\Delta_{_I}\varphi)^2$ just because $S_J^2\varphi(x) \ge \lambda$ everywhere on J, so we use the range/obstacle condition as before to obtain $B(x_I,\lambda_I) \le \max\{|x_I|, |\Delta_{_I}\varphi|\} \le \langle|\varphi|\rangle_{_I}$. Finally, (5.5.9) becomes

$$|J|B(x,\lambda) \le \sum_{I\in\mathcal{D}_N(J)} \int_I |\varphi| = \int_J |\varphi|.$$

This finishes the proof of the claim

$$B \le \mathbf{B}.$$

□

5.5.2 Finding the Candidate for $\mathbf{B}(x, \lambda)$

We introduce

$$\mathbf{b}(\tau) \stackrel{\text{def}}{=} \mathbf{B}(\tau, 1).$$

Using homogeneity, we write

$$\sqrt{\lambda}\mathbf{b}(\tau) = \sqrt{\lambda}\mathbf{B}\left(\frac{x}{\sqrt{\lambda}}, 1\right) = \mathbf{B}(x,\lambda), \quad \text{where } \tau = \frac{x}{\sqrt{\lambda}}.$$

Then $\mathbf{b}\colon \mathbb{R} \to [0,\infty)$, $\mathbf{b}$ is even in τ, and from (5.5.8):

$$\mathbf{b}(0) \le \mathbf{b}(\tau), \qquad \forall\, \tau. \tag{5.5.10}$$

Moreover, $\mathbf{b}$ satisfies

$$\mathbf{b}(\tau) = |\tau|, \qquad \forall\, |\tau| \ge 1. \tag{5.5.11}$$

We are looking for a candidate B for $\mathbf{B}$. We will assume now that $\mathbf{B}$ is smooth. We will find the candidate under this assumption, and later we will prove that thus found function is indeed $\mathbf{B}$. Using again Taylor's formula, the infinitesimal version of (5.5.7) is

$$\mathbf{B}_{xx} - 2\mathbf{B}_\lambda \geq 0. \tag{5.5.12}$$

In terms of $\mathbf{b}$, this becomes

$$\mathbf{b}''(\tau) + \tau\mathbf{b}'(\tau) - \mathbf{b}(\tau) \geq 0. \tag{5.5.13}$$

Since $\mathbf{b}$ is even, we focus next only on $\tau \geq 0$.

Let symbol Φ denote the following function:

$$\Phi(\tau) \stackrel{\text{def}}{=} \int_0^\tau e^{-y^2/2}\, dy.$$

Put

$$\Psi(\tau) = \tau\Phi(\tau) + e^{-\tau^2/2}, \qquad \forall\, \tau \geq 0.$$

The general solution of the differential equation

$$z''(\tau) + \tau z'(\tau) - z(\tau) = 0, \qquad \tau \geq 0$$

is

$$z(\tau) = C\Psi(\tau) + D\tau.$$

Note that

$$\Psi'(\tau) = \Phi(\tau), \qquad \Psi''(\tau) = e^{-\tau^2/2}. \tag{5.5.14}$$

Since $\mathbf{b}(\tau) = \tau$ for $\tau \geq 1$, see (5.5.11), a reasonable candidate for our function $\mathbf{b}$ is one already proposed by B. Bollobas [**15**]:

$$b(\tau) \stackrel{\text{def}}{=} \begin{cases} \frac{\Psi(\tau)}{\Psi(1)}, & 0 \leq \tau < 1 \\ \tau, & \tau \geq 1. \end{cases} \tag{5.5.15}$$

In other words, a candidate for $\mathbf{B}$ is

$$B(y, \lambda) = \begin{cases} \sqrt{\lambda}\frac{\Psi\left(\frac{|y|}{\sqrt{\lambda}}\right)}{\Psi(1)}, & \sqrt{\lambda} \geq |y|, \\ |y|, & \sqrt{\lambda} \leq |y|. \end{cases} \tag{5.5.16}$$

Our first goal will be to go from differential inequality (5.5.12) to its finite difference version (5.4.5).

LEMMA 5.5.2 *The function B defined in* (5.5.16) *satisfies the finite difference main inequality* (*the analog of* (5.5.7)):

$$2B(y, \lambda) \le B(y - a, \lambda - a^2) + B(y + a, \lambda - a^2). \tag{5.5.17}$$

We already saw in Lemma 5.4.11 that under some extra assumptions of convexity one can derive the finite difference inequalities from their differential form (infinitesimal form). Unfortunately, this approach will not work for function B defined in (5.5.16). This function exactly misses the extra property (5.4.29) of Lemma 5.4.11. In fact, we deal now with convexity paradigm rather than concavity conditions of Lemma 5.4.11, so the right analog of property (5.4.29) for B in the above formula would be

$$\lambda \to B(y, \lambda) \text{ is a concave function for every fixed } y.$$

But it is obvious that our candidate B does not have this property. This is why the proof of Lemma 5.5.2 requires direct calculations. This requires splitting the proof into several cases. One of them was considered in [**15**], but other cases were only mentioned there.

Proof of Lemma 5.5.2 By symmetry, we can think that $x \ge 0$.

Case 1) will be when both points $(x \pm \tau, 1 - \tau^2)$ lie in

$$\Omega_{par} \stackrel{\text{def}}{=} \{(x, \lambda) \in \mathbb{R}^2 \colon \lambda \ge x^2\}.$$

Clearly, then $(x, 1)$ will be also in Ω_{par}. We follow [**15**] in this case.

Notice that $B(x, 1) = \max(\frac{\Psi(|x|)}{\Psi(1)}, |x|) = \frac{\Psi(|x|)}{\Psi(1)}$ if $(x, 1) \in \Omega_{par}$. Put

$$X(x, \tau) \stackrel{\text{def}}{=} \frac{|x + \tau|}{(1 - \tau^2)^{1/2}}, \ \tau \in [-1, 1], \ x \in [0, 1). \tag{5.5.18}$$

In our case, (5.5.17) can be rewritten as ($\tau \stackrel{\text{def}}{=} a/\sqrt{\lambda}, \ x = y/\sqrt{\lambda}$):

$$2\Psi(x) \le (\Psi(X(x, \tau)) + \Psi(X(x, -\tau)))\sqrt{1 - \tau^2}, \tag{5.5.19}$$

which is correct for $\tau = 0$. Next, without loss of generality, we assume that $\tau \ge 0$. The inequality is true for $\tau = 0$. Let us check that

$$\frac{\partial}{\partial \tau}(\Psi(X(x, \tau)) + \Psi(X(x, -\tau))) \ge 0. \tag{5.5.20}$$

Consider the case $x - \tau \ge 0$. Notice that

$$\frac{\partial}{\partial a} X(x, \tau) = \frac{1}{\sqrt{1 - \tau^2}} + X(x, \tau)\frac{a}{1 - \tau^2};$$

$$\frac{\partial}{\partial a} X(x, -\tau) = -\frac{1}{\sqrt{1 - \tau^2}} + X(x, -\tau)\frac{a}{1 - \tau^2}.$$

Using the fact that $\Psi'(s) = \Phi(s)$, $\Psi(s) = s\Psi'(s) + e^{-s^2/2}$, we get the equality

$$\frac{\partial}{\partial \tau}(\Psi(X(x,\tau)) + \Psi(X(x,-\tau))) = \frac{1}{\sqrt{1-\tau^2}}(\Phi(X(x,\tau)) - \Phi(X(x,-\tau)))$$
$$+ \frac{\tau}{1-\tau^2}\left[\Psi(X(x,\tau)) - \exp(-(X(x,\tau))^2/2)\right.$$
$$\left.+\Psi(X(x,-\tau)) - \exp(-(X(x,-\tau))^2/2)\right]$$

Therefore,

$$\frac{\partial}{\partial \tau}\left((\Psi(X(x,\tau)) + \Psi(X(x,-\tau)))\sqrt{1-\tau^2}\right)$$
$$= (\Phi(X(x,\tau)) - \Phi(X(x,-\tau)))$$
$$- \frac{\tau}{(1-\tau^2)^{1/2}}(e^{-X(x,\tau)^2/2} + e^{-X(x,-\tau)^2/2}).$$

But $\frac{\tau}{(1-\tau^2)^{1/2}} = \frac{1}{2}(X(x,\tau) - X(x,-\tau))$, so to prove (5.5.20) one needs to check the following inequality:

$$\frac{1}{X(x,\tau) - X(x,-\tau)} \int_{X(x,-\tau)}^{X(x,\tau)} e^{-s^2/2} ds \geq \frac{1}{2}(e^{-X(x,\tau)^2/2} + e^{-X(x,-\tau)^2/2}). \tag{5.5.21}$$

This inequality holds because in our case 1), we have $X(x,-\tau) \in [0,1], X(x,\tau) \in [0,1]$, and function $s \to e^{-s^2/2}$ is concave on the interval $[-1,1]$. (It is easy to verify that for every concave function on an interval, its integral average over the interval is at least its average over the endpoints of the interval.)

If $x - \tau \leq 0$, then $\frac{\partial}{\partial \tau} X(x,-\tau) = \frac{1}{\sqrt{1-\tau^2}} + X(x,-\tau)\frac{\tau}{1-\tau^2}$. Repeating the previous calculations verbatim, eventually one will need to show the following inequality:

$$\Phi(X(x,\tau)) + \Phi(X(x,-\tau)) \geq \frac{X(x,\tau) + X(x,-\tau)}{2}$$
$$\times \left(e^{-X(x,\tau)^2/2} + e^{-X(x,-\tau)^2/2}\right),$$

which is also true. Indeed, we want to show that $\Phi(a) + \Phi(b) \geq \frac{a+b}{2}(e^{-a^2/2} + e^{-b^2/2})$ for all $a, b \in [0,1]$. If $a = b$, then the inequality follows because $a \mapsto \Phi(a) - ae^{-a^2/2}$ at $a = 0$ is true, and its derivative is $a^2 e^{-a^2/2} \geq 0$. In general, consider the map

$$a \mapsto \Phi(a) + \Phi(b) - \frac{a+b}{2}(e^{-a^2/2} + e^{-b^2/2}) \quad \text{for} \quad a \in [b,1].$$

The derivative of this map is $\frac{1}{2}(e^{-a^2/2} - e^{-b^2/2}) + \frac{a+b}{2} \cdot ae^{-a^2/2}$, which, at point $a = b$, has a nonnegative sign. Differentiating again, we obtain $\frac{e^{-a^2/2}}{2}(1 - a^2)(a + b) \geq 0$. This finishes the proof of case 1).

Case 2): when $(x, 1) \notin \Omega_{par}$. Then notice that $x \mapsto B(x, 1)$ is convex as a maximum of two convex functions. Therefore,

$$\frac{1}{2}\left(B(x + \tau, 1 - \tau^2) + B(x - \tau, 1 - \tau^2)\right) \geq B(x, 1 - \tau^2) = B(x, 1).$$

Case 3): Now suppose that $(x \pm \tau, 1 - \tau^2)$ are not in Ω_{par} and $(x, 1)$ is in Ω_{par}. We remind the reader that we are considering only $x \geq 0$. Since $\tau \mapsto |x + \tau| + |x - \tau|$ is increasing as τ increases, it suffices to consider the case when $(x - \tau, 1 - \tau^2)$ is such that $(x - \tau)^2 = 1 - \tau^2$, i.e., the left point is on the parabola. Then we need to show that

$$2\frac{\Psi(x)}{\Psi(1)} \leq |x - \tau| + x + \tau. \tag{5.5.22}$$

Clearly $0 \leq \tau \leq 1$. Consider the case when $0 \leq x \leq \tau$. From $(x - \tau)^2 = 1 - \tau^2$, we obtain that $\tau - \sqrt{1 - \tau^2} \stackrel{\text{def}}{=} F(\tau) \geq 0$, so $\tau \geq \frac{1}{\sqrt{2}}$, and the inequality (5.5.22) simplifies to

$$F(\tau) \leq \Psi^{-1}(\Psi(1)\tau), \quad 1 \geq \tau \geq \frac{1}{\sqrt{2}}.$$

The left-hand side is convex and the right-hand side is concave (as an inverse of increasing convex function). Since at $\tau = 1$ and $\tau = \frac{1}{\sqrt{2}}$, the inequality holds, then it holds on the whole interval $[1/\sqrt{2}, 1]$.

If $x \geq \tau$, the inequality (5.5.22) becomes $\frac{\Psi(x)}{\Psi(1)} \leq x$, which is correct if $x \geq 1$. Indeed, consider $g(s) = \frac{\Psi(s)}{\Psi(1)} - s$. Then $g(1) = 0$, $g'(1) < 0$, and $g''(t) \geq 0$. Also $\lim_{s\to\infty} \frac{g(s)}{s} = \frac{\int_0^\infty e^{-t^2/2}dt}{\Phi(1)+\exp(-1/2)} - 1 = -0.1428\cdots < 0$. This implies that $g(t) \leq 0$ for all $t \geq 1$.

Case 4a). Next we consider the case when $(x, 1)$ is in Ω_{par}, $(x + \tau, 1 - \tau^2)$ is not in Ω_{par}, $(x - \tau, 1 - \tau^2)$ is in Ω_{par} and it has nonnegative first coordinate, i.e., $x - \tau \geq 0$ (the remaining case with negative first coordinate will be treated in Case 4b)).

First, consider the case when $x - \tau = 0$, i.e., the first coordinate of the left point is in zero. Since the right point is outside (below) of the parabola, we have $\frac{x+\tau}{\sqrt{1-\tau^2}} = \frac{2\tau}{\sqrt{1-\tau^2}} \geq 1$. The latter means that $\tau \in [\frac{1}{\sqrt{3}}, 1]$. Then we need to show that

$$2\Psi(\tau) = 2\Psi(x) \le \sqrt{1-\tau^2}\,\Psi\left(\frac{x-\tau}{\sqrt{1-\tau^2}}\right) + \Psi(1)(x+\tau)$$
$$= \sqrt{1-\tau^2} + 2\Psi(1)\tau.$$

The left-hand side of the inequality is convex. The right-hand side of the inequality is concave. Inequality clearly holds for the endpoint cases, i.e., $\tau = 1$ and $\tau = \frac{1}{\sqrt{3}}$. Therefore, it holds in general.

Notice that if $(x-\tau)^2 = 1-\tau^2$ then we are in Case 3). So if we show that the map $\tau \mapsto B(x+\tau, 1-\tau^2) + B(x-\tau, 1-\tau^2)$ is concave when $1 \ge \frac{x-\tau}{\sqrt{1-\tau^2}} \ge 0$ (left point is in Ω_{par} with nonnegative first coordinate), $x \le 1$ (the point $(x,1)$ is in Ω_{par}), and $\frac{x+\tau}{\sqrt{1-a^2}} \ge 1$ (the right point is not in Ω_{par}) then this will prove Case 4a) completely because the concave function dominates the number $2B(x,1)$ at the endpoints of an interval. We have

$$B(x+\tau, 1-\tau^2) + B(x-\tau, 1-\tau^2) = \sqrt{1-\tau^2}\,\Psi\left(\frac{x-\tau}{\sqrt{1-\tau^2}}\right)$$
$$+ \Psi(1)(x+\tau).$$

The second term is linear in a. Its first derivative is

$$-\Phi\left(\frac{x-\tau}{\sqrt{1-\tau^2}}\right) - \frac{\tau}{\sqrt{1-\tau^2}}\exp\left(-\left[\frac{x-\tau}{\sqrt{1-\tau^2}}\right]^2/2\right) + \Psi(1).$$

Its second derivative is

$$\frac{\tau(\tau+\tau x^2-2x)}{(1-\tau^2)^{5/2}}\exp\left(-\left[\frac{x-\tau}{\sqrt{1-\tau^2}}\right]^2/2\right).$$

The map $\tau \mapsto \tau + \tau x^2 - 2x$ is increasing in a. Let us increase τ. Two scenarios can occur: 1) $x-\tau = 0$ or 2) $\frac{x-\tau}{\sqrt{1-\tau^2}} = 1$. In the first case, we get $\tau + \tau x^2 - 2x = x(x^2-1) \le 1$ since $0 \le x \le 1$. In the second case, the condition $\tau \in [0,1]$ implies

$$\tau + \tau x^2 - 2x = -\tau - 2\sqrt{1-\tau^2} + \tau^2(\sqrt{1-\tau^2}+\tau)^2$$
$$= \tau(\tau-1) + 2\sqrt{1-\tau^2}(\tau^3-1) \le 0.$$

Thus, in all cases, we obtain $\tau + \tau x^2 - 2x \le 0$, therefore, this finishes the proof of the case 4a).

Case 4b). It remains to show that if the right point already left Ω_{par} but the left point is in Ω_{par} with negative first coordinate, then (5.5.17) still holds. Then the required inequality amounts to

$$2\Psi(x) \le \sqrt{1-\tau^2}\,\Psi\left(\frac{\tau-x}{\sqrt{1-\tau^2}}\right) + \Psi(1)(x+\tau),$$

where $|\sqrt{1-\tau^2}-\tau| \le x \le a \le 1$ (notice that the latter inequality simply means that $\frac{x+\tau}{\sqrt{1-\tau^2}} \ge 1$, i.e., the right point is not in Ω_{par}, and $\frac{\tau-x}{\sqrt{1-\tau^2}} \le 1$, the left point is in Ω_{par} with negative first coordinate). It is the same as to show

$$\Psi\left(\frac{\tau-x}{\sqrt{1-\tau^2}}\right)+\Psi(1)\left(\frac{\tau-\left(\frac{2\Psi(x)}{\Psi(1)}-x\right)}{\sqrt{1-\tau^2}}\right)\ge 0 \tag{5.5.23}$$

for all $0\le x\le 1$ if $\max\left\{x, \frac{\sqrt{2-x^2}-x}{2}\right\}\le\tau\le\frac{x+\sqrt{2-x^2}}{2}$.

Let us show that the derivative in a of the left-hand side of (5.5.23) is nonnegative. If this is the case, then we are done because by increasing a we can reduce the inequality to an endpoint case, which is already verified. Ψ is increasing, and since $x\tau\le 1$, function $\tau\mapsto\Psi\left(\frac{\tau-x}{\sqrt{1-\tau^2}}\right)$, $\tau\in[f,1]$ is increasing as a composition of two increasing functions. Here we have used the fact that

$$\frac{\partial}{\partial\tau}\left(\frac{\tau-x}{\sqrt{1-\tau^2}}\right)=\frac{1-\tau x}{(1-\tau^2)^{3/2}}.$$

To check the monotonicity of the map $\tau\mapsto\frac{\tau-\left(\frac{2\Psi(x)}{\Psi(1)}-x\right)}{\sqrt{1-\tau^2}}$, it is enough to verify that $\tau\left(\frac{2\Psi(x)}{\Psi(1)}-x\right)\le 1$. The latter inequality follows from the following two simple inequalities:

$$\Psi(x)\ge\frac{\Psi(1)x}{2},\quad 0\le x\le 1, \tag{5.5.24}$$

$$\left(\frac{x+\sqrt{2-x^2}}{2}\right)\left(\frac{2\Psi(x)}{\Psi(1)}-x\right)\le 1,\quad 0\le x\le 1. \tag{5.5.25}$$

Indeed, to verify (5.5.24) notice that

$$\frac{d}{dx}\frac{\Psi(x)}{x}=\frac{x\Phi(x)-\Psi(x)}{x^2}=-\frac{e^{-\frac{x^2}{2}}}{x^2}<0, \tag{5.5.26}$$

and therefore, $\frac{\Psi(x)}{x}\ge\Psi(1)\ge\frac{\Psi(1)}{2}$.

To verify (5.5.25), it is enough to show that

$$\frac{\Psi(x)}{\Psi(1)x}\le\frac{1}{x^2+x\sqrt{2-x^2}}+\frac{1}{2},\quad x\in[0,1].$$

If $x = 1$ we have equality. Taking derivative of the mapping $x \to \frac{\Psi(x)}{\Psi(1)x} - \frac{1}{x^2+x\sqrt{2-x^2}} - \frac{1}{2}$ in x, we obtain

$$\frac{2}{x^2}\left(-\frac{e^{-\frac{x^2}{2}}}{2\Psi(1)} + \frac{x + \frac{1-x^2}{\sqrt{2-x^2}}}{(x+\sqrt{2-x^2})^2}\right) \geq 0.$$

To prove the last inequality, it is the same as to show that $\frac{\sqrt{2-x^2}+x(2-x^2)}{x\sqrt{2-x^2}+1-x^2} \leq \Psi(1)e^{\frac{x^2}{2}}$. For the exponential function, we use the estimate $e^{\frac{x^2}{2}} \geq 1 + \frac{x^2}{2}$. We estimate $\sqrt{2-x^2}$ from above in the numerator by $\sqrt{2}(1 - \frac{x^2}{4})$, and we estimate $\sqrt{2-x^2}$ from below in the denominator by $(1-\sqrt{2})(x-1)+1$ (as $x \to \sqrt{2-x^2}$ is concave). Thus, it would be enough to prove that

$$\frac{\sqrt{2}(1-\frac{x^2}{4}) + x(2-x^2)}{\sqrt{2}x(1-x)+1} \leq \Psi(1)\left(1+\frac{x^2}{2}\right), \quad 0 \leq x \leq 1.$$

If we further use the estimates $\Psi(1) \geq \frac{29}{28}$ and $\frac{41}{29} \leq \sqrt{2} \leq \frac{17}{12}$ (for denominator and numerator correspondingly), then the last inequality would follow from

$$\frac{29}{240} \cdot \frac{246x^4 - 486x^3 + 233x^2 - 12x - 8}{29 + 41x - 41x^2} \leq 0.$$

The denominator has the positive sign. The negativity of $246x^4 - 486x^3 + 233x^2 - 12x - 8 \leq 0$ for $0 \leq x \leq 1$ follows from the Sturm's algorithm, which shows that the polynomial does not have roots on $[0, 1]$. Since at point $x = 0$ it is negative, it is negative on the whole interval. □

5.5.3 Finding B

Since it is easy to verify that B satisfies the range condition $B(x, \lambda) \leq \max\{|f|, \sqrt{\lambda}\}$, we have then that B is a subsolution of (5.5.17), and so, by Theorem 5.5.1

$$B \leq \mathbf{B}.$$

Now we want to prove the opposite inequality

$$\mathbf{B} \leq B. \tag{5.5.27}$$

LEMMA 5.5.3 *Let even functions* $\mathbf{b}$ *and* b *defined on* $[-1, 1]$ *satisfy* $\mathbf{b}(1) = b(1) = 1$, *and* $b'' + xb' - b = 0$, $b \in C^2$, $\mathbf{b}$ *being a convex function such that* $\mathbf{b}'' + x\mathbf{b}' - \mathbf{b} \geq 0$ *on* $(-1, 1)$ *in the sense of distributions. Then* $\mathbf{b} \leq b$.

PROOF If b were in C^2 as well, then this would be very easy. In fact, consider $a(x) \stackrel{\text{def}}{=} \mathbf{b}(x) - b(x)$. At endpoints, it is zero, and $a'' + xa' - a \geq 0$. Assume that function a is strictly positive somewhere, then it should have a maximum, where it is positive. Let it be x_0. Then $a(x_0) > 0, a'(x_0) = 0$. So $a''(x_0) \geq a(x_0) > 0$. Then x_0 cannot be maximum, so we come to a contradiction.

If b is not C^2, we still consider $a(x) \stackrel{\text{def}}{=} \mathbf{b}(x) - b(x)$, which is still a continuous function on $[-1, 1]$ equal to 0 at the endpoints. If it is positive somewhere, it should have a positive maximum, let s_0 be a point of maximum.

Since b is assumed to be convex, function a' is of bounded variation, and as such it is the sum of f and g, where f is a continuous function and g is a jump function. Notice that all jumps are positive, as they came only from b and g is continuous everywhere except the countable set of jump points.

As a' is a function of bounded variation, it has one-sided limits at any interior point. Let $a'(s_0\pm)$ be right and left limits correspondingly. Since all the jumps are positive, we have

$$a'(s_0+) \geq a(s_0-).$$

But s_0 is a point of maximum of a, so $a'(s_0-) \geq 0$, $a'(s_0+) \leq 0$. All these inequalities may happen only if $a'(s_0+) = a'(s_0-) = 0$. But this means that s_0 is not a jump point.

By continuity at s_0, a' is small near s_0, but $a(s_0) > 0$, so we can choose a small neighborhood of s_0, where $|sa'(s)| < \frac{1}{2}a(s)$.

Since $a'' + sa' - a \geq 0$, in this neighborhood of s_0, we have

$$a'' \geq a - sa' > \tfrac{1}{2}a \geq 0$$

in the sense of distributions. But a convex function cannot have maximum strictly inside an interval. We come to a contradiction.

Lemma is proved. □

We found the Bellman function $\mathbf{B}$, the formula is given in the following theorem.

THEOREM 5.5.4

$$\mathbf{B}(x, \lambda) = \begin{cases} \sqrt{\lambda}\frac{\Psi\left(\frac{|x|}{\sqrt{\lambda}}\right)}{\Psi(1)}, & x^2 \leq \lambda, \\ |x|, & x^2 \geq \lambda. \end{cases} \tag{5.5.28}$$

Let us introduce an obstacle function defined on $\mathbb{R}^2$.

$$O(x,\lambda) \stackrel{\text{def}}{=} \begin{cases} |x|, & x^2 \geq \lambda \\ \infty, & x^2 < \lambda. \end{cases} \tag{5.5.29}$$

THEOREM 5.5.5 *Function* $\mathbf{B}$ *is the largest function satisfying the finite difference inequality* (5.5.17) *such that it is majorized by the obstacle function* $O(x,\lambda)$*:*

$$\mathbf{B}(x,\lambda) \leq O(x,\lambda). \tag{5.5.30}$$

Moreover,

$$\mathbf{B}(x,\lambda) = \max\left(\sqrt{\lambda}\frac{\Psi(\frac{x}{\sqrt{\lambda}})}{\Psi(1)}, |x|\right). \tag{5.5.31}$$

PROOF We already proved in Theorem 5.5.4 that this $\mathbf{B}$ is the largest function satisfying (5.5.17) and having the obstacle:

$$\mathbf{B}(x,\lambda) \leq \max(|x|, \sqrt{\lambda}).$$

Obstacle $O(x,\lambda)$ in (5.5.30) is bigger than $\max(|x|,\sqrt{\lambda})$, and in principle it might have had a bigger "envelope" $\mathcal{B}$ than $\mathbf{B}$. However, notice that if $\mathcal{B}$ satisfies (5.5.17), then by choosing $a^2 = \lambda$, $y = x$, we get

$$2\mathcal{B}(x,\lambda) \leq \mathcal{B}(x-\sqrt{\lambda},0) + \mathcal{B}(+\sqrt{\lambda},0) \leq |x-\sqrt{\lambda}| + |x+\sqrt{\lambda}| = 2\sqrt{\lambda}$$

when $|x| \leq \sqrt{\lambda}$. Thus, smaller obstacle function $\max(|x|,\sqrt{\lambda})$ automatically majorizes $\mathcal{B}$. We conclude that the classes of solutions of (5.5.17) majorized by O and majorized by a smaller function $\max(|x|,\sqrt{\lambda})$ coincide.

To prove (5.5.31), it is sufficient to consider $\lambda = 1$. In this case, we need to prove that $\Psi(x) \geq \Psi(1)x$ for $0 \leq x \leq 1$ and that $\Psi(x) \leq \Psi(1)x$ for $x \geq 1$. But inequality (5.5.26) guarantees that function $\Psi(x)/x$ is decreasing. □

We need the following local version of the latter theorem.

THEOREM 5.5.6 *Let a function* $V(x,\lambda)$ *defined in the strip* $\Pi = \{(x,\lambda)\colon x \in \mathbb{R},\ 0 \leq \lambda \leq 1\}$ *satisfy* (5.5.17)*, and let* O *from* (5.5.29) *be its majorant:* $V \leq O$*. Then*

$$\mathbf{B} \geq V \qquad (x,\lambda) \in \Pi, \tag{5.5.32}$$

where $\mathbf{B}$ *is the function from* (5.5.31).

PROOF Let us consider a family of functions, parametrized by T, $T \geq 1$, defined in the strip of width T as follows:

$$V_T(x,\lambda) \stackrel{\text{def}}{=} \sup_{t \geq T} \sqrt{t}\, V\left(\frac{x}{\sqrt{t}}, \frac{\lambda}{t}\right), \qquad 0 \leq \lambda \leq T.$$

It is easy to see that every function under the supremum sign satisfies (5.5.17). Also (5.5.17) is preserved under the operation of taking the supremum. Therefore, V_T satisfies main inequality (5.5.17). It obviously satisfies also $V_T \leq O$.

Make now one more transform:

$$W(x,\lambda) \stackrel{\text{def}}{=} \sup_{T \geq 1} V_T(x,\lambda).$$

The function W is defined already in the whole plane and satisfies there the main inequality (5.5.17), also $W \leq O$. By the construction, $V \leq W$ on the strip Π. Since $\mathbf{B}$ from (5.5.31) is the largest function majorized by O and satisfying (5.5.17), we conclude $\mathbf{B} \geq W \geq V$, and so (5.5.32) is proved. □

5.6 The Weak Norm of the Square Function

The following result is a corollary of Theorem 5.5.4.

THEOREM 5.6.1

$$\{t \in (0,1) \colon S^2\varphi(t) \geq \lambda\} \leq \Psi(1)\frac{\|\varphi\|_1}{\sqrt{\lambda}}, \tag{5.6.1}$$

and the constant $\Psi(1)$ is sharp, which means that it is the norm of the operator S from L^1 to $L^{1,\infty}$.

Put

$$\mathcal{F}(x,\lambda) \stackrel{\text{def}}{=} \{\varphi \in L^1(0,1) \colon \int_0^1 \varphi\, dt = x,\ \ S^2\varphi(t) \geq \lambda\ a.e.\ t \in (0,1)\}.$$

By definition of $\mathbf{B}$ and by Theorem 5.5.4, we have for any function $\varphi \in \mathcal{F}(x,\lambda)$:

$$|\{t \in (0,1) \colon S^2\varphi(t) \geq \lambda\} = 1 \leq \frac{\|\varphi\|_1}{\mathbf{B}(x,\lambda)}.$$

However, we have more. Again by definition of $\mathbf{B}$ and by Theorem 5.5.4, for any $\varepsilon > 0$, we have a function $\varphi_\varepsilon \in \mathcal{F}(0,\lambda)$ such that

$$|\{t \in (0,1)\colon S^2\varphi(t) \ge \lambda\}| = 1 \ge \frac{\|\varphi_\varepsilon\|_1 - \varepsilon}{\mathbf{B}(0,\lambda)} = \frac{1}{\mathbf{B}(0,1)} \frac{\|\varphi_\varepsilon\|_1 - \varepsilon}{\sqrt{\lambda}}.$$

Of course, this means that

$$\Psi(1) = \mathbf{B}(0,1)^{-1} \le \|S\colon L^1 \to L^{1,\infty}\|. \tag{5.6.2}$$

Now we are going to prove the opposite estimate

$$\|S\colon L^1 \to L^{1,\infty}\| \le \Psi(1). \tag{5.6.3}$$

Let us consider the following function, obtained from function B (5.5.16) by the change of variables:

$$U(p,q) \stackrel{\text{def}}{=} 1 - \Psi(1)B(p, 1-q^2). \tag{5.6.4}$$

We assumed here that function $B(x,\lambda)$ is extended to the half-plane $\lambda \le 0$ by formula $B(x,\lambda) = |x|$.

THEOREM 5.6.2 *Function U introduced above is the smallest function majorizing the obstacle*

$$O_2(p,q) = \mathbf{1}_{p^2+q^2\ge 1} - \Psi(1)|p|$$

such that U satisfies the main inequality

$$2U(p,q) \ge U(p-a, \sqrt{q^2+a^2}) + U(p+a, \sqrt{q^2+a^2}). \tag{5.6.5}$$

PROOF The fact that U satisfies (5.6.5) follows immediately because B was proved to satisfy (5.5.17); see Lemma 5.5.2.

Let us prove the obstacle claim: $U \ge O_2$. Let $\lambda = 1 - q^2$. Then by (5.5.17), we know that $B(p, 1-q^2) \le |p|$ if $p^2 + q^2 \ge 1$, and so

$$U(p,q) \ge 1 - \Psi(1)|p|, \qquad \text{if } p^2 + q^2 \ge 1.$$

So to see that $U(p,q) \ge \mathbf{1}_{p^2+q^2\ge 1}(p,q) - \Psi(1)|p|$, one needs to check this inequality for $p^2 + q^2 \le 1$, where it is equivalent to

$$\Psi(1)B(p, 1-q^2) \le \Psi(1)|p| + 1. \tag{5.6.6}$$

But

$$B(0, 1-q^2) = \sqrt{1-q^2}\frac{\Psi(\frac{0}{\sqrt{1-q^2}})}{\Psi(1)} = \sqrt{1-q^2}\frac{1}{\Psi(1)}.$$

And

$$B(\sqrt{1-q^2}, 1-q^2) = \sqrt{1-q^2}.$$

Function $p \to B(p, 1-q^2)$ is convex for a fixed q. Thus, if $p^2 + q^2 \le 1$, then

$$B(p, 1-q^2) \le \frac{1}{\Psi(1)}\left(1 - \frac{|p|}{\sqrt{1-q^2}}\right)\sqrt{1-q^2} + \frac{|p|}{\sqrt{1-q^2}}\sqrt{1-q^2} \le \frac{1}{\Psi(1)} + |p|,$$

which is (5.6.6).

Function U is obtained by the change of variable in function B. So, Theorem 5.5.6 shows that function U from (5.6.4) is the smallest majorizing O_2. □

Now we can prove (5.6.3). In fact, using Theorems 5.4.6 and 5.6.2, we obtain

$$|\{x \in (0,1) \colon \varphi^2(x) + S^2\varphi(x) \ge 1\} \le \Psi(1)\|\varphi\|_1. \tag{5.6.7}$$

This, of course, gives us (5.6.3). But it gives us also a bit more. Consider the sublinear operator

$$T\varphi \stackrel{\text{def}}{=} (\varphi^2 + S^2\varphi)^{1/2}.$$

Pointwise, $T\varphi$ is bigger than $S\varphi$. Hence, its weak norm is at least the weak norm of S. Combining (5.6.2) and (5.6.7), we conclude that these norms are equal, i.e., we proved the following theorem.

THEOREM 5.6.3

$$\|T \colon L^1 \to L^{1,\infty}\| = \|S \colon L^1 \to L^{1,\infty}\| = \Psi(1).$$

REMARK 5.6.4 The smaller the obstacle, the smaller can be the smallest function majorizing this obstacle and satisfying main inequality (5.6.5). But sometimes, relaxing the obstacle does not help. For example, the functions U, which we get for two obstacles are the same. This is illustrated in the next Theorem, where $O_1 \le O_2$.

THEOREM 5.6.5 *Function* $U(p,q) = 1 - \Psi(1)B(p, 1-q^2)$ *introduced above is the smallest function majorizing the obstacle*

$$O_1(p,q) = \mathbf{1}_{q\ge 1} - \Psi(1)|p|$$

such that U *satisfies main inequality* (5.6.5).

PROOF The fact that $U(p,q) = 1 - \Psi(1)B(p, 1-q^2)$ is the smallest function majorizing $O_2 = \mathbf{1}_{p^2+q^2\ge 1} - \Psi(1)|p|$ follows from Theorem 5.6.2. So, of course, it majorizes a smaller obstacle $O_1 = \mathbf{1}_{q\ge 1} - \Psi(1)|p|$. But why is it the smallest one?

Suppose this is not the case, and O_1 is majorized by function $\tilde{U}$ such that $\tilde{U} \leq U$, with a strict inequality somewhere.

Let us then show that the main inequality (5.6.5) allows us to prove $\tilde{U} \geq O_2$. Then by Theorem 5.6.2, we have $\tilde{U} \geq U$, and we come to a contradiction.

In fact, fix a point (p, q) with $0 \leq q \leq 1$, $p \geq 0, p^2 + q^2 \geq 1$. Consider the points $(p-a, 1)$, $(p+a, 1)$ such that $1 = q^2 + a^2$, $a \geq 0$. Notice that $a \leq p$. Then (5.6.5) implies

$$\begin{aligned} 2\tilde{U}(p,q) &\geq \tilde{U}(p-a,1) + \tilde{U}(p+a,1) \\ &\geq O_1(p-a,1) + O_1(p+a,1) \\ &= 2 - \Psi(1)(|p-a| + p + a) = 2 - \Psi(1)(p-a+p+a) \\ &= 2 - 2\Psi(1)p = 2 - 2\Psi(1)|p| = 2O_2(p,q). \end{aligned}$$

We are done. □

5.7 Saturation of Estimates by Extremal Sequences

As it often happens in extremal problems, we can observe in our situation that a certain saturation of estimates takes place. Let us introduce a new Bellman function, which, on the first glance, is not the same as $\mathbf{B}$ from (5.5.6):

$$\mathbf{B}_=(x,\lambda) \stackrel{\text{def}}{=} \inf\{\|\varphi\|_{L^1(J)} : S_J^2\varphi(x) = \lambda \text{ for a. e. } x \in J\}.$$

Some of the obvious properties of $\mathbf{B}_=$ are

- Domain: $\Omega_{\mathbf{B}_=} \stackrel{\text{def}}{=} \{(x,\lambda) : f \in \mathbb{R};\ \lambda \geq 0\}$;
- $\mathbf{B}_=$ does not depend on interval J;
- $\mathbf{B}_=$ is increasing in λ and even in f;
- Homogeneity: $\mathbf{B}_=(tf, t^2\lambda) = |t|\mathbf{B}(x,\lambda)$;
- Range/Obstacle Condition: $|f| \leq \mathbf{B}_=(x,\lambda) \leq \max\{|f|, \sqrt{\lambda}\}$;
- Main Inequality:

$$2\mathbf{B}_=(x,\lambda) \leq \mathbf{B}_=(x-a,\lambda-a^2) + \mathbf{B}_=(x+a,\lambda-a^2), \quad \forall\, |a| \leq \sqrt{\lambda}. \tag{5.7.1}$$

- $\mathbf{B}$ is convex in x and even, and so it is easy to see that $\mathbf{B}_=$ is minimal at $x = 0$:

$$\mathbf{B}_=(0,\lambda) \leq \mathbf{B}_=(x,\lambda), \quad \forall\, x, \tag{5.7.2}$$

REMARK 5.7.1 All the properties above are obvious except, maybe, the main inequality. We leave the proof of it to the reader as an exercise.

Notice that these properties repeat verbatim those of function $\mathbf{B}$. However, formally, these two functions seem to be different. In fact, they are the same.

THEOREM 5.7.2

$$\mathbf{B}_{=} = \mathbf{B}.$$

PROOF By definition $\mathbf{B} \leq \mathbf{B}_{=}$. But we saw in Theorem 5.5.1 that any function satisfying (5.7.1) and obstacle/range assumption above is majorized by $\mathbf{B}$. Hence, $\mathbf{B}_{=} = \mathbf{B}$. □

Now we prove a "duality" theorem. Since nothing depends on the choice of interval J, we shall work with $J = (0,1)$.

THEOREM 5.7.3 *Let $q \leq 1$ then*

$$\sup\{|\{x \in (0,1)\colon S^2\varphi \geq 1 - q^2\}| - \Psi(1)\|\varphi\|_1, \int_0^1 \varphi = p\}$$
$$= 1 - \Psi(1) \inf\{\|\varphi\|_1\colon S^2\varphi = 1 - q^2 \ a.\,e.\,(0,1), \int_0^1 \varphi = p\}.$$

PROOF Consider function $O_2(p,q) = \mathbf{1}_{p^2+q^2 \geq 1} - \Psi(1)|p|$. By Theorem 5.4.5, the smallest function U satisfying (5.6.5) and majorizing O_2 is given by the formula

$$U(p,q) = \sup\{|\{x \in (0,1)\colon S^2\varphi \geq 1 - q^2\}| - \Psi(1)\|\varphi\|_1, \int_0^1 \varphi = p\}. \tag{5.7.3}$$

On the other hand, Theorem 5.6.2 claims that this function U is given by

$$U(p,q) = 1 - \Psi(1)\mathbf{B}(p, 1 - q^2). \tag{5.7.4}$$

We also know from Theorem 5.7.2 that

$$\mathbf{B}(p, 1-q^2) = \inf\{\|\varphi\|_1\colon S^2\varphi = 1 - q^2 \ a.\,e.\,(0,1)\}.$$

Combining this fact with (5.7.3) and (5.7.4), we get the claim of Theorem 5.7.3. □

5.8 An Obstacle Problem Associated with the Chang–Wilson–Wolff Theorem

Consider an obstacle function

$$O \stackrel{\text{def}}{=} O_3(p, q; \lambda) = \mathbf{1}_{[\lambda,\infty)}(p)\mathbf{1}_{[0,1]}(q). \tag{5.8.1}$$

The heat envelope of this O, or, in other words, the smallest function U satisfying main inequality (5.6.5) and majorizing this O is

$$U(p, q; \lambda) = \sup\left\{\int_0^1 \mathbf{1}_{[\lambda,\infty)}(\varphi(x)) \cdot \mathbf{1}_{[0,1]}(\sqrt{q^2 + S^2\varphi(x)})dx, \int_0^1 \varphi = p\right\}. \tag{5.8.2}$$

Consider the following function:

$$B(\lambda) \stackrel{\text{def}}{=} \sup\left\{\left|\{t \in (0,1) \colon \varphi(t) \geq \lambda\}\right| \colon \|S\varphi\|_\infty \leq 1, \int_0^1 \varphi = 0\right\}. \tag{5.8.3}$$

This function was considered in Section 1.9 (see also [**135**]) and it was partially found there. In particular, it was shown in Section 1.9 that there exists a positive constant c (whose exact value remains unknown to us) such that $B(\lambda) = c\Phi(\lambda)$ for all $\lambda \geq \sqrt{3}$ where Φ is the Gaussian "error function," i.e., $\Phi(\lambda) = \displaystyle\int_\lambda^\infty e^{-y^2/2}\,dy$.

Also in Section 1.9 of Chapter 1, function B was found for $\lambda \in (0,1)$. Briefly put, this means that we know B exactly for $\lambda \leq 1$, know it up to an absolute constant factor for $\lambda \geq \sqrt{3}$, and do not have any clear idea about what B may be between 1 and $\sqrt{3}$.

THEOREM 5.8.1 $B(\lambda) = U(0, 0; \lambda)$.

PROOF By definitions of functions U and B in (5.8.2) and (5.8.1), we see immediately that $B(\lambda) \leq U(0, 0; \lambda)$. In fact, if φ satisfies $\|S\varphi\|_{L^\infty(0,1)} \leq 1$ and $q = 0$, then $\sqrt{q^2 + S^2\varphi(t)} \leq 1$ a. e. on $(0, 1)$, and the second factor in (5.8.2) is 1. So the supremum involved in the definition of $U(0, 0; \lambda)$ is at least as big as the supremum involved in the definition of B.

Now consider φ, which almost gives us the supremum in the definition of $U(0, 0; \lambda)$ (gives us the supremum up to small constant η). The second factor $\mathbf{1}_{[0,1]}(S\varphi(t))$ will be present and can be just 0 making integrand smaller than $\mathbf{1}_{[\lambda,\infty)}(\varphi(t))$, and, thus, making integral smaller than $|\{t \in (0,1)\colon \varphi(t) \geq \lambda\}|$. Let us modify this φ in this case. Namely, let $\mathcal{F}$ be the collection of maximal dyadic subintervals of $(0,1)$ such that

$$S_J^2\varphi(t) \stackrel{\text{def}}{=} \sum_{I\colon J\in\mathcal{D}(I)} (\Delta_I\varphi)^2 \frac{\mathbf{1}_I(t)}{|I|} > 1.$$

Consider a new function

$$\psi(t) \stackrel{\text{def}}{=} \begin{cases} \varphi(t), & t \in (0,1) \setminus \bigcup\{J\colon J \in \mathcal{F}\} \\ \langle\varphi\rangle_J, & t \in J,\ J \in \mathcal{F}. \end{cases}$$

Then 1) $S^2\psi(t) \leq 1$ on $(0,1) \setminus \bigcup\{J\colon J \in \mathcal{F}\}$; 2) If $\hat{J}$ is a parent of $J \in \mathcal{F}$, we see that $S\psi(t) = S_{\hat{J}}\varphi(t)$ if $t \in J$, and by maximality of chosen intervals, we get that $S^2\psi(t) \leq 1$ on any $J \in \mathcal{F}$.

Hence, $S\psi \leq 1$ almost everywhere, and it is easy to see that we did not change the integral: $\int_0^1 \psi = \int_0^1 \varphi = 0$. So ψ participates in supremum that defines $B(\lambda)$.

On the other hand, as $\psi(t) = \varphi(t)$, where $S\varphi(t) \leq 1$. Therefore,

$$\mathbf{1}_{[\lambda,\infty)}(\varphi(t))\mathbf{1}_{[0,1]}(S\varphi(t)) = \mathbf{1}_{[\lambda,\infty)}(\psi(t))\mathbf{1}_{[0,1]}(S\varphi(t)) \leq \mathbf{1}_{[\lambda,\infty)}(\psi(t)).$$

Taking the integral from zero to one, we conclude from this that

$$U(0,0;\lambda) - \eta \leq B(\lambda).$$

As $\eta > 0$ is arbitrary, we finished proving that $U(0,0;\lambda) = B(\lambda)$. □

We already mentioned that in Section 1.9, we proved that $\mathbf{B}(\lambda) = c\Phi(\lambda)$ for $\lambda \geq \sqrt{3}$, where $\Phi(\lambda) = \int_\lambda^\infty e^{-y^2/2}dy$. Here we deduce from Theorem 5.4.2 a simple but weaker estimate on $\mathbf{B}(\lambda)$ for large λ.

Notice that function $U(p,q) = e^{p-q^2/2}$ does satisfy the main inequality (5.6.5). Then by Theorem 5.4.2, we can justify the second inequality below for any test function φ

$$e^{-\|S\varphi\|_\infty^2/2} \int_0^1 e^\varphi dt \leq \int_0^1 e^{\varphi(t) - S^2\varphi(t)/2} dt \leq e^{\int_0^1 \varphi(t)dt} \leq 1, \tag{5.8.4}$$

if $\int_0^1 \varphi = 0$. In this case, we can replace φ by $\frac{\lambda}{\|S\varphi\|_\infty}\varphi$ to rewrite this inequality in the following form:

$$\int_0^1 e^{\lambda\varphi/\|S\varphi\|_\infty} dt \leq e^{\lambda^2/2}.$$

This of course gives

$$|\{t \in (0,1) \colon \varphi(t) \geq \lambda\}| \leq e^{-\lambda^2/2}, \ \forall \varphi \colon \|S\varphi\|_\infty \leq 1. \tag{5.8.5}$$

So

$$B(\lambda) \leq e^{-\lambda^2/2}.$$

However, by the formula $B(\lambda) = c\Phi(\lambda)$, $\lambda \geq \sqrt{3}$ from Section 1.9, which we noted before, function B decreases faster at infinity, even though $c > 1$. It would be important for certain questions in probability theory to find B for all λ.

5.9 Strong Weighted Estimate of the Square Function

We use an approach quite different from that in [**71**], where such sharp estimate was first obtained. Notice that more general sharp weighted estimates for square function in $L^p(w)$ were obtained by Lerner [**105**], [**109**].

We would like to find the best functions C_1, C_2 such that

$$\frac{1}{|J|}\sum_{I\in D(J)} |\Delta_I w^{-1}|^2 \langle w\rangle_I |I| \leq C_1([w]_{A_2})\langle w^{-1}\rangle_J. \tag{5.9.1}$$

$$\frac{1}{|J|}\sum_{I\in D(J)} |\Delta_I (\varphi w^{-1})|^2 \langle w\rangle_I |I| \leq C_2([w]_{A_2})\langle \varphi^2 w^{-1}\rangle_J. \tag{5.9.2}$$

Formulas for these best functions also can be obtained using our Bellman function approach below. But we will restrict our goal to finding just sharp estimates.

REMARK 5.9.1 We are working either with the operator $S\colon L^2(w) \to L^{2,\infty}(w)$ or with the operator $S\colon L^2(w) \to L^2(w)$. However, it is more convenient to work with isomorphic objects: $S_{w^{-1}}\colon L^2(w^{-1}) \to L^{2,\infty}(w)$ or $S_{w^{-1}}\colon L^2(w^{-1}) \to L^2(w)$, here $S_{w^{-1}}$ denotes the product $SM_{w^{-1}}$, where $M_{w^{-1}}$ is the operator of multiplication.

5.9.1 Sharp Estimate for C_1

DEFINITION 5.9.2 Recall that for a smooth function B of d real variables $(x_1, \ldots, x_d)$, we denote by $d^2B(x)$ the second differential form of B, namely,

$$d^2B(x) = (H_B(x)dx, dx)_{\mathbb{R}^d},$$

where vector $dx = (dx_1, \dots, dx_d)$ is an arbitrary vector in $\mathbb{R}^d$, and $H_B(x)$ is the $d \times d$ matrix of the second derivatives of B (Hessian matrix) at point $x \in \mathbb{R}^d$.

For brevity, we write A_2 in this section, but we mean the dyadic class A_2^d.

THEOREM 5.9.3 *$C_1([w]_{A_2}) \le A[w]_{A_2}^2$ and the power 2 cannot be improved.*

We introduce the following function of $x \in \mathbb{R}^2$

$$\mathbf{B}(x) \stackrel{\text{def}}{=} \mathbf{B}(x;Q) \stackrel{\text{def}}{=} \sup \frac{1}{|J|} \sum_{I \in D(J)} |\Delta_I w^{-1}|^2 \langle w \rangle_I |I|, \tag{5.9.3}$$

where supremum is taken over all $w \in A_2$, $[w]_{A_2} \le Q$, such that

$$\langle w \rangle_J = x_1, \langle w^{-1} \rangle_J = x_2.$$

Recall that by scaling argument, our function does not depend on J but depends on $Q = [w]_{A_2}$. Notice also that function $\mathbf{B}$ is defined in the domain Ω_Q, where

$$\Omega_Q \stackrel{\text{def}}{=} \{x \in \mathbb{R}^2 : 1 \le x_1 x_2 \le Q\}.$$

Function $\mathbf{B}_Q$ is the Bellman function of our problem. In particular, it is very easy to observe that to prove the estimate in Theorem 5.9.3 is equivalent to proving $\mathbf{B}(x;Q) \le AQ^2 x_2$, and the sharpness in Theorem 5.9.3 is just the claim that $\sup_{x \in \Omega_Q} \frac{1}{x_2} \mathbf{B}(x;Q) \ge cQ^2 > 0$.

REMARK 5.9.4 Therefore, Theorem 5.9.3 can be proved by finding the explicit formula for $\mathbf{B}$. To do that, we obviously need to solve an infinite dimensional optimization problem of finding the (almost) best possible $w \in A_2$ (the reader is reminded that it is a dyadic class) such that $\langle w \rangle_J = x_1, \langle w^{-1} \rangle_J = x_2$. This can be done, but here we adapt a slightly different approach to proving Theorem 5.9.3.

PROOF Below, A, a are positive absolute constants. Instead of finding precisely $\mathbf{B}$, we will find another function $B = B(x;Q)$ such that the following properties are satisfied:

- B is defined in Ω_{4Q},
- $0 \le B(x) \le AQ^2 x_2$,
- $B(x) - \frac{B(x^+)+B(x^-)}{2} \ge a(x_2^+ - x_2^-)^2 x_1$,
  if $x \in \Omega_Q$, $\quad x^\pm = (x_1^\pm, x_2^\pm) \in \Omega_Q$, $\quad$ and $x = \frac{x^+ + x^-}{2}$,
- $B(x) = 0$ if $x \in \Omega_Q$ is such that $x_1 x_2 = 1$.

For example, it is not difficult to check that $\mathbf{B}(\cdot\,; 4Q)$ satisfies the first, the third, and the fourth properties. We leave this as an exercise to the reader.

The second property is not easy at all, it can be observed when the complicated formula for $\mathbf{B}$ is written down.

Here we will write down an explicit (and rather easy) form of *some* B that satisfies all four properties. This will not be $\mathbf{B}$. However, let us observe that if the existence of such a B is proved, the inequality in Theorem 5.9.3 gets proved. In fact, fix $J \in D(I)$, and introduce $x_J = (\langle w \rangle_I, \langle w^{-1} \rangle_J)$. Of course $\{x_J\}_{J \in D(I)}$ is a martingale. Now we compose this martingale with B (notice that $x_J \in \Omega_Q$, so $B(x_J)$ is well defined). The resulting object is not a martingale anymore, but it is a supermartingale, moreover, by the third property

$$\langle w \rangle_J |\Delta_J w^{-1}|^2 \, |J| \le |J| B(x_J) - |J_+| B(x_{J_+}) - |J_-| B(x_{J_-}).$$

Now the reader who went through Chapter 1 or 3 knows what happens next: we use the telescopic nature of the sum in the right-hand side to observe that the summation in all $J \in D(I)$ cancels all the terms except $|I| B(x_I)$, which, by the second property is at most $A\, Q^2 \langle w^{-1} \rangle_I |I|$. Hence, we obtained

$$\sum_{J \in D(I)} |\Delta_J w^{-1} J^2 \langle w \rangle_I |J| \le A\, Q^2 \langle w^{-1} \rangle_I |I|.$$

We leave as an exercise for the reader to explain, where we used the positivity of B in this reasoning.

The last estimate is precisely inequality (5.9.1), and Theorem 5.9.3 gets proved (apart from the sharpness) as soon as any function B as above is proved to exist.

The construction of a certain B with abovementioned three properties is split into two steps.

5.9.2 Step 1: The Reduction to Nonlinear Ordinary Differential Equation

First of all, we wish to find a smooth B in the domain Ω_Q such that the following quadratic forms inequality holds in Ω_Q:

$$-\tfrac{1}{2} d^2 B \ge x_1 (dx_2)^2. \tag{5.9.4}$$

We will be searching for homogeneous B: $B(\tau^{-1} x_1, \tau x_2) = \tau B(x)$. Choosing $\tau = x_1$, we get

$$B(x) = \frac{1}{x_1} f(x_1 x_2)$$

for $f(t) \stackrel{\text{def}}{=} B(1,t)$. Then (5.9.4) becomes

$$\begin{bmatrix} t^2 f'' - 2tf' + 2f, & tf'' \\ tf'', & f'' + 2 \end{bmatrix} \le 0.$$

Nonpositivity of this 2×2 matrix follows from nonpositivity of one of its corner elements and vanishing of its determinant. So it is enough to satisfy for all $t \in [1, Q]$:

$$\begin{aligned} & f'' + 2 \le 0, \\ & 2(f'' + 2)(-tf' + f) + 2t^2 f'' = 0. \end{aligned} \tag{5.9.5}$$

Since $f'' \le -2$, we can see from the equality above that $-tf' + f \le 0$. Put $g = f(t)/t$. Then we know that $-t^2 g' = f - tf' \le 0$, so g is increasing. Also $tg'' + 2g' = f'' \le -2$, hence $g'' \le 0$ as g was noticed to be increasing.

We have equation

$$t(-g'g'' + g'') - 2(g')^2 = 0.$$

This is a first-order nonlinear equation on $h \stackrel{\text{def}}{=} g'$ of which we know that $h \ge 0, h' \le 0$:

$$t(-hh' + h') - 2h^2 = 0.$$

We separate the variables and get

$$\frac{1-h}{h^2} h' = \frac{2}{t}. \tag{5.9.6}$$

We saw that $h' = g''$ is negative and t here is positive, so $h \ge 1$, and the condition $f - tf' \le 0$ is the same as $h \ge 0$. Thus, any solution $h \ge 1$ of (5.9.6) gives us the desired result.

We want to solve this for $t \in [1, Q]$:

$$-\log h - \frac{1}{h} = 2\log t + c \quad \Leftrightarrow \quad -\frac{1}{h} e^{-\frac{1}{h}} = -t^2 C, \quad C > 0.$$

Recall that Lambert W function (which is multivalued) solves the equation $z = W(z)e^{W(z)}$. Thus, we must have $W(-t^2 C) = -\frac{1}{h(t)}$. The condition $-1 \le -\frac{1}{h(t)} \le 0$ requires that $-1 \le W \le 0$, and since h is decreasing we conclude that W is increasing, and this defines the single-valued solution $W_0(y)$ on the interval $[-1/e, 0]$ such that $W_0(-1/e) = -1$, $W_0(0) = 0$ and $W_0(y)$ is increasing. So $h(t) = -\frac{1}{W_0(-t^2 C)}$. The condition $-1/e \le -t^2 C \le 0$ for $t \in [1, Q]$ gives us the range for constant C i.e., $0 < C \le \frac{1}{Q^2 e}$. Going back to the functions f and B, we obtain:

$$f(t) = -t \int_1^t \frac{dt}{W_0(-t^2 C)} + t f(1), \tag{5.9.7}$$

And thus,

$$B(x) = -x_2 \int_1^{x_1 x_2} \frac{dt}{W_0(-t^2 C)} + x_2 f(1). \tag{5.9.8}$$

Let us calculate the bound on B. As we have already seen, constant C is such that $0 < C \leq \frac{1}{Q^2 e}$, so the minimal B corresponds to the choice $C = \frac{1}{Q^2 e}$. Also by the fourth property of B on page 403, B has to be 0 if $x_1 x_2 = 1$. Hence $f(1) = 0$. Then

$$\begin{aligned} B^Q(x) &\stackrel{\text{def}}{=} x_2 \int_1^{x_1 x_2} \frac{dt}{W_0(-\frac{t^2}{Q^2 e})} \\ &\leq x_2 \int_1^{x_1 x_2} \frac{Q^2 e}{t^2} = eQ^2 x_2 \left(1 - \frac{1}{x_1 x_2}\right), \qquad 1 \leq x_1 x_2 \leq Q. \end{aligned} \tag{5.9.9}$$

Here we used the fact that Lambert function $W_0(t) \leq t$ for $t \in [-1/e, 0]$. Actually one can get better estimates of the integral above by using the series expansion for W_0, i.e.,

$$W_0(t) = \sum_{n=1}^{\infty} \frac{(-n)^{n-1}}{n!} t^n, \quad |t| < \frac{1}{e},$$

but this can change only the constant e in front of Q^2.

5.9.3 Step 2: From Infinitesimal Inequality on $d^2 B^Q$ to the Third Property on page 403 for B^Q

The function B^Q defined in (5.9.9) is not function B with four properties formulated on page 403 at the beginning of the proof of this theorem. However, let us prove that $B \stackrel{\text{def}}{=} B^{4Q}$ has all these four properties. The first and the fourth properties follow just by definition, and the second property is because we just proved in (5.9.9) that

$$B(x) = B^{4Q}(x) \leq 4eQ^2 x_2.$$

To prove the third property, let us fix

$$x = (x_1, x_2) \in \Omega_Q, \quad x^{\pm} = (x_1^{\pm}, x_2^{\pm}) \in \Omega_Q$$

such that $x = \frac{x^+ + x^-}{2}$. Introduce a vector function $X = (X_1, X_2)$ defined on $[-1, 1]$:

$$X_i(t) = \frac{1+t}{2} x_i^+ + \frac{1-t}{2} x_i^- .$$

Then $X(\pm 1) = x_\pm$, $X(0) = x$. Compose the vector function X and function $B = B^{4Q}$, namely, put

$$b(t) \stackrel{\text{def}}{=} B(X(t)).$$

It is important to notice that b is well defined because

$$\forall t \in [-1, 1] \;\; X(t) \in \Omega_{4Q}.$$

The latter is an elementary geometric observation saying that if three points $X_\pm, X$ belong to Ω_Q, and $X = \frac{X_+ + X_-}{2}$, then the whole segment with endpoints $X_\pm$ lie in Ω_{4Q}. Now we differentiate twice function b. The chain rule gives us immediately

$$b''(t) = \big(H_B(X(t))(x_+ - x_-), (x_+ - x_-)\big)_{\mathbb{R}^2},$$

where H_B denotes as always the Hessian matrix of function B. Therefore, the use of (5.9.4) gives us

$$-b''(t) \geq 2X_1(t)(x_2^+ - x_2^-)^2.$$

On the other hand,

$$B(x) - \frac{B(x_+) + B(x_-)}{2} = b(0) - \frac{b(1) + b(-1)}{2} = -\frac{1}{2} \int_{-1}^{1} b''(t)(1 - |t|)\, dt$$

Notice that the integrand is always nonnegative by the previous display formula. By the same formula, the integrand is at least $x_1 = X_1(0)$ for $t \in [0, 1/2]$ because on this interval, $X_1(t) \geq \frac{1}{2} X_1(0)$ by the obvious geometric reason. Combining that we obtain

$$B(x) - \frac{B(x_+) + B(x_-)}{2} \geq \frac{3}{8} x_1 (x_2^+ - x_2^-)^2.$$

We established all four properties for B, and we have already shown that this is enough to prove the inequality in Theorem 5.9.3. The sharpness is not difficult to see for a weight with one singular point, see **[71]** for example. □

5.9.4 Proving the Instance of $T1$ Theorem by Bellman Function Technique

For function C_2 introduced on page 402, we have the following inequality.

THEOREM 5.9.5 *$C_2 \leq AQ^2$, and the power 2 is the best possible.*

Let us deduce this result from Theorem 5.9.3. This an example of the so-called weighted $T1$ theorem.

We use the notation $h_{_I}$ for a standard Haar function supported on a dyadic interval I. Now consider the same type of Haar basis but in weighted $L^2(\sigma)$: functions $h^\sigma_{_I}$ are orthogonal to constants in $L^2(\sigma)$, normalized in $L^2(\sigma)$, assume constant value on each child of I, and are supported on I.

We need a couple of lemmas, the first one in fact is Lemma 4.2.5 from Chapter 3.

LEMMA 5.9.6 *The following holds $h_{_I} = \alpha^\sigma_I h^\sigma_{_I} + \beta^\sigma_I \frac{\mathbf{1}_I}{\sqrt{I}}$ with*

$$\alpha^\sigma_I = \frac{\langle\sigma\rangle^{1/2}_{_{I+}}\langle\sigma\rangle^{1/2}_{_{I-}}}{\langle\sigma\rangle^{1/2}_{_I}}, \quad \beta^\sigma_I = \frac{\langle\sigma\rangle_{_{I+}} - \langle\sigma\rangle_{_{I-}}}{\langle\sigma\rangle_{_I}} \tag{5.9.10}$$

PROOF This is a direct calculation, we need to define two constants $\alpha^\sigma_I, \beta^\sigma_I$ and we have two conditions: $\|h^\sigma_{_I}\|_{L^2(\sigma)} = 1$ and $(h^\sigma_{_I}, \mathbf{1})_{L^2(\sigma)} = 0$. □

The second lemma follows from the chain rule.

LEMMA 5.9.7 *Let $\Phi(x')$,$B(x'')$ be smooth functions of $x' = (x_0, x_1, \ldots, x_n)$, $x'' = (x_{n+1}, \ldots, x_m)$ correspondingly. Then the second differential form of the composition function*

$$\mathcal{B}(x_1, \ldots, x_n, x_{n+1}, \ldots, x_m) \stackrel{\text{def}}{=} \Phi(B(x_{n+1}, \ldots, x_m), x_1, \ldots, x_n)$$

is given by the following formula:

$$d^2\mathcal{B} = d^2_B\Phi + \frac{\partial\Phi}{\partial x_0} d^2 B.$$

REMARK 5.9.8 We understand the left-hand side as $(H_{\mathcal{B}}(x)dx, dx)_{\mathbb{R}^m}$, where

$$x \stackrel{\text{def}}{=} (x_1, \ldots, x_n, x_{n+1}, \ldots, x_m),$$
$$dx = (dx_1, \ldots, dx_n, dx_{n+1}, \ldots, dx_m).$$

We understand $d^2_B\Phi$ in the right-hand side as

$$(H_\Phi(x_1, \ldots, x_n, B(x''))dy, dy)_{\mathbb{R}^{n+1}},$$

where

$$dy = (dx_1, \ldots, dx_n, dB), \quad dB \stackrel{\text{def}}{=} \nabla B(x'') \cdot dx''.$$

Now we are ready to prove Theorem 5.9.5.

The proof of Theorem 5.9.5 The sum we want to estimate in (5.9.2)

$$\frac{1}{|I|}\sum_{J\in\mathcal{D}(I)}(\langle\varphi w^{-1}\rangle_{J+}-\langle\varphi w^{-1}\rangle_{J-})^2\langle w\rangle_J|J|$$

is of course

$$\Sigma=\frac{2}{|I|}\sum_{J\in D(I)}(\varphi w^{-1},h_J)^2\langle w\rangle_J.$$

We can plug the decomposition of Lemma 5.9.6 with $\sigma=w^{-1}$ and take into account that for dyadic lattice obviously $\alpha_J^\sigma\le 2\langle\sigma\rangle_J^{1/2}$. Then we obtain

$$\Sigma\le\frac{8}{|I|}\sum_{J\in D(I)}(\varphi w^{-1},h_J^{w^{-1}})^2\langle w^{-1}\rangle_I\langle w\rangle_J$$
$$+\frac{2}{|I|}\sum_{J\in D(I)}\langle\varphi w^{-1}\rangle_J^2(\beta_J^{w^{-1}})^2\langle w\rangle_J|J|\stackrel{\text{def}}{=}\Sigma_1+\Sigma_2.$$

The system $\{h_J^{w^{-1}}\}_{J\in D(I)}$ is orthonormal in $L^2(w^{-1})$, and $\langle w^{-1}\rangle_J\langle w\rangle_J\le Q$. Hence, immediately we have

$$\Sigma_1\le 8Q\|\varphi\|^2_{L^2(w^{-1})}.$$

Let $\langle\cdot\rangle_{J,w^{-1}}$ mean the average with respect to measure $\mu\stackrel{\text{def}}{=}w^{-1}(x)dx$ and

$$\gamma_J\stackrel{\text{def}}{=}\left(\langle w^{-1}\rangle_{J+}-\langle w^{-1}\rangle_{J-}\right)^2\langle w\rangle_J.$$

We are left to estimate Σ_2. To do that let us rewrite Σ_2:

$$\Sigma_2=\frac{2}{|I|}\sum_{J\in D(I)}\left(\frac{\langle\varphi w^{-1}\rangle_J}{\langle w^{-1}\rangle_J}\right)^2\gamma_J|J|=\frac{2}{|I|}\sum_{J\in D(I)}\left(\langle\varphi w^{-1}\rangle_{J,w^{-1}}\right)^2\gamma_J|J|,$$

Let $\sigma\in A_2$ with $Q=[\sigma]_{A_2}$. We are going to prove now that with some absolute constant A

$$\frac{1}{|I|}\sum_{J\in D(I)}\left(\langle\varphi\sigma\rangle_{J,\sigma}\right)^2\gamma_J|J|\le AQ^2\langle\varphi^2\sigma\rangle_I. \tag{5.9.11}$$

To finish the proof of Theorem 5.9.5, we need to check (5.9.11).

To prove (5.9.11), we construct a special function of four real variables $\mathcal{B}(x)=\mathcal{B}(x;Q)$, $x=(x_1,x_2,x_3,x_4)$ that possesses the following properties:

(1) $\mathcal{B}$ is defined in a non-convex domain Ω_{4Q}, where

$$\Omega_Q\stackrel{\text{def}}{=}\{(x_1,x_2,x_3,x_4)\in\mathbb{R}^4_+: x_3^2\le x_4x_2,\ 1\le x_1x_2\le Q\},$$

(2) $0 \le \mathcal{B} \le x_4$,
(3) if $x, x_\pm$ belong to Ω_Q, and $x = \frac{x_+ + x_-}{2}$,

$$\mathcal{B}(x) - \frac{\mathcal{B}(x_+) + \mathcal{B}(x_-)}{2} \ge aQ^{-2}\frac{f^2}{v^2}x_1(x_2^+ - x_2^-)^2.$$

As soon as such a function is constructed, estimate (5.9.11) and Theorem 5.9.5 follow immediately. In fact, we repeat our telescopic consideration. We set the vector martingale $x_I \stackrel{\text{def}}{=} (x_1^I, x_2^I, x_3^I, x_4^I)$, where

$$x_1^I = \langle\sigma^{-1}\rangle_I,\ x_2^I = \langle\sigma\rangle_I,\ x_3^I = \langle\varphi\sigma\rangle_I,\ x_4^I = \langle\varphi^2\sigma\rangle_I.$$

It is obvious that vector martingale $\{x_I\}_{I\in D(J)}$ is always inside Ω_Q, and so the superposition of this martingale and $\mathcal{B}$ is well defined, so we consider $\mathcal{B}(x_I)$. Then, the property 3) claims that $\{\mathcal{B}(x_I)\}_{I\in D(J)}$ is a supermartingale, and moreover,

$$|J|\frac{(x_3^J)^2}{(x_2^J)^2}x_1^J(x_2^{J_+} - x_2^{J_-})^2 \le AQ^2(|J|\mathcal{B}(x_J) - |J_+|\mathcal{B}(x_{J_+}) - |J_+|\mathcal{B}(x_{J_-})).$$

We use the telescopic nature of the term in the right-hand side, and summing these terms for all $J \in \mathcal{D}(I)$, we then notice that all of them will get cancelled, except $AQ^2|I|\mathcal{B}(x_I)$, which is bounded by $AQ^2|I|x_4^I = AQ^2|I|\langle\varphi^2\sigma\rangle_I$. So we just proved (5.9.11) provided that the existence of $\mathcal{B}$ is validated.

Now we will write the explicit formula for $\mathcal{B}$. Exactly as in Theorem 5.9.3 we first construct, by an explicit formula, an auxiliary function $\mathcal{B}^Q$. Here it is

$$\mathcal{B}^Q(x) \stackrel{\text{def}}{=} x_4 - \frac{x_3^2}{x_2 + aQ^{-2}B^Q(x)}, \tag{5.9.12}$$

where B^Q was defined in (5.9.9). It is clear that it satisfies property 2) on page 410. It "almost" satisfies property 1), but it is defined only in Ω_Q, not in a larger domain Ω_{4Q}. As to the property 3), it does satisfy its infinitesimal version: at any point $x \in \Omega_Q$, we have

$$-d^2B^Q \ge aQ^{-2}\frac{x_3^2}{x_2}x_1(dx_2)^2. \tag{5.9.13}$$

Let us prove (5.9.13). Consider

$$\Phi(s, x_2, x_3, x_4) \stackrel{\text{def}}{=} x_4 - \frac{x_3^2}{x_2 + s}.$$

By direct calculation, one can see that it is concave in $\mathbb{R}_+^4$, so $d^2\Phi \le 0$. Hence, we know that $d^2_{B^Q}\Phi \le 0$. Now we see that

$$\mathcal{B}^Q(x) = \Phi(aQ^{-2}B^Q(x), x_2, x_3, x_4),$$

and by Lemma 5.9.7

$$d^2\mathcal{B}^Q = d^2_{B^Q}\Phi + aQ^{-2}\frac{\partial\Phi}{\partial s}\cdot(d^2B^Q) \le \frac{ax_3^2}{Q^2(x_2+aQ^{-2}B^Q(x))^2}(d^2B^Q). \tag{5.9.14}$$

But in Theorem 5.9.3 we proved that $B^Q \le 16eQ^2x_2$; hence, by choosing $a = \frac{1}{16e}$ we guarantee that $x_2 + aQ^{-2}B^Q(x) \le 2x_2$. We also proved in Theorem 5.9.3 that $-d^2B^Q \ge x_1(dx_2)^2$. Combining these facts with inequality (5.9.14), we obtain

$$-d^2\mathcal{B}^Q \ge \frac{1}{64Q^2}\frac{x_3^2}{x_2^2}x_1(dx_2)^2,$$

which is precisely (5.9.13).

We need a function $\mathcal{B}$ with property 3) and defined in the domain Ω_{4Q}. So let us put $\mathcal{B} \stackrel{\text{def}}{=} \mathcal{B}^{4Q}$. Exactly as on page 406, we can prove now that not only (5.9.13) holds with $a = \frac{1}{64e}$, but also we have with some small absolute positive a_0

$$\mathcal{B}(x) - \frac{\mathcal{B}(x_+)+\mathcal{B}(x_-)}{2} \ge a_0Q^{-2}\frac{x_3^2}{x_2^2}x_1(x_2^+ - x_2^-)^2, \tag{5.9.15}$$

for all triple of points x, x_+, x_- from Ω_Q such that $x = \frac{x_+ + x_-}{2}$. □

Remark 5.9.9 The reader should pay attention to the following formula:

$$\mathcal{B}(x) = x_4 - \frac{x_3^2}{x_2 + aQ^{-2}B_Q(x)}, \tag{5.9.16}$$

By formula (5.9.16), we transfer the claim of Theorem 5.9.3 to Theorem 5.9.5. A Bellman function of Theorem 5.9.3 was used as "a lego piece" to construct a Bellman function for Theorem 5.9.5. Joining "the lego pieces construction" is a frequent tool for building complicated Bellman functions from "simple" pieces. A good example of that approach can be found in [**130**]. In the current section we had another instance of this "lego construction." This consruction proved for us the $T1$ theorem for the weighted square function operator.

5.10 Weak Weighted Estimates for the Square Function

As in Section 5.1.1, our measure space will be $(X, \mathfrak{A}, dx)$, where σ-algebra $\mathfrak{A}$ is generated by a standard dyadic filtration $\mathcal{D} = \cup_k\mathcal{D}_k$ on $\mathbb{R}$. We considered the weak weighted estimate for the martingale transform. In Section 5.1.1 of this chapter, the endpoint exponent was $p = 1$, and critical weights belong to A_1 class.

Now we are going to consider weak weighted estimates for the dyadic square function. The endpoint exponent is now $p = 2$. The weak estimate of the square function in $L^2(w)$, $w \in A_1$ is well known, see, e.g., [**200**]. The critical weights belong now to A_2 class. For subcritical exponents ($1 < p < 2$) and supercritical ones ($p > 2$) see [**101**], [**51**], and [**98**].

The fact that the endpoint exponent is now at $p = 2$ can be explained by the bilinear nature of square function transform.

Recall that the symbol $\mathrm{ch}(J)$ denotes the dyadic children of J. Recall that the martingale difference operator Δ_J was defined as follows:

$$\Delta_J f \stackrel{\text{def}}{=} \sum_{\ell \in \mathrm{ch}(J)} \mathbf{1}_\ell (\langle f \rangle_\ell - \langle f \rangle_J).$$

For our case of dyadic lattice on the line, we have that $|\Delta_J f|$ is constant on J, and

$$\Delta_J f = \frac{1}{2}[(\langle f \rangle_{J_+} - \langle f \rangle_{J_-})\mathbf{1}_{J_+} + (\langle f \rangle_{J_-} - \langle f \rangle_{J_+})\mathbf{1}_{J_-}].$$

The square function operator is

$$Sf(t) = \left(\sum_{J \in \mathcal{D}} |\Delta_J f|^2 \mathbf{1}_J(t) \right)^{1/2}.$$

In this section, we work only with the dyadic A_2 classes of weights, but we skip the word dyadic, because we consider here only dyadic operators. We consider a positive function $w(x)$, and as before we call it A_2 weight if

$$Q \stackrel{\text{def}}{=} [w]_{A_2} \stackrel{\text{def}}{=} \sup_{J \in \mathcal{D}} \langle w \rangle_J \langle w^{-1} \rangle_J < \infty. \tag{5.10.1}$$

We are going to consider restricted weak estimate, when the operator is applied to $\varphi = \mathbf{1}_E$, but only for set E being itself a dyadic interval. We are interested in the following estimate:

$$\frac{1}{|I|} w \left\{ t \in I \colon \sum_{J \in \mathcal{D}(I)} |\Delta_J w^{-1}|^2 \mathbf{1}_J(t) > \lambda \right\} \leq C_{1w} \frac{\langle w^{-1} \rangle_I}{\lambda}. \tag{5.10.2}$$

Namely, we are interested in understanding the dependence of C_{1w} on $[w]_{A_2}$. We also provide some information on the dependence on $[w]_{A_2}$ of the constant in the full weak estimate for the square function operator:

$$\frac{1}{|I|} w \left\{ t \in I \colon \sum_{J \in D(I)} |\Delta_J (\varphi w^{-1})|^2 \mathbf{1}_J(t) > \lambda \right\} \leq C_{2w} \frac{\langle \varphi^2 w^{-1} \rangle_I}{\lambda}. \tag{5.10.3}$$

Here φ runs over all functions such that $\operatorname{supp} \varphi \subset I$ and $\varphi \in L^2(I, w)$, $w \in A_2$.

We wish to compare these estimates for the weak type to the similar estimates of strong type: (5.9.1), (5.9.2) from Section 5.9.

Of course, there are several obvious estimates: 1) weak constants are smaller than strong constants:

$$C_{1w}([w]_{A_2}) \leq C_1([w]_{A_2}), \qquad C_{2w}([w]_{A_2}) \leq C_2([w]_{A_2})$$

just by Chebyshev inequality; also 2) test function estimates are trivially at least as good as the full estimates:

$$C_{1w}([w]_{A_2}) \leq C_{2w}([w]_{A_2}), \qquad C_1([w]_{A_2}) \leq C_2([w]_{A_2}).$$

Above we proved a converse inequality:

$$C_2([w]_{A_2}) \leq A(C_1([w]_{A_2}) + [w]_{A_2}). \tag{5.10.4}$$

REMARK 5.10.1 Sharp weak type estimates for square function operator are quite difficult. The dependence on $[w]_{A_2}$ drops dramatically with respect to the strong weighted estimates, for which sharp dependence on $[w]_{A_2}$ was found in Section 5.9 of the current chapter. This is in big contrast with what happens for singular integrals T of Calderón–Zygmund type. We know that for $w \in A_2$, the strong norm $\|T \colon L^2(w) \to L^2(w)\|$ is equivalent to

$$\|T \colon L^2(w) \to L^{2,\infty}(w)\| + \|T^* : L^2(w^{-1}) \to L^{2,\infty}(w^{-1})\|,$$

see **[76, 149]** for example. This is completely false for square function operator.

REMARK 5.10.2 The reader can see that (5.10.2) is a particular case of (5.10.3) for a special choice of test function $\varphi = \mathbf{1}_I$. The same remark holds for (5.9.1) and (5.9.2) of Section 5.9. We already noted above that test estimate (5.9.1) (plus its symmetric one, where w is replaced by w^{-1}) implies the strong estimate (5.9.2). This is the essence of the weighted $T1$ theorem (testing condition theorem in the terminology of E. Sawyer).

Unfortunately there is no $T1$ principle for weak type estimates. See papers [**1, 35, 37**].

REMARK 5.10.3 Let M_a be an operator of multiplication by function a. Notice $\varphi \to \varphi w^{-1}$ is the isometry between $L^2(w^{-1})$ and $L^2(w)$. Hence, by using this isometry, one reduces the problem of estimating the norm of the square function operator $S\colon L^2(w) \to L^{2,\infty}(w)$ to the estimate of the norm of $S_{w^{-1}} \stackrel{\text{def}}{=} SM_{w^{-1}}$ from $L^2(w^{-1})$ to $L^{2,\infty}(w)$.

5.10.1 Sharp Dependence on $[w]_{A_2}$ in Weak Testing Estimate

THEOREM 5.10.4 $aQ \le \sup\{C_{1w}(Q)\colon [w]_{A_2} \le Q\} \le AQ$.

We will prove the upper estimate, the lower one is well known just for one-point-singularity weights, see e.g., [**51, 98, 101**].

As always in this section, A_2 means dyadic A_2. We introduce the following function of three real variables x_1, x_2, λ:

$$\mathbf{B}(x,\lambda;Q) \stackrel{\text{def}}{=} \sup \frac{1}{|I|} w \left\{ t \in I\colon \sum_{J \in D(I)} |\Delta_J w^{-1}|^2 \mathbf{1}_J(t) > \lambda \right\}, \tag{5.10.5}$$

where the supremum is taken over all $w \in A_2$, $[w]_{A_2} \le Q$ such that

$$x_1 = \langle w \rangle_I, \quad x_2 = \langle w^{-1} \rangle_I.$$

Notice that by scaling argument, our function does not depend on I but depends on $Q = [w]_{A_2}$.

REMARK 5.10.5 Ideally, we want to find the formula for this function. Notice that this is similar to solving a problem of "isoperimetric" type, where the solution of certain nonlinear partial differential equation is a common tool; extensive literature can be found, e.g., in the bibliographies of [**10, 11, 83, 84**].

5.10.1.1 Properties of B *and the Main Inequality.*

(1) B is defined in

$$\Omega \stackrel{\text{def}}{=} \Omega_Q \stackrel{\text{def}}{=} \{(x_1,x_2,\lambda)\colon 1 \le x_1 x_2 \le Q, x_i > 0,\ 0 \le \lambda < \infty\}.$$

(2) If three points $P = (x,\lambda)$, $P_\pm = (x^\pm, \lambda^\pm)$ belong to Ω, and $x = \frac{1}{2}(x^+ + x^-)$, $\lambda = \min(\lambda^+, \lambda^-)$, then the main inequality holds

$$\mathbf{B}\left(x, \lambda + (x_2^+ - x_2^-)^2\right) - \frac{\mathbf{B}(P_+) + \mathbf{B}(P_-)}{2} \ge 0.$$

(3) $\mathbf{B}$ is decreasing in λ.
(4) Homogeneity: $\mathbf{B}(x_1\tau, x_2/\tau, \lambda/\tau^2) = \tau\mathbf{B}(x_1, x_2, \lambda)$, $\tau > 0$.
(5) Obstacle condition: for all points (x, λ) such that

$$10 \leq x_1 x_2 \leq Q, \qquad 0 \leq \lambda \leq \delta\, x_2^2,$$

for a positive absolute constant δ, one has $\mathbf{B}(x, \lambda) = x_1$.
(6) The boundary condition $\mathbf{B}(x, \lambda) = 0$ if $x_1 x_2 = 1$.
(7) By scale invariance, $\mathbf{B}$ does not depend on interval I.

REMARK 5.10.6 We will not be using this fact, but it is not difficult to see that

$$\mathbf{B}(x, \lambda) = x_1, \qquad \text{if } \lambda < \frac{4x_2}{x_1}(x_1 x_2 - 1).$$

All the properties above are very simple consequences of the definition of $\mathbf{B}$. However, let us explain a little about the second and the fifth properties. The second property is the consequence of the scale invariance of $\mathbf{B}$. We consider data $P_\pm$ and find weights $w_\pm$ that almost supremizes $\mathbf{B}(P_\pm)$. By property (7) of $\mathbf{B}$, we can think that $w_\pm$ lives on $I_\pm$.

The next step is to consider the concatenation of w_+ and w_-, we call it w_c, it is supported on I and is defined as usual:

$$w(t) \stackrel{\text{def}}{=} \begin{cases} w_+(t), & \text{on } t \in I_+ \\ w_-(t), & \text{on } t \in I_-. \end{cases}$$

Clearly, this new weight is a competitor for giving the supremum for date P on I. But it is only a competitor, the real supremum in (5.10.5) is bigger. This implies the second property above (*the main inequality*).

Now let us explain the fifth property above, we call it *the obstacle condition*. Let us consider a special weight w in I: it is one constant on I_- and just another constant on I_+. Moreover, we wish to have $\langle w^{-1}\rangle_{I_+} = 4\langle w^{-1}\rangle_{I_-}$. Notice that then $\delta\langle w^{-1}\rangle_I \leq |\Delta_J w^{-1}|$ with $\delta = \frac{3}{5}$.

Now it is obvious that if $\lambda \leq \delta^2\langle w^{-1}\rangle_I^2$, then

$$\{t \in J\colon S^2_{w^{-1}}(\mathbf{1}_I) \geq \lambda\} = I,$$

and therefore,

$$\frac{1}{|I|} w\{t \in J\colon S^2_{w^{-1}}(\mathbf{1}_I) \geq \lambda\} = \langle w\rangle_I.$$

Notice now that w is just one admissible weight, and that we have to take supremum over all such admissible weights.

We get the fifth property above (= the obstacle condition): $\mathbf{B}(x, \lambda) = x_1$ for those points (x, λ) in the domain of definition of $\mathbf{B}$, where the corresponding

w with $\langle w\rangle_I = x_1, \langle w^{-1}\rangle_I = x_2$ exists. It is obvious that for all sufficiently large Q and for any pair (x_1, x_2) such that $10 \le x_1x_2 \le Q$, one can construct a just "two-valued" w as above with $[w]_{A_2} \le Q$ (we recall that we deal only with dyadic A_2 weights).

Notice that the main inequality above transforms into a partial differential inequality if considered infinitesimally (and if we tacitly assume that $\mathbf{B}$ is smooth):

$$-\frac{1}{2}d^2_{x_1,x_2}\mathbf{B} + \frac{\partial \mathbf{B}}{\partial \lambda}(dx_2)^2 \ge 0. \tag{5.10.6}$$

We are not going to find $\mathbf{B}$ defined in (5.10.5). But instead we will construct smooth $\mathcal{B}$ that satisfies all the properties above except for two of them: (2) and (6). But instead of satisfying property (2) function $\mathcal{B}$ will satisfy a slightly weaker inequality

$$\mathcal{B}\left(x, \lambda + c(x_2^+ - x_2^-)^2\right) - \frac{\mathcal{B}(P_+) + \mathcal{B}(P_-)}{2} \ge 0. \tag{5.10.7}$$

with some positive constant c. Its infinitesimal version looks as follows:

$$-\frac{1}{2}d^2_{x_1,x_2}\mathcal{B} + c\frac{\partial \mathcal{B}}{\partial \lambda}(dx_2)^2 \ge 0. \tag{5.10.8}$$

The function $\mathcal{B}$ will satisfy even slightly stronger properties, for example, the obstacle condition (the fifth property) will be satisfied with 1 instead of 10:

$$\mathcal{B}(x, \lambda) = x_1, \qquad \forall (x, \lambda) \in \Omega_Q : \lambda \le \delta\, x_2^2. \tag{5.10.9}$$

Here δ is some positive absolute constant (it will not depend on Q).

Using our usual telescopic sums consideration it will be very easy to prove the following.

THEOREM 5.10.7 *Suppose there exists a smooth function $\mathcal{B}$ on Ω_Q satisfying the main inequality* (5.10.8) *and the obstacle condition* (5.10.9). *And suppose $\mathcal{B}$ also satisfies*

$$\mathcal{B}(x, \lambda) \le A\, Q \frac{x_2}{\lambda}. \tag{5.10.10}$$

Then the constant C_{1w} in (5.10.2) *is at most $A\,Q$.*

PROOF It is enough to prove the theorem for w that is constant on dyadic intervals of $\mathcal{D}_N(I)$ with estimates that are uniform in N.

For any $J \in \mathcal{D}(I)$, we put

$$\lambda(J_\pm) \stackrel{\text{def}}{=} \lambda - c \sum_{\substack{L:\, L \in \mathcal{D}(J) \\ I \subset L}} |\Delta_I w^{-1}|^2 \mathbf{1}_L .$$

Notice that $J_\pm \in \mathcal{D}_N(I)$ satisfy $\lambda(J_\pm) > 0$ if and only if $(Sw^{-1})^2 > \lambda$ on their dyadic father J. Also the latter condition is equivalent to the following statement about this father:

$$\lambda(J) < c(\langle w^{-1}\rangle_{J_+} - \langle w^{-1}\rangle_{J_-})^2. \tag{5.10.11}$$

Let us call by $\mathcal{I}$ the collection of intervals $J \in \mathcal{D}_{N-1}(I)$ with the property (5.10.11).

Now combine (5.10.11) with an inequality

$$c(\langle w^{-1}\rangle_{J_+} - \langle w^{-1}\rangle_{J_-})^2 \le \delta \langle w^{-1}\rangle_J^2,$$

which is obvious if, e.g., $\delta = 4c$. Hence, we use the obstacle condition (5.10.9) to conclude that on intervals $J \in \mathcal{I}$ we have

$$\mathcal{B}(\langle w\rangle_J, \langle w^{-1}\rangle_J, \lambda(J)) = \langle w\rangle_J . \tag{5.10.12}$$

Next we use (5.10.10) and the main inequality (5.10.8), then we can write

$$\begin{aligned} \frac{AQ\langle w^{-1}\rangle_I}{\lambda}|I| &\ge \mathcal{B}(\langle w\rangle_I, \langle w^{-1}\rangle_I, \lambda)|I| \\ &\ge |I_+|\mathcal{B}(\langle w\rangle_{I_+}, \langle w^{-1}\rangle_{I_+}, \lambda(I_+)) \\ &\quad + |I_-|\mathcal{B}(\langle w\rangle_{I_-}, \langle w^{-1}\rangle_{I_-}, \lambda(I_-)) \end{aligned} \tag{5.10.13}$$

We can now apply the main inequality to the both terms in the right-hand side. We do this repeatedly for $N-1$ times.

After that, we come to the following inequality

$$\begin{aligned} \frac{AQ\langle w^{-1}\rangle_I}{\lambda}|I| &\ge \mathcal{B}(\langle w\rangle_I, \langle w^{-1}\rangle_I, \lambda)|I| \\ &\ge \sum_{J \in \mathcal{D}_{N-1}(I)} |J|\mathcal{B}(\langle w\rangle_J, \langle w^{-1}\rangle_J, \lambda(J)). \end{aligned} \tag{5.10.14}$$

Now combine (5.10.12) and (5.10.14) (and also the fact that $\mathcal{B} \ge 0$):

$$\sum_{J \in \mathcal{I}} |J|\langle w^{-1}\rangle_J \le \frac{AQ\langle w^{-1}\rangle_I}{\lambda}|I|. \tag{5.10.15}$$

Therefore, we proved

$$\frac{1}{|J|}w\left\{t \in J\colon \sum_{I \in D(J)} |\Delta_I w^{-1}|^2 \chi_I(t) > \lambda\right\} \le AQ\frac{\langle w^{-1}\rangle_J}{\lambda},$$

which is (5.10.2). □

5.10.1.2 Formula for the Function $\mathcal{B}$: Monge–Ampère Equation with a Drift. The homogeneity of $\mathcal{B}$ allows us to write it in the form

$$\mathcal{B}(x,\lambda) = \frac{1}{\sqrt{\lambda}}\Theta\left(x_1\sqrt{\lambda},\ \frac{x_2}{\sqrt{\lambda}}\right), \quad (x,\lambda) \in \Omega_Q. \tag{5.10.16}$$

for a certain function Θ.

Let us choose Θ to be

$$\Theta(\gamma,\tau) \stackrel{\text{def}}{=} \min\left(\gamma,\ Qe^{-\tau^2/2}\int_0^\tau e^{s^2/2}ds\right).$$

We are going to show that then function $\mathcal{B}$ given by (5.10.16) satisfies all assumption of Theorem 5.10.7 for this choice of Θ.

Let

$$H \stackrel{\text{def}}{=} \{(\gamma,\tau) > 0\colon 1 \le \gamma\tau \le Q\}.$$

If $(x,\lambda) \in \Omega_Q$, then $(x_1\sqrt{\lambda},\ \frac{x_2}{\sqrt{\lambda}}) \in H$.

In terms of Θ, the main inequality (5.10.8) with $c = \frac{1}{8}$ turns into

$$\frac{1}{\sqrt{1+\frac{(\Delta\tau)^2}{8}}}\Theta\left(\sqrt{1+\frac{(\Delta\tau)^2}{8}}\,\frac{\gamma_- + \gamma_+}{2},\ \frac{1}{\sqrt{1+\frac{(\Delta\tau)^2}{8}}}\,\frac{\tau_- + \tau_+}{2}\right) \ge \frac{\Theta(\gamma_-,\tau_-) + \Theta(\gamma_+,\tau_+)}{2}, \tag{5.10.17}$$

where $(\gamma_-,\tau_-),(\gamma_+,\tau_+) \in H, 0 < \tau_- < \tau_+, \Delta\tau \stackrel{\text{def}}{=} \tau_+ - \tau_-$.

Since we try to find a smooth function Θ, the infinitesimal version of (5.10.17) appears. It is a sort of Monge–Ampère relationship with a drift. Namely, using Taylor expansion up to the terms of the second order with respect to $\Delta\gamma \stackrel{\text{def}}{=} \gamma_+ - \gamma_-$, $\Delta\tau$, we notice that all terms of order less than two get cancelled. The terms of the second order give us the following matrix inequality:

$$\begin{bmatrix} \Theta_{\gamma\gamma}, & \Theta_{\gamma\tau} \\ \Theta_{\gamma\tau}, & \Theta_{\tau\tau} + \Theta + \tau\Theta_\tau - \gamma\Theta_\gamma \end{bmatrix} \leq 0. \tag{5.10.18}$$

This is a very interesting condition because it can be interpreted geometrically.

DEFINITION 5.10.8 We call a smooth function Θ given in the domain $\Omega \subset \mathbb{R}^2$ *tacitly concave* if there exists a second-order ordinary linear differential operator $\mathcal{L} = \frac{\partial^2}{\partial\tau^2} + a(\tau)\frac{\partial}{\partial\tau} + b(\tau)$ such that given any solution ϕ, $\mathcal{L}\phi = 0$, function $\gamma = \phi_\Theta(\tau)$, where $\phi_\Theta(\tau) \stackrel{\text{def}}{=} \Theta(\phi(\tau), \tau)$, is the supersolution, namely, $\mathcal{L}\phi_\Theta \leq 0$.

The geometric meaning is the following: the family of curves $\gamma = \phi(\tau)$, such that $\mathcal{L}\phi = 0$, depends on two arbitrary constants, hence the family of such curves passing through a given point (τ_0, γ_0) generically points in any direction. Suppose (this is not always the case), that the family of curves $\gamma = \phi(\tau)$ such that $\mathcal{L}\phi = 0$ can be "straightened up," meaning that there exists the change of variables

$$\begin{cases} T = T(\tau, \gamma), \\ \Gamma = \Gamma(\tau, \gamma), \end{cases} \tag{5.10.19}$$

such that the family of curves mentioned above becomes the family of straight lines $\Gamma = \Gamma(T)$, $\Gamma'' = 0$. Then we can apply this change of variable to Θ and notice that functions $\Gamma = \Gamma(\tau, \Theta(\gamma, \tau))$, $\gamma = \phi(\tau)$, are concave as functions of T.

This means that we changed the variables in such a way that the new function

$$\Psi(T, \Gamma) \stackrel{\text{def}}{=} \Gamma(\tau(T, \Gamma), \Theta(\gamma(T, \Gamma), \tau(T, \Gamma)))$$

is concave in new coordinates (T, Γ).

LEMMA 5.10.9 *Let*

$$M_\Theta \stackrel{\text{def}}{=} \begin{bmatrix} \Theta_{\gamma\gamma}, & \Theta_{\gamma\tau} \\ \Theta_{\gamma\tau}, & \Theta_{\tau\tau} + b(\tau)\Theta + a(\tau)\Theta_\tau - b(\tau)\gamma\Theta_\gamma \end{bmatrix} \leq 0. \tag{5.10.20}$$

Function Θ satisfying (5.10.20) *is tacitly concave with respect to*

$$\mathcal{L} = \frac{\partial^2}{\partial\tau^2} + a(\tau)\frac{\partial}{\partial\tau} + b(\tau).$$

Namely, the following holds: On any curve $\gamma = \phi(\tau)$ lying in the domain H and such that $\mathcal{L}\phi = 0$ the following holds:

$$\mathcal{L}\phi_\Theta \leq 0, \tag{5.10.21}$$

where $\phi_\Theta(\tau) = \Theta(\phi(\tau), \tau)$. *Moreover, if the family of curves* $\gamma = \phi(\tau)$, $\mathcal{L}\phi = 0$ *passes through an arbitrary point in an arbitrary direction, then* (5.10.20) *is equivalent to* (5.10.21) *to hold for any* ϕ *such that* $\mathcal{L}\phi = 0$.

Proof Let ϕ be a smooth function of τ, and $(\phi(\tau), \tau)$ lies in H for $\tau \in I$. It is easy to check that

$$\mathcal{L}\phi_\Theta - \Theta_\gamma(\phi, \tau)\mathcal{L}\phi = (M_\Theta(\phi, \tau)e_\phi, e_\phi),$$

where

$$e_\phi \stackrel{\text{def}}{=} \begin{bmatrix} \phi'(\tau) \\ 1 \end{bmatrix}.$$

Therefore, since $\mathcal{L}\phi = 0$, we see that

$$\mathcal{L}\phi_\Theta = (M_\Theta e_\phi, e_\phi) \tag{5.10.22}$$

and, thus, $\mathcal{L}\phi_\Theta \le 0$ if $M_\Theta \le 0$. Inequality (5.10.21) is proved.

But if the family of curves $\gamma = \phi(\tau)$, $\mathcal{L}\phi = 0$, passes through an arbitrary point in an arbitrary direction, the negativity of matrix M_Θ follows. □

In particular, any function Θ satisfying (5.10.18) is tacitly concave with respect to $\mathcal{L} = \frac{\partial^2}{\partial\tau^2} + \tau\frac{\partial}{\partial\tau} + 1$.

This hints at a possibility to have a change of variables

$$(\gamma, \tau) \to (\Gamma, T)$$

such that condition (5.10.18) transforms to a simple concavity. First we need the following simple lemma.

Lemma 5.10.10 *Consider the following change of variable:* $T = \int_0^\tau e^{s^2/2} ds$. *Then* $\phi''(\tau) + \tau\phi'(\tau) + \phi(\tau) \le 0$ *if and only if* $(e^{\tau^2/2}\phi(\tau))_{TT} \le 0$ *and* $\phi''(\tau) + \tau\phi'(\tau) + \phi(\tau) = 0$ *if and only if* $(e^{\tau^2/2}\phi(\tau))_{TT} = 0$.

Proof The proof is a direct differentiation. □

This lemma reveals that the right change of variable as in (5.10.19) should look like follows:

$$\begin{cases} \Gamma \stackrel{\text{def}}{=} \gamma e^{\tau^2/2}, \\ T \stackrel{\text{def}}{=} \int_0^\tau e^{s^2/2} ds, \qquad (\gamma, \tau) \in \mathbb{R}^1_+ \times \mathbb{R}^1_+. \end{cases} \tag{5.10.23}$$

In the new coordinates, the family of curves $\gamma = \phi(\tau)$ such that $\phi'' + \tau\phi' + \phi = 0$ becomes a family of straight lines $\Gamma = CT + D$. (Notice that both families depend on two arbitrary constants.)

The following holds true:

$$\begin{aligned}&\left(e^{\tau^2/2}\Theta(\phi(\tau),\tau)\right)_{TT} \le 0,\\ &\gamma=\phi(\tau)\,,\ \phi''+\tau\phi'+\phi=0, \qquad (\gamma,\tau)\in G.\end{aligned} \tag{5.10.24}$$

which is the concavity of $e^{\tau^2/2}\Theta(\gamma,\tau)$ in a new coordinate T along the line $\Gamma = CT + D$. Let us rewrite two functions in the new coordinates:

$$\Phi(\Gamma,T) \stackrel{\text{def}}{=} \Theta(\gamma,\tau)\,,\ U(T) \stackrel{\text{def}}{=} e^{\tau^2/2}.$$

Denote

$$O \stackrel{\text{def}}{=} \{(\Gamma,T)\colon (\gamma,\tau)\in H\}.$$

Then (5.10.24) transforms into

$$\forall C,D\in\mathbb{R}, \qquad (U(T)\Phi(CT+D,T))_{TT}\le 0. \tag{5.10.25}$$

This is just a concavity of $U(T)\Phi(\Gamma,T)$ on O of course. Notice that neither H nor O are convex, so we should understand (5.10.25) as a local concavity in O: just the negativity of its second differential form

$$d^2_{\Gamma,T}\big(U(T)\Phi(\Gamma,T)\big)\le 0, \qquad (\Gamma,T)\in O.$$

So we reduce the question to finding a concave function in new coordinates. Now we choose the following concave function:

$$U(T)\Phi(\Gamma,T) \stackrel{\text{def}}{=} \min(\Gamma, KT),$$

where the constant $K = K(Q)$ will be chosen momentarily. We recall to the reader that function Θ has to satisfy a certain obstacle condition.

If we write down now $\Theta(\gamma,\tau) = \Phi(\Gamma,T)$ in the old coordinates, we get exactly function Θ from (5.10.16) (we need to define constant K yet), namely,

$$\Theta(\gamma,\tau) \stackrel{\text{def}}{=} \min\left(\gamma, Ke^{-\tau^2/2}\int_0^\tau e^{s^2/2}ds\right). \tag{5.10.26}$$

Recall that we put

$$\mathcal{B}(x,\lambda) = \frac{1}{\sqrt{\lambda}}\Theta(x_1\sqrt{\lambda}, \frac{x_2}{\sqrt{\lambda}}) \tag{5.10.27}$$

and we are going to apply Theorem 5.10.7 to it. But we need to choose K to satisfy all the assumptions of Theorem 5.10.7.

First of all, it is now very easy to understand why the form of the domain $H = \{1 \le \gamma\tau \le Q\}$ plays the role. In fact, by choosing

$$K = AQ$$

with some absolute constant A, we guarantee that in this domain our function Θ satisfies the obstacle condition

$$\Theta(\gamma, \tau) = \gamma \text{ as soon as } \tau \geq a_0 > 0, \tag{5.10.28}$$

where a_0 is an absolute positive constant. In fact, for all sufficiently small τ, $e^{-\tau^2/2} \int_0^\tau e^{s^2/2} ds \asymp \tau$, and therefore, for all sufficiently small τ (smaller than a certain absolute constant)

$$\Theta(\gamma, \tau) \stackrel{\text{def}}{=} \min(\gamma, K\tau).$$

The obstacle condition (5.10.9) requires that $\min(\gamma, K\tau) = \gamma$ if $\tau \geq a_0 > 0$. But on the upper hyperbola $\gamma = Q/a_0$ for $\tau = a_0$. Hence, at this point on the upper hyperbola, we have to have $\min(\gamma, K\tau) = \min(Q/a_0, Ka_0) = Q/a_0$. We see that the smallest possible K we can choose to satisfy the obstacle condition is $K \asymp Q$.

Secondly, function Θ satisfies the infinitesimal condition (5.10.18) by construction. But we need to check that the main inequality (5.10.17) is satisfied as well.

This can be done by the following simple lemma.

LEMMA 5.10.11 *Inequality* (5.10.17) *for function* Θ *built above holds if the following inequality is satisfied for* $\varphi(\tau) \stackrel{\text{def}}{=} e^{-\tau^2/2} \int_0^\tau e^{s^2/2} ds$:

$$\frac{1}{\sqrt{1 + \frac{(\Delta\tau)^2}{8}}} \varphi\left(\frac{\tau_1 + \tau_2}{2\sqrt{1 + \frac{(\Delta\tau)^2}{8}}}\right) \geq \frac{\varphi(\tau_1) + \varphi(\tau_2)}{2}, \qquad \forall 0 < \tau_1 \leq \tau_2 \leq \tau_0, \tag{5.10.29}$$

with some absolute positive small constant τ_0.

PROOF Lemma is easy because we can immediately see that the main inequality (5.10.17) is stable under the operation of taking minimum. □

As soon as (5.10.29) is checked, Theorem 5.10.7 gets proved. There are many ways to prove (5.10.29); we choose the proof that imitates (with some changes) the proof of Barthe and Maurey of the similar statement, see [**11**].

PROOF Consider the new function

$$U(p, q) \stackrel{\text{def}}{=} \frac{1}{q} \varphi\left(\frac{p}{q}\right), \tag{5.10.30}$$

given in the domain $\{(p,q)\colon p \geq 0, 0 \leq \frac{p}{q} \leq \tau_0\}$. Here τ_0 is a small positive number to be chosen below. Then (5.10.29) follows from

$$U(p,q) \geq \tfrac{1}{2}U\left(p+a, \sqrt{q^2 - \frac{a^2}{100}}\right) + \tfrac{1}{2}U\left(p-a, \sqrt{q^2 - \frac{a^2}{100}}\right). \tag{5.10.31}$$

Notice that we need to prove it only for

$$a \leq p \leq \tau_0 q, \tag{5.10.32}$$

and so

$$\sqrt{q^2 - \frac{a^2}{50}} \Big/ \sqrt{q^2 - \frac{a^2}{100}} \leq 0.9. \tag{5.10.33}$$

Notice also that infinitesimally (5.10.29) and (5.10.31) are satisfied, this is very easy to see because φ is strictly concave on small interval $[0, \tau_0]$.

Without loss of generality assume $a \geq 0$. Consider the process

$$X_t = U\left(p + W(t), \sqrt{q^2 - \frac{t}{50}}\right), \quad 0 \leq t \leq q^2/4.$$

Here $W(t)$ is the standard Brownian motion starting at zero. The infinitesimal version of (5.10.31) shows that

$$\frac{1}{2}U_{pp} - \frac{1}{100}\frac{U_q}{q} \leq 0.$$

It follows from Itô's formula together with the last observation that X_t is a supermartingale. Let τ be the stopping time

$$\tau = \frac{q^2}{4} \wedge \inf\{t \geq 0\colon W(t) \notin (-a,a)\}.$$

It follows from the fact that X_t is a supermartingale that

$$\begin{aligned}
U(p,q) &= X_0 \geq \mathbb{E}X_\tau = \mathbb{E}U(p + W(\tau), \sqrt{q^2 - \tau/50}) \\
&= P(W(\tau) = -a)\mathbb{E}(U(p-a, \sqrt{q^2 - \tau/50})|W(\tau) = -a) \\
&\quad + P(W(\tau) = a)\mathbb{E}(U(p+a, \sqrt{q^2 - \tau/50})|W(\tau) = a) \\
&\quad + P(|W(\tau)| < a, \tau = \frac{q^2}{4})\mathbb{E}(U(p+a, \sqrt{q^2 - \tau/50})|\, W(\tau)| = a).
\end{aligned}$$

Notice that the last probability is very small if τ_0 is chosen to be small. In fact, using (5.10.32), we see

$$\left(\frac{q^2}{4}\right)^2 P(|W(\tau)| < a) \le \left(\frac{q^2}{4}\right)^2 P\left(\tau = \frac{q^2}{4}\right)$$
$$\le \mathbb{E}\tau^4 \le C\mathbb{E}|W(\tau)|^4 \le Ca^4 \le C(\tau_0 q)^4.$$

Also clearly $P(W(\tau) = -a) = P(W(\tau) = -a)$, and by the last observation this probabilities are at least $\frac{1}{2.2}$. So we obtained the following:

$$\begin{aligned} 2.2U(p,q) &\ge \mathbb{E}\left(U(p-a, \sqrt{q^2 - \tau/50})|W(\tau) = -a\right) \\ &+ \mathbb{E}\left(U(p+a, \sqrt{q^2 - \tau/50})|W(\tau) = a\right) \\ &\ge U\left(p-a, \sqrt{q^2 - \mathbb{E}(\tau/50|W(\tau) = -a)}\right) \\ &+ U\left(p+a, \sqrt{q^2 - \mathbb{E}(\tau/50|W(\tau) = a)}\right) \\ &= U\left(p-a, \sqrt{q^2 - a^2/50}\right) + U\left(p+a, \sqrt{q^2 - a^2/50}\right). \end{aligned}$$

Notice that we have used $\mathbb{E}(\tau|W(\tau) = a) = \mathbb{E}(\tau|W(\tau) = -a) = a^2$, and the fact that the map $t \mapsto U(p, \sqrt{t})$, $t \in [q^2/4, q^2]$, is convex together with Jensen's inequality. The convexity follows from the fact that

$$t \to \frac{1}{\sqrt{t}}\varphi\left(\frac{p}{\sqrt{t}}\right)$$

is convex if $t \in [q^2/4, q^2]$. This is easy to check by direct calculation, putting $x = \frac{p}{\sqrt{t}}$ we get

$$\left[\frac{1}{\sqrt{t}}\varphi\left(\frac{p}{\sqrt{t}}\right)\right]''_{tt} = \frac{3}{4}\frac{1}{t^{5/2}}[(1 - 2x^2 - x^2(1-x^2))\varphi(x) + 2x - x^3] > 0,$$

if $x = \frac{p}{\sqrt{t}} \le 2\frac{p}{q} \le 2\tau_0$ is sufficiently small.

Now we use (5.10.33) to conclude that

$$\begin{aligned} 2.2U(p,q) &\ge U(p-a, \sqrt{q^2 - a^2/50}) + U(p+a, \sqrt{q^2 - a^2/50}) \\ &\ge 1.1[U(p-a, \sqrt{q^2 - a^2/100}) + U(p+a, \sqrt{q^2 - a^2/100})]. \end{aligned}$$

To obtain the last inequality, we use (5.10.30) and the following consideration. Put

$$x_1 = \frac{p \pm a}{\sqrt{q^2 - \frac{a^2}{100}}}, \quad x_2 = \frac{p \pm a}{\sqrt{q^2 - \frac{a^2}{50}}}$$

and notice that

$$\frac{\varphi(x_2)}{\varphi(x_1)} \geq 0.99 \geq 1.1 \times 0.9 \geq 1.1\frac{x_1}{x_2} = 1.1\sqrt{q^2 - \frac{a^2}{50}}\Big/\sqrt{q^2 - \frac{a^2}{100}}.$$

Here we used again (5.10.32). □

5.10.2 Full Weak Weighted Estimate of Square Function

We promised to provide information on full weak estimate for square function, namely on the asymptotics of constant $C([w]_{A_2})$ from (5.10.3).

We are going to present now the following result proved by C. Domingo-Salazar, M. Lacey, G. Rey [**51**].

THEOREM 5.10.12 *Let $w \in A_2$, then the norm of the square function operator $S\colon L^2(w) \to L^{2,\infty}(w)$ is bounded by $C[w]_{A_2}^{1/2}\log^{1/2}(1+[w]_{A_\infty})$, where C is an absolute constant.*

The claim of the theorem is equivalent to the following inequality that should be proved for an arbitrary function $\varphi \in L^2((0,1),w)$:

$$w\{x \in (0,1)\colon S\varphi(x) > 1\} \leq C[w]_{A_2}\log(1+[w]_{A_\infty})\,\|\varphi\|_w^2. \qquad (5.10.34)$$

Before proving it let us make several remarks.

REMARK 5.10.13 In Section 5.10.1, we studied the case of special functions φ, namely, we proved that if $\varphi = w^{-1}\mathbf{1}_I$, $I \in \mathcal{D}$, then inequality (5.10.34) can be strengthened. By this we mean that for such special test functions, the estimate above has the right-hand side $C[w]_{A_2}\|\varphi\|_w^2$. There is no logarithmic blow-up.

REMARK 5.10.14 Exponent $p = 2$ is critical for Theorem 5.10.12. By this we mean that one can quite easily deduce from this theorem the result for $p > 2$: if $w \in A_p$, then the norm of the square function operator $S\colon L^p(w) \to L^{p,\infty}(w)$ is bounded by $C[w]_{A_p}^{1/2}\log^{1/2}(1+[w]_{A_\infty})$, where C is an absolute constant. For that reduction to the case $p = 2$, the reader can look at [**51**]. It is important to note that for $1 \leq p < 2$, and $w \in A_p$, [**51**] proves the estimate $C[w]_{A_p}^{1/2}$ for the weak norm of the square function operator.

The first step in proving Theorem 5.10.12 is to reduce it to the weak estimate of *sparse square function*.

5.10.2.1 *Sparse Square Function Operators.*

DEFINITION 5.10.15 A family $\mathcal{S}$ of intervals of $\mathcal{D}$ is called ε-sparse if the following condition is satisfied

$$\sum_{\substack{I\in\mathcal{S}\\ I\subsetneq J}} |I| \le \varepsilon |J|, \quad \forall J\in\mathcal{S}. \tag{5.10.35}$$

DEFINITION 5.10.16 Sparse square function operator is defined for each sparse family $\mathcal{S}$ as follows:

$$S^{\mathrm{sp}}\varphi \stackrel{\mathrm{def}}{=} S^{\mathrm{sp}}_{\mathcal{S}}\varphi \stackrel{\mathrm{def}}{=} \left(\sum_{I\in\mathcal{S}} \langle\varphi\rangle_I^2 \mathbf{1}_I\right)^{1/2}.$$

THEOREM 5.10.17 *For any $\varepsilon > 0$ and any $\varphi \in L^1$, there exists a constant $C = C(\varepsilon)$ independent of φ and a sparse family $\mathcal{S}$ (depending on ε and on φ) such that pointwise almost everywhere*

$$S\varphi \le C S^{\mathrm{sp}}\varphi.$$

PROOF It is well known (see e.g., [**200**]) that the square function operator is weakly bounded in unweighted L^1. Let us call A the norm of the operator S from L^1 to $L^{1,\infty}$. Fix ε and let $C = 100A/\varepsilon$. We start with interval $I_0 = (0,1)$, put $\mathcal{S}_0 \stackrel{\mathrm{def}}{=} \{I_0\}$, and define the first generation of stopping intervals $\mathcal{S}_1$ as follows: $Q \in \mathcal{S}_1$ if it is the maximal interval in I_0 such that

$$S^{I_0}_Q\varphi \stackrel{\mathrm{def}}{=} \left(\sum_{\substack{I\in\mathcal{D}\\ Q\subset I\subset I_0}} (\Delta_I\varphi)^2\right)^{1/2} > C\langle|\varphi|\rangle_{I_0};$$

The second generation of stopping intervals $\mathcal{S}_2$ will be nested inside the first generation $\mathcal{S}_1$. For every $I \in \mathcal{S}_1$, we define its subintervals from $\mathcal{D}$ by the same rule as before, but with I playing the rôle of I_0. Namely, we define the first generation of stopping intervals $\mathcal{S}_2$ inside $I \in \mathcal{S}_1$ as follows: $Q \in \mathcal{S}_2$ if it is the maximal interval in I such that

$$S^{I}_Q\varphi \stackrel{\mathrm{def}}{=} \left(\sum_{\substack{J\in\mathcal{D}\\ Q\subset J\subset I}} (\Delta_J\varphi)^2\right)^{1/2} > C\langle|\varphi|\rangle_I;$$

We continue the construction of generations of intervals $\mathcal{S}_3, \mathcal{S}_4, \ldots$ recursively, and we put $\mathcal{S} \stackrel{\mathrm{def}}{=} \cup_{k=0}^{\infty}\mathcal{S}_k$.

Notice by the fact that operator S and dyadic maximal operator M are weakly bounded in unweighted L^1 and from our choice of constant

C at the beginning of the proof, we get that $\sum_{Q\in\mathcal{S}_1}|Q| \le \frac{\varepsilon}{50}|I_0|$, and similarly,

$$\sum_{\substack{Q\in\mathcal{S}_{k+1}\\ Q\subset I}} |Q| \le \frac{\varepsilon}{50}|I|, \quad \forall I\in\mathcal{S}_k.$$

Obviously, and with a good margin, we obtained that $\mathcal{S}$ is ε-sparse.

Now to see the pointwise estimate of the theorem, let us notice that given $x\in I_0$, which is not an endpoint of any dyadic interval, we will be able to find the tower of intervals $\cdots \subsetneq I_k \subsetneq I_1 \subsetneq I_0$ such that x is contained in all of them and such that $I_k\in\mathcal{S}_k$. This tower may degenerate to just one interval I_0 or it can be an infinite tower. But the set of points for which the tower is infinite has the Lebesgue measure zero. This is clear from the fact that $\mathcal{S}$ is sparse. In any case,

$$S^2\varphi(x) = \sum_{\substack{I\in\mathcal{D}\\ I_1\subsetneq I\subset I_0}} (\Delta_I\varphi)^2 + \sum_{\substack{I\in\mathcal{D}\\ I_2\subsetneq I\subset I_1}} (\Delta_I\varphi)^2 + \sum_{\substack{I\in\mathcal{D}\\ I_3\subsetneq I\subset I_2}} (\Delta_I\varphi)^2 + \cdots$$

But then, using our stopping criterion, we see that the last expression is bounded by

$$C^2\big(\langle|\varphi|\rangle_{I_0}^2 + \langle|\varphi|\rangle_{I_1}^2 + \langle|\varphi|\rangle_{I_2}^2 + \cdots\big),$$

which proves the theorem.

□

5.10.2.2 Weak Weighted Estimate of S^{sp}. We are going to prove

$$w\{x\in(0,1)\colon S^{\mathrm{sp}}\varphi(x) > 1\} \le C[w]_{A_2}\log(1+[w]_{A_\infty})\,\|\varphi\|_w^2. \quad (5.10.36)$$

The combination of this inequality with sparse domination Theorem 5.10.17 gives us Theorem 5.10.12.

We need several lemmas. Everywhere below, we deal with dyadic classes A_p and we omit the superscript d in A_p^d, also everywhere below all intervals are dyadic.

LEMMA 5.10.18 *Let $w\in A_2$. Consider the operator of averaging over one interval: $f\to\langle f\rangle_I\mathbf{1}_I$. Its norm as an operator in $L^2(w)$ is bounded by $[w]_{A_2}^{1/2}$.*

PROOF It is the same as to prove that for any nonnegative function f, we have

$$\langle f\rangle_I^2 w(I) \le [w]_{A_2}\|f\|_w^2. \quad (5.10.37)$$

In the next chain, we use Cauchy inequality.

$$\langle f\rangle_I^2 w(I) = \langle fww^{-1}\rangle_I^2 w(I) = \left(\frac{1}{w^{-1}(I)}\int_I fww^{-1}\right)^2 \frac{w^{-1}(I)w(I)}{|I|^2} w^{-1}(I)$$
$$\le \frac{w^{-1}(I)w(I)}{|I|^2} w^{-1}(I) \frac{1}{w^{-1}(I)}\int_I f^2 w^2 w^{-1} = \frac{w^{-1}(I)w(I)}{|I|^2}\int_I f^2 w,$$

and we are done proving (5.10.37). □

LEMMA 5.10.19 *Let $w \in A_2$ and $\{g_Q\}_{Q\in\mathcal{F}}$ be an arbitrary collection of nonnegative measurable functions, where g_Q is supported on interval Q and belong to L^2. Then*

$$\left\| \left[\sum_{Q\in\mathcal{F}} \langle g_Q\rangle_Q^2 \mathbf{1}_Q\right]^{1/2} \right\|_{L^2(w)} \le A[w]_{A_2}^{1/2} \left\| \left[\sum_{Q\in\mathcal{F}} g_Q^2\right]^{1/2} \right\|_{L^2(w)}.$$

PROOF One just has to apply Lemma 5.10.18 to each function g_Q. □

LEMMA 5.10.20 *Let $w \in A_\infty$. Then there exists an absolute constant a, $a > 0$, such that for $\delta = a/[w]_{A_\infty}$ the following reverse Hölder inequality holds:*

$$\langle w^{1+\delta}\rangle_I^{\frac{1}{1+\delta}} \le 2\langle w\rangle_I.$$

PROOF The Bellman function proof is due to V. Vasyunin [**184**]. The stopping time proof can be found in the paper of T. Hytönen, C. Pérez [**75**]. □

LEMMA 5.10.21 *Let G be a measurable subset of interval I. Let $w \in A_\infty$. Then there exists an absolute constant b, $b > 0$,*

$$\frac{w(G)}{w(I)} \le 2\left(\frac{|G|}{|I|}\right)^{\frac{1}{b[w]_\infty}}.$$

PROOF Denote $r = 1 + \delta$, where δ is from Lemma 5.10.20. As always r' denotes the dual exponent.

$$\frac{w(G)}{|I|} = \langle w\mathbf{1}_G\rangle_I \le \langle w^r\rangle_I^{1/r}\frac{|G|^{1/r'}}{|I|^{1/r'}} \le 2\langle w\rangle_I \cdot \frac{|G|^{1/r'}}{|I|^{1/r'}} = 2\frac{|G|^{1/r'}}{|I|^{1/r'}} \cdot \frac{w(I)}{|I|},$$

which implies the lemma because clearly $r' \asymp [w]_{A_\infty}$. □

LEMMA 5.10.22 *Let M be the dyadic maximal operator, and w be in dyadic A_2. Then the norm of M from $L^2(w)$ to $L^{2,\infty}(w)$ is bounded by $[w]_{A_2}^{1/2}$.*

Proof Given a test function $\varphi \geq 0$, let $\{I\}$ be the maximal dyadic intervals for which $M\varphi > 1$. Then

$$\sum_I w(I) \leq \int \sum \frac{w(I)}{|I|}\varphi \mathbf{1}_I\, dx = \int \sum_I \frac{w(I)}{|I|}\varphi \mathbf{1}_I\, w^{-1/2} w^{1/2} dx$$

$$\leq \left(\sum_I \left(\frac{w(I)}{|I|}\right)^2 w(I)\right)^{1/2} \|f\|_w \leq [w]_{A_2}^{1/2} \left(\sum_I w(I)\right)^{1/2} \|\varphi\|_w$$

Hence,

$$\big(w\{x\colon M\varphi > 1\}\big)^{1/2} = \left(\sum_I w(I)\right)^{1/2} \leq [w]_{A_2}^{1/2}\|\varphi\|_w,$$

which is precisely what the lemma claims. □

Remark 5.10.23 One can skip the word "dyadic" everywhere in the statement of this lemma. Also one can generalize the statement to $\mathbb{R}^n$. Then lemma remains true, only the estimate becomes by $C_n[w]_{A_2}^{1/2}$. See [**20**].

Now let $\mathcal{S}$ be an ε-sparse system of dyadic intervals, and S^{sp} be a corresponding sparse square function. We split $\mathcal{S} = \mathcal{S}^0 \cup \cup_{m=0}^{\infty} \mathcal{S}_m$, where $\mathcal{S}^0$ consists of intervals of $\mathcal{S}$ such that $\langle\varphi\rangle_I > 1$, and $\mathcal{S}_{m+1}$ consists of intervals of $\mathcal{S}$ such that

$$2^{-m-1} < \langle\varphi\rangle_I \leq 2^{-m}, \quad m = 0, 1, \ldots.$$

We denote

$$S_m^{\mathrm{sp}}\varphi \stackrel{\mathrm{def}}{=} \left(\sum_{I\in\mathcal{S}_m} \langle\varphi\rangle_I^2 \mathbf{1}_I\right)^{1/2}.$$

We are going to estimate measures

$$W_0 \stackrel{\mathrm{def}}{=} w\{x \in I_0\colon (S_0^{\mathrm{sp}}\varphi)^2 > 1\},\ W_1 \stackrel{\mathrm{def}}{=} w\left\{x \in I_0\colon \sum_{m=1}^{m_0} (S_m^{\mathrm{sp}}\varphi)^2 > 1\right\},$$

$$W_2 \stackrel{\mathrm{def}}{=} w\left\{x \in I_0\colon \sum_{m=m_0+1}^{\infty} (S_m^{\mathrm{sp}}\varphi)^2 > 1\right\}.$$

We will choose m_0 a bit later.

The estimate of W_0 is easy. The sum $(S_0^{\mathrm{sp}}\varphi)^2$ is supported on intervals where the dyadic maximal function of φ is bigger than one. The set, where the dyadic maximal function is bigger than one has w-measure bounded by $[w]_{A_2}\|\varphi\|_w^2$ by Lemma 5.10.22.

To estimate W_1, we first introduce for each $I \in \mathcal{S}_m$, $1 \le m \le m_0$,

$$E_m(I) \stackrel{\text{def}}{=} I \setminus \bigcup_{\substack{J \subsetneq I \\ J \in \mathcal{S}_m}} J.$$

Notice that

$$\langle \varphi \mathbf{1}_{E_m(I)} \rangle_I \ge \tfrac{1}{4} \langle \varphi \rangle_I \tag{5.10.38}$$

for any $I \in \mathcal{S}_m$ if $\varepsilon \le \frac{1}{4}$. In fact, if J^* mean the maximal intervals from $\mathcal{S}_m$ lying strictly inside I then

$$\int_I \varphi \mathbf{1}_{E_m(I)} = \int_I \varphi - \sum_{J^*} \int_{J^*} \varphi \ge 2^{-m}|I|\big(\tfrac{1}{2} - \varepsilon\big) \ge 2^{-m-2}|I|,$$

if $\varepsilon \le \frac{1}{4}$. Thus, (5.10.38) holds.

Now we use Chebyshev inequality. For that, we first estimate

$$\begin{aligned}\int_{I_0} (S_m^{\mathrm{sp}} \varphi)^2 w dx &= \int_{I_0} \sum_{I \in \mathcal{S}_m} \big[\langle \varphi \rangle_I^2 \mathbf{1}_I\big] w dx \\ &\le 16 \int_{I_0} \sum_{I \in \mathcal{S}_m} \big[\langle \varphi \mathbf{1}_{E_m(I)} \rangle_I^2 \mathbf{1}_I\big] w dx \\ &\le 16 [w]_{A_2} \int_{I_0} \sum_{I \in \mathcal{S}_m} \big[(\varphi \mathbf{1}_{E_m(I)})^2\big] w dx \\ &\le 16 [w]_{A_2} \int_{I_0} \varphi^2 \, w dx = 16 [w]_{A_2} \|\varphi\|_w^2 .\end{aligned}$$

The first inequality uses (5.10.38), the second one is just Lemma 5.10.19, the last inequality follows because by construction the sets $E_m(I)$ are disjoint when I runs over $\mathcal{S}_m$. Now Tchebyshoff inequality gives us

$$W_1 \le 16 m_0 [w]_{A_2} \|\varphi\|_w^2 .$$

We are left to estimate W_2. Notice that the support of $S_m^{\mathrm{sp}} \varphi$ is in $\cup Q^*$, where we denote by Q^* the maximal dyadic intervals in the family $\mathcal{S}_m$ (we drop the index m in the notation of these intervals).

We use the notation

$$b_m \stackrel{\text{def}}{=} 2^{2m} (S_m^{\mathrm{sp}} \varphi)^2 .$$

Notice that b_m is supported on $\cup Q^*$ and that

$$b_m \le \sum_{I \in \mathcal{S}_m} \mathbf{1}_I . \tag{5.10.39}$$

For every such Q^* we denote by $G_j(Q^*)$ the subset of Q^*, where $b_m \geq j$. From the ε-sparseness of $\mathcal{S}_m$, one immediately concludes that

$$\frac{|G_j(Q^*)|}{|Q^*|} \leq e^{-Cj}, \tag{5.10.40}$$

where $C = \log \frac{1}{\varepsilon}$. By Lemma 5.10.21, we get

$$\frac{w(G_j(Q^*))}{w(Q^*)} \leq 2e^{-cj/[w]_{A_\infty}}, \tag{5.10.41}$$

To estimate W_2, we will use (5.10.41) and a rather rude union estimate.

$$\begin{aligned}
& w\left\{x \in Q^* : \sum_{m=m_0+1}^{\infty} (S_m^{\mathrm{sp}}\varphi)^2 > 1\right\} \\
& \quad = w\left\{x \in Q^* : \sum_{m=m_0+1}^{\infty} (S_m^{\mathrm{sp}}\varphi)^2 > \sum_{m=m_0+1}^{\infty} 2^{m_0-m}\right\} \\
& \quad = w\left\{x \in Q^* : \sum_{m=m_0+1}^{\infty} b_m > \sum_{m=m_0+1}^{\infty} 2^{m_0+m}\right\} \\
& \quad \leq \sum_{m=m_0+1}^{\infty} w\{x \in Q^* : b_m > 2^{m_0+m}\} \leq 2w(Q^*) \sum_{m=m_0+1}^{\infty} e^{-\frac{c2^{m_0}}{[w]_{A_\infty}}\cdot 2^m}
\end{aligned}$$

On the other hand,

$$w(\cup Q^*) \leq A[w]_{A_2} 2^{2m} \|\varphi\|_w^2.$$

This is the sharp weak type estimate for the maximal function because $\cup Q^*$ is contained in the set where (dyadic) maximal operator $M\varphi$ is bigger than 2^{-m-1}. In the next chain of inequalities, A denote absolute constants that can change from line to line.

$$\begin{aligned}
W_2 &= w\left\{x \in \cup Q^* : \sum_{m=m_0+1}^{\infty} (S_m^{\mathrm{sp}}\varphi)^2 > 1\right\} \\
&\leq A[w]_{A_2} \sum_{m=m_0+1}^{\infty} 2^{2m} e^{-\frac{c2^{m_0}}{[w]_{A_\infty}}\cdot 2^m} \\
&\leq A[w]_{A_2} \int_{m_0}^{\infty} 2^{2x} e^{-c\frac{2^{m_0}}{[w]_{A_\infty}}\cdot 2^x}\, dx \\
&\leq A[w]_{A_2} \int_{2^{m_0}}^{\infty} y^2 e^{-c\frac{2^{m_0}}{[w]_{A_\infty}}\cdot y} \frac{dy}{y} = A[w]_{A_2} \int_{2^{m_0}}^{\infty} y e^{-c\frac{2^{m_0}}{[w]_{A_\infty}}\cdot y} dy \\
&= A[w]_{A_2} \left(\frac{[w]_{A_\infty}}{2^{m_0}}\right)^2 \int_{\frac{2^{2m_0}}{[w]_{A_\infty}}}^{\infty} u e^{-u} du.
\end{aligned}$$

Gathering together, the estimates for W_0, W_1, W_2 and choose $2^{m_0} \asymp 1 + [w]_{A_\infty}$, we get

$$w\{x \in I_0 \colon S^{\mathrm{sp}}\varphi > 3\}$$
$$\leq A\left([w]_{A_2} + [w]_{A_2}\log(1+[w]_{A_\infty}) + [w]_{A_2}\int_{[w]_{A_\infty}}^{\infty} ue^{-u}du\right)\|\varphi\|_w^2. \tag{5.10.42}$$

Inequality (5.10.36) is finally proved.

This proves that the norm of S^{sp} as the operator from $L^2(w)$ to $L^{2,\infty}(w)$ is at most $A[w]_{A_2}^{1/2}\log^{1/2}(1+[w]_{A_\infty})$. Theorem 5.10.12 is completely proved.

REMARK 5.10.24 The reader may suspect that by choosing m_0 differently, e.g., by choosing $m_0 \approx \log(1+[w]_{A_\infty})/\log\log(e+[w]_{A_\infty})$, one can optimize the estimate of $W_1 + W_2$. However, it is easy to see that the choice of $m_0 \approx \log(1+[w]_{A_\infty})$ above was already optimal.

REMARK 5.10.25 In the reasoning above, one can change L^2 to L^p and A_2 to A_p, $p > 2$. Then just minor modifications are needed to prove that the norm of S^{sp} as the operator from $L^p(w)$ to $L^{p,\infty}(w)$ is at most $[w]_{A_p}^{1/2}\log^{1/2}(1+[w]_{A_\infty})$. As it is shown in **[101]** for $1 \leq p < 2$ one can drop the term W_1 completely and get that the norm of S^{sp} as the operator from $L^p(w)$ to $L^{p,\infty}(w)$ is at most $[w]_{A_p}^{1/2}$.

REMARK 5.10.26 The Bellman function technique of the previous section gave us a better estimate in this particular case $\varphi = w^{-1}$:

$$w\{x \in I_0 \colon S^{\mathrm{sp}}w^{-1} > 3\} \leq A[w]_{A_2}\int_{I_0} w^{-1}\,dx. \tag{5.10.43}$$

5.11 Restricted Weak Weighted Estimate of the Square Function

THEOREM 5.11.1 *Let $w \in A_2$ and $\mathcal{S}$ be a collection of sparse dyadic intervals in $\mathcal{D}(I_0)$. Let S^{sp} be the sparse square function operator built on this collection. Then the restricted weak type of the operator $S^{\mathrm{sp}}_{w^{-1}}$ from $L^2(w^{-1})$ to $L^{2,\infty}(w)$ is bounded by $A[w]_{A_2}^{1/2}$, where A is an absolute constant.*

Now we can use sparse domination Theorem 5.10.17. Then this theorem gives us the following one.

THEOREM 5.11.2 *Let $w \in A_2$. Let S be the dyadic square function operator. Then the restricted weak type of the operator $S_{w^{-1}}$ from $L^2(w^{-1})$ to $L^{2,\infty}(w)$ is bounded by $A[w]_{A_2}^{1/2}$, where A is an absolute constant.*

This result strengthens Theorem 5.10.4. Indeed, Theorem 5.10.4 claims that the norm of $S_{w^{-1}}\mathbf{1}_I$ in weak space $L^{2,\infty}(w)$ is bounded by

$$A[w]_{A_2}^{1/2}\|\mathbf{1}_I\|_{w^{-1}},$$

where A does not depend on w or interval I. Theorem 5.11.2 claims the same estimate $A[w]_{A_2}^{1/2}\|\mathbf{1}_E\|_{w^{-1}}$ of the norm of $S_{w^{-1}}\mathbf{1}_E$ in weak space $L^{2,\infty}(w)$, where E is an arbitrary measurable set, and not just an interval.

The reader can find the proofs of these results in [**77**].

References

[1] I. Assani, Z. Buczolich, R. D. Mauldin, *An L^1 counting problem in ergodic theory*, J. Anal. Math., **95** (2005), 221–241.

[2] K. Astala, *Area distortion of quasiconformal mappings*, Acta Math., **173** (1994), 37–60.

[3] A. Baernstein, S. Montgomery-Smith, *Some conjectures about integral means of ∂f and $\overline{\partial} f$*, Complex analysis and differential equations, Proc. of the Marcus Wallenberg symposium in honor of Matts Essén, Uppsala, Sweden, 1997, 92–109.

[4] R. Bañuelos, *The foundational inequalities of D. L. Burkholder and some of their ramifications*, Illinois J. Math., **54** (2010), no. 3, 789–868.

[5] R. Bañuelos, P. Janakiraman, *L^p-bounds for the Beurling–Ahlfors transform*, Trans. Amer. Math. Soc., **360** (2008), 3603–3612.

[6] R. Bañuelos, P. J. Méndez-Hernández, *Space-time Brownian motion and the Beurling–Ahlfors transform*, Indiana Univ. Math. J., **52** (2003), 981–990.

[7] R. Bañuelos, A. Osękowski, *Burkholder inequalities for submartingales, Bessel processes and conformal martingales*, Amer. J. Math., **135** (2013), no. 6, 1675–1698.

[8] R. Bañuelos, A. Osękowski, *On the Bellman function of Nazarov, Treil and Volberg*, Math. Z., **278** (2014), no. 1–2, 385–399.

[9] R. Bañuelos, G. Wang, *Sharp inequalities for martingales with applications to the Beurling–Ahlfors and Riesz transforms*, Duke Math. J., **80** (1995), 575–600.

[10] F. Barthe, N. Huet, *On Gaussian Brunn–Minkowski inequalities*, Stud. Math., **191** (2009), 283–304.

[11] F. Barthe, B. Maurey, *Some remarks on isoperimetry of Gaussian type*, Ann. de l'Institut Henri Poincaré (B) Prob. Stat., **36** (2000), no. 4, 419–434.

[12] C. Bennett, R. Sharpley, *Interpolation of operators*, Pure and Applied Mathematics, **129**, Academic Press, Boston, MA, 1988.

[13] V. Bentkus, D. Dzindzalieta, *A tight Gaussian bound for weighted sums of Rademacher random variables*, Bernoulli, **21** (2015), no. 2, 1231–1237.

[14] S. G. Bobkov, F. Götze, *Discrete isoperimetric and Poincaré-type inequalities*, Prob. Theory Relat. Fields, **114** (1999), 245–277.

[15] B. Bollobás, *Martingale inequalities*, Math. Proc. Cambridge Philos. Soc., **87** (1980), no. 3, 377–382.

[16] A. Borichev, P. Janakiraman, A. Volberg, *On Burkholder function for orthogonal martingales and zeros of Legendre polynomials*, Amer. J. Math., **135** (2013), no. 1, 207–236.

[17] A. Borichev, P. Janakiraman, A. Volberg, *Subordination by conformal martingales in L^p and zeros of Laguerre polynomials*, Duke Math. J., **162** (2013), no. 5, 889–924.

[18] N. Boros, L. Székelyhidi, A. Volberg, *Laminates meet Burkholder functions*, J. Math. Pures Appl., **100** (2013), no. 5, 687–700.

[19] St. Buckley, *Summation condition on weights*, Mich. Math. J., **40** (1993), no. 1, 153–170.

[20] St. Buckley, *Estimates for operator norms on weighted spaces and reverse Jensen inequalities*, Trans. Amer. Math. Soc., **340** (1993), no. 1, 253–272.

[21] D. Burkholder, *Martingale transforms*, Ann. Math. Stat., **37** (1966), 1494–1504.

[22] D. Burkholder, *A geometrical characterization of Banach spaces in which martingale difference sequences are unconditional*, Ann. Prob., **9** (1981), no. 6, 997–1011.

[23] D. Burkholder, *Boundary value problems and sharp estimates for the martingale transforms*, Ann. Prob., **12** (1984), no. 3, 647–702.

[24] D. Burkholder, *Martingales and Fourier analysis in Banach spaces*, Probability and Analysis (Varenna, 1985), Lecture Notes in Math., **1206** (1986), 61–108.

[25] D. Burkholder, *An extension of classical martingale inequality*, Probability Theory and Harmonic Analysis (Cleveland, OH, 1983), Monogr. Textbook Pure Appl. Math., **98**, 21–30. Marcel Dekker, New York, 1986.

[26] D. Burkholder, *Sharp inequalities for martingales and stochastic integrals*, Colloque Paul Lévy sur les Processus Stochastiques (Palaiseau, 1987), Astérisque, **157–158** (1988), 75–94.

[27] D. Burkholder, *A proof of the Peczyński's conjecture for the Haar system*, Stud. Math., **91** (1988), no. 1, 79–83.

[28] D. Burkholder, *Differential subordination of harmonic functions and martingales*, Harmonic Analysis and Partial Differential Equations (El Escorial, 1987), Lecture Notes in Math., **1384** (1989), 1–23.

[29] D. Burkholder, *Explorations of martingale theory and its applications*, École d'Été de Probabilités de Daint-Flour XIX–1989, Lecture Notes in Math., **1464** (1991), 1–66.

[30] D. Burkholder, *Strong differential subordination and stochastic integration*, Ann. Prob., **22** (1994), no. 2, 995–1025.

[31] D. Burkholder, *Martingales and singular integrals in Banach spaces*, Handbook of the Geometry of Banach Spaces, **I**, Ch. 6, 233–269. North-Holland, Amsterdam, 2001.

[32] D. L. Burkholder, R. F. Gundy, *Extrapolation and interpolation of quasi-linear operators on martingales*, Acta Math., **124** (1970), 249–304.

[33] A. Carbonaro, O. Dragičević, *Bellman function and linear dimension-free estimates in a theorem of Bakry*, J. Funct. Anal., **265** (2013), no. 7, 1085–1104.

[34] A. Carbonaro, O. Dragičević, *Functional calculus for generators of symmetric contraction semigroups*, Duke Math. J., **166** (2017), no. 5, 937–974.

[35] M. Carro, *From restricted weak type to strong type estimates*, J. London Math. Soc., **70** (2004), no. 3, 750–762.

[36] M. Carro, C. Domingo-Salazar, *Weighted weak-type (1, 1) estimates for radial Fourier multipliers via extrapolation theory*. J. Anal. Math., **138** (2019), no. 1, 83–105.

[37] M. Carro, L. Grafakos, *Weighted weak type (1, 1) estimates via Rubio de Francia extrapolation*, J. Funct. Anal., **269** (2015), no. 5, 1203–1233.

[38] S.-Y. A. Chang, J. M. Wilson, T. H. Wolff, *Some weighted norm inequalities for the Schrödinger operator*, Comment. Math. Helv., **60** (1985), no. 1, 217–246.

[39] M. Christ, *A $T(b)$ theorem with remarks on analytic capacity and the Cauchy integral*, Colloq. Math., **60/61** (1990), no. 2, 601–628.

[40] J. A. Clarkson, *Uniformly convex spaces*, Trans. Amer. Math. Soc., **40** (1936), no. 3, 396–414.

[41] E. A. Coddington, *An introduction to ordinary differential operators*, Dover, New York, 1989.

[42] E. A. Coddington, N. Levinson, *The theory of ordinary differential operators*, McGraw-Hill, New York, 1955.

[43] J. Conde-Alonso, A. Culiuc, F. Di Plinio, Yumeng Ou, *A sparse domination principle for rough singular integrals*, Anal. PDE, **10** (2017), no. 5, 1255–1284.

[44] J. Conde-Alonso, G. Rey, *A pointwise estimate for positive dyadic shifts and some applications*, Math. Ann., **365** (2016), no. 3–4, 1111–1135.

[45] D. Cruz-Uribe, J. M. Martell, C. Pérez, *Weights, extrapolation and the theory of Rubio de Francia*, Operator Theory: Advances and Applications, **215**, Birkhäuser/Springer, Basel, 2011.

[46] D. Cruz-Uribe, C. Pérez, *Sharp two-weight, weak-type norm inequalities for singular integral operators*, Math. Res. Lett., **6** (1999), no. 3–4, 417–427.

[47] L. Dalenc, Y. Ou, *Upper bound for multi-parameter iterated commutators*, Publ. Mat., **60** (2016), no. 1, 191–220.

[48] G. David, J.-L. Journé, *A boundedness criterion for generalized Calderón–Zygmund operators*, Ann. Math., **120** (1984), no. 2, 371–397.

[49] G. David, J.-L. Journé, *Opérateurs de Calderón-Zygmund, fonctions para-accrétives et interpolation*, Rev. Mat. Iberoam., **1** (1985), no. 4, 1–56 (in French).

[50] B. Davis, *On the L^p norms of stochastic integrals and other martingales*, Duke Math. J., **43** (1976), 697–704.

[51] C. Domingo-Salazar, M. Lacey, G. Rey, *Borderline weak-type estimates for singular integrals and square functions*, Bull. Lond. Math. Soc., **48** (2016), no. 1, 63–73.

[52] O. Dragicevic, S. Treil, A. Volberg, *A theorem about three quadratic forms*. Int. Math. Res. Notices, IMRN, 2008, Art. ID rnn 072, 9 pp.

[53] O. Dragičević, A. Volberg, *Sharp estimates of the Ahlfors–Beurling operator via averaging of martingale transform*, Michi. Math. J., **51** (2003), 415–435.

[54] O. Dragičević, A. Volberg, *Bellman function, Littlewood–Paley estimates, and asymptotics of the Ahlfors–Beurling operator in $L^p(\mathbb{C})$, $p \to \infty$*, Indiana Univ. Math. J., **54** (2005), 971–995.

[55] O. Dragičević, A. Volberg, *Bellman function and dimensionless estimates of classical and Ornstein–Uhlenbeck Riesz transforms*, J. Oper. Theory, **56** (2006), 167–198.

[56] J. Duoandikoetxea, *Fourier analysis*, Graduate Studies in Mathematics, **29**, American Mathematical Society, Providence, RI, 2001.

[57] J. Duoandikoetxea, J. L. Rubio de Francia, *Maximal and singular integral operators via Fourier transform estimates*, Invent. Math., **84** (1986), no. 3, 541–561.

[58] I. Ekelan, R. Temam, *Convex analysis and variational problems*, North-Holland, Amsterdam, 1976.

[59] P. Enflo, *Banach spaces which can be given an equivalent uniformly convex norm*, Israel J. Math., **13** (1972), 281–288.

[60] L. C. Evans, R. F. Gariepy, *Measure theory and Fine properties of functions*, Studies in Advanced Mathematics, CRC Press, Boca Raton, FL, 1992. ISBN: 0-8493-7157-0

[61] J. Garnett, *Bounded analytic functions*, Springer, New York, 2007 (first edition Academic Press, 1981).

[62] S. Geiss, S. Montgomery-Smith, E. Saksman, *On singular integral and martingale transforms*, Trans. Amer. Math. Soc., **362** (2010), 553–575.

[63] I. I. Gikhman, A. V. Skorokhod, *The theory of stochastic processes I, II, III*, Springer-Verlag, 2004–2007.

[64] L. Grafakos, J. M. Martell, F. Soria, *Weighted norm inequalities for maximally modulated singular integral operators*, Math. Ann., **331** (2005), no. 2, 359–394.

[65] R. R. Hall, *On a conjecture of Littlewood*, Math. Proc. Cambridge Philos. Soc., **78** (1975), 443–445.

[66] O. Hanner, *On the uniform convexity of L^p and l^p*, Ark. Mat., **3** (1956), no. 3, 239–244.

[67] L. H. HARPER, *Optimal numberings and isoperimetric problems on graphs*, J. Combin. Theory, **1** (1996), 385–393.

[68] S. HOFMANN, *Banach spaces of analytic functions*, Dover, Mineola, NY, 2007.

[69] B. HOLLENBECK, I. VERBITSKY, *Best constants for the Riesz projections*, J. Funct. Anal., **175** (2000), 370–392.

[70] I. HOLMES, P. IVANISVILI, A. VOLBERG, *The sharp constant in the weak (1,1) inequality for the square function: A new proof*, arXiv:1710.01346, 1–17, Rev. Mat. Iberoam., 2019, (online first), European Mathematical Society, doi: 10.4171/rmi/1147.

[71] S. HUKOVIC, S. TREIL, A. VOLBERG, *The Bellman functions and sharp weighted inequalities for square functions*, Complex Analysis, Operators, and Related Topics, Oper. Theory Adv. Appl., **113** (2000), 97–113, Birkhäuser, Basel.

[72] T. HYTÖNEN, *The sharp weighted bound for general Calderón–Zygmund operators*, Ann. Math. (2), **175** (2012), no. 3, 1473–1506.

[73] T. HYTÖNEN, *Representation of singular integrals by dyadic operators, and the A_2 theorem*, Expo. Math., **35** (2017), no. 2, 166–205.

[74] T. HYTÖNEN, H. MARTIKAINEN, *Non-homogeneous Tb theorem and random dyadic cubes on metric measure spaces*, J. Geom. Anal., **22** (2012), no. 4, 1071–1107.

[75] T. HYTÖNEN, C. PÉREZ, *Sharp weighted bounds involving A_∞*, Anal. PDE, **6** (2013), no. 4, 777–818.

[76] T. HYTÖNEN, C. PÉREZ, S. TREIL, A. VOLBERG, *Sharp weighted estimates for dyadic shifts and the A_2 conjecture*, J. Reine Angew. Math., **687** (2014), arXiv:1010.0755, 43–86.

[77] P. IVANISVILI, P. MOZOLYAKO, A. VOLBERG, *Strong weighted and restricted weak weighted estimates of the square function*, arXiv:1804.06869, 1–18.

[78] P. IVANISVILI, F. NAZAROV, A. VOLBERG, *Hamming cube and martingales*, Comptes Rendus de L'Acad. Sci. Paris, **355** (2017), no. 10, 1072–1076.

[79] P. IVANISVILI, F. NAZAROV, A. VOLBERG, *Square functions and the Hamming cube: Duality*, Discrete Anal., (2018), no. 1, arXiv:1608.4021, 1–18.

[80] P. IVANISHVILI, N. OSIPOV, D. STOLYAROV, V. VASYUNIN, P. ZATITSKIY, *Bellman function for extremal problems in BMO*, Trans. Amer. Math. Soc., **368** (2016), no. 5, arXiv:1205.7018, 3415–3468.

[81] P. IVANISHVILI, N. OSIPOV, D. STOLYAROV, V. VASYUNIN, P. ZATITSKIY, *Bellman function for extremal problems in BMO II: Evolution*, Mem. Amer. Math. Soc., **255** (2018), no. 1220, arXiv:1510.01010, 1–148.

[82] P. IVANISVILI, D. M. STOLYAROV, P. B. ZATITSKIY, *Bellman vs Beurling: Sharp estimates of uniform convexity for L^p spaces*, Algebra i Analiz, **27** (2015), no. 2, 218–231 (Russian); English translation in: St. Petersburg Math. J., **27** (2016), no. 2, 333–343.

[83] P. IVANISVILI, A. VOLBERG, *Bellman partial differential equation and the hill property for classical isoperimetric problems*, arXiv:1506.03409, 1–30.

[84] P. IVANISVILI, A. VOLBERG, *Isoperimetric functional inequalities via the maximum principle: The exterior differential systems approach*, in V. P. Havin memorial volume, 50 years with Hardy Spaces, 281–305, Oper. Theory Adv. Appl., **261**, Birkhuser/Springer, Cham, 2018 (preprint, arXiv: 1511.06895, 1–16).

[85] T. IWANIEC, *Extremal inequalities in Sobolev spaces and quasiconformal mappings*, Z. Anal. Anwendungen, **1** (1982), 1–16.

[86] R. C. JAMES, *Some self-dual properties of normed linear spaces*, Symposium on Infinite-Dimensional Topology (Louisiana State University, Baton Rouge, LA., 1967), 159–175, Ann. Math. Stud., **69**, Princeton University Press, Princeton, NJ, 1972.

[87] P. JANAKIRAMAN, *Orthogonality in complex martingale spaces and connections with the Beurling-Ahlfors transform*, Illinois J. Math., **54** (2010), no. 4, 1509–1563.

[88] F. John, L. Nirenberg, *On functions of bounded mean oscillation*, Comm. Pure Appl. Math., **14** (1961), 415–426.

[89] I. Karatzas, S. Shreve, *Brownian motion and stochastic calculus*, Graduate Texts in Mathematics, Springer, New York, 1991.

[90] S. Konyagin, A. Volberg, *On measures with the doubling condition*, Izv. Akad. Nauk SSSR Ser. Mat., **51** (1987), no. 3, 666–675 (Russian); English translation in: Math. USSR-Izv., **30** (1988), no. 3, 629–638.

[91] V. Kovač, *Applications of the Bellman function technique in multilinear and nonlinear harmonic analysis*, Thesis, UCLA, 2011, 1–111.

[92] V. Kovač, *Bellman function technique for multilinear estimates and an application to generalized paraproducts*, Indiana Univ. Math. J., **60** (2011), no. 3, 813–846.

[93] V. Kovač, *Boundedness of the twisted paraproduct*, Rev. Mat. Iberoam., **28** (2012), no. 4, 1143–1164.

[94] V. Kovač, *Uniform constants in Hausdorff-Young inequalities for the Cantor group model of the scattering transform*, Proc. Amer. Math. Soc., **140** (2012), no. 3, 915–926.

[95] N. Krylov, *Optimal control of diffusion processes*, Springer, Berlin 1980.

[96] M. Lacey, *An elementary proof of an A_2 bound*, Israel J. Math., **217** (2017), no. 1, 181–195.

[97] M. Lacey, *Two-weight inequality for the Hilbert transform: A real variable characterization, II*, Duke Math. J., **163** (2014), no. 15, 2821–2840.

[98] M. Lacey, K. Li, *On $A_p - A_\infty$ type estimates for square functions*, Math. Z., **284** (2016), 1211–1222.

[99] M. Lacey, S. Petermichl, M. Reguera, *Sharp A_2 inequality for Haar shift operators*, Math. Ann., **348** (2010), no. 1, 121–141.

[100] M. Lacey, E. Sawyer, C.-Y. Shen, I. Uriarte-Tuero, *Two-weight inequality for the Hilbert transform: A real variable characterization, I*, Duke Math. J., **163** (2014), no. 15, 2795–2820.

[101] M. Lacey, J. Scurry, *Weighted weak type estimates for square function*, arXiv:1211. 4219.

[102] M. Ledoux, *Concentration of measure and logarithmic Sobolev inequalities*, Séminare de Probabilités XXXIII, Lecture Notes in Math., **1709**, Springer, Berlin, 1999.

[103] M. Ledoux, *Isoperimetry and Gaussian analysis*, Lectures on Probability Theory and Statistics, Lecture Notes in Math., **1648** (1996), 165–294.

[104] O. Lehto, *Remarks on the integrability of the derivatives of quasiconformal mappings*, Ann. Acad. Sci. Fenn., Series A, **371** (1965), 1–8.

[105] A. Lerner, *Sharp weighted norm inequalities for Littlewood–Paley operators and singular integrals*, Adv. Math., **226** (2011), 3912–3926.

[106] A. Lerner, *Mixed $A_p - A_r$ inequalities for classical singular integrals and Littlewood–Paley operators*, J. Geom. Anal., **23** (2013), no. 3, 1343–1354.

[107] A. Lerner, *A simple proof of the A_2 conjecture*, Int. Math. Res. Notices, **2013** (2013), no. 14, 3159–3170.

[108] A. Lerner, *On an estimate of Calderón–Zygmund operators by dyadic positive operators*, J. Anal. Math., **121** (2013), no. 1, 141–161.

[109] A. Lerner, *On sharp aperture-weighted estimates for square functions*, J. Fourier Anal. Appl., **20** (2014), no. 4, 784–800.

[110] A. Lerner, F. Nazarov, S. Ombrosi, *On the sharp upper bound related to the weak Muckenhoupt–Wheeden conjecture*, arXiv:1710.07700, 1–17.

[111] A. Lerner, S. Ombrosi, C. Pérez, *A_1 bounds for Calderón–Zygmund operators related to a problem of Muckenhoupt and Wheeden*, Math. Res. Lett., **16** (2009), no. 1, 149–156.

[112] A. Lerner, F. Di Plinio, *On weighted norm inequalities for the Carleson and Walsh–Carleson operators*, J. London Math. Soc., **90** (2014), no. 3, 654–674.

[113] L. D. López-Sánchez, J. M. Martell, J. Parcet, *Dyadic harmonic analysis beyond doubling measures*, Adv. Math., **267** (2014), 44–93.

[114] J. Marcinkiewicz, *Quelque théorèmes sur les séries orthogonales*, Ann. Soc. Polon. Math., **16** (1937), 84–96 (pages 307–318 of the Collected Papers).

[115] H. Martikainen, *Representation of bi-parameter singular integrals by dyadic operators*, Adv. Math., **229** (2012), no. 3, 1734–1761.

[116] A. Melas, *The Bellman functions of dyadic-like maximal operators and related inequalities*, Adv. Math., **192** (2005), no. 2, 310–340.

[117] P. A. Meyer, *Démonstration probabiliste de certain inégalités de Littlewood–Paley*, Seminaire de Probabilitées X (University of Strasbourg, 1974/75), Lecture Notes in Math., **511**, Springer-Verlag, Berlin, 1982.

[118] P. W. Millar, *Martingale integrals*, Trans. AMS, **133** (1968), 145–166.

[119] C. Muscalu, T. Tao, C. Thiele, *A Carleson theorem for a Cantor group model of the scattering transform*, Nonlinearity, **16** (2003), no. 1, 219–246.

[120] N. Nadirashvili, V. Tkachev, S. Vladut, *Nonlinear elliptic equations and nonassociative algebras*, Mathematical Surveys and Monographs, **200** American Mathematical Society, Providence, RI, 2014.

[121] A. Naor, G. Schechtman, *Remarks on non linear type and Pisier's inequality*, J. Reine Angew. Math., **552** (2002), 213–236.

[122] F. Nazarov, A. Reznikov, S. Treil, A. Volberg, *A Bellman function proof of the L^2 bump conjecture*, J. Anal. Math., **121** (2013), 255–277.

[123] F. Nazarov, A. Reznikov, V. Vasyunin, A. Volberg, *A Bellman function counterexample to the A_1 conjecture: The blow-up of the weak norm estimates of weighted singular operators*, arXiv:1506.04710v1, 1–23.

[124] F. Nazarov, A. Reznikov, V. Vasyunin, A. Volberg, *On weak weighted estimates of the martingale transform and a dyadic shift*, Anal. PDE **11** (2018), no. 8, arXiv:1612.03958, 2089–2109.

[125] F. Nazarov, A. Reznikov, A. Volberg, *The proof of A_2 conjecture in a geometrically doubling metric space*, Indiana Univ. Math. J., **62** (2013), no. 5, 1503–1533, arXiv:1106.1342, 1–23.

[126] F. Nazarov, X. Tolsa, A. Volberg, *On the uniform rectifiability of AD-regular measures with bounded Riesz transform operator: The case of codimension* 1, Acta Math., **213** (2014), no. 2, 237–321.

[127] F. Nazarov, S. Treil, *The hunt for Bellman function: Applications to estimates of singular integral operators and to other classical problems in harmonic analysis*, Algebra i Analiz, **8** (1996), no. 5, 32–162 (Russian); English translation in: St. Petersburg Math. J., **8** (1997), no. 5, 721–824.

[128] F. Nazarov, S. Treil, A. Volberg, *Cauchy integral and Calderón–Zygmund operators on nonhomogeneous spaces*, Int. Math. Res. Notices, **1997** (1997), no. 15, 703–726.

[129] F. Nazarov, S. Treil, A. Volberg, *Weak type estimates and Cotlar inequalities for Calderón–Zygmund operators on nonhomogeneous spaces*, Int. Math. Res. Notices, **1998** (1998), no. 9, 463–487.

[130] F. Nazarov, S. Treil, A. Volberg, *The Bellman functions and two-weight inequalities for Haar multipliers*, J. Amer. Math. Soc., **12** (1999) 909–928.

[131] F. Nazarov, S. Treil, A. Volberg, *Bellman function in stochastic control and harmonic analysis*, Systems, Approximation, Singular Integral Operators, and Related Topics (Bordeaux, 2000), Oper. Theory Adv. Appl., **129** (2001), 393–423, Birkhäuser, Basel.

[132] F. Nazarov, S. Treil, A. Volberg, *The Tb-theorem on non-homogeneous spaces*, Acta Math., **190** (2003), 151–239.

[133] F. Nazarov, S. Treil, A. Volberg, *Two weight inequalities for individual Haar multipliers and other well localized operators*, Math. Res. Lett., **15** (2008), no. 4, 583–597.

[134] F. Nazarov, S. Treil, A. Volberg, *Two weight estimate for the Hilbert transform and corona decomposition for non-doubling measures*, arXiv:1003.1596v1.

[135] F. Nazarov, V. Vasyunin, A. Volberg, *On Bellman function associated with Chang–Wilson–Wolff theorem*, Algebra and Analysis.

[136] F. Nazarov, A. Volberg, *Bellman function, two weighted Hilbert transforms and embeddings of the model spaces K_θ*, Dedicated to the memory of Thomas H. Wolff, J. Anal. Math., **87** (2002), 385–414.

[137] F. Nazarov, A. Volberg, *Heating of the Beurling operator and estimates of its norm*, Algebra and Analysis, **15** (2003), no. 4, 142–158 (Russian); English translation in: St. Petersburg Math. J., **15** (2004), no. 4, 563–573.

[138] C. J. Neugebauer, *Inserting A_p-weights*, Proc. Amer. Math. Soc., **87** (1983), no. 4, 644–648.

[139] A. A. Novikov, *On stopping times for Wiener processes*, Theory Prob. Appl., **16** (1971), 449–456.

[140] A. Osekowski, *On the best constant in the weak type inequality for the square function of a conditionally symmetric martingale*, Stat. Prob. Lett., **79** (2009), no. 13, 1536–1538.

[141] A. Osekowski, *Sharp Martingale and Semimartingale Inequalities*, Monografie Matematyczne, **72**, Birkhäuser, Basel, 2012.

[142] A. Osekowski, *Some sharp estimates for the Haar system and other bases in $L^1(0,1)$*, Math. Scand., **115** (2014), 123–142.

[143] A. Osekowski, *A splitting procedure for Bellman functions and the action of dyadic maximal operator in L^p*, Mathematika, **61** (2015), 199–212.

[144] A. Osekowski, *Weighted square function inequalities*, Publ. Mat., **62** (2018), 75–94.

[145] Y. Ou, *Multi-parameter singular integral operators and representation theorem*, Rev. Mat. Iberoam., **33** (2017), no. 1, arXiv:1410.8055, 325–350.

[146] C. Pérez, *Weighted norm inequalities for general maximal operators*, Conference on Mathematical Analysis (El Escorial, 1989), Publ. Mat., **35** (1991), no. 1, 169–186.

[147] C. Pérez, *A remark on weighted inequalities for general maximal operators*. Proc. Amer. Math. Soc., **119** (1993), no. 4, 1121–1126.

[148] C. Pérez, *On sufficient conditions for the boundedness of the Hardy-Littlewood maximal operator between weighted L_p-spaces with different weights*. Proc. London Math. Soc. (3), **71** (1995), no. 1, 135–157.

[149] C. Pérez, S. Treil, A. Volberg, *On A_2 conjecture and corona decomposition of weights*, arXiv:1006.2630, 1–39.

[150] S. Petermichl, *A sharp bound for weighted Hilbert transform in terms of classical A_p characteristic*, Amer. J. Math., **129** (2007), 1355–1375.

[151] S. Petermichl, *The sharp weighted bound for the Riesz transforms*. Proc. Amer. Math. Soc., **136** (2008), no. 4, 1237–1249.

[152] S. Petermichl, A. Volberg, *Heating the Beurling operator: Weakly quasiregular maps on the plane are quasiregular*, Duke Math. J., **112** (2002), 281–305.

[153] S. Pichorides, *On the best values of the constants in the theorems of M. Riesz, Zygmund and Kolmogorov*, Stud. Math., **XLIV** (1972), 165–179.

[154] G. Pisier, *Martingales with values in uniformly convex spaces*, Israel J. Math., **20** (1975), no. 3–4, 326–350.

[155] G. Pisier, *Probabilistic methods in the geometry of Banach spaces*, Probability and analysis (Varenna, Italy, 1985), Lecture Notes in Math. **1206** (1986), 167–241, Springer-Verlag.

[156] A. V. Pogorelov, *Extrinsic geometry of convex surfaces*, Translations of Mathematical Monographs, **35**, American Mathematical Society, Providence, RI, 1973.

[157] M. C. Reguera, *On Muckenhoupt–Wheeden conjecture*, Adv. Math., **227** (2011), no. 4, 1436–1450.

[158] M. C. Reguera, C. Thiele, *The Hilbert transform does not map $L^1(Mw)$ to $L^{1,\infty}(w)$*, Math. Res. Lett., **19** (2012), no. 1, arXiv:1011.1767, 1–7.

[159] A. Reznikov, *Sharp weak type estimates for weights in the class $A_{p_1 p_2}$*, Rev. Mat. Iberoam., **29** (2013), no. 2, 433–478.

[160] A. Reznikov, A. Beznosova, *Equivalent definitions of dyadic Muckenhoupt and reverse Hölder classes in terms of Carleson sequences, weak classes, and comparability of dyadic $L \log L$ and A_∞ constants*, Rev. Mat. Iberoam., **30** (2014), no. 4, 1191–1236.

[161] A. Reznikov, S. Treil, A. Volberg, *A sharp estimate of weighted dyadic shifts of complexity* 0 *and* 1, arXiv:1104.5347.

[162] A. Reznikov, V. Vasyunin, A. Volberg, *Extremizers and the Bellman function for the weak type martingale inequality*, arXiv:1311.2133.

[163] A. Reznikov, V. Vasyunin, A. Volberg, *An observation: Cut-off of the weight w does not increase the A_{p_1,p_2}-norm of w*, arXiv:1008.3635v1.

[164] L. C. G. Rogers, D. Williams, *Diffusions, Markov processes and martingales*, Vol. 1, 2, Cambridge University Press, 2000.

[165] E. Sawyer, *Two-weight norm inequalities for certain maximal and integral operators*, Lecture Notes Math., **908** (1982), 102–127, Springer-Verlag, Berlin–Heidelberg–New York.

[166] L. A. Shepp, *A first passage problem for the Wiener process*, Ann. Math. Stat., **38** (1967), 1912–1914.

[167] M. Sion, *On general minimax theorems*, Pac. J. Math., **8** (1958), 171–176.

[168] L. Slavin, *Bellman function and BMO*, PhD thesis, Michigan State University, 2004.

[169] L. Slavin, A. Stokolos, *The Bellman PDE and its solution for the dyadic maximal function*, 2006, 1–16.

[170] L. Slavin, A. Stokolos, and V. Vasyunin, *Monge-Ampère equations and Bellman functions: The dyadic maximal operator*, C. R. Math. Acad. Sci. Paris, **346** (2008), no. 9–10, 585–588.

[171] L. Slavin, V. Vasyunin, *Sharp results in the integral-form John–Nirenberg inequality*, Trans. Amer. Math. Soc., **363** (2011), arXiv:0709.4332, 4135–4169.

[172] L. Slavin, V. Vasyunin, *Sharp L^p estimates on* BMO, Indiana Univ. Math. J., **61** (2012), no. 3, 1051–1110.

[173] E. Stein *Singular integrals and differentiability properties of functions*, Princeton University Press, Princeton, NY, 1986.

[174] D. M. Stolyarov, P. B. Zatitskiy, *Theory of locally concave functions and its applications to sharp estimates of integral functionals*. Adv. Math., **291** (2016), arXiv:1412.5350, 228–273.

[175] S. J. Szarek, *On the best constants in the Khintchine inequality*, Stud. Math., **18** (1976), 197–208.

[176] G. Szegö, *Orthogonal polynomials*, Fourth edition, American Mathematical Society, Colloquium Publications, **XXIII**, American Mathematical Society, Providence, RI, 1975.

[177] M. Talagrand, *Isoperimetry, logarithmic Sobolev inequalities on the discrete cube, and Margulis' graph connectivity*, Geom. Funct. Anal., **3**, (1993), no. 3, 295–314.

[178] C. Thiele, S. Treil, A. Volberg, *Weighted martingale multipliers in the non-homogeneous setting and outer measure spaces*, Adv. Math., **285** (2015), 1155–1188.

[179] X. Tolsa, *Painlevé's problem and the semiadditivity of analytic capacity*, Acta Math., **190** (2003), no. 1, 105–149.

[180] X. Tolsa, *Analytic capacity, the Cauchy transform, and non-homogeneous Calderón–Zygmund theory*, Progress in Mathematics, Birkhäuser, Basel, 2013.

[181] S. Treil, *Sharp A_2 estimates of Haar shifts via Bellman function*, Recent trends in analysis, Theta Ser. Adv. Math., **16** (2013), Theta, Bucharest, arXiv:1105.2252v1, 187–208.

[182] S. Treil, A. Volberg, *Entropy conditions in two weight inequalities for singular integral operators*, Adv. Math., **301** (2016), 499–548.

[183] V. Vasyunin, *The sharp constant in the John–Nirenberg inequality*, POMI no. 20, 2003.

[184] V. Vasyunin, *The sharp constant in the reverse Hölder inequality for Muckenhoupt weights*, Algebra i Analiz, **15** (2003), no. 1, 73–117 (Russian); English translation in: St. Petersburg Math. J., **15** (2004), no. 1, 49–79.

[185] V. Vasyunin, *Sharp constants in the classical weak form of the John-Nirenberg inequality*, POMI no. 10, 2011, www.pdmi.ras.ru/preprint/2011/eng-2011.html.

[186] V. Vasyunin, *Cincinnati lectures on Bellman functions*, edited by L. Slavin, arXiv:1508.07668, 1–33.

[187] V. Vasyunin, A. Volberg, *The Bellman functions for a certain two-weight inequality: A case study*, Algebra i Analiz, **18** (2006), no. 2, 24–56 (Russian); English translation in: St. Petersburg Math. J., **18** (2007), 201–222.

[188] V. Vasyunin, A. Volberg, *Bellman functions technique in harmonic analysis*, 2009, 1–86, www.sashavolberg.wordpress.com

[189] V. Vasyunin, A. Volberg, *Monge–Ampère equation and Bellman optimization of Carleson embedding theorems*, Amer. Math. Soc. Transl. Ser. 2, **226** (2009), 195–238.

[190] V. Vasyunin, A. Volberg, *Burkholder's function via Monge-Ampère equation*, Illinois J. Math., **54** (2010), no. 4, 1393–1428.

[191] V. Vasyunin, A. Volberg, *Sharp constants in the classical weak form of the John–Nirenberg inequality*, Proc. London Math. Soc. (3), **108** (2014), no. 6, 1417–1434.

[192] A. Volberg, *The proof of the nonhomogeneous T1 theorem via averaging of dyadic shifts*, Algebra i Analiz, **27** (2015), no. 3, 75–94; English translation in: St. Petersburg Math. J., **27** (2016), no. 3, 399–413.

[193] A. Volberg, *Bellman approach to some problems in Harmonic Analysis*, Séminaires des Equations aux derivées partielles, Ecole Politéchnique, 2002, exposé XX, 1–14.

[194] A. Volberg *Calderón–Zygmund capacities and operators on nonhomogeneous spaces*, CBMS Regional Conference Series in Mathematics, **100**. Published for the Conference Board of the Mathematical Sciences, Washington, DC; by the American Mathematical Society, Providence, RI, 2003.

[195] A. Volberg, P. Zorin-Kranich, *Sparse domination on non-homogeneous spaces with an application to A_p weights*, Rev. Mat. Iberoam., **34** (2018), no 3, arXiv:1606.03340, 1401–1414.

[196] G. Wang, *Sharp square function inequalities for conditionally symmetric martingales*, Trans. Amer. Math. Soc., **328** (1991), no. 1, 393–419.

[197] G. Wang, *Some sharp inequalities for conditionally symmetric martingales*, PhD thesis, University of Illinois at Urbana-Champaign, 1989.

[198] G. N. Watson, *A treatise on the theory of Bessel functions*, Reprint of the second (1944) edition, Cambridge Mathematical Library, Cambridge University Press, Cambridge, 1995.

[199] A. D. Wentzel, *Course in the theory of stochastic processes*, Moscow, Nauka, Fizmatgiz, 1996.

[200] M. Wilson, *Weighted Littlewood–Paley theory and exponential-square integrability*, Lecture Notes in Math., **1924**, Springer-Verlag, New York, 2007.

Index